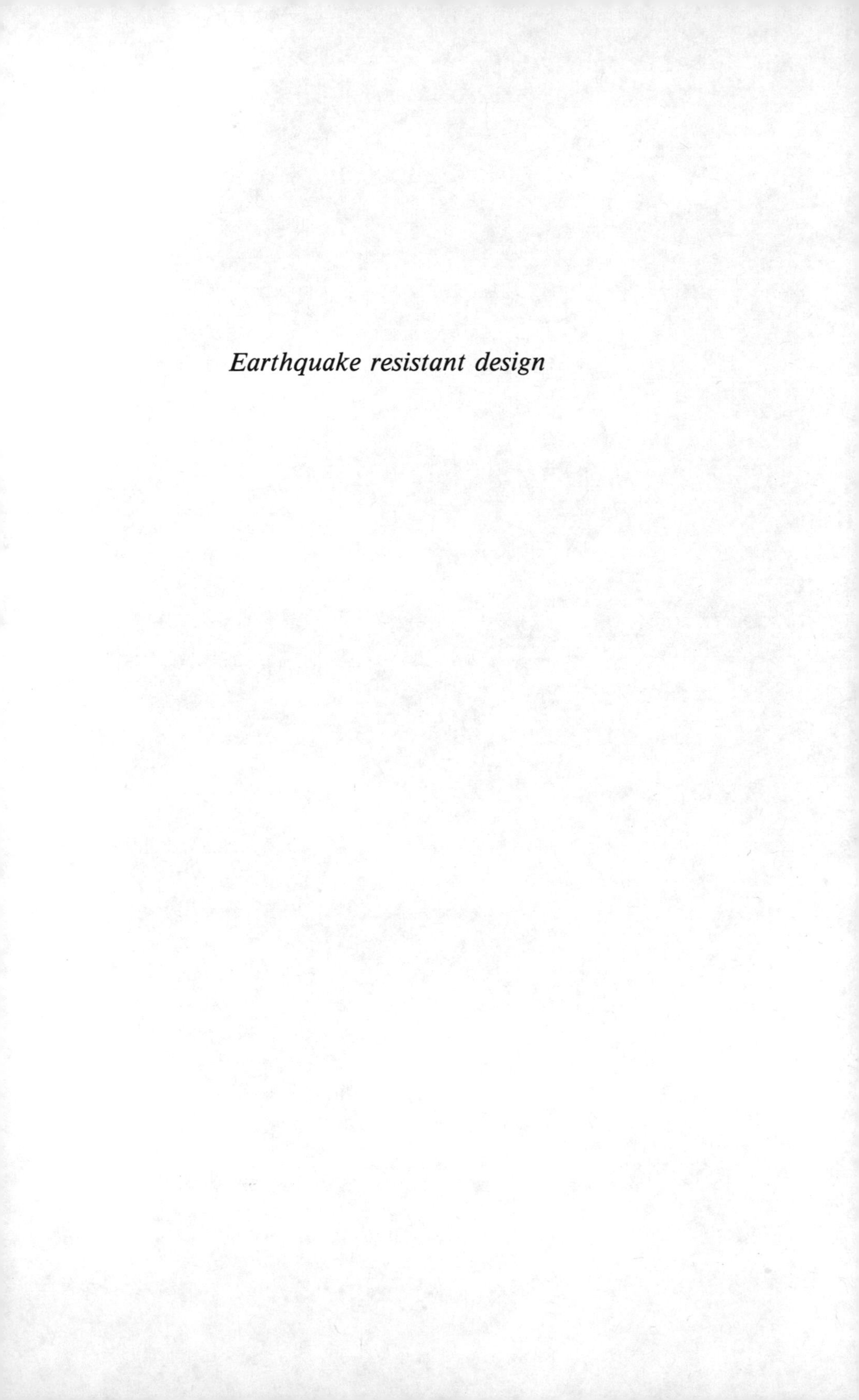

Earthquake resistant design

Earthquake resistant design

For engineers and architects

Second Edition

David J. Dowrick

A Wiley–Interscience Publication

JOHN WILEY & SONS

Chichester · New York · Brisbane · Toronto · Singapore

Reprinted October 1977
Reprinted May 1978
Reprinted October 1978

Library of Congress Cataloging-in-Publication Data:

Dowrick, D. J.
Earthquake resistant design.

'A Wiley-Interscience Publication.'
Includes bibliographical references and index.
1. Earthquake resistant design. I. Title.
TA 658.44.D67 1987 624.1'762 87-2030

ISBN 0 471 91503 3

British Library Cataloguing in Publication Data:

Dowrick, D. J.
Earthquake resistant designs : for engineers and architects.—2nd ed.
1. Earthquake resistant design
I. Title
624.1'762 TA658.44

ISBN 0 471 91503 3

Typeset by Dobbie Typesetting Service, Plymouth, Devon
Printed and bound in Great Britain by Bath Press Ltd

Dedicated
to
my wife Gulielma
and to all people who
live in earthquake areas

Contents

Foreword xv

Preface to the First Edition xvii

Preface to the Second Edition xix

Introduction 1
The design process for earthquake resistant construction

Chapter 1 **Seismic risk—the causes, strength, and effects of earthquakes**
1.1 Seismic risk and hazard 3
1.2 The cause and strength of earthquakes 4
1.3 The social and economic consequences of earthquakes 7
1.3.1 Earthquake consequences and their acceptability 7
1.3.2 Economic consequences of earthquakes 9
1.4 Theory of dynamics and seismic response 13
1.4.1 Introduction 13
1.4.2 Single-degree-of-freedom systems 15
1.4.3 Multi-degree-of-freedom systems 22
1.4.4 Non-linear inelastic earthquake response 28

Chapter 2 **Seismic activity in a regional setting**
2.1 Introduction 31
2.2 Seismotectonics—seismicity and geology 31
2.2.1 Introduction 31
2.2.2 Global seismotectonics 32
2.2.3 Regional seismotectonics 37
2.2.4 Faulting 40
2.3 Earthquake distribution in space, size, and time 54
2.3.1 Introduction 54
2.3.2 Spatial distribution of earthquakes—maps 55
2.3.3 Earthquake distribution in time and size 59
2.3.4 Seismic gaps and seismic quiescence 61
2.3.5 Models of earthquake-occurrence processes 64
2.4 The nature and attenuation of ground motions 67
2.4.1 Earthquake source models 67
2.4.2 The characteristics of strong ground motion 72
2.5 Defining design events 87

Chapter 3 **Determination of site characteristics**
3.1 Introduction 94

3.2 Local geology and soil conditions 94
3.3 Site investigations and soil tests 100
3.3.1 Introduction 100
3.3.2 Field determination and tests of soil characteristics . 100
3.3.3 Laboratory tests relating to dynamic behaviour of soils 107

Chapter 4 **Design earthquakes**
4.1 Introduction 113
4.2 Probability distributions of ground motion criteria 113
4.3 Determining design earthquakes 118
4.3.1 Introduction 118
4.3.2 Sources of accelerograms and response spectra 120
4.3.3 Response spectra as design earthquakes 122
4.3.4 Accelerograms as design earthquakes 130
4.4 Faults—risk and design considerations 135
4.4.1 Introduction 135
4.4.2 Probability of occurrence of fault displacement 135
4.4.3 Designing for fault movements 138

Chapter 5 **Earthquake resistant design philosophy—choice of form and materials**
5.1 Introduction 142
5.2 Criteria for earthquake resistant design 142
5.2.1 Function, cost, and reliability 142
5.2.2 Reliability criteria 143
5.3 Principles of reliable seismic behaviour—form, material, and failure modes 148
5.3.1 Introduction 148
5.3.2 Simplicity and symmetry 148
5.3.3 Length in plan 150
5.3.4 Shape in elevation 150
5.3.5 Uniform and continuous distribution of strength and stiffness 151
5.3.6 Appropriate stiffness 153
5.3.7 Choice of construction materials 156
5.3.8 Failure mode control 157
5.4 Specific structural forms for earthquake resistance 162
5.4.1 Moment-resisting frames 163
5.4.2 Framed tube structures 163
5.4.3 Structural walls (shear walls) 163
5.4.4 Concentrically braced frames 164
5.4.5 Eccentrically braced frames 164
5.4.6 Hybrid structural systems 165
5.5 Energy isolating and dissipating devices 165

5.5.1 Introduction 165
5.5.2 Isolation from seismic motion.................... 166
5.5.3 Base isolation using flexible bearings............ 168
5.5.4 Isolation using flexible piles and energy dissipators . 170
5.5.5 Isolation using uplift............................. 173
5.5.6 Energy dissipators for base-isolated structures...... 176
5.5.7 Energy dissipators for non-isolated structures 177

Chapter 6 **Seismic response of soils and structures**

6.1 Introduction.. 182
6.2 Seismic response of soils............................. 182
6.2.1 Dynamic properties of soils 182
6.2.2 Site response to earthquakes 190
6.3 Seismic response of soil–structure systems 207
6.3.1 Introduction 207
6.3.2 Dynamic analysis of soil–structure systems 208
6.3.3 Soil models for dynamic analysis 209
6.3.4 Useful results from soil–structure interaction studies 224
6.4 Aseismic design of foundations 231
6.4.1 Introduction 231
6.4.2 Shallow foundations 233
6.4.3 Deep box foundations.............................. 234
6.4.4 Caissons.. 235
6.4.5 Piled foundations................................. 235
6.4.6 Foundations in liquefiable ground 245
6.5 Aseismic design of earth-retaining structures 248
6.5.1 Introduction 248
6.5.2 Seismic earth pressures 249
6.6 Seismic response of structures 256
6.6.1 Elastic seismic response of structures............. 256
6.6.2 Non-linear seismic response of structures 258
6.6.3 Mathematical models of non-linear seismic behaviour............................... 265
6.6.4 Level of damping in different structures.......... 266
6.6.5 Periods of vibration of structures................. 267
6.6.6 Interaction of frames and infill panels 269
6.6.7 Methods of seismic analysis for structures 272

Chapter 7 **Concrete structures**

7.1 Introduction.. 288
7.2 *In situ* reinforced concrete design and detail 288
7.2.1 Introduction 288
7.2.2 Seismic response of reinforced concrete 289
7.2.3 Reliable seismic behaviour of concrete structures ... 289
7.2.4 Reinforced concrete structural walls 302

7.2.5 *In situ* concrete design and detailing—general requirements 313
7.2.6 Foundations 317
7.2.7 Retaining walls 320
7.2.8 Walls 320
7.2.9 Columns 327
7.2.10 Beams 332
7.2.11 Beam–column joints in moment-resisting frames ... 334
7.2.12 Slabs 335
7.2.13 Staircases 335
7.2.14 Upstands and parapets 335
7.3 Structural precast concrete detail 339
7.3.1 Introduction 339
7.3.2 Connections between bases and precast columns.... 340
7.3.3 Connections between precast columns and beams... 341
7.3.4 Connections between precast floors and walls 343
7.3.5 Connections between adjacent precast floor and roof units 345
7.3.6 Connections between adjacent precast wall units ... 347
7.4 Precast concrete cladding detail 350
7.5 Prestressed concrete design and detail
7.5.1 Introduction 352
7.5.2 Official recommendations for seismic design of prestressed concrete 352
7.5.3 Seismic response of prestressed concrete 352
7.5.4 Factors affecting ductility of prestressed concrete members 353
7.5.5 Detailing summary for prestressed concrete 355

Chapter 8 **Masonry structures**
8.1 Introduction 361
8.2 Seismic response of masonry 361
8.3 Reliable seismic behaviour of masonry structures 363
8.3.1 Introduction 363
8.3.2 Structural form for masonry buildings 364
8.3.3 Structural form for reinforced masonry 364
8.4 Design and construction details for reinforced masonry..... 366
8.4.1 Minimum reinforcement 366
8.4.2 Horizontal continuity 366
8.4.3 Grouting 368
8.4.4 Hollow concrete blocks 368
8.4.5 Supervision of construction 368
8.5 Construction details for structural infill walls 369

Chapter 9 **Steel structures**
9.1 Introduction 371

9.2 Seismic response of steel structures 371
9.3 Reliable seismic behaviour of steel structures 374
9.3.1 Introduction 374
9.3.2 Material quality of structural steel 375
9.4 Steel beams 377
9.4.1 Moment–curvature relationships for steel beams under monotonic loading 377
9.4.2 Behaviour of steel beams under cyclic loading 380
9.4.3 Design of steel beams 380
9.5 Steel columns 382
9.5.1 Monotonic and hysteretic behaviour of steel columns 382
9.5.2 Design of steel columns 383
9.6 Steel frames with diagonal braces 387
9.6.1 Concentrically braced steel frames 387
9.6.2 Eccentrically braced steel frames 389
9.7 Steel Connections 392
9.7.1 Introduction 392
9.7.2 Behaviour of steel connections under cyclic loading . 393
9.7.3 Deformation behaviour of steel panel zones 396
9.7.4 Design of steel connections for seismic loading 396
9.8 Composite construction 399
9.9 Steel structural walls 399

Chapter 10 **Timber structures**
10.1 Introduction 402
10.2 Seismic response of timber structures 402
10.3 Reliable seismic behaviour of timber structures 407
10.4 Foundations of timber structures 407
10.5 Timber-sheathed walls (shear walls) 408
10.6 Timber horizontal diaphragms 410
10.7 Timber moment-resisting frames and braced frames 413
10.8 Connections in timber construction 414
10.9 Fire resistance of timber construction 416

Chapter 11 **Earthquake resistance of services and equipment**
11.1 Seismic response and design criteria 419
11.1.1 Introduction 419
11.1.2 Earthquake motion—accelerograms 419
11.1.3 Design norms—design earthquakes 420
11.1.4 The response spectrum design method 420
11.1.5 Comparison of design requirements for buildings and equipment 421
11.1.6 Equipment mounted in buildings 422
11.1.7 Material behaviour 423

11.1.8 Costs of providing earthquake resistance for equipment 424
11.2 Seismic analysis and design procedures for equipment 424
11.2.1 Design procedures using dynamic analysis 425
11.2.2 Design procedures using equivalent-static analysis 426
11.3 Aseismic protection of equipment 433
11.3.1 Introduction 433
11.3.2 Rigidly mounted equipment 434
11.3.3 Equipment mounted on isolating or energy-absorbing devices 436
11.3.4 Light fittings 436
11.3.5 Ductwork 437
11.3.6 Pipework 437

Chapter 12 **Architectural detailing for earthquake resistance**
12.1 Introduction 443
12.2 Non-structural infill panels and partitions 444
12.2.1 Introduction 444
12.2.2 Integrating infill panels with the structure 445
12.2.3 Separating infill panels from the structure 445
12.2.4 Separating infill panels from intersecting services 447
12.3 Cladding, wall finishes, windows, and doors 448
12.3.1 Introduction 448
12.3.2 Cladding and curtain walls 448
12.3.3 Weather seals 448
12.3.4 Wall finishes 449
12.3.5 Windows 449
12.3.6 Doors 449
12.4 Miscellaneous architectural details 450
12.4.1 Exit requirements 450
12.4.2 Suspended ceilings 450
12.4.3 Landscape elements 453
12.4.4 Window-washing rigs 453

Appendix A **Earthquake resistance of specific structures**
A.1 Earthquake resistance of bridges 455
A.1.1 Introduction 455
A.1.2 Reliable seismic behaviour of bridges 455
A.1.3 Seismic analysis of bridges 459
A.1.4 Strength and ductility of bridges 461
A.1.5 Superstructure forces on abutments 461
A.1.6 Movement joints and horizontal linkage systems 462
A.1.7 Holding-down devices 463
A.1.8 Energy-isolating and dissipating devices for bridges 463
A.2 Chimneys and towers 463

A.2.1 Introduction 463
A.2.2 Reliable seismic behaviour of chimneys and towers . 463
A.2.3 Seismic analysis of chimneys and towers 464
A.2.4 Framed chimneys and tower structures 465
A.2.5 Free-standing chimneys and stack-like towers 465
A.2.6 Inverted pendulum structures 469
A.2.7 Energy-isolating and dissipating devices for chimneys and towers 472
A.3 Low-rise commercial–industrial buildings 472
A.3.1 Introduction 472
A.3.2 Reliable seismic behaviour of low-rise commerical–industrial buildings 472
A.3.3 Seismic analysis 473
A.3.4 Foundations for low-rise commercial–industrial buildings 473
A.3.5 Connections in low-rise commercial–industrial buildings 473
A.4 Low-rise housing 474
A.4.1 Introduction 474
A.4.2 Symmetry in plan of low-rise housing 477
A.4.3 Apertures in walls 477
A.4.4 The strength and stiffness of walls 477
A.4.5 Horizontal continuity 478
A.4.6 Foundations for low-rise housing 478
A.4.7 Roofs of heavy construction 478
A.4.8 Chimneys and decorative panels 479
A.5 Improving the earthquake resistance of existing structures .. 479

Appendix B **Miscellaneous information**
B.1 Modified Mercalli intensity scale 484
B.2 Quality of reinforcement for concrete 485
B.3 Statistical methods for probability studies 490
B.3.1 Introduction 490
B.3.2 Definitions of some statistical terms 490
B.3.3 Establishing relationships from data of seismicity observations 492
B.3.4 Estimating goodness of fit—correlation 494
B.3.5 Confidence limits of regression estimates 499
B.4 Determining probabilities of ground-motion criteria 503
B.4.1 The basic method 503
B.4.2 Probabilistic enhancement from attentuation variability 507

Index 511

Foreword

Earthquakes are one of nature's greatest hazards to life on this planet; throughout historic time they have caused the destruction of countless cities and villages on nearly every continent. They are the least understood of the natural hazards and in early days were looked upon as supernatural events. Possibly for this reason earthquakes have excited concern which is out of proportion to their actual hazard. Certainly the average annual losses due to wind and flood exceed those due to earthquakes in many parts of the world, and all of these represent lesser life hazards than are accepted daily in our streets and highways. Nevertheless, the totally unexpected nearly instantaneous devastation of a major earthquake has a unique psychological impact which demands serious consideration by modern society.

The hazards imposed by earthquakes are unique in many respects, and consequently planning to mitigate earthquake hazards requires a unique engineering approach. An important distinction of the earthquake problem is that the hazard to life is associated almost entirely with man-made structures. Except for earthquake-triggered landslides, the only earthquake effects that cause extensive loss of life are collapses of bridges, buildings, dams, and other works of man. It is this fact that has led to the great emphasis placed on earthquake prediction in one of the world's great seismic regions—The People's Republic of China. With even a few hours of advance notice, people can be evacuated from buildings and houses into open fields where loss of life can be almost completely avoided. Apparently such a prediction was effective in saving hundreds or possibly thousands of lives during the Haicheng, China, earthquake of February 1975.

However, it is evident that even a successful prediction cannot eliminate the earthquake hazard; even if all the people are evacuated safely, the structures which largely determine the standard of living of the community remain, and their destruction could be a disastrous loss to the regional economy. This aspect of earthquake hazard can be countered only by the design and construction of earthquake resistant structures, and therefore a completely successful earthquake prediction programme could not eliminate the need for effective earthquake engineering. On the other hand, with effective application of earthquake engineering knowledge, the collapse of structures and the resulting life hazard can be avoided; this would greatly reduce the value of any earthquake prediction programme.

Earthquake hazard poses a unique engineering design problem in that an intense earthquake constitutes the most severe loading to which most civil engineering structures might possibly be subjected, and yet the probability that any given structures will ever be affected by a major earthquake is very low.

The optimum engineering approach to this combination of conditions is to design the structure so as to avoid collapse in the most severe possible earthquake—thus ensuring against loss of life—but accepting the possibility of damage, on the basis that it is less expensive to repair or replace the small number of structures which will be hit by a major earthquake than to build all structures strong enough to avoid damage. Clearly this design concept presents the structural engineer with a most challenging problem: to provide an economical design which is susceptible to earthquake damage but which is essentially proof against collapse in the greatest possible earthquake.

Another unique feature of the earthquake excitation provides the key to the solution of this design problem. In contrast to the other loads considered in structural design—wind, gravity, hydrodynamic, etc.—the intensity of the earthquake loading depends on the properties of the structure. Thus adequate earthquake resistance may be provided either by the traditional approach of increasing strength or by the unique seismic design concept of reducing stiffness and thereby reducing the forces to be resisted. This additional approach to earthquake design imposes a greater need for understanding of structural behaviour in earthquake engineering than in any other field of civil engineering design. Seemingly minor changes in the framing system or in design details may have an overwhelming influence on the seismic performance; and merely adding more materials—though it will directly increase costs—will not guarantee satisfactory performance.

It is because understanding is the key to good earthquake design, and because the quality of design has such a profound influence on the earthquake performance of structures, that this earthquake design manual by David Dowrick will occupy an important place in engineering design offices throughout the world.

RAY W. CLOUGH
Berkeley, California

Preface to the First Edition

This book was originally written for the guidance of architects and engineers employed in the international practice of the Ove Arup Partnership. Because it is not written specifically for application in any one country it should be of assistance to designers in any part of the world. Much of the text should also be of interest to students of architecture and engineering, the elements of Chapter 4 being particularly recommended at that stage.

In preparing the text, valuable advice has been obtained from generous people in many parts of the world, some of whom are acknowledged below. Because of the enormous scope of the book brevity has been essential, with referencing to source material. With the rapid advance in the understanding of much of the subject matter it is hoped that the text will be revised from time to time, and suggestions for its improvement will be welcomed.

Grateful acknowledgements for their assistance in preparing parts of the text is made to my colleagues C. H. I. Balmond, R. J. Bentley, J. C. Blanchard, A. K. Denney, M. V. Harley, P. Parlour, C. P. Wade, and R. T. Whittle. The Ove Arup Partnership and the author wish to thank Professor N. A. Mowbray of Auckland University and Professor R. Park of Canterbury University for their kindness, advice, and encouragement during part of the preparation of this document. Amongst the many architects, engineers, and seismologists who also gave helpful advice, the author is particularly indebted to Professor T. Paulay, Canterbury University, Mr. R. Granwell, Professor R. Shepherd, and Professor P. W. Taylor, Auckland University; Mr O. A. Glogau, New Zealand Ministry of Works; Professor G. W. Housner, California Institute of Technology; Dr A. G. Brady, US Geological Survey, San Fransisco; Professor R. V. Whitman, Massachusetts Institute of Technology; and Mr J. Lord, Seismic Engineering Associates, Los Angeles.

The author is grateful to the Literary Executor of the late Sir Ronald A. Fisher, FRS, to Dr Frank Yates, FRS, and to Longman Group Ltd, London, for permisison to reprint Table III from their book *Statistical Tables for Biological, Agricultural and Medical Research*. (6th edition, 1974.)

The author is especially grateful to Professor R. W. Clough of Berkeley for his careful scrutiny of the manuscript and his constructive Foreword.

Preface to the Second Edition

In the decade or so since the writing of the first edition of this book much progress has been made in understanding earthquakes and in how to build more safely. In some areas of study great developments have occurred, such as in seismotectonics, hazard analysis, and design earthquakes, base isolation, and microcomputers for everyone, and there has been wider recognition of the importance of structural form. However, one of the great difficulties for designers arises simply from the enormous volume of literature being produced on each of the many specialisms with the overall subject area. Hopefully, this book will help some of us to find our way better through this maze.

This book was written from the standpoint of a designer trying to keep a broad perspective on the total process, starting from the nature of the loading through to the details of construction. To this end the successful overall format of the first edition has been retained, although with some reorganization of the sequence of the sub-sections. I have attempted to give the book as international a flavour as possible, although I have inevitably drawn more heavily on information from the literature that I know best.

Our greatest current need for progress lies in improving the reliability of seismic response, i.e. in failure mode control, of all types of construction, not least in the housing of the poorer people in the world. This requires (1) better modelling of both the design earthquake and structural response, (2) better understanding of the roles of ductilility and energy dissipation, (3) better collaboration between engineers and architects, and (4) simpler methods of analysis and detailing rules.

Finally I must express my gratitude to all those fellow workers upon whose works I have drawn, and I also acknowledge all the valuable work to which I have not been able to refer, through the sheer enormity of the task. Special thanks are owed to my wife Gulielma for typing the manuscript, thus easing my burden so much.

Introduction

This book is intended to help architects and engineers carry out good earthquake-resistant design expeditiously. Earthquake-resistant design is such a wide and immature subject that there exists a genuine difficulty in deciding what design criteria and analytical methods should be applied to any one project.

The principal objectives are as follows

(1) To discuss the chief aspects of seismic risk evaluation and earthquake-resistant design.
(2) To evaluate various alternative design techniques.
(3) To give guidance on topics where no generally accepted method is currently available.
(4) To suggest procedures to be adopted in earthquake regions having no official zoning or lateral force regulations.
(5) To indicate the more important specialist literature.

The general principles of this book apply to the whole range of building construction and civil engineering, while the more detailed sections relate to the structural rather than the heavy civil engineering industry.

Whereas an attempt has been made to provide guidance in the more important areas of design, the coverage can scarcely be exhaustive, even in the long term.

THE DESIGN PROCESS FOR EARTHQUAKE RESISTANT CONSTRUCTION

Earthquakes provide architects and engineers with a number of important design criteria foreign to the normal design process. As some of these criteria are fundamental in determining the form of the structure it is crucial that adequate attention is given to earthquake considerations at the correct stage in the design. To this end a simplified flow chart of the design process for earthquake resistant structures is shown in Figure (i).

Although the real interrelationships between all the factors shown in the diagram are obviously much more complex than indicated, the overall sequence is correct. All factors 1 to 9 are related when evaluating the level of seismic risk, as the risk depends not only on the possible earthquake loadings but also on the capacity of the construction to avoid damage.

The *design brief* (Box 1 of Figure (i)) for different projects is developed by the designers with varying amounts of input from the owner, and varying aspects and degrees of detail of the brief are subject to the owner's agreement.

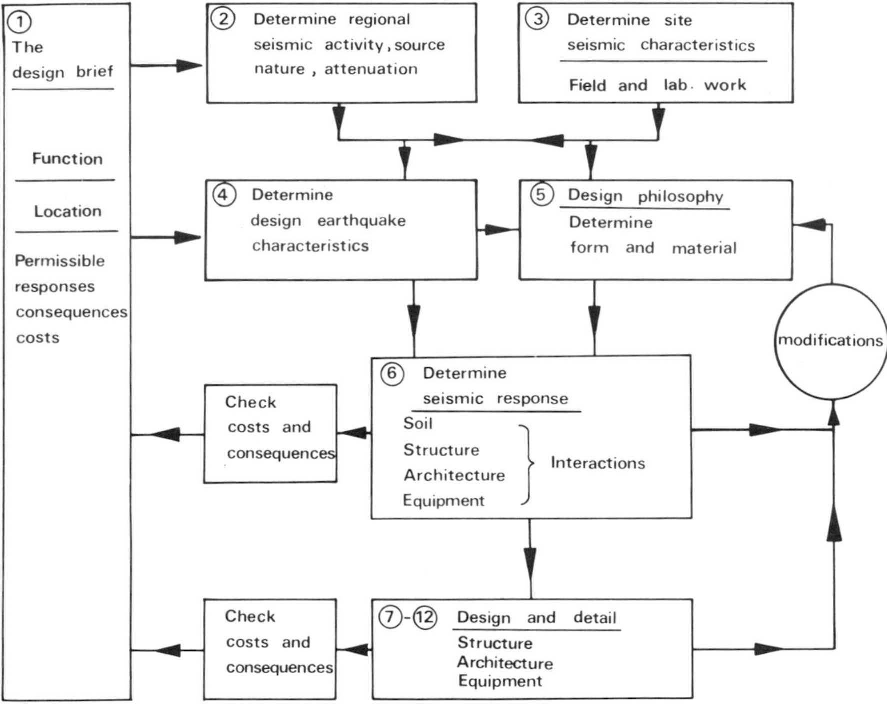

Figure (i) Simplified flow chart for the design of earthquake resistant construction (non-seismic factors omitted for clarity)

Few owners wish to be involved in deciding the acceptable level of risk, but, in any case, it is important that the owner should be informed of the risks consequent to the available options. Even when the design is done according to a good local code, high risks may still exist.

The various stages of the design process are now discussed, with the sequence of numbers in Figure (i) corresponding to the numbers of the relevant chapters.

Chapter 1

Seismic risk: the causes, strength, and effects of earthquakes

1.1 SEISMIC RISK AND HAZARD

In normal English usage the word *risk* means *exposure to the chance of injury or loss*. It is noted that the word *hazard* is almost synonymous with *risk*, and the two words are used in the risk literature with subtle variations which can be confusing.

Fortunately an authoritative attempt has been made to overcome this difficulty through the publication by the Earthquake Engineering Research Institute's Committee on Seismic Risk of a glossary[1] of standard terms for use in this subject. Their terminology will be used in this book.

Thus the definition of *seismic risk* is *the probability that social or economic consequences of earthquakes will equal or exceed specified values at a site, at several sites, or in an area, during a specified exposure time*. Risk statements are thus given in quantitative terms.

Seismic hazard, on the other hand, is *any physical phenomenon (e.g. ground shaking, ground failure) associated with an earthquake that may produce adverse effects on human activities*. Thus hazards may be either purely descriptive terms or quantitatively evaluated, depending on the needs of the situation.

It follows that seismic risk is an outcome of seismic hazard as described by relationships of the form

$$\text{Seismic risk} = (\text{Seismic hazard}) \times (\text{Vulnerability})(\text{Value}) \qquad (1.1)$$

where *Vulnerability* is the amount of damage, induced by a given degree of hazard, and expressed as a fraction of the *Value* of the damaged item under consideration. Referring to Figure 1.1, the Monetary Seismic Risk to a building could be evaluated by taking the Seismic Hazard to be the MM intensity of the appropriate probability of occurrence, the Vulnerability would then be taken as the damage ratio on the appropriate curve for that intensity, and the Value would be the Replacement Cost. It is noted that Seismic Hazard in equation (1.1) is qualitative and not quantitative in nature, i.e. it is only used for finding the Vulnerability and the probability level.

For design or risk assessment purposes the assessment of seismic hazard consists of the following basic steps;

(1) Definition of the nature and locations of earthquake sources;
(2) Magnitude–frequency relationships for the sources;
(3) Attenuation of ground motion with distance from source;
(4) Determination of ground motions at the site having the required probability of exceedance.

Because seismic risk and hazard statements are essentially forecasts of future situations they are inherently *uncertain*. Seismic hazard assessments are attempts to forecast the likely future seismic activity rates and strengths, based on knowledge of the past and present, and significant uncertainties arise partly because the processes involved are not fully understood and partly because relevant data are generally scarce and poor in quality. For reasonable credibility considerable knowledge of both *historical seismicity* and *geology* need to be used, together with an appropriate analysis of the uncertainties. *Seismicity* is defined as the frequency of occurrence of earthquakes per unit area in a given region, and is illustrated in non-numerical terms by the seismicity map of the world presented in Chapter 2 in Figure 2.1. Where available, other geophysical or seismological knowledge, such as strain-release studies, may also be helpful, particularly in evaluating regional seismic activity patterns. Once both the estimated future seismic-activity rates and the acceptable risks are known, appropriate earthquake loadings for the proposed structure may be determined, e.g. loadings with mean recurrence intervals of say 100 or 1000 years, depending on the consequences of failure.

Because of the difficulties involved in seismic hazard evaluation, earthquake design criteria in different areas of the world vary, from well codified to inadequate or non-existent. Hence, depending on the location and nature of the project concerned, seismic risk evaluation ranging from none through arbitrary to thorough-going may be required.

The whole of this book is essentially to do with the explicit or implicit management of seismic risk, and hence the forcgoing bricf introduction to risk and hazard will be expanded upon in the subsequent text.

1.2 THE CAUSE AND STRENGTH OF EARTHQUAKES

An earthquake is a spasm of ground shaking caused by a sudden release of energy in the earth's lithosphere (i.e. the crust plus part of the upper mantle). This energy arises mainly from stresses built up during tectonic processes, which consist of interaction between the crust and the interior of the earth. In some parts of the world earthquakes are associated with volcanic activity. For example, in Guatemala such earthquakes occur in swarms, with an average duration of three to four months, the largest having a magnitude normally under 6.5. These events are of shallow focus and cause considerable damage within a radius of about 30 km from the epicentre. Human activity also sometimes modifies crustal stresses enough to trigger small or even moderate earthquakes, such as the periodic swarms of minor tremors resulting from mining in the Midlands of England, or the sometimes larger events induced by the impounding of large amounts of water behind dams, such as the earthquakes associated with the construction of the Koyna dam in central India in 1967.[2]

While the design provisions of this book apply to all earthquakes regardless of origin, any discussion of earthquakes themselves is generally confined to events derived from the main cause of seismicity, i.e. tectonic activity. During such earthquakes the release of crustal stresses is believed generally to involve the fracturing of the rock along a plane which passes through the point of origin (the *hypocentre* or *focus*) of the event (Figure 2.22). Sometimes, especially in larger shallower earthquakes, this rupture plane, called a *fault*, breaks through to the ground surface, where it is known as a *fault trace* (Figure 2.31).

The cause and nature of earthquakes is the subject of study of the science of *seismology*, and further background may be obtained from the classical books by Richter[3] and Eiby[4].

Unfortunately for non-seismologists at least, understanding the general literature related to earthquakes is impeded by the difficulty of finding precise definitions of fundamental seismological terms. For assistance in the use of this book, definitions of some basic terms are set out below. Further definitions may be found elsewhere in this book or in the references given above.

The *strength* of an earthquake is not an official technical term, but is used in the normal language sense of 'How strong was that earthquake?' Earthquake strength is defined in two ways: first the strength of shaking at any given place (called the *intensity*) and second, the total strength (or size) of the event itself (called *magnitude*, seismic *moment*, or *moment magnitude*). These entities are described below.

Intensity[1] is *a qualitative or quantitative measure of the severity of seismic ground motion at a specific site*. Over the years, various subjective scales of what is often called *felt intensity* have been devised, notably the Mercalli, the Rossi and Forel and the MSK scales. The most widely used in the English speaking world is the Modified Mercalli scale (commonly denoted MM), which has twelve grades denoted by Roman numerals I–XII. A detailed description of this intensity scale is given in Appendix B.1.

Quantitative instrumental measures of intensity include engineering parameters such as peak ground acceleration, peak ground velocity, the Housner spectral intensity, and response spectra in general. Because of the high variability of both subjective and instrumental scales, the correlation between these two approaches to describing intensity is inherently weak (Figure 2.24).

Magnitude is a quantitative measure of the size of an earthquake, related indirectly to the energy released, which is independent of the place of observation. It is calculated from amplitude measurements on seismograms, and is on a logarithmic scale expressed in ordinary numbers and decimals. The most commonly used magnitude scale is that devised by and named after Richter, and is denoted M or M_L. It is defined as

$$M_L = \log A - \log A_o \qquad (1.2)$$

where A is the maximum recorded trace amplitude for a given earthquake at a given distance as written by a Wood-Anderson instrument and A_o is that for a particular earthquake selected as standard.

The Wood-Anderson seismograph ceases to be useful for shocks at distances beyond about 1000 km, and hence Richter magnitude is now more precisely called *local magnitude* (M_L) in order to distinguish it from magnitude measured in the same way but from recordings on long-period instruments, which are suitable for more distant events. These latter magnitudes are measured from surface wave impulses and are hence denoted by M_s. Because *surface wave magnitudes* are not always reliable, Gutenburg proposed what he called 'unified magnitude', denoted m or m_b, which is dependent on body waves, and is now generally named *body wave magnitude* (m_b). This magnitude scale is particularly appropriate for events with a focal depth greater than *c*. 45 km.

There are significant discrepancies between values of magnitude determined by the above methods at the bottom and top ends of the scale (See Tables 1.1 and 2.3), so that it is often important to know the type of magnitude being used. In this book the local (i.e. Richter) magnitude is commonly used and is denoted by M.

The greatest magnitude events that have been instrumentally recorded (up to 1984) were the Colombia–Equador earthquake of 31 January 1906 and the Sanriku Japan earthquake in 1933, both of which reached $M = 8.9$. Probably the largest known event was the pre-instrumental Lisbon earthquake of 1755, which may have been of magnitude 9.0[4]. In fact these values are about the maximum that can be achieved on the magnitude scale, which system tends to saturate at about $M = 9$ regardless of the amount of energy released (Figure 2.18).

Magnitude in Multiple Events

A complication in the understanding of earthquakes which has been recognized in recent years is the occurrence of multiple events, which consist of two or more separate fault ruptures (not necessarily on the same fault) for which the ground shaking overlaps. Such events thus appear superficially to be one event, and added difficulties arise in assigning magnitudes and calculating the stress drop on the fault(s) during the earthquake as compared with single or simple events. The magnitudes of numbers of past events may be re-determined if examination of their records shows them to be multiple events. For example, it has been suggested that the 1971 San Fernando California earthquake may have been a double event.[5]

Seismic moment measures the size of an earthquake directly from the energy released, through the expression[6]

$$M_o = GAD \tag{1.3}$$

where G is the shear modulus of the medium (and is usually taken as 3×10^{11} dyne/cm^2), A is the area of the dislocation or fault surface, and D is the

Table 1.1. Seismic moments, magnitudes, and moment magnitudes for some well-known Californian earthquakes[7]

Event	M_o $\times 10^{25}$ (dyn. cm)	M_L	M_s	M_m
San Francisco, 18 April 1906	400	–	8.25	7.7
Long Beach, 10 March 1933	2	6.3	6.25	6.2
El Centro, 18 May 1940	30	6.4	6.7	7.0
Kern County, 21 July 1952	200	7.2	7.7	7.5
San Fernando, 9 February 1971	20	6.4	6.6	6.6
Point Mugu, 21 February 1973	0.1	5.9	5.2	5.3

average displacement of slip on that surface. Seismic moment is a modern alternative to magnitude, which avoids the shortcomings of the latter but is not so readily determined. Up to 1985 seismic moment had generally only been used by seismologists.

Moment magnitude is a relatively recent magnitude scale from Hanks and Kanamori[7] which overcomes the above-mentioned saturation problem of other magnitude scales by incorporating seismic moment into its definition, such that moment magnitude

$$M_m = \tfrac{2}{3}\log M_o - 10.7 \tag{1.4}$$

A comparison of the values of seismic moment, plus local and surface wave magnitudes, with moment magnitude for some Californian earthquakes is given in Table 1.1. It appears that this moment magnitude scale has much to recommend it, and it is therefore to be hoped that values on this scale become available for many earthquakes in the near future so that they may become normal usage for other than seismologists. (Note that M_m is adopted here as the symbol for moment magnitude, instead of bold type **M** used by Hanks and Kanamori,[7] because of the difficulty of writing **M** by hand, and confusions between **M** and M easily arise.)

1.3 THE SOCIAL AND ECONOMIC CONSEQUENCES OF EARTHQUAKES

1.3.1 Earthquake consequences and their acceptability

The physical consequences of earthquakes for human beings are generally viewed under two headings;

(1) Death and injury to human beings;
(2) Damage to the constructed and natural environments.

These physical effects in turn are considered as to their social and economic consequences, both of which aspects are taken into account when the acceptable consequences are being decided, i.e. the acceptable seismic risk.

Both financially and technically it is possible only to *reduce* these consequences for large earthquakes. The basic design aims are therefore confined (a) to the reduction of loss of life in any earthquake, either through collapse or through secondary damage such as falling debris or earthquake-induced fire, and (b) to the reduction of damage and loss of use of the constructed environment.

Obviously some facilities demand greater earthquake resistance than others, because of their greater social and/or financial significance. It is important to determine in the design brief not only the more obvious intrinsic value of the structure, its contents, and function or any special parts thereof, but also the survival value placed upon it by the owner.

In some countries the greater importance to the community of some types of facility is recognized by regulatory requirements, such as in New Zealand, where various public buildings are designed for higher earthquake forces than other buildings. Some of the most vital facilities to remain functional after destructive earthquakes are dams, hospitals, fire and police stations, government offices, bridges, radio and telephone services, schools, energy sources, or, in short, anything vitally concerned with preventing major loss of life in the first instance and with operation of emergency services afterwards.

In some cases the owner may be aware of the consequences of damage to his property but may do nothing about it. It is worth noting that even in earthquake-conscious California, it was only since the destruction of three hospitals and some important bridges in the San Fernando earthquake of 1971 that there have been statutory requirements for extra protection of various vital structurcs.

The consequences of damage to structures housing intrinsically dangerous goods or processes is another category of consideration, and concerns the potential hazards of fire, explosion, toxicity, or pollution represented by installations such as liquid petroleum gas storage facilities or nuclear power or nuclear weapon plants. These types of consequences often become difficult to consider objectively, as strong emotions are provoked by the thought of them. Acknowledging the general public concern about the integrity of nuclear power plants, the authorities in the United Kingdom decided in the 1970s that future plants should be designed against earthquakes, although that country is one of low seismicity and aseismic design is not generally required.

Since the 1960s, with the growing awareness of the high seismic risks associated with certain classes of older buildings, programmes for strengthening or replacement of such property have been introduced in various parts of the world, notably for pre-earthquake code buildings of lightly reinforced or unreinforced masonry construction in New Zealand and the USA. While the substantial economic consequences of the loss of many such buildings in earthquakes are, of course, apparent, the main motivating force behind these risk-reduction programmes has been social, i.e. the general attempt to reduce loss of life and injuries to people, plus the desire to save buildings or monuments of historical and cultural importance.

While individual owners, designers, and third parties are naturally concerned specifically about the consequences of damage to their own proposed or existing property, the overall effects of a given earthquake are also receiving increasing attention. Government departments, emergency services, and insurance firms all have critical interests in the physical and financial overall effects of large earthquakes on specific areas. In the case of insurance companies, they need to have a good estimate of their likely losses in any single large catastrophe event[8] so that they can arrange sufficient reinsurance if they are over-exposed to seismic risk. Disruption of lifelines such as transport, water, and power systems obviously greatly hampers rescue and rehabilitation programmes.

1.3.2 Economic consequences of earthquakes

The economic consequences of earthquakes occur both before and after the event. Those arising before the event include protection provisions such as earthquake resistance of new and existing facilities, insurance premiums, and provision of earthquake emergency services. Insurance companies themselves need to reinsure against large earthquake losses, as mentioned in the previous section.

Post-earthquake economic consequences include;

(1) Cost of death and injury;
(2) Cost of damage;
(3) Losses of production and markets;
(4) Insurance claims.

The direct cost of damage depends on the nature of building or other type of facility, its individual vulnerability, and the strength of shaking or other seismic hazard to which it is subjected. Figure 1.1 shows the estimated damage costs as functions of MM intensity for several different types of commercial building in New Zealand built from 1966 to 1976. It is stressed that these curves are based on comparatively few data, especially for higher intensities, and relate to average damage levels for each class of building.

In a study of earthquake damage in the Greece–Turkey region Ambraseys and Jackson[9] developed relationships for the total damage to houses as functions of epicentral intensity and of magnitude. Figure 1.2 shows the percentage of houses destroyed or rendered uninhabitable within the isoseismal MMI = VI + as a function of magnitude for a population density of $50/km^2$.

The curves in Figure 1.2 are for different construction types, indicated by symbols defined as;

R = Rubble masonry;
A = Adobe;
B = Brick with reinforced concrete floors;
T = Timber framed;
S = Stone masonry;
s = Rubble or adobe, reinforced with timber.

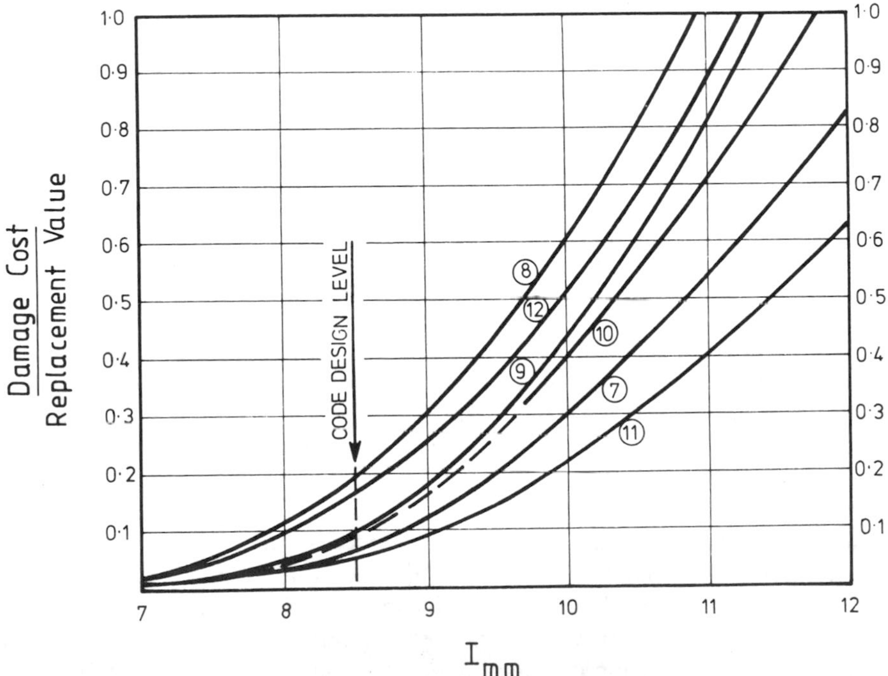

Figure 1.1 Estimated average vulnerability (damage ratios) of New Zealand commercial buildings built between 1966 and 1976[8]

A fuller description of the construction types is given in the original paper[9], where the difference in vulnerability of the non-engineered construction (types R and A) is contrasted with the remainder, which all have basic engineering content.

During the briefing and budgeting stages of a design, the cost of providing earthquake resistance will have to be considered, at least implicitly, and sometimes explicitly, such as for the upgrading of older structures. The cost will

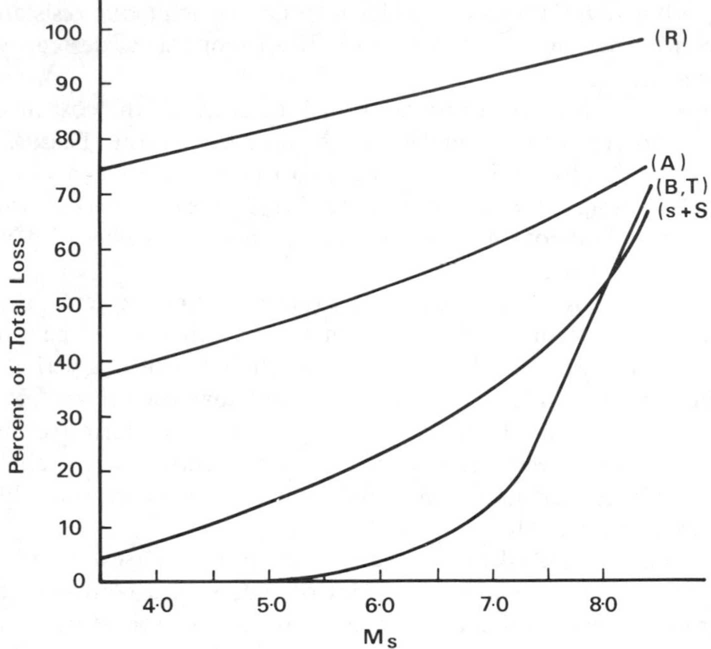

Figure 1.2 Percentage of total loss of houses within isoseismal MMI = VI + for Greece–Turkey region: population normalized at 50/km^2 (after Ambraseys *et al.*[9])

depend on such things as the type of project, site conditions, the form of the structure, the seismic activity of the region, and statutory design requirements. The capital outlay actually made may in the end be determined by the wealth of the client and his or her attitude to the consequences of earthquakes, and insurance to cover losses.

Unfortunately it is not possible to give simple guides on costs, although it would not be misleading to say that most engineering projects designed to the fairly rigorous Californian or New Zealand regulations would spend a maximum of 10 percent of the total cost on earthquake provisions, with 5 percent as an average figure.

The cost of seismic upgrading of older buildings varies from as little as about 10 percent to more than the replacement cost, depending on the nature of the building, the level of earthquake loadings used, and the amount of non-structural upgrading that is done at same time as the strengthening. It is sad to record that many fine old buildings have been replaced rather than strengthened, despite it often being much cheaper to strengthen than to replace. In such cases factors or desires, often than short-term economic or cultural/historical, have governed the decisions of the owners.

Where the client simply wants the minimum total cost satisfying local regulations, the usual cost-effectiveness studies comparing different forms and materials will apply. For this a knowledge of good earthquake-resistant forms will, of course, hasten the determination of an economical design, whatever the material chosen.

In some cases, however, a broader economic study of the cost involved in prevention and cure of earthquake damage may be fruitful. These costs can be estimated on a probability basis and a cost-effectiveness analysis made to find the relationship between capital expenditure on earthquake resistance on the one hand, and the cost of repairs and loss of income together with insurance premiums on the other.

For example, Elms *et al.*,[10] have found that in communal terms the capital cost savings of neglecting aseismic design and detailing would be more than offset by the increased economic losses in earthquakes over a period of time in any part of New Zealand. It is not clear just how low the seismic activity rate needs to be in order for it to be cheaper in the long term for any given community to omit specific seismic resistance provisions. The availability or not of private sector earthquake insurance in such circumstances would be part of the economic equation.

Hollings[11] has discussed the earthquake economics of several engineering projects. In the case of a 16-storey block of flats with a reinforced concrete ductile frame it was estimated that the cost of incorporating earthquake resistance against collapse and subsequent loss of life was 1.4 percent of the capital cost of building, while the cost of preventing other earthquake damage was reckoned as a further 5.0 percent, a total of 6.4 percent. The costs of insurance for the same building were estimated as 4.5 percent against deaths and 0.7 percent against damage, a total of 5.2 percent. Clearly a cost-conscious client would be interested in outlaying a little more capital against danger from collapse, thus reducing the life insurance premiums, and he or she might well consider offsetting the danger of damage mainly with insurance.

Loss of income due to the building being out of service was not considered in the preceding example. In a hypothetical study of a railway bridge, Hollings showed that up to 18 percent of the capital cost of the bridge could be spent in preventing the bridge going out of service, before this equalled the cost of complete insurance cover.

In a study of Whitman *et al.*,[12] an estimate was made of the costs of providing various levels of earthquake resistance for typical concrete apartment buildings of different heights, as illustrated in Figure 1.3. Until further studies of this type have been done, results such as shown in the figure should be used qualitatively rather than quantitatively.

It is most important that at an early stage the owner should be advised of the relationship between strength and risk so that he can agree to what he is buying. Where stringent earthquake regulations must be followed the question of insurance versus earthquake resistance may not be a design consideration: but it can still be important, for example for designing non-structural partitions to be expendable or if a 'fail-safe' mechanism is proposed for the structure. Where there are loose earthquake regulations or none at all, insurance can be a

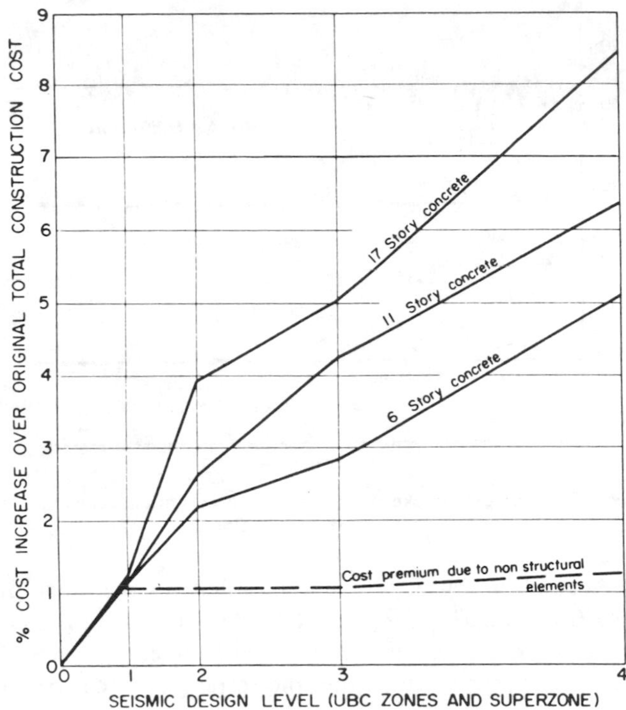

Figure 1.3 Effect on cost of a seismic design of typical concrete apartment buildings in Boston (after Whitman *et al.*[12])

much more important factor, and the client may wish to spend little on earthquake resistance and more on insurance.

However, in some cases insurance may be more expensive, or unavailable, for facilities of high seismic vulnerability. For example, the latter is now the case for older masonry buildings in some high seismic risk areas of New Zealand, i.e. those built prior to the introduction of that country's earthquake loadings code in 1935.

1.4 THEORY OF DYNAMICS AND SEISMIC RESPONSE

1.4.1 Introduction

In Section 1.3 above we have discussed the effects of earthquakes in terms of the social and economic consequences, which are largely the result of the shaking or dynamic response of soils and structures. As this direct effect of *dynamic response* underlies most of the rest of this book it is convenient to introduce this subject in this first chapter.

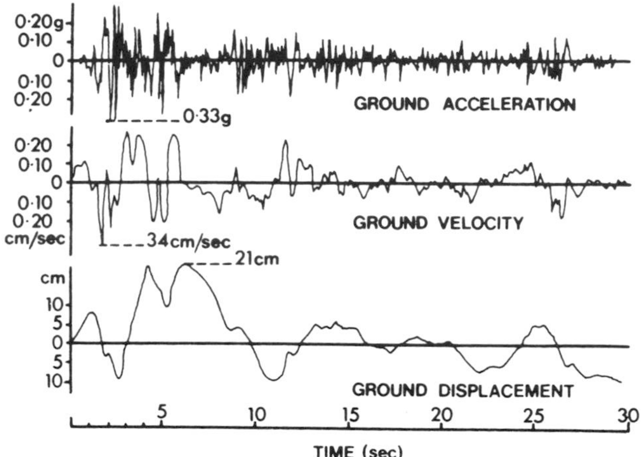

Figure 1.4 The 1940 El Centro earthquake ground motion: acceleration, velocity, and displacement in north–south direction

Earthquake ground motion is typically like that for the much-quoted 1940 Imperial Valley (El Centro) earthquake (Figure 1.4), where the accelerations in particular can be seen to change rapidly with time.

As dynamic loading varies with time, the response of the structure (soil or man-made) also varies with time, and hence a full dynamic analysis involves determining the responses at each of a series of time intervals throughout the motion induced by the loading. Two different assumptions are generally used in specifying the deflected shape of the system being analysed, the lumped mass approach and the generalized co-ordinate approach. In both cases the number of displacement components required to specify the position of all significant mass particles in the system is called the number of degrees of freedom of the system. Only the lumped mass system will be considered here. As it assumes that the mass is concentrated into a number of discrete parts, it is the simpler approach, and although consequently less accurate than the generalized co-ordinate approach it is satisfactory for most structural frames, and for simple models of soil systems.

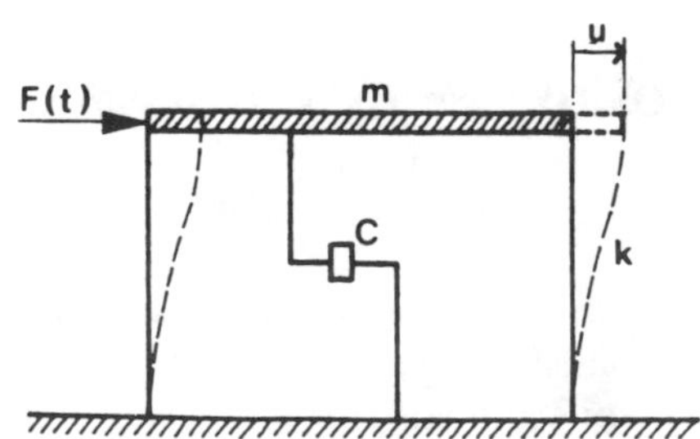

Figure 1.5 Idealized single-degree-of-freedom system

Fuller treatments of the following theory have been given by Clough and Penzien,[13] Biggs,[14] and Newmark and Rosenblueth.[15]

1.4.2 Single-degree-of-freedom systems

To find the displacement history of a structure it is necessary to solve the equations of motion of the system. There is one such equation of dynamic equilibrium for each degree of freedom. A common representation of a single-degree-of-freedom system is as shown in Figure 1.5.

In Figure 1.5 $F(t)$ is a force varying with time, k is the total spring constant of resisting elements, c is the damping coefficient (the damping force is usually taken as proportional to the velocity of the mass for ease of computation), and u is the displacement.

Generally, the equation of dynamic equilibrium is

$$F_{\mathrm{I}} + F_{\mathrm{D}} + F_{\mathrm{S}} = F(t) \tag{1.5}$$

where the inertia force

$$F_{\mathrm{I}} = m\ddot{u}$$

the damping force

$$F_{\mathrm{D}} = c\dot{u}$$

the elastic force

$$F_{\mathrm{S}} = ku$$

Thus

$$m\ddot{u} + c\dot{u} + ku = F(t) \tag{1.6}$$

For the case of earthquake excitation (Figure 1.6), the only external loading is in the form of an applied motion at ground level, $u_{\mathrm{g}}(t)$, i.e. the total acceleration of the mass m is

$$\ddot{u}_{\mathrm{t}} = \ddot{u} + \ddot{u}_{\mathrm{g}}$$

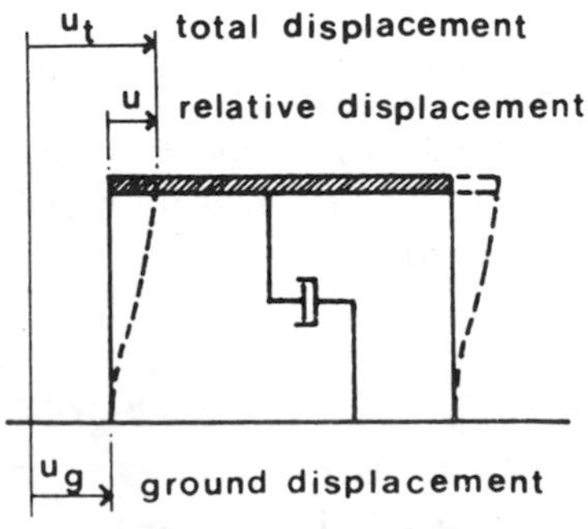

Figure 1.6 Single-degree-of-freedom system subjected to ground motion

Therefore

$$F_I = m\ddot{u}_t = m\ddot{u} + m\ddot{u}_g$$

and

$$F(t) = 0 = F_I + F_D + F_S$$

Therefore

$$m\ddot{u} + m\ddot{u}_g + c\dot{u} + ku = 0$$

or

$$m\ddot{u} + c\dot{u} + ku = F_{eff}(t) \tag{1.7}$$

where

$$F_{eff}(t) = -m\ddot{u}_g$$

is the effective load resulting from the ground motion, and is equivalent to the product of the mass of the structure and the ground acceleration.

1.4.2.1 Free vibrations (undamped)

In order to solve the equation of motion (equation 1.7), first consider the case of free vibration, $F(t) = 0$, with zero damping, i.e.

$$m\ddot{u} + ku = 0$$

or

$$\ddot{u} + \omega^2 u = 0 \tag{1.8}$$

where $\omega = \sqrt{(k/m)}$ is the circular frequency of this free vibration system.

The solution of equation (1.8) is

$$u = A \sin \omega t + B \cos \omega t$$

Solving for A and B,

$$u = \dot{u}_o/\omega \sin \omega t + u_o \cos \omega t \tag{1.9}$$

where u_o and $\dot{u}_o$ are the initial displacement and velocity, respectively. The resulting simple harmonic motion is shown in Figure 1.7.

The period of the above motion is

$$T = 2\pi/\omega$$

and the amplitude is

$$R = \sqrt{\left\{\left(\frac{\dot{u}_o}{\omega}\right)^2 + u_0^2\right\}}$$

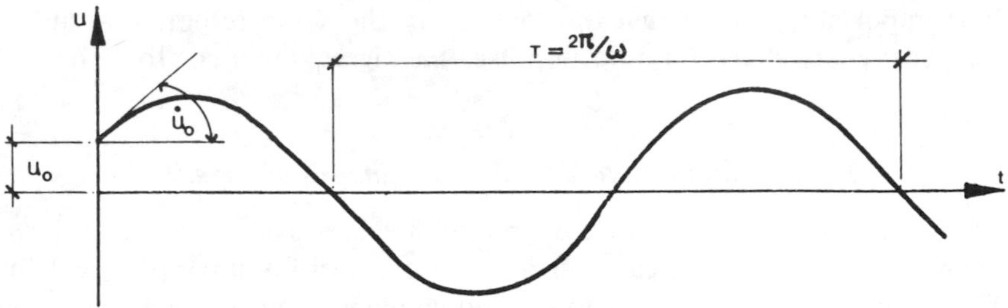

Figure 1.7 Undamped simple harmonic motion of single-degree-of-freedom system given initial displacement and velocity

1.4.2.2 Response to short impulse (undamped)

An approximate solution for the response to a very short-duration loading is easily derived from the free vibration results above. If the length of impulse $t_1 \ll T$, the period of vibration, it can be taken that $u_o \approx 0$ and from impulse-momentum

$$m\Delta\dot{u} = \int F dt$$

therefore

$$\dot{u}_o = \frac{\int F dt}{m}$$

Using these values of u_o and $\dot{u}_o$ in equation (1.9)

$$u \approx \frac{\int F dt}{m\omega} \sin \omega t \tag{1.10}$$

This loading and response system is shown in Figure 1.8.

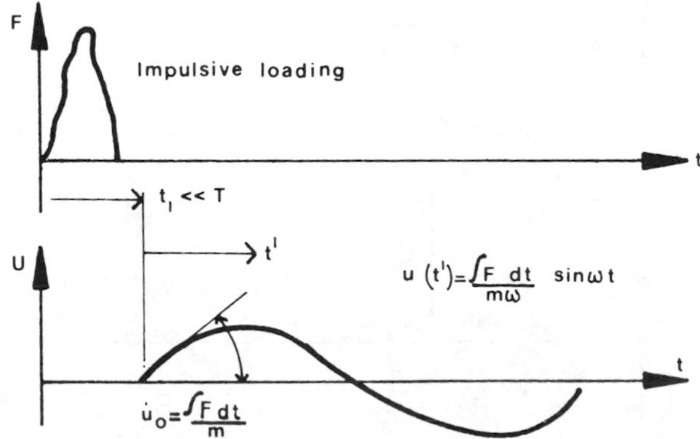

Figure 1.8 Response of single-degree-of-freedom system to impulsive undamped loading

It is important to note that approximately the same response would be developed by any short duration impulses having equal values for $\int F dt$.

1.4.2.3 Response to arbitrary loading (undamped)

To find the response of a single-degree-of-freedom system to an arbitrary loading, the latter can be treated as a series of short impulses (Figure 1.9).

The displacement response due to any individual increment of loading ending at time τ and of duration $d\tau$, can be written down in the form of equation (1.10) as

$$du = \frac{F(\tau)}{m\omega} \sin\omega t' \ d\tau$$

or

$$du = \frac{F(\tau)}{m\omega} \sin\ \omega(t-\tau) d\tau$$

where

$$t' = t - \tau$$

The total response to the arbitrary loading is the sum of all the impulses of duration $d\tau$, i.e.

$$u(t) = \int_0^t \frac{F(\tau)}{m\omega} \sin\ \omega(t-\tau)\ d\tau \tag{1.11}$$

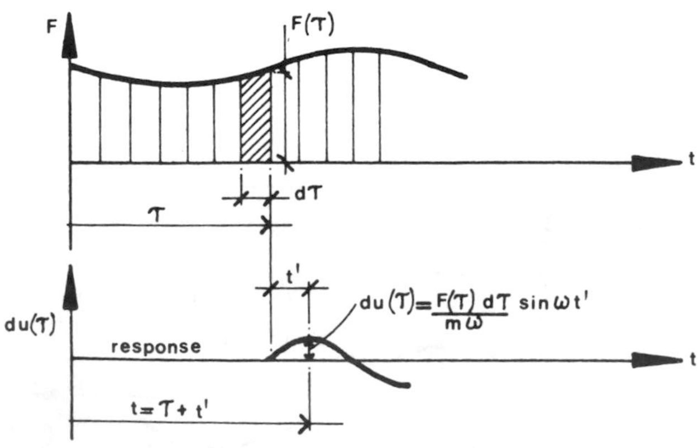

Figure 1.9 Response of single-degree-of-freedom system to impulsive undamped loading

This is an exact expression called the Duhamel integral. Because it depends upon the principle of superposition it is applicable to linear structures only.

1.4.2.4 Solution of equations of motion (damped)

The solution of the damped problem is similar to that of undamped systems, and only the key results will be given here.

1.4.2.5 Free vibrations (damped)

The equation of motion for a damped system may be written

$$\ddot{u} + 2\xi\omega\dot{u} + \omega^2 u = 0 \tag{1.12}$$

from which (for moderate damping)

$$u = \mathrm{e}^{-\xi\omega t}\left(\frac{\dot{u}_o + \xi\omega u_o}{\omega_D}\sin\omega_D t + u_o\cos\omega_D t\right) \tag{1.13}$$

where

$$\xi = \frac{c}{2m\omega}$$

is the damping ratio, and

$$\omega_D = \omega\sqrt{(1-\xi^2)}$$

is the damped circular frequency.

1.4.2.6 Response to short impulse (damped)

The free vibration response is

$$u = \frac{\int F\,dt}{m\omega_D}\mathrm{e}^{-\xi\omega t}\sin\,\omega_D t \tag{1.14}$$

1.4.2.7 Response to arbitrary loading (damped)

The damped form of the Duhamel integral becomes

$$u(t) = \int_o^t \frac{F(\tau)}{m\omega_D}\mathrm{e}^{-\xi\omega(t-\tau)}\sin\,\omega_D(t-\tau)\,d\tau \tag{1.15}$$

To evaluate this response to an arbitrary loading history, such as would occur in an earthquake, many numerical integration processes are available, some of which are discussed by Clough and Penzien.[13]

1.4.2.8 Earthquake response

The response of damped single-degree-of-freedom structures to earthquake motion comes from the above as follows. Equation (1.15) can be rewritten in terms of the ground acceleration $\ddot{u}_g(\tau) = F(\tau)/m$, taking $\omega \approx \omega_D$, which is reasonable for small damping. Thus

$$u(t) = \frac{1}{\omega} \int_0^t \ddot{u}_g(\tau) e^{-\xi\omega(t-\tau)} \sin \omega(t-\tau)\, d\tau \tag{1.16}$$

Denoting the integral in the above equation by the response function

$$V(t) = \int_0^t \ddot{u}_g(\tau) e^{-\xi\omega(t-\tau)} \sin \omega(t-\tau)\, d\tau \tag{1.17}$$

The earthquake deflection response of a lumped mass system becomes

$$u(t) = \frac{1}{\omega} V(t) \tag{1.18}$$

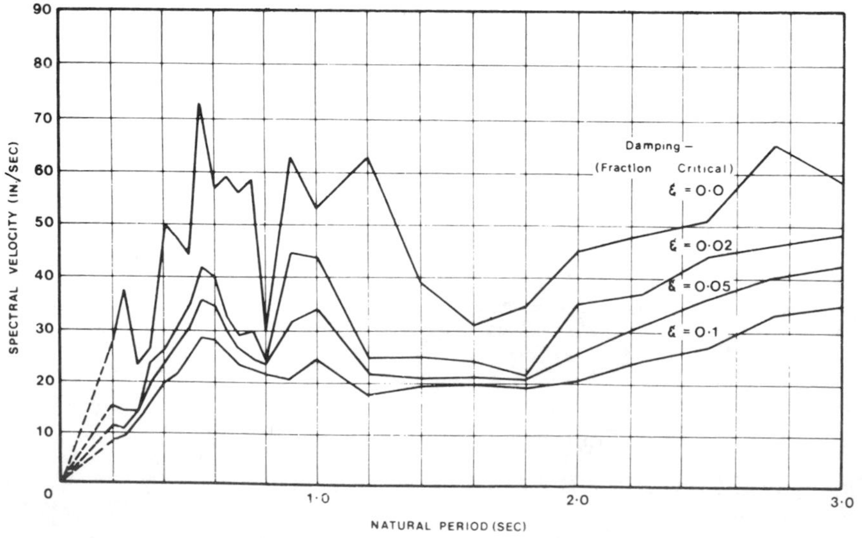

Figure 1.10 Velocity response spectra for the El Centro earthquake of 18 May 1940 (north–south)

The forces generated in the structure may best be found in terms of the effective acceleration

$$\ddot{u}_e(t) = \omega^2 u(t) \tag{1.19}$$

The effective earthquake force on the structure follows simply as

$$\begin{aligned} Q(t) &= m\ddot{u}_e(t) \\ &= m\omega^2\, u(t) \end{aligned}$$

Therefore

$$Q(t) = m\omega V(t) \tag{1.20}$$

Thus the effective earthquake force (or base shear) is found in terms of the mass of the structure, its circular frequency, and the response function $V(t)$ expressed in equation (1.17). Equations (1.18) and (1.20) describe the earthquake response at any time t for a single-degree-of-freedom structure, and solutions to these earthquakes depend upon the evaluation of equation (1.17) as discussed under equation (1.15).

1.4.2.9 Response spectra

To obtain the entire history of forces and displacements during an earthquake using the above equations in clearly a tedious and costly procedure. For many structures it will suffice to evaluate only the maximum responses. From equations

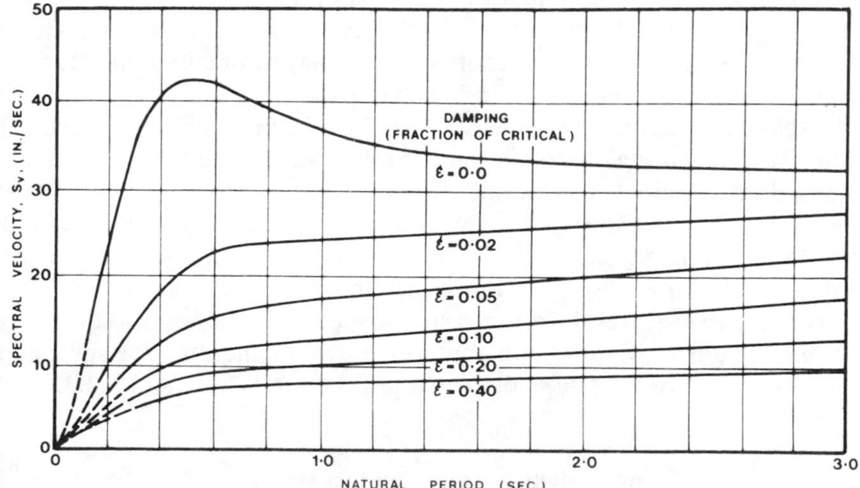

Figure 1.11 Averaged velocity response spectra, based on the spectral intensity of the 1940 El Centro earthquake

(1.18) and (1.20) this means finding the maximum value of response function $V(t)$. This maximum value is called the spectral velocity, S_v, or more accurately the spectral pseudo-velocity because it is not exactly the maximum velocity of a damped system. The spectral velocity is

$$S_v = \left\{ \int_0^t \ddot{u}_g(\tau) e^{-\xi\omega(t-\tau)} \sin \omega(t-\tau) \, d\tau \right\}_{max} \tag{1.21}$$

From before it follows that the maximum displacement or spectral displacement

$$S_d = \frac{S_v}{\omega} \tag{1.22}$$

and the spectral acceleration (or spectral pseudo-acceleration)

$$S_a = \omega S_v \tag{1.23}$$

From these relationships, the maximum earthquake displacement response

$$u_{max} = S_d \tag{1.24}$$

and the maximum effective earthquake force or base shear is

$$Q_{max} = MS_a \tag{1.25}$$

If equation (1.21) is evaluated for single-degree-of-freedom structures of varying natural periods, a maximum velocity response curve (called a response spectrum) can be plotted. A family of curves is usually calculated for any given excitation, showing the effect of variation in the amount of damping. For example, the ground acceleration history shown in Figure 1.4 results in the velocity response spectra shown in Figure 1.10 and the acceleration response spectra shown in Figure 6.29, the latter form being that most commonly used. The maximum responses of a single-degree-of-freedom structure may be obtained directly from the spectra and equations (1.24) and (1.25).

The velocity spectrum of Figure 1.10 is for the ground motion of a specific earthquake recorded at a specific site, and the sharp discontinuities in the spectral curves indicate local resonances only. In any case the period of vibration of a structure cannot be known with enough certainty to design with either a peak spectral value or an adjacent trough. For general design purposes an averaged spectrum as shown in Figure 1.11 will therefore be more appropriate.

To obtain a realistic result from a response spectrum analysis, it is necessary to use a spectrum derived from an appropriate ground motion. The subject of selection of response spectra is discussed in Sections 1.2.4.1 and 4.3.3.

1.4.3 Multi-degree-of-freedom systems

In the dynamic analysis of most structures it is necessary to assume that the mass is distributed in more than one discrete lump. For most buildings the mass

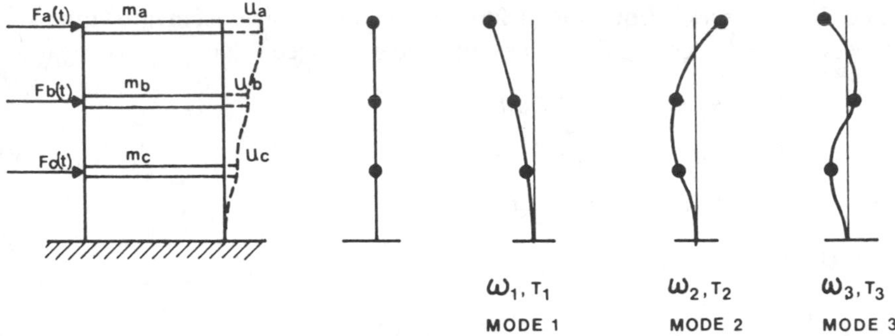

Figure 1.12 Multi-degree-of-freedom system subjected to dynamic loading

is assumed to be concentrated at the floor levels, and to be subjected to lateral displacements only. To illustrate the corresponding multi-degree-of-freedom analysis, consider a three-storey building (Figure 1.12). Each storey mass represents one degree-of-freedom, each with an equation of dynamic equilibrium,

$$\begin{aligned} F_{\mathrm{Ia}} + F_{\mathrm{Da}} + F_{\mathrm{Sa}} &= F_{\mathrm{a}}(t) \\ F_{\mathrm{Ib}} + F_{\mathrm{Db}} + F_{\mathrm{Sb}} &= F_{\mathrm{b}}(t) \\ F_{\mathrm{Ic}} + F_{\mathrm{Dc}} + F_{\mathrm{Sc}} &= F_{\mathrm{b}}(t) \end{aligned} \tag{1.26}$$

The inertia forces in equation (1.26) are simply

$$\begin{aligned} F_{\mathrm{Ia}} &= m_{\mathrm{a}}\ddot{u}_{\mathrm{a}} \\ F_{\mathrm{Ib}} &= m_{\mathrm{b}}\ddot{u}_{\mathrm{b}} \\ F_{\mathrm{Ic}} &= m_{\mathrm{c}}\ddot{u}_{\mathrm{a}} \end{aligned} \tag{1.27}$$

In matrix form

$$\begin{Bmatrix} F_{\mathrm{Ia}} \\ F_{\mathrm{Ib}} \\ F_{\mathrm{Ic}} \end{Bmatrix} = \begin{bmatrix} m_{\mathrm{a}} & 0 & 0 \\ 0 & m_{\mathrm{b}} & 0 \\ 0 & 0 & m_{\mathrm{c}} \end{bmatrix} \begin{Bmatrix} \ddot{u}_{\mathrm{a}} \\ \ddot{u}_{\mathrm{b}} \\ \ddot{u}_{\mathrm{c}} \end{Bmatrix} \tag{1.28}$$

Or more generally

$$\mathbf{F}_{\mathbf{I}} = \mathbf{M}\ddot{\mathbf{u}} \tag{1.29}$$

where $\mathbf{F}_{\mathbf{I}}$ is the inertia force vector, $\mathbf{M}$ is the mass matrix and $\ddot{\mathbf{u}}$ is the acceleration vector. It should be noted that the mass matrix is of diagonal form for a lumped-sum system, giving no coupling between the masses. In more generalized shape co-ordinate systems, coupling generally exists between the co-ordinates, complicating the solution. This is a prime reason for using the lumped-mass method.

The elastic forces in equation (1.26) depend on the displacements, and using stiffness influence coefficients they may be expressed

$$\begin{aligned} F_{Sa} &= k_{aa}u_a + k_{ab}u_b + k_{ac}u_c \\ F_{Sb} &= k_{ba}u_a + k_{bb}u_b + k_{bc}u_c \\ F_{Sc} &= k_{ca}u_a + k_{cb}u_b + k_{cc}u_c \end{aligned} \tag{1.30}$$

In matrix form

$$\begin{Bmatrix} F_{Sa} \\ F_{Sb} \\ F_{Sc} \end{Bmatrix} = \begin{bmatrix} k_{aa} & k_{ab} & k_{ac} \\ k_{ba} & k_{bb} & k_{bc} \\ k_{ca} & k_{cb} & k_{cc} \end{bmatrix} \begin{Bmatrix} u_a \\ u_b \\ u_c \end{Bmatrix} \tag{1.31}$$

Or more generally

$$\mathbf{F}_S = \mathbf{K}\mathbf{u} \tag{1.32}$$

where $\mathbf{F}_S$ is the elastic force vector, $\mathbf{K}$ is the stiffness matrix, and $\mathbf{u}$ is the displacement vector. The stiffness matrix $\mathbf{K}$ generally exhibits coupling and will best be handled by a standard computerized matrix analysis.

By analogy with the expressions (1.30), (1.31), and (1.32) in equation (1.26) may be expressed

$$\mathbf{F}_D = \mathbf{C}\dot{\mathbf{u}} \tag{1.33}$$

where $\mathbf{F}_D$ is the damping force vector, $\mathbf{C}$ is the damping matrix and $\dot{\mathbf{u}}$ is the velocity vector. In general it is not practicable to evaluate $\mathbf{C}$, and damping is usually expressed in terms of damping coefficients.

Using equations (1.29), (1.32), and (1.33), the equations of dynamic equilibrium (1.26) may be written generally as

$$\mathbf{F}_I + \mathbf{F}_D + \mathbf{F}_S = \mathbf{F}(t) \tag{1.34}$$

which is equivalent to

$$\mathbf{M}\ddot{\mathbf{u}} + \mathbf{C}\dot{\mathbf{u}} + \mathbf{K}\mathbf{u} = \mathbf{F}(t) \tag{1.35}$$

The matrix equation (1.35) for a multi-degree system is identical in form to the single-degree-of-freedom equation (1.6)

1.4.3.1 Vibration frequencies and mode shapes

As the dynamic response of a structure is dependent upon the frequency (or period T) and the displaced shape, the first step in the analysis of a multi-degree system is to find its free vibration frequencies and mode shapes. In free vibration there is no external force and damping is taken as zero. The equations of motion (1.35) become

$$\mathbf{M}\ddot{\mathbf{u}} + \mathbf{K}\mathbf{u} = 0 \tag{1.36}$$

But in free vibration the motion is simple harmonic

$$\mathbf{u} = \hat{\mathbf{u}}\sin \omega t$$

Therefore

$$\ddot{\mathbf{u}} = -\omega^2 \hat{\mathbf{u}}\sin \omega t \tag{1.37}$$

where $\hat{\mathbf{u}}$ represents the amplitude of vibration.

Substituting in equation (1.36)

$$\mathbf{K}\hat{\mathbf{u}} - \omega^2 \mathbf{M}\hat{\mathbf{u}} = 0 \tag{1.38}$$

Equation (1.38) is an eigenvalue equation and is readily solved for ω by standard computer programs. Its solution for a system having N degrees of freedom yields a vibration frequency ω_n and a mode shape vector ϕ_n represents the *relative* amplitudes of motion for each of the displacement components in mode n. It should be noted that equation (1.38) cannot be solved for absolute values of ϕ, as the amplitudes are arbitrary in free vibration. Figure 1.12 shows the shapes of the three normal modes of a typical three-storey building.

An important simplification can be made in the equations of motion because of the fact that each mode has an independent equation of exactly equivalent form to that for a single-degree-of-freedom system. Because of the orthogonality properties of mode shapes, equation (1.35) can be written

$$\ddot{Y}_n + 2\xi_n\omega_n\dot{Y}_n + \omega_n^2 Y_n = \frac{\phi_n^T \mathbf{F}(t)}{\phi_n^T \mathbf{M}\phi_n} \tag{1.39}$$

where Y_n is a generalized displacement in mode n, leading to the actual displacement (see equation 1.44) and ϕ_n^T is the row mode shape vector corresponding to the column vector ϕ_n.

1.4.3.2 *Earthquake response analysis by mode superposition*

The dynamic analysis of a multi-degree-of-freedom system can therefore be simplified to the solution of equation (1.38) for each mode, and the total response is then obtained by superposing the modal effects.

In terms of excitation by earthquake ground motion $\ddot{u}_g(t)$, equation (1.39) becomes

$$\ddot{Y}_n + 2\xi_n\omega_n\dot{Y}_n + \omega_n^2 Y_n = \frac{L_n}{\phi_n^T \mathbf{M}\phi_n}\ddot{u}_g(t) \tag{1.40}$$

where the earthquake participation factor

$$L_n = \phi_n^T \mathbf{M}\hat{\mathbf{I}} \tag{1.41}$$

in which $\hat{\mathbf{I}}$ is a unit column vector of dimension N.

The response of the nth mode at any time t demands the solution of equation (1.40) for Y_n. This may be done by evaluating the Duhamel integral (Section 1.4.2.3):

$$Y_n(t) = \frac{L_n}{\phi_n^T \mathbf{M}\phi_n} \cdot \frac{1}{\omega_n} \int_o^t \ddot{u}_g(\tau) e^{-\xi_n\omega_n(t-\tau)} \sin \omega_n(t-\tau)\, d\tau \tag{1.42}$$

This displacement of floor (or mass) i at time t is then obtained by superimposing the response of all modes evaluated at this time t:

$$u_i = \sum_{n=1}^{N} \phi_{in} Y_n(t) \tag{1.43}$$

where ϕ_{in} is the relative amplitude of displacement of mass i in mode n. It should be noted that in structures with many degrees-of-freedom most of the vibrational energy is absorbed in the lower modes, and it is usually sufficiently accurate to superimpose the effects of only the first few modes.

The earthquake forces in the structure may then be expressed in terms of the effective accelerations

$$\ddot{Y}_{neff}(t) = \omega_n^2\, Y_n(t) \tag{1.44}$$

from which the acceleration at any floor i is

$$\ddot{u}_{ineff}(t) = \omega_n^2 \phi_{in} Y_n(t) \tag{1.45}$$

and the earthquake force at any floor i at time t is

$$q_{in}(t) = m_i \omega_n^2 \phi_{in} Y_n(t) \tag{1.46}$$

Superimposing all the modal contributions, the earthquake forces in the total structure may be expressed in matrix form as

$$\mathbf{q}(t) = \mathbf{M}\phi\omega^2 \mathbf{Y}_n(t) \tag{1.47}$$

where ϕ is the square matrix of relative amplitude distributions in each mode and ω^2 is the diagonal matrix of ω^2 for each of the n modes.

From equations (1.43) and (1.47) the entire history of displacement and force response can be defined for any multi-degree-of-freedom system, having first determined the modal response amplitudes of equation (1.43).

1.4.3.3 Response spectrum analysis for multi-degree systems

As with single-degree-of-freedom structures considerable simplification of the analysis is achieved if only the maximum response to each mode is considered rather than the whole response history. If the maximum value $Y_{n\,\max}$ of the Duhamel equation (1.42) is calculated, the distribution of maximum displacements in that mode is

$$u_{n\,\max} = \phi_n Y_{n\,\max} = \phi_n \frac{L_n}{\phi_n^T \mathbf{M}\phi_n} \frac{S_{\mathrm{vn}}}{\omega_n} \tag{1.48}$$

and the distribution of maximum earthquake forces in that mode is

$$\mathbf{q}_{n\,\max} = \mathbf{M}\phi_n \omega_{\mathrm{n}}^2 Y_{n\,\max} = \mathbf{M}\phi_n \frac{L_n}{\phi_{\mathrm{n}}^2 \mathbf{M}\phi_n} S_{an} \tag{1.49}$$

In equations (1.48) and (1.49), S_{vn} and S_{an} are the spectral velocity and spectral acceleration for mode n, and are as defined in Section 1.4.2.9.

Equations (1.48) and (1.49) enable the maximum response in each mode to be determined. As the modal maxima do not necessarily occur at the same time, nor necessarily have the same sign, they cannot be combined to give the precise total maximum response. The best that can be done in a response spectrum analysis is to combine the modal responses on a probability basis. Various approximate formula for superposition are used, the most common being the Square-Root-of-Sum-of-Squares (SRSS) procedure. As an example the maximum deflection at the top of a three-storey structure (three masses) would be

$$u_{\mathrm{a}\,\max} \approx \sqrt{(u_{\mathrm{a}\,1\,\max}^2 + u_{\mathrm{a}\,2\,\max}^2 + u_{\mathrm{a}\,3\,\max}^2)} \tag{1.50}$$

This approximation is usually, but not necessarily, conservative.

Considerable savings in computation are made by the further approximation of using the responses of only the first few modes in this equation. Usually the first three to six modes are all that need to be included, as most of the energy of vibration is absorbed in these modes.

A preferred alternative to the above SRSS method which reduces the errors in modal combination is the Complete Quadratic Combination (CQC) method[16], in which a typical displacement component is

$$u_{\mathrm{k}} = \sqrt{(\sum_i \sum_j u_{\mathrm{k}i}\ \rho_{ij} u_{\mathrm{k}j})} \tag{1.51}$$

and a typical force component is

$$f_{\mathrm{k}} = \sqrt{(\sum_i \sum_j f_{\mathrm{k}i}\ \rho_{ij} f_{\mathrm{k}j})} \tag{1.52}$$

where $u_{\mathrm{k}i}$ is a typical component of the modal displacement vector, $U_{\mathrm{i},\max}$ and $f_{\mathrm{k}i}$ is a typical force component which is produced by the modal displacement vector, $U_{i,\max}$. The cross-modal coefficients, ρ_{ij}, are functions of the duration and frequency content of the loading and of the modal frequencies and damping ratios of the structure.

If the duration of the earthquake is long compared with the periods of the structure, if the earthquake spectrum is smooth over a wide range of frequencies, and for constant modal damping, ξ, the modal cross-correlation coefficient, ρ_{ij} may be approximated as

$$\rho_{ij} = \frac{8\xi^2(1+r)r^{3/2}}{(1-r_2)^2 + 4\xi^2 r(1+r)^2} \tag{1.53}$$

where $r = \omega_j/\omega_i$.

If the frequencies are well separated, the off-diagonal terms approach zero and the CQC method approaches the SRSS. It is noted[16] that the dynamic analysis computer program TABS, from the University of California, Berkeley, was modified to use the CQC modal combination method instead of the SRSS method.

Most analyses utilizing response spectra take the spectral velocities S_{vn} for all modes from a single-degree-of-freedom spectrum. This approximation is reasonable for uniform and regular structures, but for irregular structures with larger changes of stiffness more general forms of analysis are advisable. Apart from the obvious possibility of using a full modal analysis or direct integration technique, it is possible to create response spectra for systems of more than one degree-of-freedom. Penzien[17] has used two degree-of-freedom response spectra developed for analysing buildings having large setbacks of the facade at high level, but these have not been widely used.

1.4.4. Non-linear inelastic earthquake response

The importance of and difficulties involved in carrying out realistic non-linear analyses is discussed in Sections 6.6 and 6.8. A very brief summary of the processes involved follows. Further discussions of non-linear analysis may be found in the literature.[13,15]

The mode superposition techniques discussed above are necessarily limited to the study of linear material behaviour only. To analyse the effects of non-linear inelastic response during strong-motion earthquakes, a step-by-step procedure is necessary as outlined below.

1.4.4.1 Step-by-step integration

A number of step-by-step integration procedures are possible. Generally the response history is divided into very short time increments, during each of which the structure is assumed to be linearly elastic. Between each interval the properties of the structure are modified to match the current state of deformation. Therefore the non-linear response is obtained as a sequence of linear responses of successively differing systems. One method of step-by-step integration is now described. In each time increment the following computations are made.

(1) The stiffness of the structure for that increment is computed, based on the state of displacement existing at the beginning of the increment.

(2) Changes of displacement are computed assuming the accelerations to vary linearly during the interval.
(3) These changes of displacement are added to the displacement state of the beginning of the interval to give the displacements at the end of the interval.
(4) Stresses appropriate to the total displacements are computed.

In the above procedure the equations of motion must be integrated in their original form during each time increment. For this purpose equation (1.35) may be written

$$\mathbf{M}\Delta\ddot{\mathbf{u}} + \mathbf{C}\Delta\dot{\mathbf{u}} + \mathbf{K}(t)\Delta\mathbf{u} = \Delta\mathbf{F}(t) \tag{1.54}$$

where $\mathbf{K}(t)$ is the stiffness matrix for the time increment beginning at time t, and $\Delta\mathbf{u}$ is the change in displacement during the interval. The determination of $\mathbf{K}$ for each increment is the most demanding part of the analysis, as all the individual member stiffnesses must be found each time for their current state of deformation.

REFERENCES

1. EERI Committee on Seismic Risk, Haresh C. Shah, Chairman. 'Glossary of terms for probabilistic seismic-risk and hazard analysis', *Earthquake Spectra*, 1, No. 1, 33–40 (1984).
2. Chopra, A. K., and Chakrabarti, P., 'The Koyna earthquake and damage to Koyna dam', *Bull. Seism. Soc. Amer.*, **63**, No. 2, 381–97 (1973).
3. Richter, C. F., *Elementary Seismology*, Freeman, San Francisco (1958).
4. Eiby, G. A., *Earthquakes*, Heinemann Educational Books, Auckland (1980).
5. Heaton, T. H., 'The 1971 San Fernando earthquake: A double event?' *Bull. Seism. Soc. Amer.*, **72**, No. 6, Part A, 2037–62 (1982).
6. Wyss, M., and Brune, J., 'Seismic moment, stress and source dimensions', *Jnl Geoph. Research*, **73**, 4681–94 (1968).
7. Hanks, T. C., and Kanamori, H., 'A moment magnitude scale', *Jnl Geophys. Res.*, **84**, B5, 2348–50 (1979).
8. Dowrick, D. J., 'An earthquake catastrophe model with particular reference to central New Zealand', *Bull NZ Nat. Soc. for Earthq. Eng.*, **16**, No. 3, 213–21 (1983).
9. Ambraseys, N. N., and Jackson, J. A., 'Earthquake hazard and vulnerability in the north eastern Mediterranean: the Corinth earthquake sequence of February–March 1981', *Disasters*, **5**, No. 4, 355–68 (1981).
10. Elms, D. G., and Silvester, D., 'Cost effectiveness of code base shear requirements for reinforced concrete frame structures', *Bull. NZ Nat. Soc. for Earthq. Eng.*, **11**, No. 2, 85–93 (1978).
11. Hollings, J. P., 'The economics of earthquake engineering', *Bull. NZ Soc. for Earthq. Eng.*, **4**, No. 2, 205–21 (1971).
12. Whitman, R. V., Biggs, J. M., Brennan, J., Cornell, C. A., Neufville, R. de, and Vanmarcke, E. H., 'Seismic design analysis', *Structures Publication No. 381*, Massachusetts Institute of Technology (1974).
13. Clough, R. W., and Penzien, J., *Dynamics of Structures*, McGraw-Hill, New York (1975).
14. Biggs, J. M., *Introduction to Structural Dynamics*, McGraw-Hill, New York (1964).

15. Newmark, N. M., and Rosenblueth, E., *Fundamentals of Earthquake Engineering*, Prentice-Hall, Englewood Cliffs, NJ (1971).
16. Wilson, E. L., Der Kiureghian, A., and Bayo, E. P., 'A replacement for the SRSS method in seismic analysis', *Earthquake Engineering and Structural Dynamics*, **9**, 187–94 (1981).
17. Penzien, J., 'Earthquake response of irregularly shaped buildings', *Proc. 4th World Conf. on Earthq. Eng., Santiago*, **2**, A3, 75–90 (1969).

Chapter 2

Seismic activity in a regional setting

2.1 INTRODUCTION

The seismic hazard at any given site obviously depends on the seismic activity of the region. Hence studies of this topic are primary requirements for the preparation of earthquake loadings codes, for determining the earthquake loadings for projects requiring special study, for areas where no earthquake codes exist, or for regional and town planning purposes. Background information may be obtained from a variety of sources such as local officials, engineers, seismologists, geologists, local building regulations, published papers, maps, and reference books. The available data are often inadequate or imprecise and are hence best evaluated by people with specialist knowledge, while some further pitfalls in seismic hazard analysis have been described by Smith[1] and Vere-Jones.[2]

For the purposes of the following discussion, regional seismic activity is seen to include the following main ingredients;

(1) Source mechanisms;
(2) Distribution of sources;
(3) Magnitudes;
(4) Intensity of shaking;
(5) Attenuation of intensity with distance;
(6) Rates of activity.

2.2 SEISMOTECTONICS—SEISMICITY AND GEOLOGY

2.2.1 Introduction

As most earthquakes arise from stress build-up due to deformation of the earth's crust, understanding of seismicity depends heavily on aspects of geology, which is the science of the earth's crust, and also calls upon knowledge of the physics of the earth as a whole, i.e. geophysics. The particular aspect of geology which sheds most light on the source of earthquakes is *tectonics*, which concerns the structure and deformations of the crust and the processes which accompany it; the relevant aspect of tectonics is now often referred to as *seismotectonics*.

Geology tells us the overall underlying level of seismic hazard which may differ from the available evidence of historical seismicity, notably in areas experiencing present day quiescent periods.

2.2.2 Global Seismotectonics

On a global scale the present-day seismicity pattern of the world is illustrated in general terms by the seismic events plotted in Figure 2.1. Most of these events can be seen to follow clearly defined belts which form a map of the boundaries of segments of the earth's crust known as *tectonic plates*. This may be seen by comparing Figure 2.1 with Figure 2.2, which is a world map of the main tectonic plates taken from the highly understandable book on the theory of continental drift by Stevens.[4] According to the latter, the earth's crust is composed of at least 15 virtually undistorted plates of lithosphere. The lithosphere moves differentially on the weaker asthenosphere which starts at the Low-Velocity Layer in the Upper Mantle at a depth of about 50 km. Boundaries of plates are of four principal types;

(1) Divergent zones, where new plate material is added from the interior of the earth;
(2) Subduction zones, where plates converge and the under-thrusting one is consumed;
(3) Collision zones, former subduction zones where continents riding on plates are colliding;
(4) Transform faults, where two plates are simply gliding past one another, with no addition or destruction of plate material.

Almost all the earthquake, volcanic, and mountain-building activity which marks the active zones of the earth's crust closely follows the plate boundaries and is related to movements between them.

Divergent boundaries are found at the oceanic sea-floor ridges, affecting scattered islands of volcanic origin, such as Iceland and Tristan da Cunha, which are located on these ridges. As these zones involve lower stress levels, they generate somewhat smaller earthquakes than the other types of plate boundary.

As can be seen in Figure 2.2, subduction zones occur in various highly populated regions, notably Japan and the western side of Central and South America. Figure 2.3 shows the cross-section of the likely structure of the subduction zone formed by the Pacific plate thrusting under the Indian–Australian plate beneath the North Island of New Zealand. The seismic cross-section corresponding to Figure 2.3 is on Figure 2.4, and gives earthquakes located under the shaded region of the key map (during a period of time when no events shallower than *c*. 40 km occurred). According to Stevens,[4] the zone of diffuse seismic activity which exists down to a depth of about 100 km is believed to be related both to *volcanic activity* (movement of magma in the crust and upper mantle and related expansion and contraction), and to *faulting* within

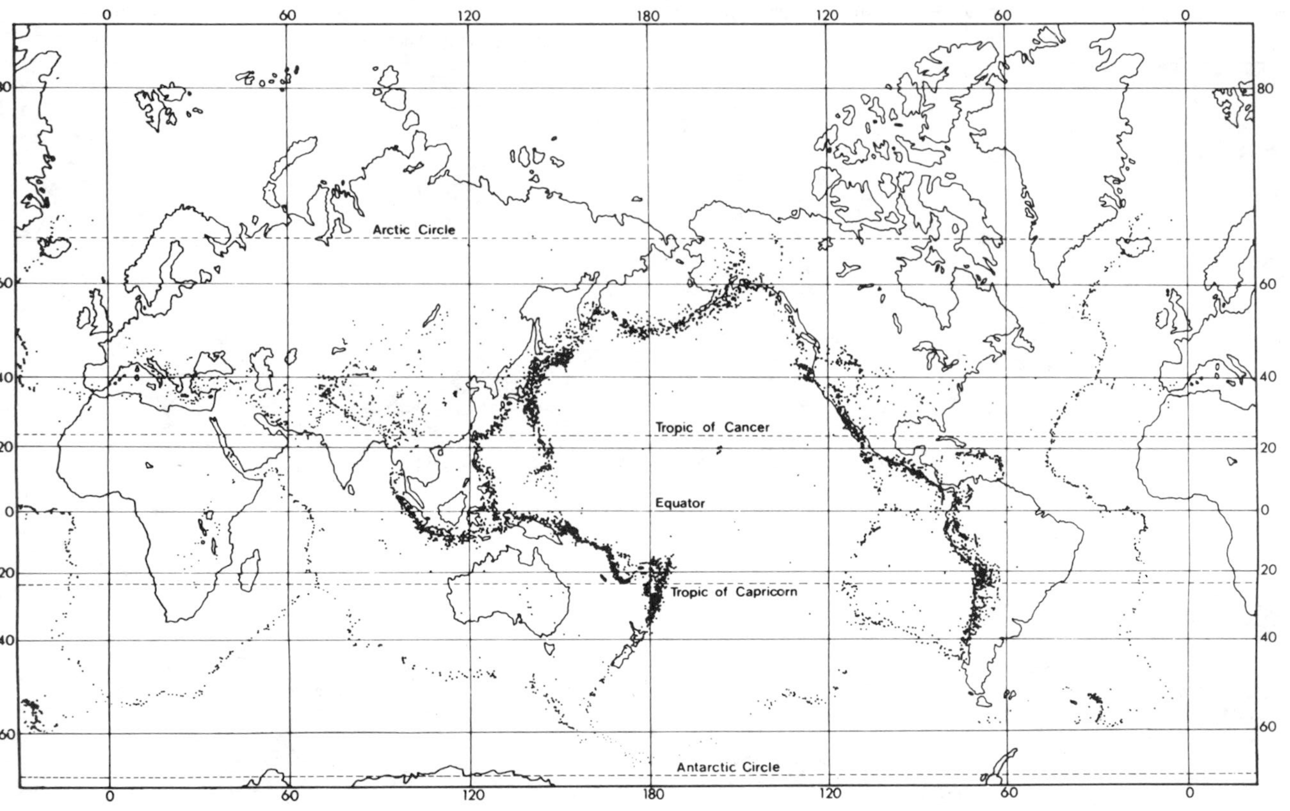

Figure 2.1 Seismicity map of the world. The dots indicate the distribution of seismic events in the mid-twentieth century (after Barazangi and Dorman[3])

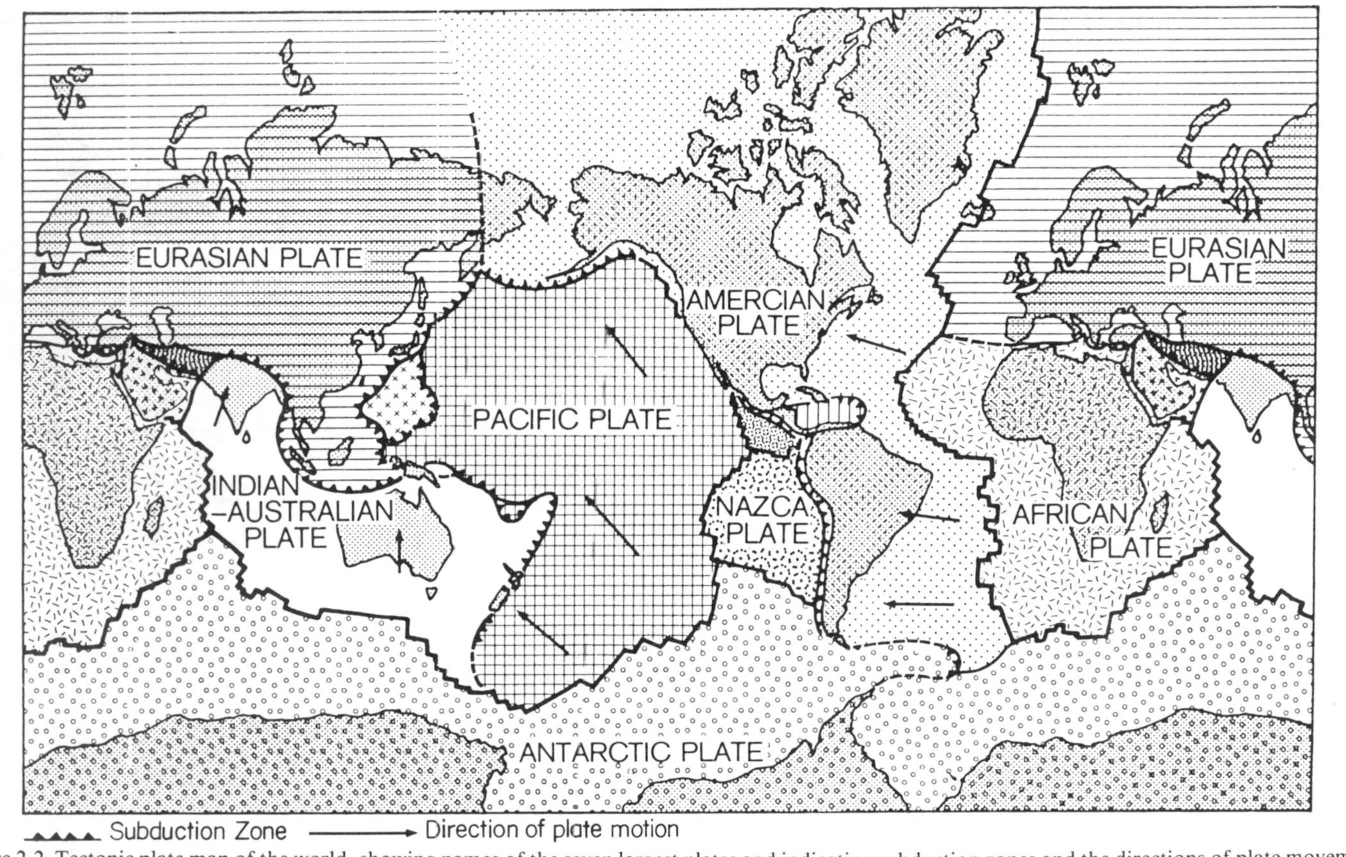

Figure 2.2 Tectonic plate map of the world, showing names of the seven largest plates and indicating subduction zones and the directions of plate movement (reproduced with permission from Stevens[4])

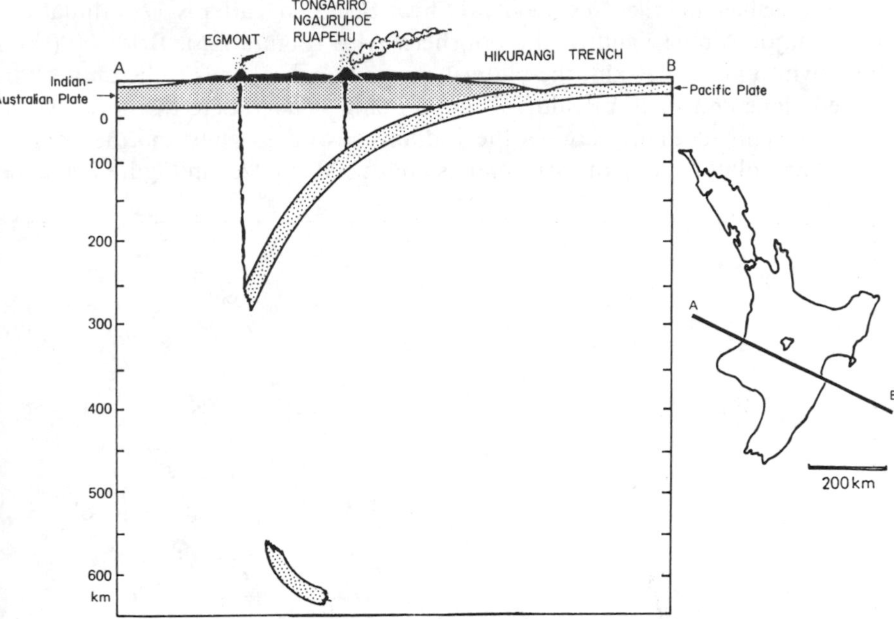

Figure 2.3 The likely structure of the subduction zone beneath the North Island of New Zealand inferred from Figure 2.4 (reproduced with permission from Stevens[4])

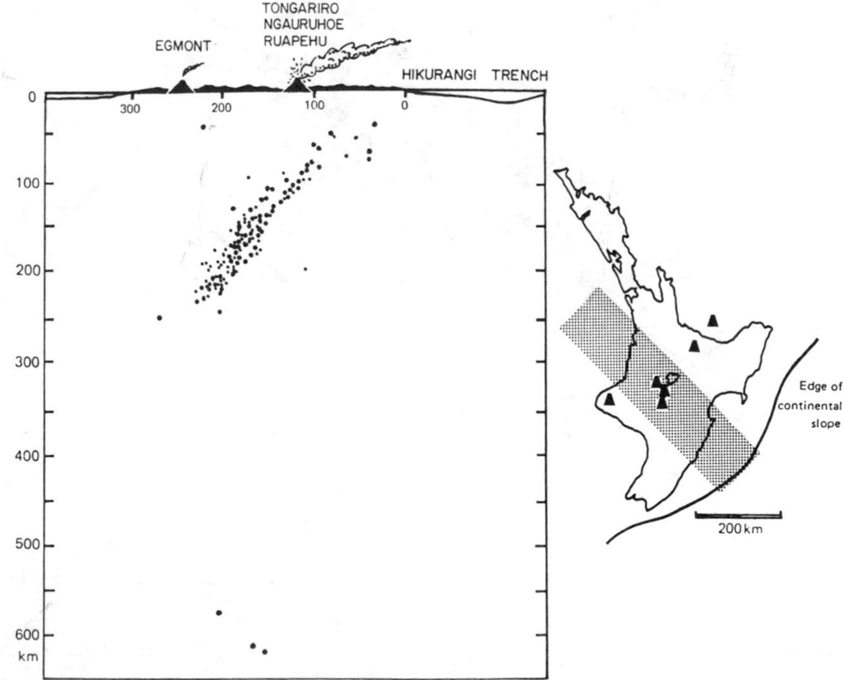

Figure 2.4 Seismic cross-section through the North Island of New Zealand, showing locations of earthquake foci (reproduced with permission from Stevens[4])

the volcanic belt and the 'New Zealand Shear Belt'. The latter is a continuation of the major Alpine Fault of the Southern Alps (Figure 2.5). Below 100 km and down to about 250 km, the pattern of earthquakes tends to lie on a well-defined plane known as a Benioff Zone, dipping 50 degrees to the north-west. This is the contact plane between the Indian–Australian plate and the Pacific plate. The isolated group of earthquakes about 600 km deep in Figure 2.4 have

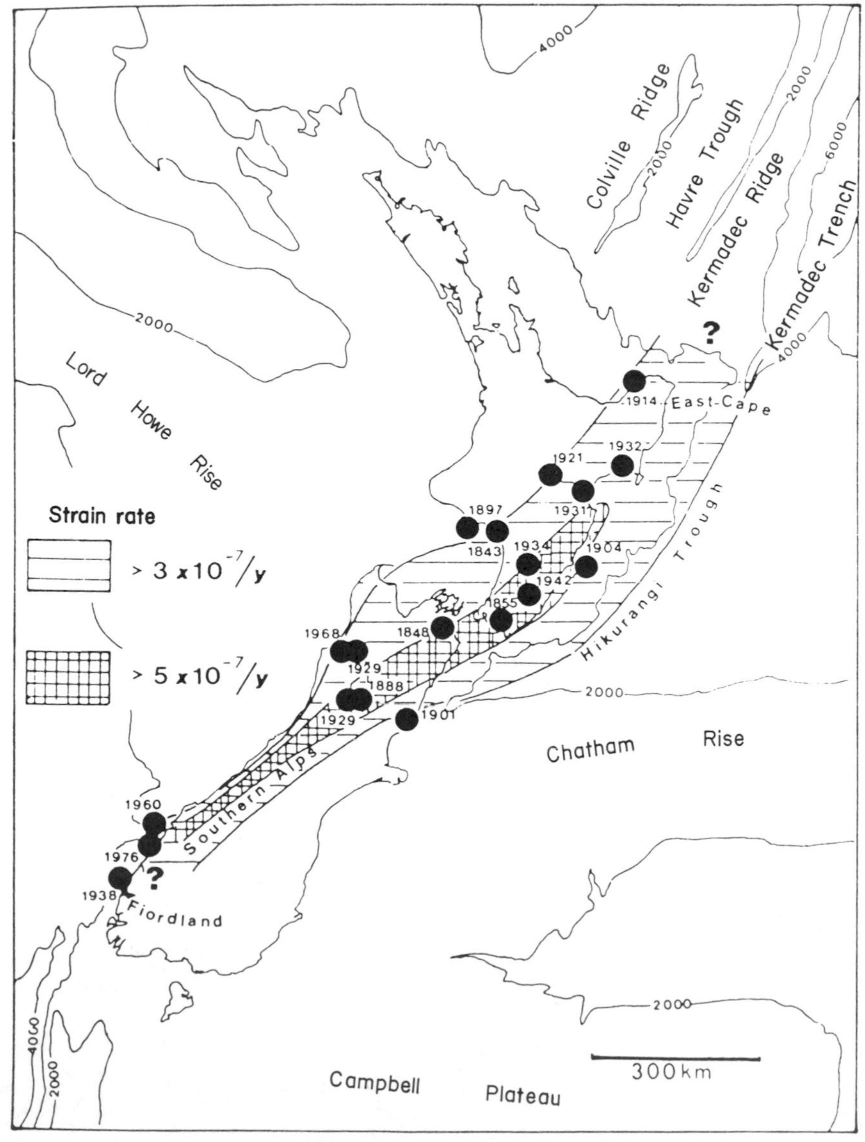

Figure 2.5 Generalized shear strain rates and large ($M \geqslant 7$) shallow ($h < 100$ km) New Zealand earthquakes (after Walcott[5])

been conjectured to be caused by a piece of lithosphere that has become detached and has moved deeper into the mantle, as illustrated in Figure 2.3.

The progressive movement of the Pacific plate, subducting under the overlying plate, caused shear stresses to develop, as illustrated by Walcott's geophysical study[5] (Figure 2.5), which relates the shear strains of the shear belt referred to above to large historical earthquakes. It is believed that the fault forming the plate boundary periodically locks together, perhaps due to a large bump on its surface, and this leads to an accumulation of shear and compressional strain until it is in part relieved by a large thrust type of earthquake. The sudden release of strain (when the resistance of the bump is overcome) signals the recommencement of movement of the subducting plate in a further cycle of aseismic slip, then another locking of the fault leads to the next earthquake.

As well as the 15 or so main plates shown in Figure 2.2, studies of seismic activity need to consider the smaller buffer plates or sub-plates which in certain areas tend to ease the relative movements of the world's giant plates. Buffer plates have been recognized in Tibet and China, the western USA, and at the complex junction of the African, Arabian, Iranian, and Eurasion plates, where eight Mediterranean buffer plates have been identified.[4]

In the foregoing discussion tectonic plates have been described as rigid, virtually undistorted plates and the world's principal zones of seismicity have been shown to be associated with the interaction between the plates. However, occasional damaging intra-plate earthquakes also occur, well within the interior of the plates that clearly are not associated with plate boundary conditions, and so far their origins are ill-understood. The high strength of earthquakes such as New Madrid, 7 February 1812, Charleston 1886, both in the USA, and Meckering 1968 in Australia with magnitudes (M_s) of 8.8, 7.5, and 6.8, respectively, combined with their rarity and the difficulty of finding convincing causal explanations, are factors that give rise to risk assessment's classical dilemma: 'Should they be ignored, or not?'

2.2.3 Regional seismotectonics

In attempting to understand and then quantify seismic activity, whether it be inter-plate or intra-plate in origin, in addition to using the above global tectonic plate data we try to relate seismicity to quantifiable deformational features such as *faulting, tilting, warping, or folding*, or to major geological structures such as *basins, grabens, platforms, and arches*, which are basement rock features. The nature, age, location, and movement history of these features need to be known, though evidence for all these data will generally not be available for any given area.

Mogi[6] has pointed out that the majority of large shallow earthquakes occur in ocean-facing slopes of deep-sea trenches, or in local depressions or troughs or ends of depressions. The magnitude and frequency of earthquakes in a given area may be estimated in broad terms from the size and strength of the fault blocks.[7] The larger and stronger the block, the larger is the maximum size of earthquake which can be generated along the boundaries of that block. Also the greater the rate of tectonic movement and the less the competency of the tectonic structures, the more rapid is the build-up of the stress needed for a fault movement, and the more frequent will be the occurrence of the maximum magnitude of earthquake for that structure.

In studies of the intra-plate area of the USA, originally because of difficulties in recognizing active faults, the concept of *tectonic provinces* has been used in estimating seismicity. The boundaries of tectonic provinces should be defined by major geological structure relevant to present-day seismogenic (earthquake-producing) mechanisms, although in the absence of such knowledge the arbitrary use of old geological structures has traditionally been necessary, and the definition and recognition of appropriate boundaries remains problematical.[8] It is clear that the tectonic province concept, at least as it is currently used, is inherently weak. As the understanding of any given tectonic province is enriched by data concerning seismogenic features within it, the tectonic province itself becomes, ironically, correspondingly redundant as an analytical tool. However, if tectonic provinces can be defined with clearly defined boundary geometry and boundary conditions, then the behaviour of internal structures may be predicted.

As it is believed that most, if not all, damaging earthquakes occur on faults, faults are the most useful seismographic feature to identify and evaluate. However, this is often not feasible, either because the rupture plane does not reach the surface or because the fault trace cannot be readily recognized. This difficulty with active fault identification is not confined to intra-plate areas, but has been well noted in the Eastern USA.[8] It is thus necessary to look for other structural features such as grabens, basins, platforms, and folds, and to try to assess their relative seismogenic characteristics.

As an example of the identification of seismogenic features in an intra-plate zone, Allen[9] notes that the faults bounding grabens in China have been, and remain, seismically very active features, such as the normal fault of the Shansi graben which gave rise to the 1556 Sian earthquake of magnitude 8 which caused more than 820 000 deaths.

Basin and graben areas in themselves are more unstable than platform areas, and might therefore be expected to exhibit more seismic activity than platforms, at least in the longer term. However, a seismotectonic study[10] of the seismically relatively quiet North Sea area (having a maximum magnitude[11] of $M=6$) found no correlation between these geological structures and known seismic events (Figure 2.6). It may be that for this area there is little or no residual difference between these two structures, or that the time period covered by the known seismic events was too short to show and differentiate their full potential for seismic activity.

In another intra-plate zoning study, Klimkiewicz *et al.*[12] identified the major basement structures of the North Central USA using geophysical and geological evidence, and had the benefit of a relatively long historical record of seismic

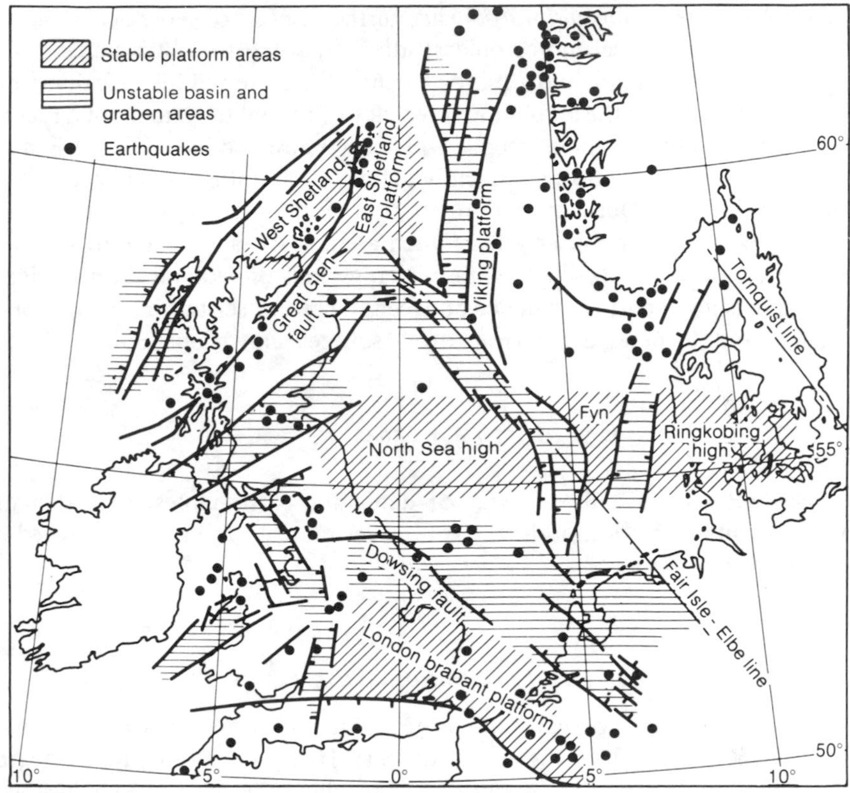

Figure 2.6 Geological structure and seismic events in the North Sea area[10] (reproduced with permission from the Institution of Civil Engineers[10])

events (200 years for larger magnitudes) which they believed to be sufficient to identify all seismogenic features. However, apart from confirming the seismic potential of the Michigan Basin they appeared to conclude that further research was necessary to understand the relative seismic potential of other geological features in their large region of study.

In a detailed analysis of the seismotectonic features of Western Montana, Waldron *et al.*[13] divided the area into two major tectonic provinces separated by a major tectonic intra-plate boundary. Because each of these three areas exhibited unique geological, tectonic, and seismic characteristics, their seismic hazard potentials were seen to be significantly different. Also one seismogenic zone was identified in both of the provinces and in the boundary zone between them. A further outcome of the study was an estimate of the maximum magnitude of earthquake for each of the above three zones, ranging from $M = 6.0$ in the Flathead Lake seismogenic zone, through $M = 6.5$ in the mountain ranges of the intra-plate boundary zone, to $M = 7.5$ in the Inter-mountain Seismic Belt of the southern tectonic province.

As noted above, *tilting and warping* are further aspects of geological structure which are helpful in seismotectonic studies. In some regions, such as New Zealand, they accompany most major earthquakes. Viewed historically, tilt is helpful in determining the amount and recency of crustal movement in a region, and is measured by the slope of beds which are known to have been originally almost horizontal. The most seismically active regions of the world are in belts of late Tertiary and Quaternary deformation, and by dating sloping beds the age of activity may be estimated. In such a study of New Zealand, Clark *et al.*[14] plotted the slopes of tilted strata of two periods of geological time (Figure 2.7). There is good correspondence between the recent seismic activity implied by this map and the evidence from other sources shown on Figure 2.5.

2.2.4 Faulting

Because faults are usually the seat of damaging earthquakes, they demand specific attention in addition to the references made to them in the preceding sections of this chapter.

2.2.4.1 Location of Active Faults

It is widely held that virtually all large earthquakes are caused by sudden displacements on faults at varying depths. However, in many damaging earthquakes no evidence of the fault trace reaching the surface soils has been found. In localities where the earthquake foci are very shallow and the surface soils are competent, such as in California, most fault planes readily reach the surface. Where the foci are deeper and/or the overburden is not stiff enough to fracture right through, surface manifestations of faults do not occur. These situations occur in New Zealand, where not all earthquake faults reach the surface, but some faults in the basement rocks beneath the sediments have been located by geophysical surveys or by instrumentally detected linear arrays of small earthquakes. In Chile no fault ruptures have broken through to the land surface in recent geological times, no doubt partly because of the depth of the subduction zone, whereas the shallower events which occur a little offshore will have fault trace expressions on the seabed, where they would obviously not be readily seen.

In some cases faults may reach the surface but are difficult to recognize, and it may not be possible to identify an active fault from surface traces prior to its next major movement. Apart from the presence of weak superficial deposits, such as in parts of New Zealand or the once glaciated areas of the eastern USA, other factors contribute to the difficulty of identifying faults, such as;

(1) Low degree of fault activity, thus creating less evidence;
(2) Erosion and sediment deposition rates that are higher than the fault slip rates;

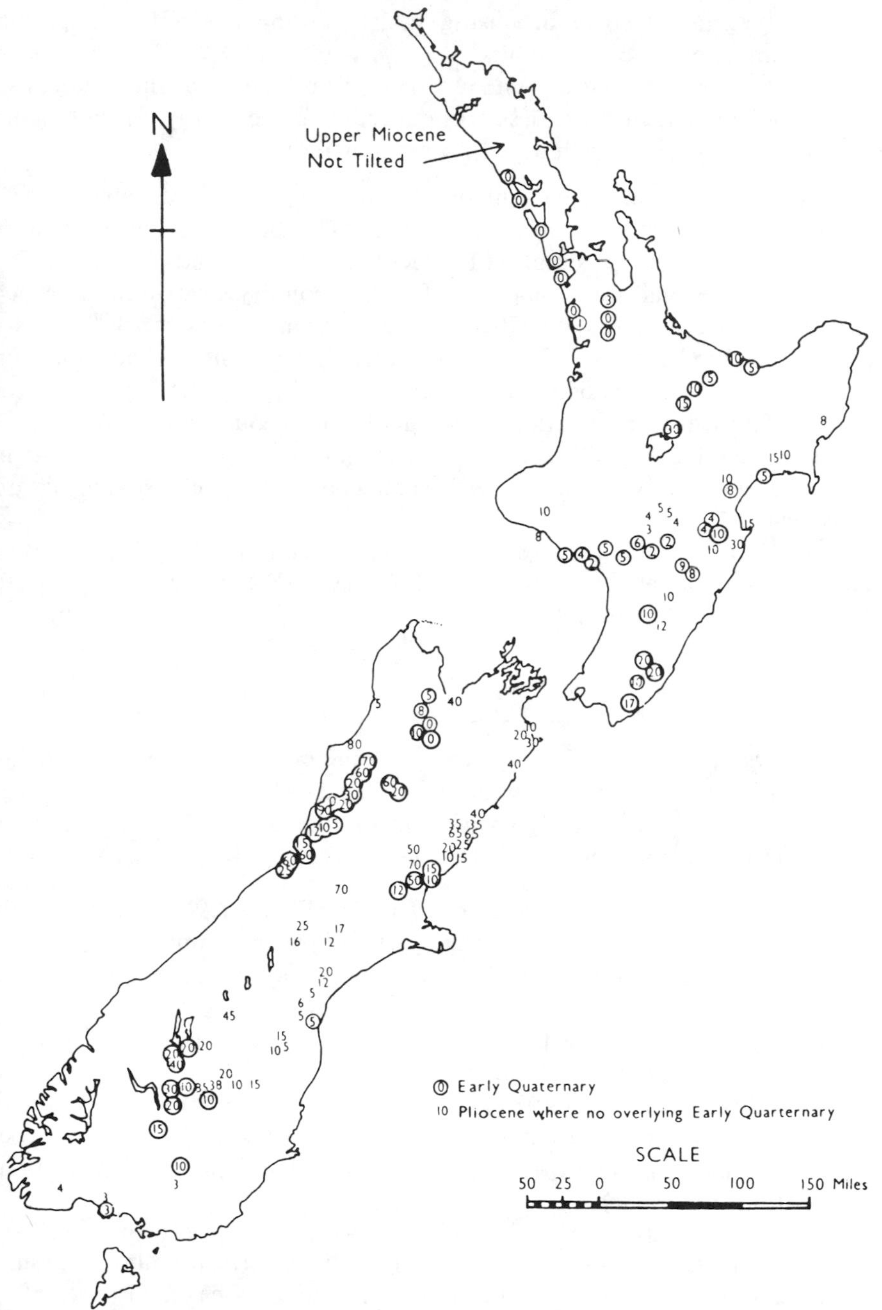

Figure 2.7 Map of New Zealand showing dip angles on Early Quaternary and Pliocene beds (after Clark *et al.*[14])

(3) Dense vegetation cover disguising faults, (although aerial photographs sometimes show up such faults[9]);
(4) Some tectonic processes result in dispersed fault zones at the surfaces so that individual features are less pronounced. Fault zones vary in width from a few metres to as much as a kilometre or more.

The locations of many active faults in different regions are, of course, shown on the appropriate geology maps, but because of the above-mentioned problems such maps are inevitably incomplete. In New Zealand by the mid-1980s geologists believed that they had found about half of that country's active faults which reach the surface, whereas in California the larger majority of such faults are now presumed to be known. Fault maps usually indicate the probable location of those hidden portions of faults, which, although not positively identified, may be inferred by interpolation or extrapolation of outcropping faults plus other evidence (Figure 2.8). The fact that the known array of active faults in a given area constitute a minimum array should be allowed for in seismic hazard assessments.

In addition to existing maps, preliminary indications of faulting may be obtained from aerial photographs and Landsat (ERTS) images, prior to investigations on the ground.

2.2.4.2 Types of Fault

It appears that the characteristics of strong ground motion in the general vicinity of the causative fault can be strongly influenced by the type of faulting. Housner[16] suggests that four types of fault should be considered in the study of destructive earthquakes;

(1) Low-angle, compressive, underthrust faults (Figure 2.9(a)). These result from tectonic sea-bed plates spreading apart and thrusting under the adjacent continental plates, a phenomenon common to much of the circum-Pacific earthquake belt;
(2) Compressive, overthrust faults (Figure 2.9(b)): compressive forces cause shearing failure forcing the upper portion upwards, as occurred in San Fernando California, in 1971 (also called reverse faults);
(3) Extensional faults (Figure 2.9(c)): this is the inverse of the previous type, extensional strains pulling the upper block down the sloping fault plane (also called normal faults);
(4) Strike-slip faults (Figure 2.9(d)): relative horizontal displacement of the two sides of the fault takes place along an essentially vertical fault plane, such as occurred at San Francisco in 1906 on the San Andreas fault (also called wrench or transcurrent faults).

Few pure examples of the above occur, most earthquake fault movements having components parallel and normal to the fault trace. Detailed discussions of faulting may be found elsewhere.[9,17-19]

Table 2.1. Active fault classification, New Zealand Geological Survey[20]

		Movement during 5000–50 000 yr BP		
		Repeated	Single	None
Movement during last 5000 yr	Repeated	I	I	I
	Single	I	II	III
	None	II	III	

		Movement during 50 000–500 000 yr BP		
		Repeated	Single	None
Movement during last 5000 yr	Repeated	I	I	I
	Single	II	III	III
	None	III		

		Movement during 50 000–500 000 yr BP		
		Repeated	Single	None
Movement during 5000–50 000 yr BP	Repeated	II	II	II
	Single	II	III	III
	None	III		

2.2.4.3 Degree of Fault Activity

Active faults include any faults which are considered capable of moving in the future. Because the amount and frequency of movement can vary enormously, it is important to be able to estimate the degree of activity likely to be exhibited by any fault in the region of interest, and various schemes have been devised for doing this. For example, the New Zealand Geological Survey[20] uses three different classes of activity depending on the frequency and times of movement during the past 500 000 years (Table 2.1). To this rating of likelihood of movement should be added estimates of the associated magnitudes of events.

With the growing understanding of source mechanisms and the use of the concept of seismic moment, a comprehensive system of fault activity classification has been proposed by Cluff *et al.*,[21] as set out in Table 2.2, based on an analysis of 150 active faults on a worldwide basis. The classification uses all the parameters of interest in fault behaviour, in six general classes of active fault and five sub-classes. Cluff *et al.* gave the following examples of different classes of fault;

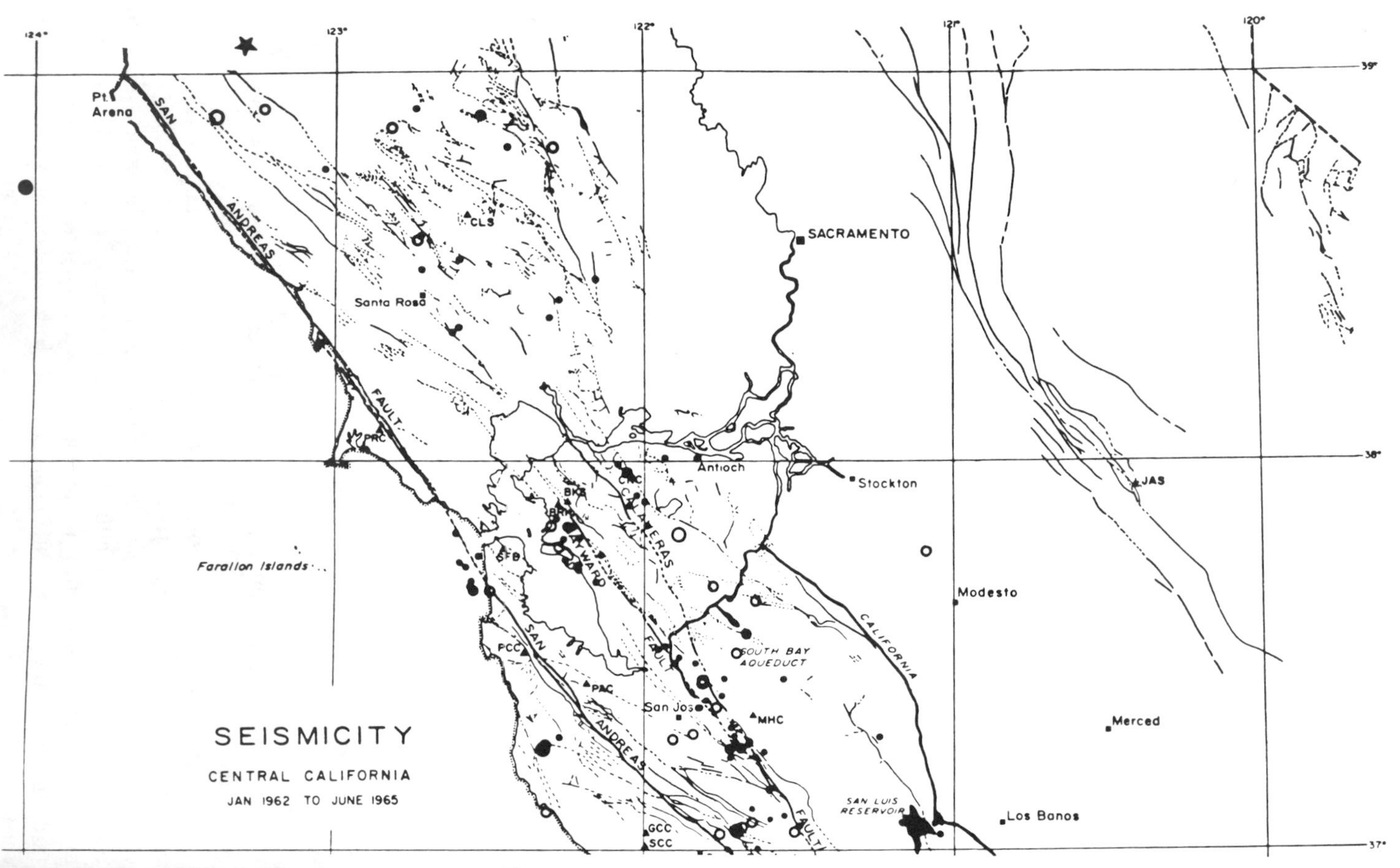
SEISMICITY
CENTRAL CALIFORNIA
JAN 1962 TO JUNE 1965
124°
123°
122°
121°
120°
39°
38°
37°
Pt. Arena
SAN ANDREAS FAULT
Santa Rosa
CLS
PRC
SACRAMENTO
Antioch
Stockton
JAS
Modesto
Merced
Los Banos
Farallon Islands
SFB
BKS
CMC
CALAVERAS
HAYWARD
FAULT
SOUTH BAY AQUEDUCT
CALIFORNIA
SAN LUIS RESERVOIR
San Jose
MHC
PAC
PCC
GCC
SCC
SAN ANDREAS
FAULT

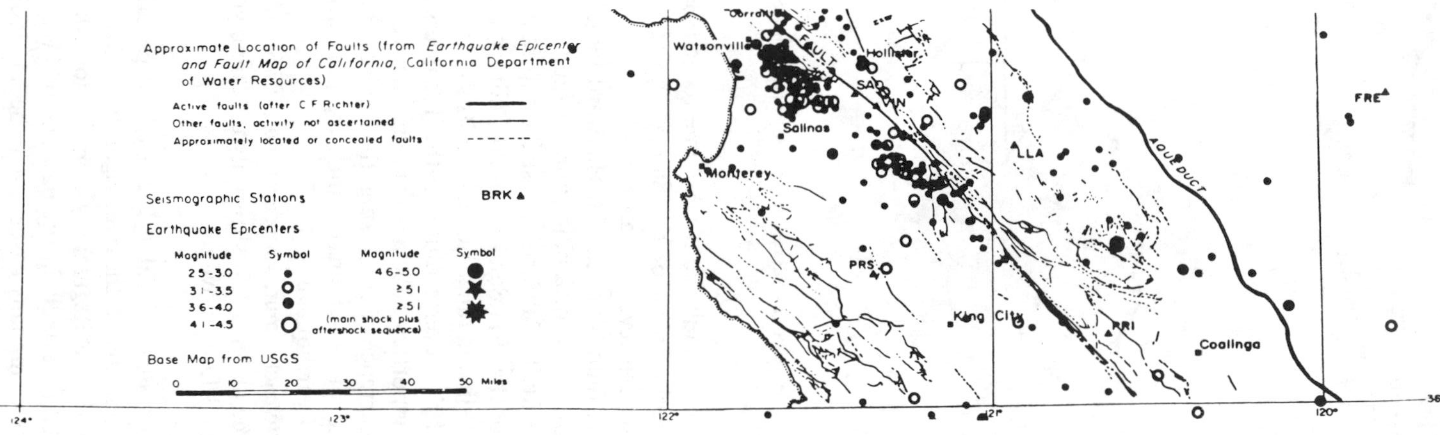

Figure 2.8 Map of California showing faults and locations of earthquakes which occurred between January 1962 and June 1965 (after Bolt,[15] reproduced by permission of Prentice-Hall, Englewood Cliffs, NJ)

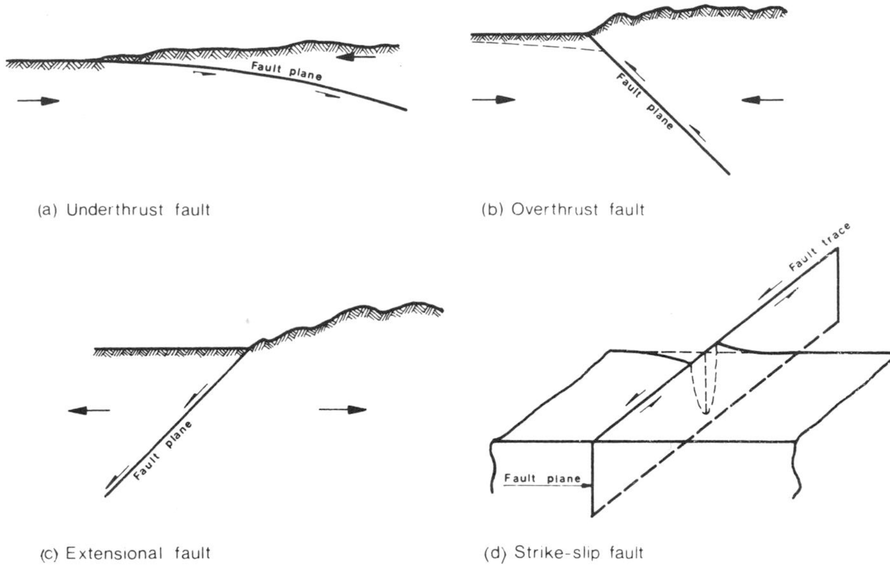

Figure 2.9 The main fault types to be considered in the study of strong ground-motion characteristics (after Housner[16])

Class 1—San Andreas fault from Cholome to San Bernardino;
Class 1B—Parkfield segment of the San Adreas fault, having smaller but more frequent events than Class 1;
Class 2—Motagua fault, Guatemala. Hayward fault, California. Wasatch fault, Utah, an intra-plate normal fault;
Class 3—Elsinore fault, California, a strike-slip fault. Sierra Madre fault, source of the 1971 San Fernando California earthquake, a reverse fault. Basin and range normal faults such as the Dixie Valley fault;
Class 4—Most of the reverse faults in the Transverse Range, California;
Class 4A—This important subclass exhibits relatively large magnitudes at long recurrence intervals, e.g. the Zenjoki fault in Japan has a low slip rate of less than 0.2 mm/yr, but produced an estimated $M_S = 7.4$ earthquake in 1847;
Class 6—The Pitaycachi fault, source of the 1887 Sonora Mexico earthquake estimated at $M_L = 7.5$, slips only 0.02 mm/yr but experienced 4 m of normal fault movement.

The classification system of Table 2.2, by incorporating the full range of possible fault behaviour, avoids the over-rigidity of active–inactive descriptions such as that used by the US Regulatory Commission which considered a fault active if it had moved repeatedly in the past 500 000 years, or once in the past 35 000 years. However, in many cases it may be impossible to obtain enough

Table 2.2. Comprehensive activity criteria for fault classification (after Cluff *et al.*[21])

Class	Slip rate (mm/yr)	Slip per event (m)	Rupture length (km)	Seismic moment (dyne-cm)	Magnitude M_S	Recurrence interval (yr)
1	$\geqslant 10$	$\geqslant 1$	$\geqslant 100$	$\geqslant 10^{26}$	$\geqslant 7.5$	$\leqslant 500$
1A (as 1, except)	$\geqslant 5$					$\leqslant 1000$
1B (as 1, except)		< 1			$\geqslant 7.0$	$\leqslant 100$
2	1–10	$\geqslant 1$	50–200	$\geqslant 10^{25}$	$\geqslant 7.0$	100–1000
2A (as 2, except)		< 1			< 7.0	< 100
2B (as 2, except)		$\geqslant 5$	$\geqslant 100$			1000
3	0.5–5	0.1–3	10–100	$\geqslant 10^{25}$	$\geqslant 6.5$	500–5000
4	0.1–1	0.01–1	1–50	$\geqslant 10^{24}$	$\geqslant 5.5$	1000–10 000
4A (as 4, except)		$\geqslant 0.5$	$\geqslant 10$	10^{25}	$\geqslant 6.5$	
5	< 1					$\geqslant 10\ 000$
6	< 0.1					$\geqslant 100\ 000$

information about the dates and rates of movement to use Table 2.2, and a simpler classification such as the type shown in Table 2.1 may be necessary.

In zones of crustal convergence or divergence, without underthrusting, where the rates of these movements are known, Anderson[22] has developed a method of computing slip rates on each fault from the rate at which seismic moment is released. The method needs to be calibrated from data from historical events on some of the faults in the region, and this proved possible for Anderson's study area of Southern California.

In order to estimate the degree of activity of a fault the mean slip rate plus the frequency and size of movements are measured by examining sections through faults, which may occur in natural geomorphological features such as marine or river terraces, cliffs, and slip faces, or in quarries, road cuttings, or other excavations or purpose-dug trenches.

A study of fault activity was carried out by the New Zealand Geological Survey of the Nevis–Cardrona Fault System in relation to hydroelectric development proposals on the Kawarau River in Central Otago, in the eastern ranges of the Southern Alps. Referring to Figure 2.5, the location is about 100 km east of the 1960 event at the southern end of the seismic gap (Section 2.3.4). A trench dug across a reverse fault at one location illustrates the complex geometrical distortions that sometimes occur in fault zones (Figure 2.10). In an analysis of this fault Beanland *et al.*[23] positively identified three movements the sequence of which is illustrated in Figure 2.11. However, there was qualitative evidence for up to six events for the same total displacement of 5.6 m. The age of the displaced surface was estimated to be *c.* 16 000 years, based on the chronology of the culminating aggradation of the last glacial advance correlated with local terrace surfaces and deposits. As this age embraced three to six fault movements it was concluded that the fault had moved at average intervals of *c.* 2500 to 7000 years, with mean displacements of *c.* 1 m to 2 m, respectively.

In addition to the above-mentioned uncertainties encountered in analysing sections through faults, sometimes there arise even greater difficulties in interpretation due to modifications to the basic fault displacement pattern with the depth of the investigations. Figure 2.12 shows a section about 400 m long by 150 m deep through the Pisa fault in New Zealand close to the Nevis–Cardrona fault discussed above. Here the general fault zone is evident from prominent fault traces (see the Backscarp, Figure 2.12) disrupting an alluvial fan *c.* 70 000 years old.[24] Three boreholes at 160 m spacings, with depths ranging from 50 to 110 m, helped to provide a stratigraphic framework, and trenches across two parts of the fault zone were dug for fault activity determination purposes. Seven reverse faults were found (A–G, Figure 2.12) with vertical displacements totalling *c.* 2.3 m, and were interpreted as antithetic faults. The main 9 m uplift of the alluvial fan was inferred to have occurred on a more important synthetic fault which was not exposed at ground surface or in the trenches, possibly because fault movement has been dissipated in the gravels.

The largest vertical displacement for a single event was estimated at 2 m, but there were insufficient data to determine the number of movements which

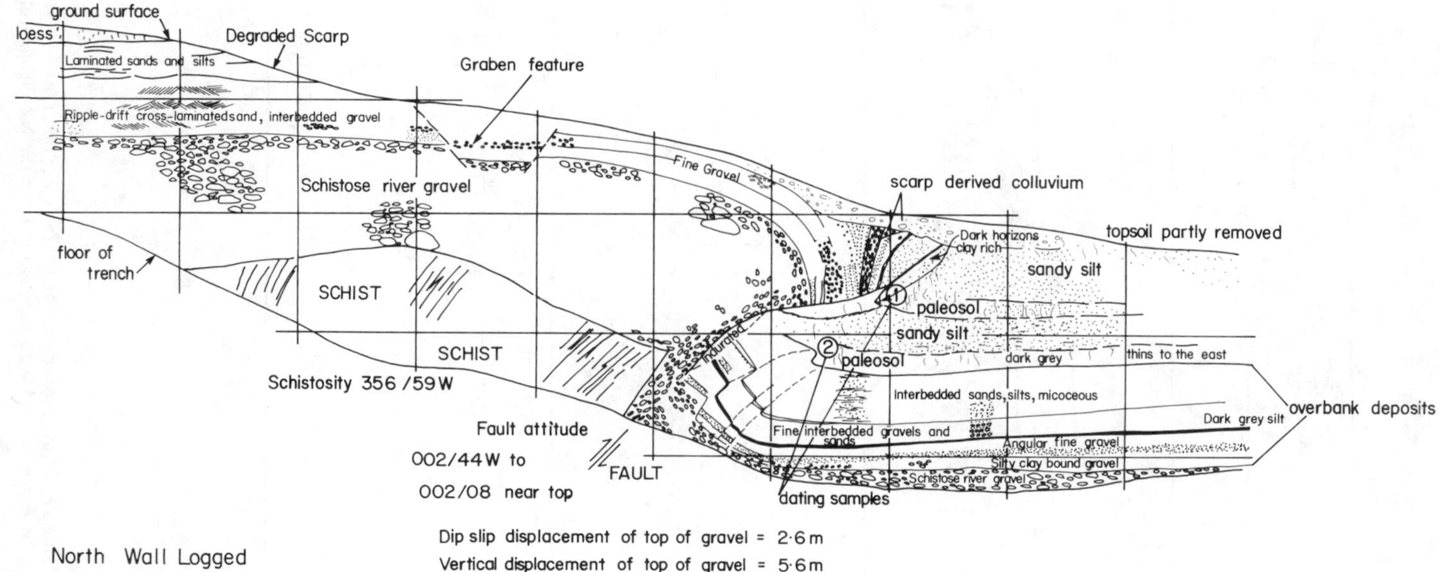

Figure 2.10 Geological log of a trench dug across part of the Nevis–Cardrona fault, Central Otago, New Zealand (after Beanland and Fellows,[23] reproduced with permission from the NZ Geological Survey)

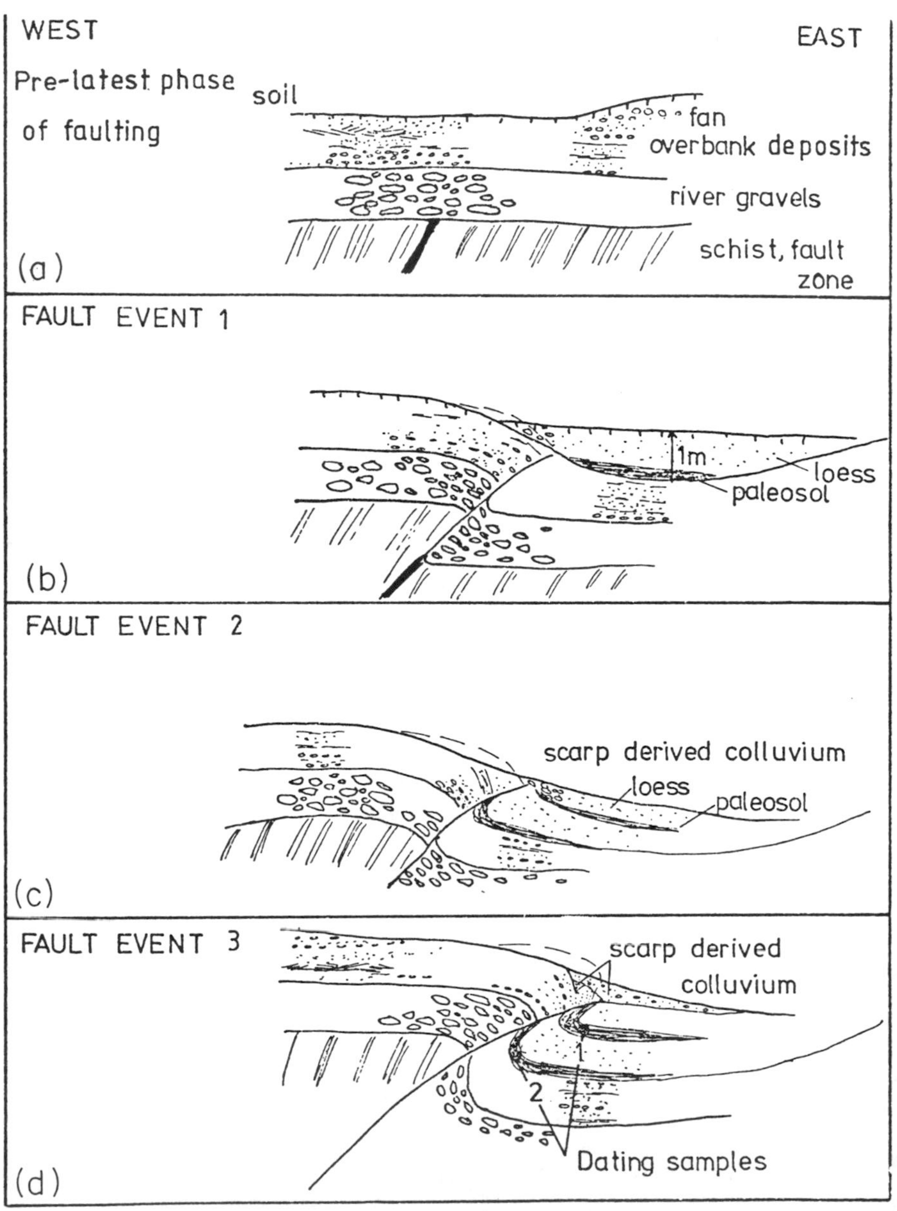

Figure 2.11 Inferred sequence of fault movements on the fault section shown in Figure 2.10 (after Beanland and Fellows,[23] reproduced with permission from the NZ Geological Survey)

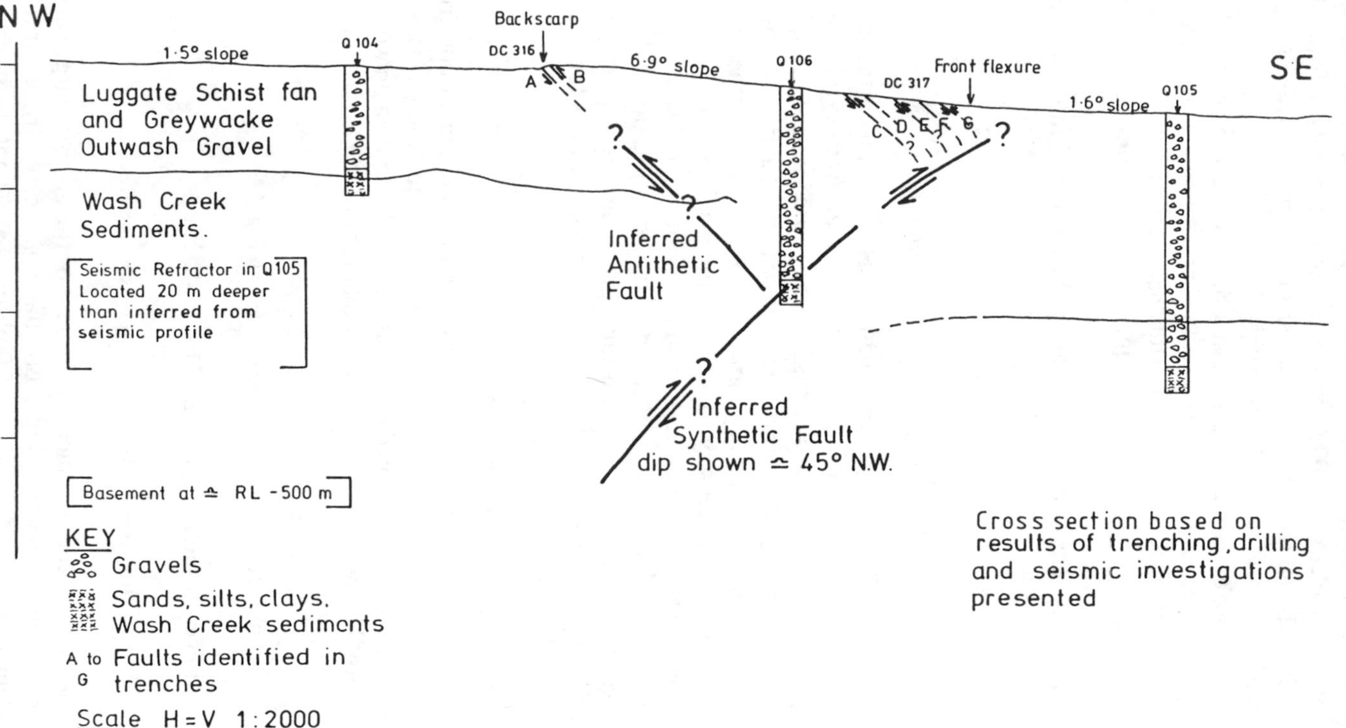

Figure 2.12 Cross-section through Lochar locality of the Pisa fault, Central Otago, New Zealand (reproduced with permission from NZ Geological Survey[24])

contributed to the 9 m uplift in the 70 000-year period concerned. However, because a related surface (*c.* 35 000 years) was not tectonically deformed, the faulting activity was thought to be 70 000–35 000 years old. This means that this Pisa fault is Class II according to Table 2.1.

It is clear from the foregoing discussion that there is likely to be considerable uncertainty in the hazard quantification data that is obtained from trenches. Unless the sites of trenches are very carefully selected their data yield will be low. The cost-effectiveness of this technique needs to be considered carefully for any given project.

Finally, it is noted that means of estimating the hazard of fault movements in probabilistic terms are discussed in Sections 2.3.5 and 4.4.2.

2.2.4.4 Faults and earthquake magnitudes

It is normal and reasonable to ask the question 'How big an earthquake will this fault generate?' Because of the difficulty of answering this question definitively, a variety of ill-defined terms have been used in answering it, e.g. 'maximum credible earthquake', or 'safe shutdown earthquake'. The problem of definition is overcome by relating magnitudes to specific recurrence intervals (Table 2.2), but it is still necessary to estimate the maximum magnitude that is possible due to geological constraints, and assigning an M_{max} value to a magnitude–frequency relationship is an important step (Section 4.2). Three methods of estimating magnitudes from fault behaviour are currently in use, all with their frailties.[8,25] These methods, plus some relationships for mid-plate events, are briefly described below.

Magnitude versus Fault Rupture Length

This is currently the most common method of estimating magnitude from a fault. As seismic moment is a function of fault area (equation (1.3)), it may be expected that magnitude is related to rupture length. From data on worldwide historical events Slemmons[26] found rough correlations for different fault types as follows;

$$\text{Normal faults:} \quad M_s = 0.809 + 1.341 \log L \tag{2.1}$$

$$\text{Reverse faults:} \quad M_s = 2.021 + 1.142 \log L \tag{2.2}$$

$$\text{Strike-slip faults:} \quad M_s = 1.404 + 1.169 \log L \tag{2.3}$$

where L is the rupture length in metres. Because of the inherent large variance in the above (or similar) relationships, Bonilla *et al.*[27] carefully reworked the worldwide data, examining five different fault types and carefully studying the variance. For their total data set of 45 faults of all fault types (which did not include any subduction zone events), they found

$$M_s(L) = 6.04 + 0.708 \log L \qquad S = 0.306 \tag{2.4}$$

and the 95 percentile magnitude

$$M_{0.95} \approx M_s(L) + t_{0.05} S \left[\frac{1}{\upsilon} + 1 \right]^{1/2} \tag{2.5}$$

where L is the fault rupture length in *kilometres*, S is the standard error, υ is the number of degrees of freedom, and t is the statistical t-test parameter (Appendix B.3). Based on the data set used in deriving equation (2.4), $\gamma = 43$, $S = 0.306$, and $t_{0.05} = 1.68$, and from equation (2.5)

$$M_{0.95} = M_s(L) + 0.52$$

Whereas there appears to be reasonable agreement between equations (2.1) to (2.3) and equation (2.4) in the magnitude range 7.0–7.5, their use at the extreme end of their data range is typically uncertain. For example, a magnitude $M_s = 8.0$ event corresponds to L values of *c.* 230, 170, 440, and 590 km for equations (2.1) to (2.4), respectively. $L = 590$ km is so large that equation (2.4) seems to be unreliable in this magnitude range. It should also be noted that equations (2.1) to (2.5) are based on worldwide data which are predominantly from inter-plate zones, and are not relevant to mid-plate events (i.e. events at least 500 km from a plate boundary). A mid-plate event of $M_s = 8.0$ would have a mean fault rupture length of only 40 km (see Table 2.3 and related text).

Magnitude versus Fault Rupture Area

Because of the above-mentioned relationship between fault rupture area and seismic moment, a stronger relationship may be expected using rupture area rather than rupture length. Indeed, even with errors in rupture area up to a factor of two, estimates of magnitudes vary only by 0.3 magnitude units according to Wyss,[28] who gives the following expression:

$$M_s = 4.15 + \log A \tag{2.6}$$

where A is the area of the fault rupture surface in square kilometres.

Table 2.3. Average source parameters for mid-plate earthquakes (after Nuttli[29])

m_b	M_s	$\log M_o$ (dyn-cm)	L (km)	W (km)	D (m)	$\Delta\sigma$ (bars)
4.5	3.35	22.2	2.1	2.1	0.011	4.2
5.0	4.35	23.2	3.8	3.8	0.033	7.0
5.5	5.35	24.2	7.0	6.4	0.11	13
6.0	6.35	25.2	13	11	0.34	23
6.5	7.35	26.2	24	18	1.1	43
7.0	8.32	27.2	45	29	3.7	82
7.2	8.53	27.6	58	36	5.8	102
7.4	8.87	28.0	75	44	9.2	129
7.5	9.00	28.2	85	49	11.5	143

Magnitude versus Fault Displacement

Various empirical relationships have been developed between magnitude and maximum observed surface displacements for historical events, and those derived by Slemmons[26] are

$$\text{Normal faults:} \quad M_s = 6.668 + 0.75 \log D \tag{2.7}$$

$$\text{Reverse faults:} \quad M_s = 6.793 + 1.306 \log D \tag{2.8}$$

$$\text{Strike-slip faults:} \quad M_s = 6.974 + 0.804 \log D \tag{2.9}$$

where D is the maximum surface displacement in metres. As well as difficulties experienced in finding and measuring the true maximum displacement in the field, the large variance in the data used to derive these expressions needs to be recognized. A special problem[8] with earthquakes in the magnitude 5 to 6 range is that many such events have been associated with no observed displacement, but regression analyses have not usually allowed for this and are thus biased in this magnitude range.

Moment Magnitude and Seismic Moment

In addition to setting bounds on magnitude M_s, Table 2.2 incorporates *seismic moment*, which depends on fault rupture area as discussed above. Anticipating a growing use of *moment magnitude*, which is derived from seismic moment using equation (1.1), this measure of earthquake size may eventually be included in the criteria of Table 2.2, perhaps replacing magnitude.

Fault Rupture Characteristics of Mid-Plate Earthquakes

As noted above, the average rupture length of mid-plate events, i.e. those at least 500 km from plate boundaries, is much shorter than for other events located near the boundaries. This was ascertained by Nuttli[29] in an empirical study of the source mechanisms and spectral characteristics of 140 continental and oceanic mid-plate earthquakes, from various parts of the world. His average relationships between m_b, M_s, M_o, and fault rupture length, width, displacement, and stress drop ($L, W, D, \Delta\sigma$) are given in Table 2.3.

2.3 EARTHQUAKE DISTRIBUTION IN SPACE, SIZE, AND TIME

2.3.1 Introduction

Earthquakes occur at irregular intervals in space, size, and time, and in order to quantify seismic hazard at any given site it is necessary to identify the patterns in the spatial, size, and temporal distributions of seismic activity in the surrounding region. Understanding the tectonic causes of earthquakes and identifying the seismogenic geological features in a region, as discussed in Section 2.2, enable the formulation of distribution patterns of potential sources. These patterns or distributions of occurrence involve the careful plotting or mapping of known historical events, and the correlation of these historical data with the models of crustal structure and deformation. Aspects of the study of the distribution of earthquakes are further discussed in the following sections.

2.3.2 Spatial distribution of earthquakes—maps

Figure 2.1 is a plot of spatial distribution of seismic events with the lowest possible level of information, but has its uses in global tectonic plate recognition. A similar world map, 'Significant earthquakes 1900–1979', produced in the USA, uses different sizes and colours of symbols to indicate the numbers of deaths and cost of damage of the events plotted. Thus a qualitative feel for the relative seismic risk of different localities can be obtained. However, for most analytical purposes the events need to be much more precisely plotted on local maps. Larger-scale seismic event maps of many areas are given in various publications, principally the reference works by Gutenberg and Richter,[30] Karnick,[31] and Lomnitz.[32]

For the assessment of a particular site the simplest map consists of seismic events only, such as that shown in Figure 2.12. This map indicates the location in plan, the order of depths, and the magnitudes of all earthquakes of magnitude 5 and greater within 300 km of Djakarta recorded from 1900 to 1972 (and uses raw data!). For preparing such maps the seismic data should be evaluated by seismologists, as lists of raw data from recording agencies are subject to significant error. Generally speaking, the earlier the event, the less accurate are the data.

A scrutiny of the depths of the events on Figure 2.13 shows that they get progressively deeper from south to north, which is explained by the fact that the Indian–Australian plate is subducting under the Pacific plate in this region (Figure 2.2). This effect is much more obvious on a vertical section, such as that shown of New Zealand in Figure 2.4. Other attempts to establish patterns in seismicity by comparing maps of events with maps of tectonic features have also been illustrated earlier (Figure 2.5, 2.6, and 2.8).

The type of seismicity map shown on Figure 2.13 is conveniently prepared on a size A2 drawing sheet, to a scale of about 25 km:10 mm. The choice of symbols poses something of a problem as there is no international convention on symbols. Various systems have been used by seismologists such as Gutenberg and Richter or the US Coast and Geodetic Survey, but these all suffer from the disadvantage that different magnitudes of earthquakes are shown by symbols which differ only in size. Such symbols cannot be easily distinguished by eye and hence are not as rapidly interpreted as they could be.

It is therefore recommended that for engineering purposes, symbols differing in shape or colour should be used. Magnitudes less than 5.0 are generally of little direct design significance, as such earthquakes cause little structural damage. Therefore, events of $M < 5.0$ have been excluded from the notation. However, in areas of low seismicity, it may be worth plotting events where $M \geqslant 4$,

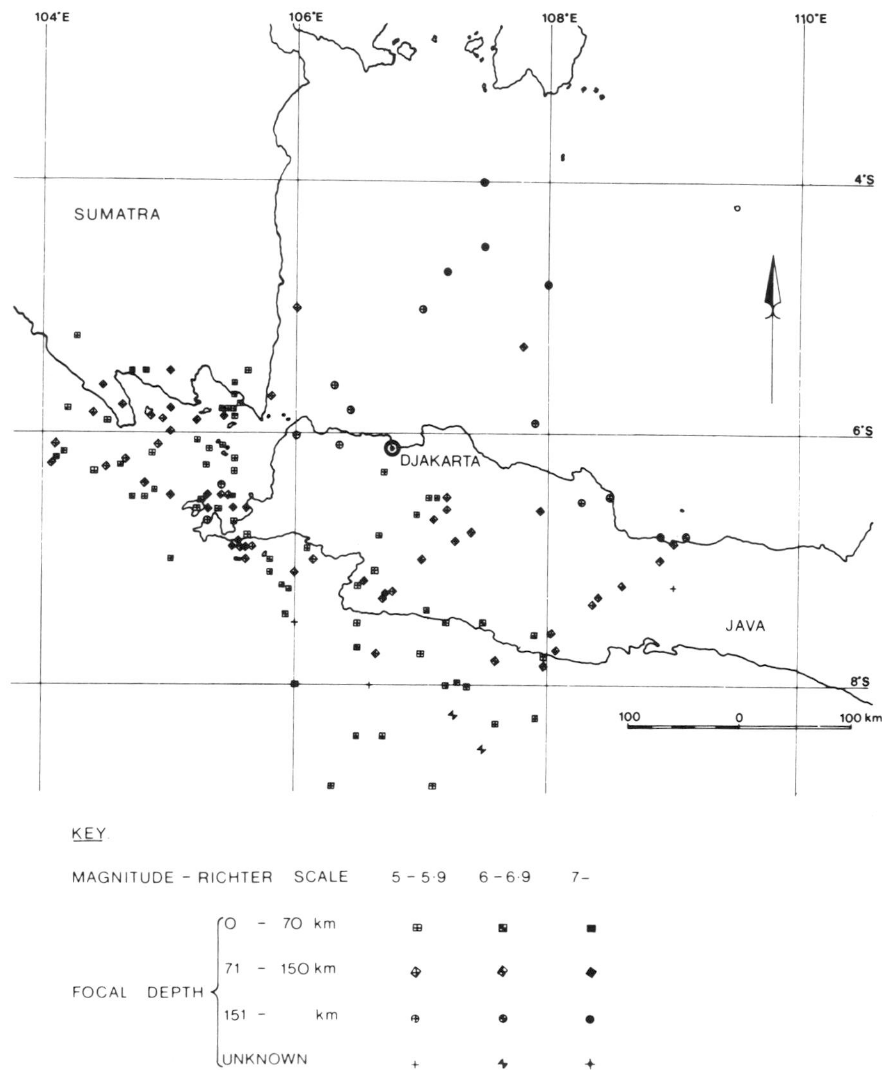

Figure 2.13 Seismic event map for Djakarta (1900–72)

in order to emphasize the pattern of seismic activity and hence help delineate the zones of greater risk. Earthquakes occurring at great depths cause little damage, and for that reason all events occurring at depths greater than 150 km have been grouped together under one symbol as these may be ignored for engineering purposes. Perhaps the deepest damaging earthquake known to have occurred was the Bucharest earthquake of 4 March 1977, which had a focal depth of about 90 km and was of magnitude $M = 7.2$.

Another method of identifying spatial patterns of seismic activity is by mapping *strain release*, which can be done by strain measurements as discussed

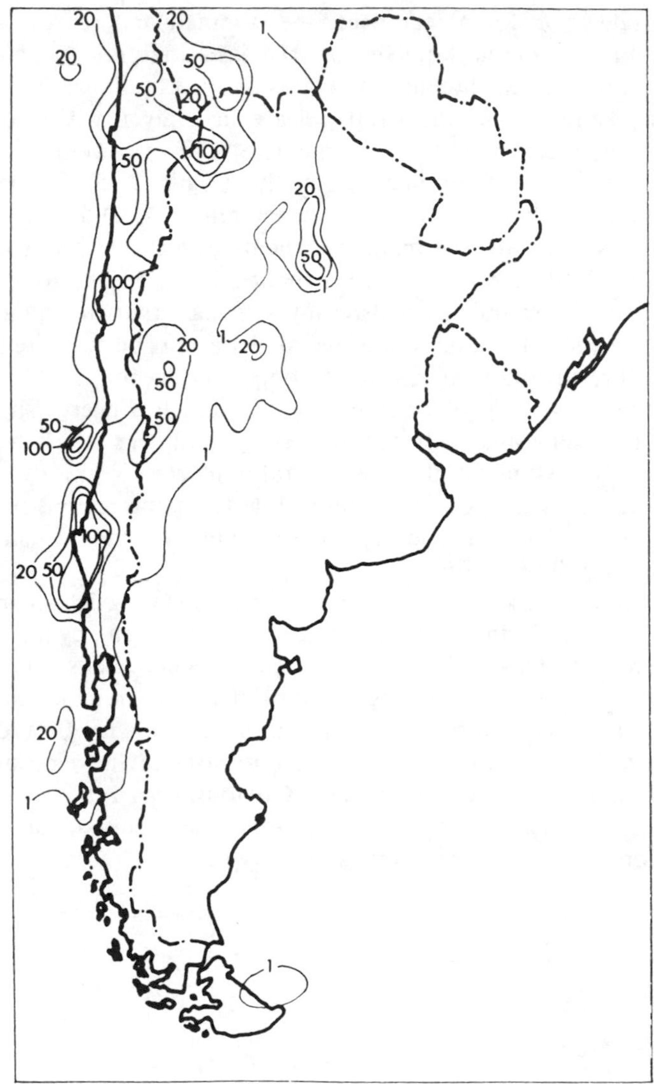

Figure 2.14 Mean annual strain release map of southern America, for the period 1920–71 (after Carmona and Castano[33])

in Section 2.2.2 above, or by converting seismic event maps into strain release, as discussed below.

The strain released during an earthquake is taken to be proportional to the square root of its energy release. The relationship between energy E, (in ergs) and magnitude M for shallow earthquakes has been given by Richter[34] as

$$\log E = 11.4 + 1.5M \tag{2.10}$$

The strain release U for a region can be summed and represented by the equivalent number of earthquakes of $M=4$ in that region, $N(U4)$. The equivalent number of earthquakes $N(U4)$ is divided by the area of the region to give a measurement of the strain release in a given period of time for that region which can be used for comparisons of one region with another (Figure 2.14), or one period of time with another (Figure 2.15). Further examples of strain energy studies may be found in references 35 and 36.

Large shocks constitute the main increments to a cumulative strain release plot. For example, a shock of $M=5.5$ is about thirteen times stronger than one of $M=4$. It therefore requires extensive low seismic activity to equal the energy release of a large shock. Nevertheless, as smaller events occur more frequently, in the study of relative strain release rates, comprehensive information is required of low-magnitude activity. The summation of many low-energy shocks in one region may be comparable with that of a few large shocks in another. Because the recording of low-magnitude events is still in its early years, this method of estimating seismicity may be either impossible for some regions or of limited value due to lack of data. However, where suitable records exist strain release plots can be very illuminating.

A plot of strain release against time is a step function to which an upper bound curve can be drawn, giving an indication of the trend in energy release for that region (Figure 2.15). Obviously, if a flattening of the curve tends to be asymptotic to a constant strain value over a significant time, then the faults in the region may have at least temporarily taken up a more stable configuration. On the other hand, a mechanical blockage of strain release may have occurred which only a pending large shock could release. Obviously, as in the case of seismic event maps, information from other sources is required for the proper interpretation of strain release curves and maps.

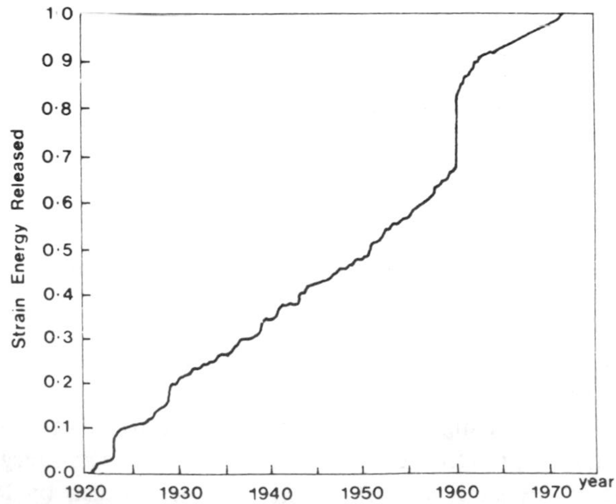

Figure 2.15 Rate of strain energy release in the portion of South America shown in Figure 2.14 (after Carmona and Castano[33])

2.3.3 Earthquake distribution in time and size

On any given fault within any given region, earthquakes occur at irregular intervals in time, and one of the basic activities in seismology has long been the search for meaningful patterns in the time sequences of earthquake occurrence. The longer the historical record, the better is the overall picture that can be obtained. In most places the useful historical record is short, often only a few decades or sometimes one or two centuries, the great exceptions being China and the Eastern Mediterranean, which both have useful records going back around 2000 years. Figure 2.16 shows the time distribution of damaging earthquakes in the latter area from the first to the eighteenth century AD, as derived by Ambraseys.[37] The long quiescent period between active periods is worth noting, and similar gaps centuries long have been found in China.[9,38]

During any given interval in time, the general underlying pattern or distribution of size of events is that first described by Gutenberg and Richter,[30] who derived an empirical relationship between magnitude and frequency of the form

$$\log N = A - bM \tag{2.11}$$

where N is the number of shocks of magnitude M and greater than M per unit time and unit area, and A and b are seismic constants for any given region (Figure 2.17).

A varies significantly from study to study while b varies from about 0.5 to 1.5 over various regions of the earth. Values of A and b derived by Kaila and Narain[39] are given in Table 2.4. An authoritative discussion of magnitude–frequency relationships in various parts of the world has been given by Evernden[40], whose values of b should be compared with those in Table 2.4.

Various researchers have tried to relate A and b; Kaila and Narain[39] obtained the relationship

$$A = 6.35b - 1.41 \tag{2.12}$$

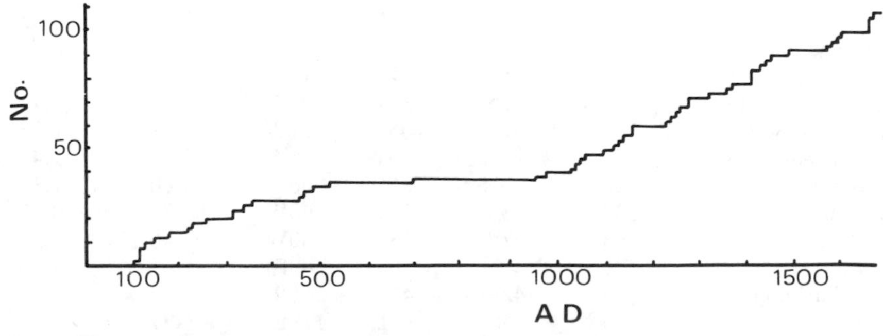

Figure 2.16 Time distribution of damaging earthquakes in the Anatolian fault zone (after Ambraseys[37]) (reprinted by permission from *Nature* **232**, pp. 375–379. Copyright © 1971 Macmillan Journals Limited)

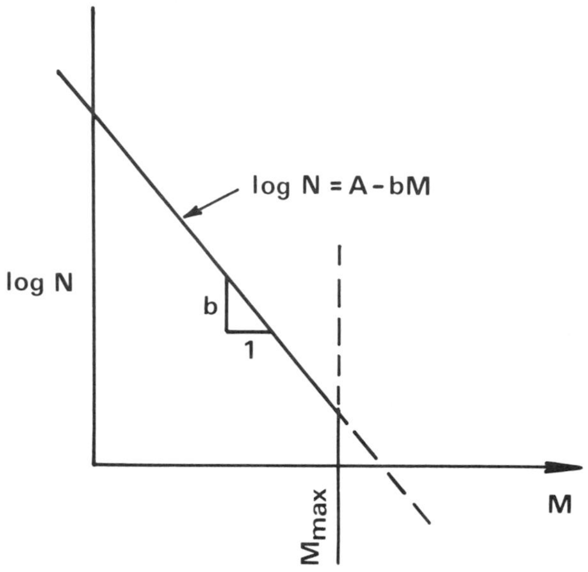

Figure 2.17 Magnitude–frequency relationship for earthquake occurrence

In order to derive this equation, values of A and b derived for various regions of the earth were plotted against each other, and the least squares line was found. The correlation coefficient or goodness of fit was 0.90.

The slope b of the least squares line has a significant seismic meaning. A decrease in b over a period of time indicates an increase in the proportion of large shocks. This may be caused by a relative increase in the frequency of large shocks, or by a relative decrease in the frequency of small ones. Some

Table 2.4. Values of A and b based on 14 years of shallow earthquake data, for various regions normalized to a $2° \times 2°$ grid (Extracted from Kaila and Narain[39])

Region	Boundary				A	b
Japan	26N	40N	132E	150E	6.86	1.22
New Guinea	13S	1N	132E	148E	7.83	1.35
New Zealand	48S	37S	164E	180E		1.04
W. Canada	47N	65N	142E	115W	5.05	1.09
W. United States	25N	47N	135W	105W	5.94	1.14
E. United States	25N	47N	105W	51W	5.79	1.38
Central America	10N	25N	120W	85W	7.36	1.45
Colombia–Peru	18S	6N	85W	60W	5.60	1.11
N. Chile	37N	18S	78W	60W	4.78	0.88
S. Chile	63S	37S	78W	60W	4.46	0.92
Mediterranean	30N	50N	20W	48E	5.45	1.10
Iran–Turkmenia	15N	42N	48E	65W	6.02	1.18
Java	13S	5S	90E	118E	5.37	0.94
E. Africa	40S	30N	20E	48E	3.80	0.87

Note: Great care must be taken to use a sufficiently large sample of earthquakes over as long a period of time as possible in order to obtain reasonably meaningful values of b.

investigators have found that periods of maximum strain release in the earth's crust (see the year 1960 in Figure 2.15) have been preceded and accompanied by a marked decrease in b. From uniaxial compression experiments in the laboratory, Scholz[41] found that the magnitude–frequency relationship for microfractures in a given rock is characterized by b decreasing when the stress level is raised. Consequently regional variations in b may indicate variations in the level of compressive stress in the earth's crust.

For any given region if enough data is available a plot of M against log N can be made, and the 'best' line for equation (2.11) can be determined using a linear regression analysis as described in Appendix B.3. Records of events that give magnitude $M<4$ go back only a few years, and it is usual to neglect these values as they may give a misleading bias to the relationship

$$\log N = A - bM \tag{2.11}$$

Although there are also arguments in favour of a quadratic form, the empirical log-linear relationship of equation (2.11) fits the data reasonably well in the lower-magnitude range, and, because of its simplicity, is the expression in general use at present. However, it is unsatisfactory at high magnitudes, as demonstrated by Chinnery and North[42] (Figure 2.18), because there is a maximum achievable magnitude, M_{max}. The latter arises both because all of the traditional magnitude scales saturate, e.g. M_s does not exceed a value of about 9.0 (Section 1.2), and because a given fault or tectonic region has physical constraints on the maximum size of event it can generate. Despite the difficulty of reliably estimating values for M_{max} (Section 2.2.4.4) it is important to have a magnitude cut-off in equation (2.11), as shown in Figure 2.17, when estimating seismic hazard, as such estimates are greatly reduced at lower probability levels (Section 4.2). Various expressions have been developed which modify equation (2.11) to incorporate M_{max}, with different transitions between the two lines, some of which are discussed by Anderson *et al.*[43]

Earthquake-recurrence relationships taking the above forms are derived to fit the seismicity of a region, which in general will comprise a number of faults. However, studies of palaeoseismicity suggest that any given fault segment generates earthquakes having a very narrow range of magnitudes which do not fit the Gutenberg–Richter or similar distributions, and an alternative 'characteristic earthquake' model has been proposed.[44]

2.3.4 Seismic gaps and seismic quiescence

In the study of earthquake occurrence in space, size, and time arise the important and difficult concepts of *seismic gaps and seismic quiescence*. In general terms a seismic gap is a segment of space experiencing a significantly long interval of time between large events (usually a longer time than might be expected for the locality concerned). As pointed out by Lomnitz,[45] some confusion about

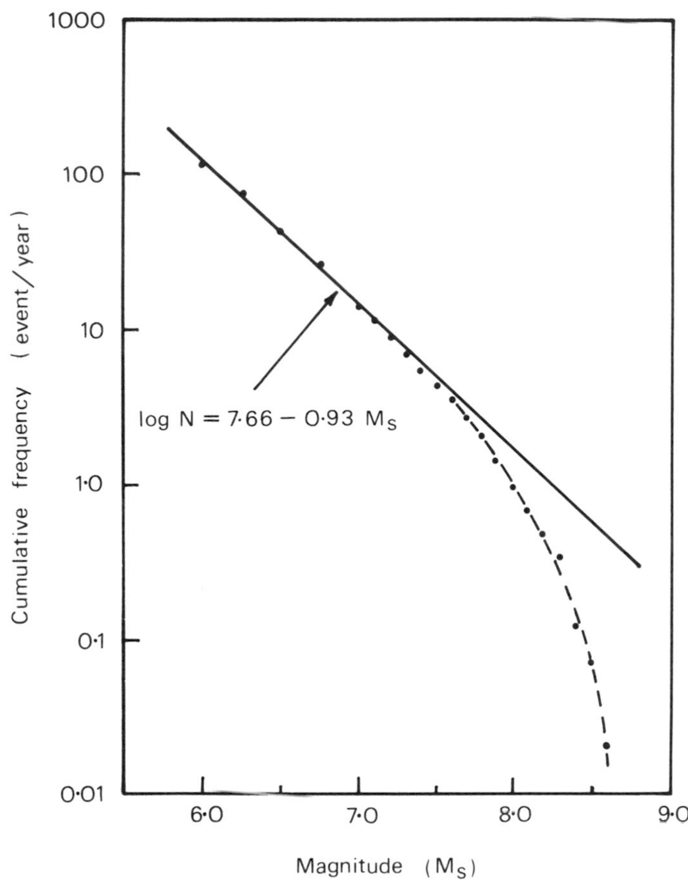

Figure 2.18 Cumulative magnitude–frequency relationship for large earthquakes from worldwide data, showing tendency for M_s to saturate in value (reproduced with permission from Chinnery and North[42]. Copyright 1975 by the AAAS)

seismic gaps occurs because of difficulties with defining the minimum duration of a gap and the minimum magnitudes concerned.

In response to Lomnitz's prompting, Haberman *et al.*[46] defined a seismic gap as *a segment of a plate boundary which has not ruptured during a large earthquake during the last three decades*. Elsewhere in the same paper, Haberman *et al.* define large earthquakes as being of $M \geqslant 7.0$. It is clear that seismic gaps are primarily spatial rather than temporal.

An example of a seismic gap is provided by examining Figure 2.5, where a seismic gap over 400 km long exists in the Southern Alps of New Zealand on the boundary between the Pacific and the Indian–Australian plates. Adams[47] found that the last major movement of this segment of the Alpine fault occurred 550 years ago, and that earthquakes of magnitude 8 recur at average intervals of 500 years, based on carbon-dating of terraces that record river downcutting across the fault. Many other seismic gaps have been noted around the world, for example by Kelleher *et al.*[48]

Seismic gaps can have background seismicity ranging from high to low,[46] but the maximum size of event conforming to background seismicity is ill-defined.[49] Quoting further definitions from Haberman *et al.*[46] permanent seismic gaps are segments of plate boundary with no clear history of significant earthquakes, and which for tectonic reasons may not produce large events. By contrast, *mature seismic gaps* are those in which one (or more) precursory event indicates that preparation for a large gap-filling earthquake may be in progress. For example the magnitude $M = 7.8$ earthquake that occurred near San Antonio, Chile, on 3 March 1985 was suspected of being a precursor to a larger event ($M \approx 8.3$) to fill the local seismic gap, which may therefore have been made *mature* by the March 1985 event. It should be noted that a *seismic gap* is simply an acknowledgement of prior history, making no statement about the physical preparation for a gap-filling event unless it becomes recognized as a *mature seismic gap*, on the basis of the precursory evidence noted above or evidence of strain build-up. *Gap-filling earthquakes*, because of their spatial extent and associated slip, significantly reduce the likelihood of another large shock in the same area during the next several decades. The possibility of predicting the approach of a gap-filling event has led to much use of the seismic gap concept in the field of earthquake prediction. Haberman *et al.*[46] define *seismic quiescence* as the state when the seismicity rate falls significantly below the *background seismicity rate* which would be the norm between major ruptures in a given seismogenic zone. In contrast to seismic gaps, seismic quiescence implies a change in physical processes occurring in the region. Finally it is noted that *precursory quiescence* has been defined[46] as temporal quiescence that occurs in the vicinity of the rupture zone of an earthquake prior to that event.

While the foregoing sequence of definitions provides a basis for the discussion of various nuances in the study of time intervals between larger earthquakes in a given region, it unfortunately does not eliminate all of the inconsistencies previously noted by Lomnitz.[45] Thus in a futher attempt to clarify the situation, Lomnitz and Nava[49] proposed the following definition of a seismic gap which attempts to include those features associated with them by scientists working in the field;

A seismic gap is a region (known to be seismically active) in which there is a lull in seismic activity (quiescence) which lasts over a period large compared with the recurrence time for events larger than a certain threshold.

This definition appears to be both comprehensive and flexible, and accords exactly with the author's general understanding of these phenomena given above in the second sentence of this section, but no doubt further constructive arguments will emerge as study of this subject continues.

2.3.5 Models of earthquake-occurrence processes

The time sequence of earthquakes in a given region is the result of an ongoing physical process, and models may be made of

(1) The physics of the phenomena producing the earthquakes; and
(2) The magnitude series of the resulting events.

In the past most effort has gone into the purely numerical approach (2), but more recently greater effort has been put into combining the above methods, such as that of Anderson *et al.*,[43] who used average slip rate as a physical control but allowed no uncertainty in it (Section 2.3.3). Two other approaches assign uncertainties to the physical information, and use the maximum entropy (likelihood) principle[50] or the Bayes' model[51] to evaluate magnitude probability distributions.

Real sequences of earthquakes in a given region are typified by those shown in Figure 2.16, with the magnitudes, the locations, and the time intervals varying randomly, and various methods of time-series analysis have been applied by different researchers[52,53] in attempts to find analytical models, including those discussed below:

(1) Evaluation of the ratio of sample variance of the number of shocks to its expected value.[53,54] This ratio, called *Poisson's index of dispersion*, equals unity for Poisson processes, is less than one for nearly periodic sequences and is greater than one when events tend to cluster.
(2) Use of the *hazard function* $h(t)$, defined so that $h(t)$ is the conditional probability that an event will take place in the time interval $(t, t+dt)$, given that no events have occurred before t. If $F(t)$ is the cumulative probability distribution of the time between the events

$$h(t) = f(t)/[1 - F(t)] \tag{2.13}$$

where $f(t) = \partial F(t)/\partial t$

For the Poisson process, $h(t)$ is a constant equal to the mean rate of the process.

Most *stochastic* models of seismicity assume that event sequences constitute Poisson processes, and the events M_i are independent and identically distributed. The Poisson process implies that the distribution of waiting times to the next event is not modified by the knowledge of the time elapsed since the last one; the distribution of waiting times is exponential (the horizontal line in Figure 2.19). In reality the data often do not conform to the Poisson renewal process, usually because of clustering, and a better model would be a more general renewal process where the expected time to the next event decreases as time goes on.

Various attempts at creating better models have been made, such as *trigger models*.[53] These treat the overall process of earthquake generation as the

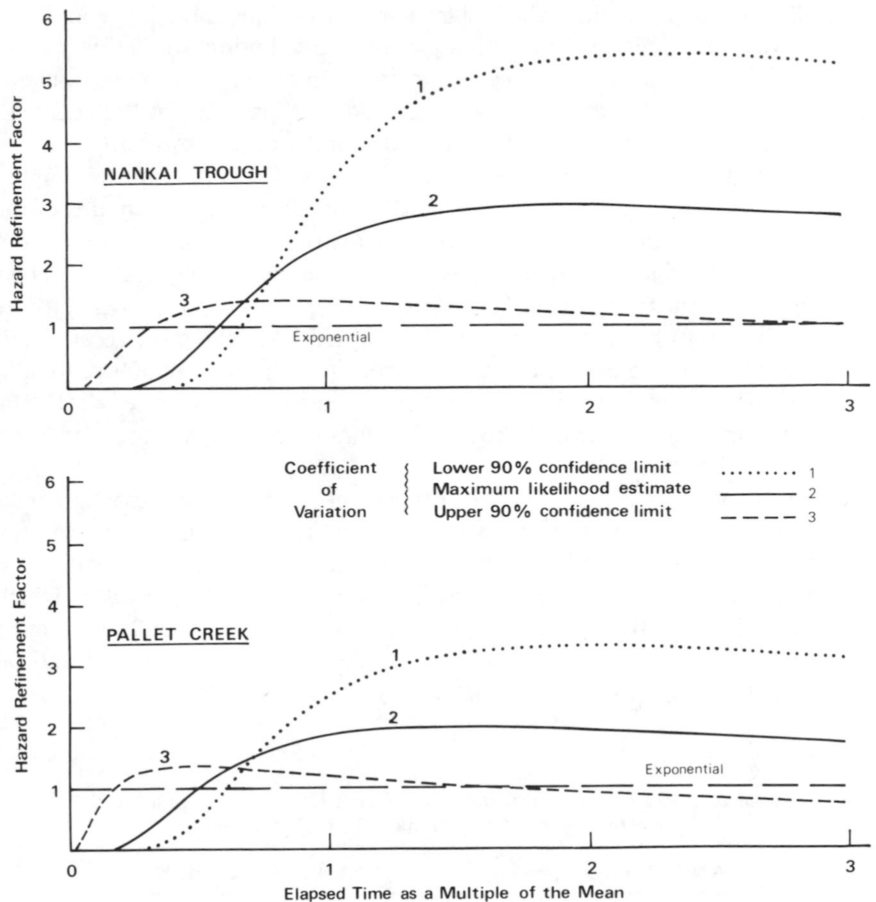

Figure 2.19 Estimates of hazard of surface fault rupture for two faults, assuming that hazard increases with time since the last event (after Rhoades and Miller,[57] reproduced with permission from the NZ Geological Survey

superposition of a number of time series, each having a different origin, where the origin times are the events of a Poisson process.

Vere-Jones[53] has also studied *branching renewal processes*, in which the intervals between cluster centres, as well as those between cluster members, constitute renewal processes. This model was shown to be valid from a study of hazard function estimated from sequences of small shocks in the Hindu-Kush.

Time-dependent models present further attempts to allow for the fact that earthquakes are not necessarily independent events. Savy *et al.*[55] have suggested a non-homenogenous Poisson process. Also a time-dependent stochastic model[56] has been proposed which assumes that an earthquake would occur when the accumulated stress reaches a certain level. Among other things, this study highlights how the Poisson model becomes less desirable as the time since the last earthquake increases.

Another approach in which the hazard varies with time, models the intervals between events (the waiting times) as a log-normal distribution, so that the hazard increases as the time since the last event increases. In a study of fault movements during the past c. 1400 years in the Nankai Trough, Japan, and at Pallet Creek, California, as shown in Table 2.5, Rhoades and Millar[57] estimated hazard functions (Figure 2.19) for these two seismic sources. The three curves in Figure 2.19 reflect the uncertainty in the data, and are based on three estimates of the coefficient of variation for the log-normal distribution, i.e. the maximum likelihood estimate and the 90 percent confidence bounds. The vertical axis shows the hazard refinement factor, i.e. the ratio of the instantaneous hazard to the mean (exponential) hazard, given the time elapsed since the most recent fault movement. For example, if the mean recurrence interval is 12 500 years and the hazard refinement factor is 2, then the instantaneous hazard rate is 2/12 500 events per year. This represents a probability of approximately 0.02 that an event will occur in the next 100 years.

As shown on Figure 2.19, the hazard refinement factor is much less than unity immediately after a fault movement has taken place, reaching unity when the elapsed time reaches about half the mean recurrence interval. After sufficient elapsed time, the hazard refinement factor rises to a fairly stable value, which in the case of the Nankai Trough is most likely to be little under 3 and almost certainly between 1 and 6, as indicated by the maximum likelihood and 90 percent confidence limit curves, respectively.

Table 2.5. Recent fault movements at Pallet Creek (California) and the western half of the Nankai Trough (Japan)[57]

NANKAI TROUGH		PALLET CREEK	
Approximate date	Holding time (yr)	Approximate date	Holding time (yr)
1946		1857	
	92		112
1854		1745	
	147		275
1707		1470	
	102		225
1605		1245	
	244		55
1361		1190	
	262		225
1099		965	
	212		105
887		860	
	203		195
684		665	
			90
		575	

The difference been the hazard functions for Nankai Trough and Pallet Creek is related to the difference in the estimated coefficients of variation (Table 2.5), and although the latter difference is not statistically significant it may be related to the difference between the tectonic settings of the two faults.[57]

While some of the above analytical models have been found to correspond well to specific sets of real seismic data, no general theory has yet been established. Difficulties in evaluating models arise because of the scarcity of data of larger events; this is, of course, an inherent problem. Also those models which do not incorporate a physical causal model will remain at best a tuned echo of the partly known outcome of real processes. However, stochastic models permit further understanding of earthquake generation and distribution to be gained, and the Poisson process model, despite its weaknesses compared with the more complex models, remains a valuable tool because of its simplicity. Vere-Jones[2] does not consider that the more complex models lead to significantly improved estimates of seismic hazard compared with those obtained from the Poisson assumption, but gives some guidance on key items where the Poisson assumption leads to significant errors, and also gives some simple means of improvement of Poisson model results. Useful reviews of the above modelling techniques will be found elsewhere.[2,50,52]

2.4 THE NATURE AND ATTENUATION OF GROUND MOTIONS

In order to obtain a complete predictive model for the ground motion at a given site, it is necessary (1) to describe fully the ground motion at the source and (2) to describe the modifications to the ground motion as it propagates from source to site, i.e. the attenuation. The nature of the sources and the attenuation are not the same for all regions, and hence the appropriate regional descriptions need to be determined from assessing the seismic hazard at a given site.

2.4.1 Earthquake source models

The subject of source models is an area of study for seismologists, the results of which are fundamental to our understanding of the nature of ground motion. From amidst the complexities of this major study area a number of key parameters are evident as being of interest to earthquake engineers, some of which have already been introduced, such as fault length, fault width, fault displacement (or slip), stress drop on a fault, and, of course, earthquake magnitude. Some regional differences in fault length have been noted in Section 2.2.4. A few further features of source models are briefly described below, and for further reading specialist text books[58–60] should be consulted.

An earthquake is the product of a displacement discontinuity sweeping across a fault surface. The shape of the rupture surface and the resistance across it are variable, such that mathematical modelling of the source process, while often qualitatively plausible, remains quantitatively promising rather than convincing. Nevertheless, various simplified models are useful predictors of gross features of ground motion and can be helpful for extrapolations in predicting design ground motions in regions with few data at the appropriate magnitudes and focal distances (Section 4.3.3.1iv).

Early work on source models concentrated on what could be learned from the kinematics only, while more recently studies have been carried out based on the fracture mechanics of cracks initiated in pre-existing stress fields on a fault plane. In this approach, called the dynamic model, components of the model such as fault slip and rupture velocity are obtained by solving a mixed boundary problem.

Aki[61] notes that there are two extreme ways of modelling a heterogeneous fault plane, as illustrated schematically in Figure 2.20. The hatching represents resistance to slip. The upper right of Figure 2.20 shows the fault plane containing an unbroken area after the earthquake, constituting a residual barrier. This has been shown by numerical studies to be possible, and the system is referred to as the *barrier model*. The alternative *asperity model* is shown on the bottom left of the figure, and illustrates the strong patches or asperities which are broken during the earthquake.

Using a barrier model of rectangular faults containing circular cracks of radius ρ, model parameters were determined[61] for several Californian earthquakes using their observed acceleration power spectra (Table 2.6). The local stress drop $\Delta\sigma$ inside the circular crack areas is relatively constant, ranging from 200 to 400 bars for all the events studied, which is qualitatively consistent with the result obtained by Hanks and McGuire[62] using data of root mean square acceleration. These stress drops may also be compared with those for mid-plate earthquakes in Table 2.3, where the greatest stress drop is 143 bars, which is the average expected value for an event of $M_s = 9.0$.

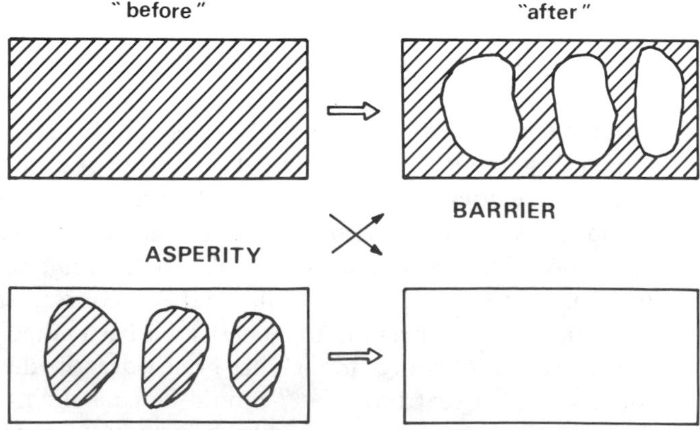

Figure 2.20 Asperities and barriers as two extreme models of heterogeneous fault planes (Aki[61]. Reproduced by permission of Earthquake Engineering Research Institute (EERI))

Table 2.6. Source parameters for some Californian earthquakes derived from a barrier source model[61]

Event	M_s	M_L	m_b	$M_o \times 10^{27}$ (dyn-cm)	L (km)	W (km)	$\Delta\sigma$ (bars)	2ρ (km)	D (m)	f_{max} (Hz)
Kern County, 1952	7.7	7.2		2.0	70	20	350	13	3.0	2.5
San Fernando, 1971	6.6	6.35	6.2	0.12	20	14	300	5	1.0	5.0
Borrego Mtn, 1968	6.7	6.8	6.1	0.063	33	11	200–300	2–3	0.4	4.0
Long Beach, 1933	6.25	6.43		0.028	30	15	220	1	0.4	4.0
Parkfield, 1966	6.5	5.5	5.9	0.014	35	15	200–300	1–2	0.4	5.0

A simple expression for estimating the stress drop comes from Brune[63]

$$\Delta\sigma = \frac{7M_o}{16r^3} \tag{2.14}$$

where the rupture surface is assumed to be circular with radius r. This equation and the previous methods give estimates of stress drop which differ for the same event by as much as an order of magnitude. Thus the maximum value of stress drop that is likely to occur in any earthquake is uncertain, but values higher than several hundred bars do not seem likely.

Rupture velocity, the velocity at which fault rupture propagates, is a basic parameter of source modelling, with estimates typically varying from about half to about equal to the shear wave velocity of the ruptured material, yielding rupture velocities $v_r \approx$ 2 to 3 km/s.

Rise time is the time required for the slip or stress change on a fault to take place, and is most simply expressed as a ramp function, so that the displacement at time t at a point x on the fault

$$u(t) = u_\infty G(t - \frac{x}{v_r}) \tag{2.15}$$

where G is a ramp function which increases linearly from zero at $t=0$ to unity at $t=T$ (where T is the rise time), and u_∞ is the final displacement. The rise time may be determined directly if the ground motion near the fault is recorded completely. Rise times computed from two theoretical models including equation (2.15) for 41 events in different parts of the world[58] range in value from 0.7 to 36 s, while Brune[60] postulates a value as low as $T=0.1$ s in estimating an upper bound for peak ground acceleration. Clearly, peak ground velocity and acceleration are dependent on the rise time.

Two *frequency parameters* f_o and f_{max} that arise in source modelling are most readily described by reference to Figure 2.21. Far-field shear wave acceleration spectra are characteristically flat at frequencies greater than the *corner frequency* f_o, which has been defined by Brune[63] as the frequency at the intersection of the low- and high-frequency asymptotes of the spectrum. Corner frequencies are calculated for both P- and S-waves, and, despite controversy[61] about the relative magnitudes of f_o (P) and f_o (S), the corner frequency is an important feature of source models. The ordinates of Figure 2.21 are logarithms of the acceleration spectral density.

The parameter f_{max}, also shown on Figure 2.21, is less well established. According to Hanks,[64] some f_{max} (often, but not necessarily much larger than f_o) almost always exist, above which spectral amplitudes diminish, often abruptly. Sound observational evidence for f_{max} comes from spectra of recordings at close epicentral distances, but the mechanism causing it is uncertain and it is unclear whether it solely represents band-limiting source phenomena or whether it is in part a function of the properties of the propagation path and recording site.[64]

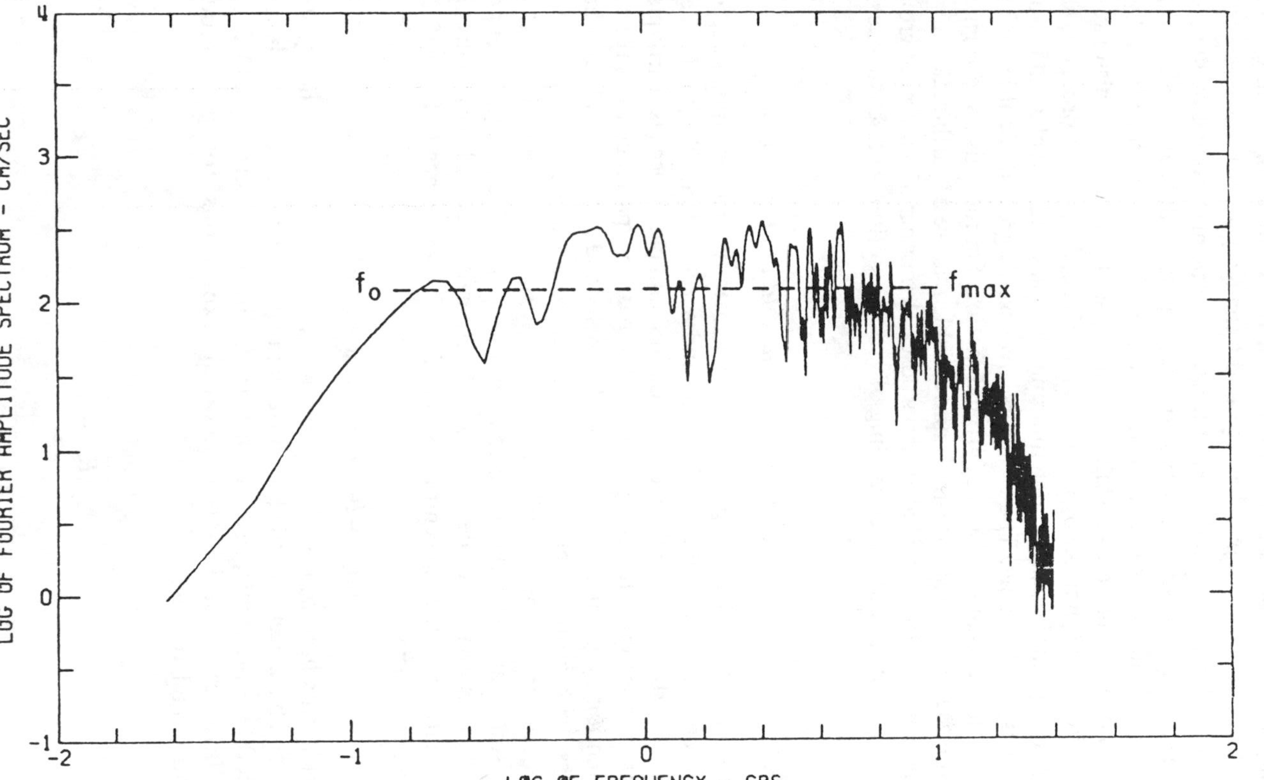

Figure 2.21 f_o and f_{max} estimated from the spectrum of the San Fernando earthquake of 9 February 1971, $M_L = 6.4$, at the Pacoima Dam site (Hanks,[64] reproduced with permission from the Seism. Soc. of America)

2.4.2 The characteristics of strong ground motion

In earthquakes the motion of any particle of the ground follows in general a complex three-dimensional path having rapidly changing accelerations, velocities, and displacements and a broad band of frequency content. Strong ground motion is measured by a large-amplitude type of seismograph called an *accelerograph*, in the form of an accelerogram, which is an acceleration history typically of the form shown in Figure 1.4, while the velocity and displacement histories are obtained after the earthquake by integration of the acceleration record. As is clear from Figure 1.4, earthquake ground motion is complex and irregular, and also no earthquakes are the same. These factors give rise to great difficulties in fully understanding the characteristics of ground motion. Seismologists and engineers are continuing to seek a single simple and reliable feature of earthquake motion that characterizes it for their analytical purposes. A number of features of ground motion contribute to our understanding of ground motion, as discussed below.

2.4.2.1 Peak ground motions and attenuation

The most obvious piece of information to be gained from an earthquake record is the maximum acceleration, or peak ground acceleration, which is 0.33 g in Figure 1.4. Partly because it is so easy to obtain, and partly because earthquake forces are proportional to acceleration, in the past this parameter has received most attention by engineers. Peak ground velocity and displacements also have their uses, with growing interest in velocity in recent years.

The characteristics of ground motion vary with the nature and size of the event at source and with the distance from the source. Traditionally the peak ground motions y have been described as a function of magnitude M and distance x from the source, either epicentral distance D or focal distance R (Figure 2.22), in the general form

$$\log y = b_1 + b_2 M + b_3 \log (x + b_4) \qquad (2.16)$$

The coefficients b_1 to b_4 vary depending on the data set to which the equation is fitted, and have been modified for different regions as more data have become available. As an example, using data from California events, in 1973 Esteva and Villaverde[65] published the following expressions for peak ground acceleration and velocity:

$$a = \frac{5600\ \mathrm{e}^{0.8M}}{(R+40)^2} \qquad (2.17)$$

$$v = \frac{32\mathrm{e}^{M}}{(R+25)^{1.7}} \qquad (2.18)$$

where a and v are in cm/s^2 and cm/s, respectively, and R is in kilometres. The two expressions are valid for focal distances in excess of 15 km.

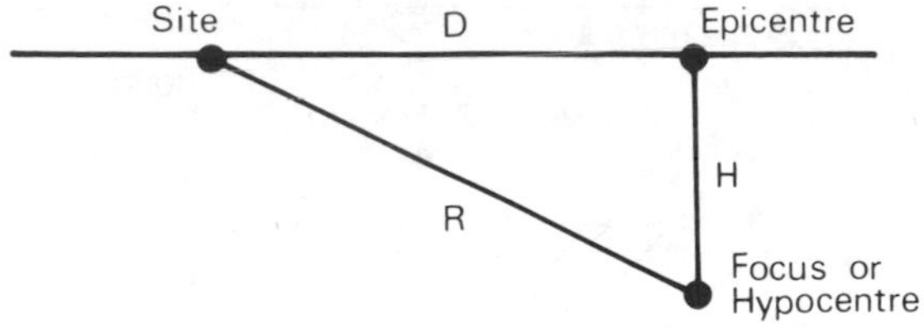

Figure 2.22 Geometric relationship between focus and site

Equations (2.17) and (2.18) should be compared with McGuire's attenuation equations[66] for a, v, and d, given in Table 2.7 together with coefficients of variation for each parameter. The coefficients are empirically derived from regression analyses of 68 earthquake records for sites in the western USA, with $5.3 \leqslant M \leqslant 7.6$ and $15 \leqslant R \leqslant 125$ km.

For any specific event, M is, of course, known, and the expected peak ground motions can be estimated as a function of R, the relationship

$$a = \frac{516\ 500}{(R+25)^{2.04}} \tag{2.19}$$

being the best-fit least squares curve found by Donovan[67] to fit the data for the San Fernando 1971 earthquake (Figure 2.23). The weakness of peak ground acceleration as a measure of earthquake strength is illustrated by the variability in this data set, e.g. at a hypocentral distance of 30 km the acceleration at the mean minus two standard deviations was about 55 cm/s^2 while at the mean plus two standard deviations the acceleration was about 280 cm/s^2. Even greater scatter must, of course, occur when the variability of source energy release mechanisms are included, i.e. when the data for more than one event of a given magnitude are included in the data set.

As mentioned above, equations (2.17) and (2.18) were stated to be invalid for focal distances less than 15 km. This was primarily because there were few or no data recorded in this relatively small area near the epicentre, known as the

Table 2.7. McGuire's attenuation expressions[66] for peak ground motions

$$y = b_1' \, 10^{b_2 M} \, (R+25)^{-b_3}$$

$$\log y = b_1 + b_2 M - b_3 \log (R+25)$$

	b_1'	b_1	b_2	b_3	Coeff. of var. of y
a (cm/s^2)	472.3	2.649	0.278	1.301	0.548
v (cm/s)	5.640	0.714	0.401	1.202	0.696
d (cm)	0.393	−0.460	0.434	0.885	0.883

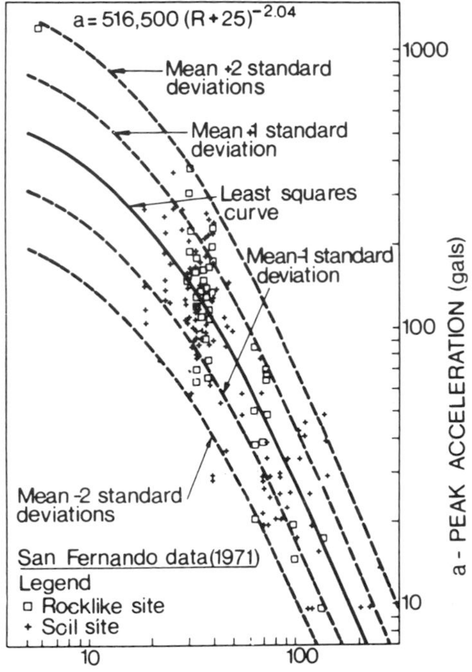

Figure 2.23 Attenuation of peak ground acceleration. Least squares and standard deviation curves for the 1971 San Fernando, California, earthquake (after Donovan[67])

near-field. Despite obtaining many reasonably close recordings in the San Fernando earthquake there was only one data point for R less than *c*. 20 km, i.e. the value in excess of 1 g at $R = 6$ km in Figure 2.22, which is the famous Pacoima Dam recording which has its own special features (Section 4.3.4.1).

An understanding of the nature of ground motion in the near-field is crucial to obtaining a comprehensive predictive model for ground motion hazard. Not only is the ground itself important for the design of structures located in such zones, but it is required for verifying our models of earthquake sources which would so greatly strengthen our hazard forecasting capability. To that end, since the mid-1970s great effort and expense has been spent on setting up two- and three-dimensional synchronized arrays of closely spaced accelerograms in highly active locations such as California, Japan, and Taiwan. An early success came from the 1979 Imperial Valley earthquake of local magnitude 6.6, which caused a fault rupture through the El Centro array in Southern California.

Taking advantage of new near-field data in a fresh look at Western USA earthquakes, Joyner and Boore[68] produced new attenuation relations for peak horizontal acceleration and velocity:

$$\log a = -1.02 + 0.249M_{\rm m} - \log r - 0.00255r + 0.26P \qquad (2.20)$$
$$r = (d^2 + 7.3^2)^{1/2} \qquad 5.0 \leqslant M_{\rm m} \leqslant 7.7$$

$$\log v = -0.67 + 0.489M_{\rm m} - \log r - 0.00256r + 0.17\,S + 0.22P \qquad (2.21)$$
$$r = (d^2 + 4.0^2)^{\frac{1}{2}} \qquad 5.3 \leqslant M_{\rm m} \leqslant 7.4$$

where a is in g, v is in cm/s, $M_{\rm m}$ is moment magnitude, d is the closest distance to surface projection of the fault rupture in km, $S=0$ at rock sites and $S=1.0$ at soil sites, and $P=0$ for 50 percentile values and $P=1.0$ for 84 percentile values.

Equations (2.20) and (2.21) have some interesting differences from the traditional attenuation relationships discussed above. First, the epicentral and focal distances D and R (Figure 2.22) have been replaced by the closest horizontal distance to the fault trace, which will improve the accuracy of the expression for points which are near the fault but not near the epicentre (as will be clear from inspection of Figure 2.26). Also because of the data points at small values d, no lower limit for d has been imposed for the validity of the equations, and at distances of less than 6 or 7 km these equations give virtually constant values of a and v (Figure 2.24). The use of such equations, of course, requires a knowledge of fault geometry in relation to the site.

Equations (2.20) and (2.21) are stated in terms of moment magnitude $M_{\rm m}$ rather than the usual $M_{\rm s}$, but for practical comparisons these two scales are numerically similar (Table 1.1). The clear statements of the valid ranges for $M_{\rm m}$ and the variability factor P ideally should always be readily available. A further important feature of these expressions is that concerning differences

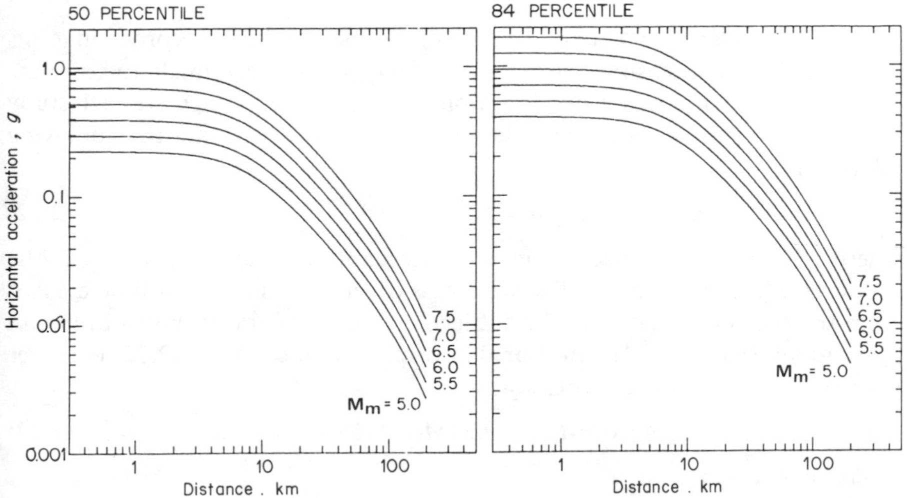

Figure 2.24 Values of peak horizontal acceleration for 50 and 84 percentiles as a function of distance and moment magnitude for western USA data using equation (2.20) (after Joyner and Boore,[68] reproduced with permission from the Seism. Soc. of America)

between soil and rock sites. Joyner and Boore[68] found that for these Western USA events the peak horizontal acceleration was not significantly different for rock and soil recording sites, but values of peak horizontal velocity and displacement were significantly greater on soil sites than on rock. At present it is not clear how universal will be this result for peak accelerations. For example, a study of Alpide Belt data by Chiaruttini and Siro[69] found the peak horizontal acceleration to be significantly greater on soil than on rock, although their data set from several countries may have been less consistent than that used for the Western USA study.

Near-field ground motion and attenuation vary regionally. For example, Ambraseys[70] found a significant difference between Europe and the Western USA. There is also the much-studied difference between Western and Eastern North America. Hermann *et al.*[71] give attenuation relationships for the latter area which show that peak ground motions are higher near the source and attenuate much more slowly at large epicentral distances in the eastern than in the western part of the North America. Another example comes from New Zealand, where Smith[72] has identified three regions in which intensity of ground motion attenuates very differently. It is apparent, though incompletely understood, that peak ground motion amplitudes and attenuation are functions of both source mechanism and the properties of the travel path. Hence because of seismotectonic similarities between Japan and New Zealand it seems likely that attenuation relations derived from the more abundant Japanese data by Katayama *et al.*[73] may be appropriate for parts of New Zealand.

2.4.2.2 Peak ground acceleration versus Modified Mercalli intensity

Because so much of the data on strength of ground shaking is recorded in terms of the subjective intensity scales, many attempts have been made to correlate intensity scales with peak ground motions. The most comprehensive such study is that of Murphy *et al.*,[74] who derived from worldwide data the following relationship:

$$\log a = 0.25\ I_{\mathrm{mm}} + 0.25 \tag{2.22}$$

where a is the peak horizontal ground acceleration in cm/s^2 and I_{mm} is the Modified Mercalli intensity. The weakness of this relationship will be evident from the enormous scatter in the subjective intensity data as shown in Figure 2.25. Analyses of the difference of the correlation in equation (2.22) on other variables resulted in the expression

$$\log a = 0.14 I_{\mathrm{mm}} + 0.24 M_{\mathrm{L}} - 0.68 \log D + \beta_{\mathrm{k}} \tag{2.23}$$

where R is in km, and

$$\begin{aligned} \beta\ &(\text{Western USA}) &= 0.60 \\ \beta\ &(\text{Japan}) &= 0.69 \\ \beta\ &(\text{Southern Europe}) &= 0.88 \end{aligned}$$

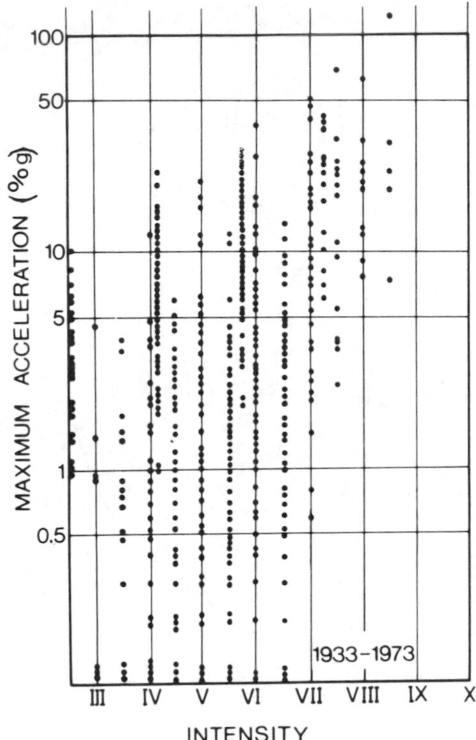

Figure 2.25 Peak ground accelerations plotted against the corresponding intensity as observed in earthquakes occurring between 1933 and 1973 (after Ambraseys[77])

The apparent regional differences in the attenuation equations are obviously strong, but whether this is the result of inherent differences in the tectonic environment of the regions, or at least in part due to consistent regional bias in the assignment of intensities, needs to be explained. However, the values of a obtained from equation (2.23) seem to be excessively high.

2.4.2.3 Directional Effects

Figure 2.26 shows an idealized distribution of intensity of ground shaking in relation to a near-vertical fault rupture, such as discussed for Californian earthquakes by Housner.[75] The traditional attenuation relationships, (equations (2.16) to (2.19)) are made to fit the mean of the data about a point source, and hence represent all the intensity contours as circles with attentuation being the same in all directions. As noted above, equations (2.20) and (2.21) allow for the effect of the line source by relating peak ground motion to distance to the fault trace, implying a contour pattern consisting of a series of straight lines parallel and equal in length to the fault trace with the ends joined by semicircles.

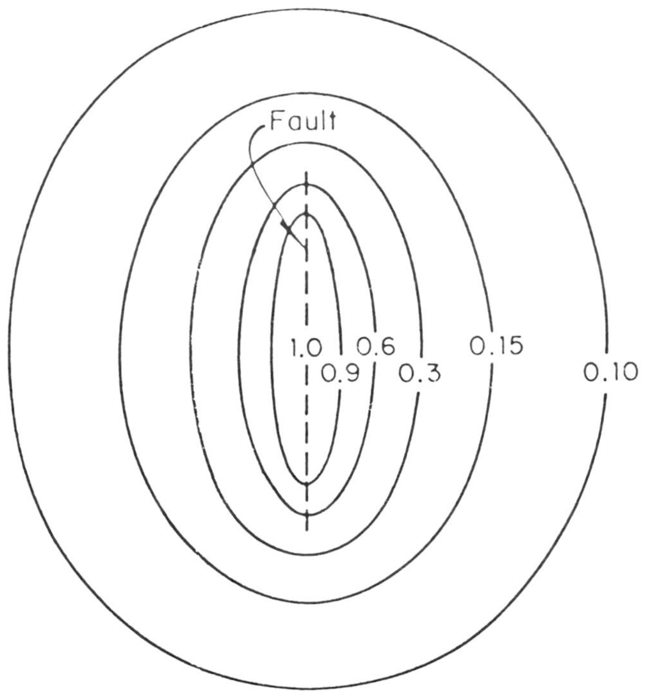

Figure 2.26 Idealized contour lines of intensity of ground shaking, normalized to unit epicentral intensity

The symmetry about the fault trace (i.e. where the fault breaks the ground surface) of the contours in Figure 2.26 clearly depends on the slope of the fault rupture surface, and an asymmetrical pattern, at least about the fault trace, could be expected from underthrust faults of the type shown in Figure 2.9(a). For shallow focal depths on faults of moderate dip angles (Figure 2.9(b) and (c)), while asymmetry exists, it will only be significant at relatively small distances from the fault trace.

Another directional feature of ground motion concerns the phenomenon of gross variation with direction from the epicentre, rather than the slowly varying contours discussed above. In a study of two 1980 events in Livermore Valley, California, of moment magnitudes 5.8 and 5.5, Boatwright and Boore[76] were able to elaborate on previous work by others, through data from 27 stations scattered around the epicentres at epicentral distances varying between 4 and 90 km. They found that peak accelerations varied by a factor of about 10, depending on the direction from the epicentre, and that peak velocities varied by a factor of about five, apparently as a result of the fault rupture process. This finding underlies the importance of understanding source mechanisms, and if such directional variation proves to be of general occurrence it will help to explain some of the large scatter in the historical attenuation data.

2.4.2.4 *Upper bounds to peak ground motion*

Expressions of the types given in equations (2.16) to (2.19) do not always give appropriate estimates of peak ground motions that may be expected at a given site, because there may be limitations on the amplitudes achievable at a given site.

From considerations of stress, frequency content, and rupture velocity, Brune[60] gives two arguments for an upper bound of about $2g$ for horizontal acceleration in solid rock near to the source. An elaboration of these results come from considering the ground motion at the surface due to an S-wave radiated vertically during the failure of the most heavily loaded asperity on a fault, whence McGarr[77] found the expression for peak horizontal acceleration;

$$a < 1.58\ \Delta\tau_i / \rho z \tag{2.24}$$

where $\Delta\tau_i$ is the stress drop on the asperity (bars), ρ is the density of the rock (gm/cm^3), and z is the depth (km).

McGarr also found upper bounds for this stress drop of

Compressional stress state:	$\Delta\tau_i = 334z$	(2.25)
Perfect strike-slip state:	$\Delta\tau_i = 112z$	(2.26)
Extensional stress state:	$\Delta\tau_i = 67z$	(2.27)

Substituting equations (2.25) to (2.27) into equation (2.24) yields upper bounds on peak horizontal accelerations as follows:

Compression state:	$a < 2.0g$
Perfect strike-slip state:	$a < 0.7g$
Extensional state:	$a < 0.4g$

If the strike-slip state of stress is other than 'perfect', the upper bound for a can lie anywhere between 0.4 and $2.0g$, depending on the ratio of horizontal to vertical principal stress.[77]

A method of estimating peak horizontal accelerations which is independent of source mechanism and location comes from considering the maximum acceleration that can be transmitted according to the strength of the soil. Consider a seismic shear wave being transmitted upwards through an elementary column of soil with forces and motions as shown in Figure 2.27(a). From dynamic equilibrium

$$T\ \Delta x\ \Delta y = \rho z\ \Delta x\ \Delta y\ \ddot{x}$$

$$\text{whence } a = \ddot{x}_{max} = \frac{\tau}{\rho z} \tag{2.28}$$

where τ is the shear stress at failure and ρ is the density.

For a saturated, normally consolidated (NC) soil, assuming the mean specific gravity $\bar{\gamma} = \gamma/2$, we find the following approximate relationship for the maximum horizontal acceleration that can be transmitted by this soil:

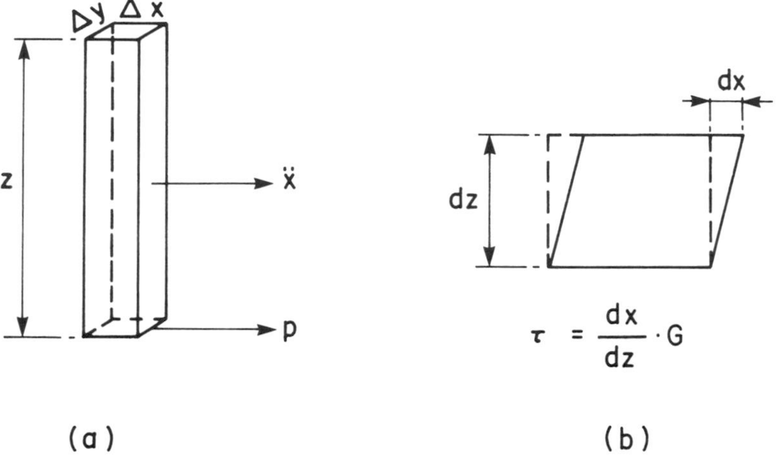

Figure 2.27 Shear behaviour of soil column assumed for calculating bounds on peak ground motion. (a) Free body; (b) shear distortion

$$\frac{a}{g} \approx \tfrac{1}{2} \frac{c_u}{p'} \tag{2.29}$$

where c_u is the undrained shear strength and p' is the effective overburden pressure.

From dynamic soil response analyses which incorporated similar limiting criteria for a variety of soils, Ambraseys[78] found the following upper bounds on horizontal accelerations:

Low plasticity NC clays: $0.1g \leqslant a \leqslant 0.15g$
High plasticity NC clays: $0.25g \leqslant a \leqslant 0.35g$
Saturated sandy clays and dense medium sands: $a \leqslant 0.6g$

Overconsolidated clays can transmit very high accelerations. In a study of the sea bed in part of the Gulf of Alaska in which the author was involved, using a method based on that outlined above for normally consolidated clay, it was estimated that 6 m below the sea bed the maximum acceleration would be $1.85g$. The soil shear strength was 215 kPa at the surface and increased in strength at a rate of 4.7 kPa per metre of depth. It is of interest that this value is nearly as high as the upper bound of $2g$ predicted above for rock considering source characteristics.

Finally, it is noted that the broad relationship of the above physical bounds on horizontal ground acceleration to bounds derived from attenuation equations with M_{max} limits is illustrated in Figure 4.1.

A simple means of obtaining upper bounds on *peak ground velocity*, v, comes from considering the shear distortion of an element of soil as shown in Figure 2.27(b), from which

$$\tau = \frac{dx}{dz} G = \frac{dx}{v_s dt} G$$

$$= \frac{v}{v_s} G$$

$$\text{i.e. } v = \frac{v_s \tau}{G} \tag{2.30}$$

where v_s is the velocity of the shear wave and G is the shear modulus for the soil. A limiting value for v is reached when the shear stress $\tau = c$ the shear strength of the soil, and using the relationship $v_s = \sqrt{(G/\rho)}$, equation (2.30) becomes

$$v_{max} = \frac{c}{\sqrt{(G\rho)}} \tag{2.31}$$

Thus the maximum horizontal particle velocity is a simple function of the mechanical properties of the local soil or rock. For a strong rock such as limestone or basalt (Table 6.3), using $E = G/3$, $E = 10^5 \text{MN/m}^2$, $E/c = 600$, $\rho = 2700\,\text{kg/m}^3$, it follows that $v_{max} \approx 1800\,\text{cm/s}$. However, it has been shown[79] that the initial velocity on a fault surface may be expressed by

$$v = cv_s \left(1 - \frac{c_r}{c}\right) \Big/ 2G \tag{2.32}$$

where c is the average strength of the material on the fault surface at rupture and c_r is the residual strength after rupture. According to Ambraseys[78], velocities near the source may reach as high as 480 cm/s, but because c_r is unlikely to be zero, and because there has been neither observable melting on fault planes nor heat-flow anomalies along re-activated faults, the upper bound for velocities in rock near a fault break should be about 150 cm/s.

Returning to equation (2.31), soft clay could have $E = 10\ \text{MN/m}^2$ and $\rho = 1600\,\text{kg/m}^3$, which gives $20 \leqslant v_{max} \leqslant 40\,\text{cm/s}$, depending on the ratio E/c. Thus equations (2.31) and (2.32) have shown that for ground conditions covering the practical range of building sites, the peak ground velocities that are physically sustainable at the site lie in the appoximate range of 20 to 150 cm/s.

Our ability to predict *peak ground displacement* suffers because most data on displacements are derived from double integrations of acceleration records, and hence numerical errors add to the uncertainties. A rough estimate of horizontal ground displacement, d, may be made from the empirical relationship given by Newmark and Rosenblueth:[80]

$$5 \leqslant \frac{ad}{v^2} \leqslant 15 \tag{2.33}$$

in which the value 5 is appropriate for large epicentral distances (say, 100 km) and the value 15 is for small epicentral distances.

2.4.2.5 Duration of strong motion

This variable is important because the amount of cumulative damage incurred by structures increases with number of cycles of loading, and also because the duration of strong motion is used in evaluating one of the measures of strength of shaking, namely the root-mean-square acceleration (discussed below). Duration of strong motion is usually defined in relation to ground accelerations and several different definitions exist. The direct approach is to equate duration to the time between the first and last accelerations on the record which exceed some arbitrary minimum value,[81] typically taken as 0.05 g for stronger events. However, it has become customary to define duration of strong motion indirectly as a function of time of the squared accelerations of the digital earthquake record. For example, Dobry *et al.*[82] defined the duration of significant shaking as the time needed to build up between 5 and 95 percent of the total Arias intensity of the record

$$I_A = \frac{\pi}{g} \int_0^T a^2(t)\mathrm{d}t \tag{2.34}$$

where T is the total length of the record.

Duration of strong motion tends to increase with both magnitude and distance from the source and may also increase from rock to soil sites. Unfortunately, widely varying expressions for such correlations exist, partly because of the inherent scatter in the data (since magnitude is not a strong definition of earthquake size) and partly because of the use of varying definitions for duration of strong motion. Both linear and exponential relationships between duration, D, and magnitude have been found. For example, as shown in Figure 2.28, Donovan[83] found the expression for all site conditions for a data set including US and Japanese events

$$D = 4 + 11(M - 5), \quad M > 5 \tag{2.35}$$

while Dobry *et al.*[82] later found for rock sites in the western USA that

$$\log D = 0.432M - 1.83 \tag{2.36}$$

The latter authors found that equation (2.36) for rock sites was a lower bound for soil sites, but that soil sites had much more scatter in the duration (Figure 2.29) such that they refrained from giving an expression for soil sites. However, Donovan's linear equation (2.35) would appear by eye to fit the mean of the rock soil data of Figure 2.29 moderately well. In line with the above observation by Dobry *et al.*, Trifunac *et al.*[84] found that the duration of strong motion on 'alluvium or otherwise soft sediment' sites was on average 10–12 s longer than on basement rock sites.

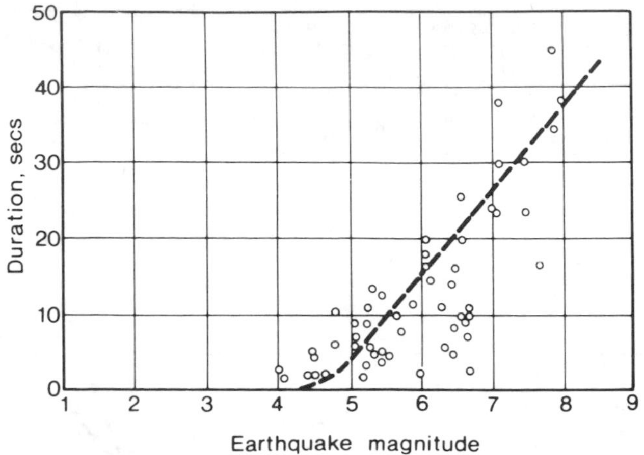

Figure 2.28 Duration of strong shaking as a function of earthquake magnitude (after Donovan,[83] reproduced by permission of Dames & Moore)

Trifunac *et al.* also give expressions for the change of duration with focal distance (i.e. 1 to 1.5 s increase in D per 10 km increase in distance), and with magnitude, but their increase in D of about 2 s for each magnitude increment is greatly at variance with equations (2.35) and (2.36), except for the latter at small values of M.

As with all earthquake variables, durations should be calculated from data for the region concerned, but at present the regional dependence of duration of strong motion needs further elucidation.

2.4.2.6 Root-mean-square acceleration

A measure of the strength of ground shaking which is frequently used in strong motion seismology is the root-mean-square acceleration, which is defined as

$$a_{\text{rms}} = \left\{ \frac{1}{T_2 - T_1} \int_{T1}^{T2} a^2(t)\mathrm{d}t \right\}^{1/2} \tag{2.37}$$

For a stationary process (Section 4.3.4.2) the location and size of the duration $(T_2 - T_1)$ over which the squared accelerations are averaged is relatively unimportant, but for a transient signal like an earthquake record, a_{rms} is obviously strongly dependent upon which portion of the record is included. Commonly, T_1 and T_2 are chosen to exclude the (arbitrarily defined) insignificant shaking and use has been made of one or other of the definitions of duration of strong motion such as noted in the preceding section. Two further approaches to duration that have been tried both take T_1 as the time of the S-wave arrival, one[85] taking $T_2 = T_1 + 10$ s and the other[84] taking $T_2 = T_1 + T_d$, where T_d, is the duration of faulting. In the latter study it

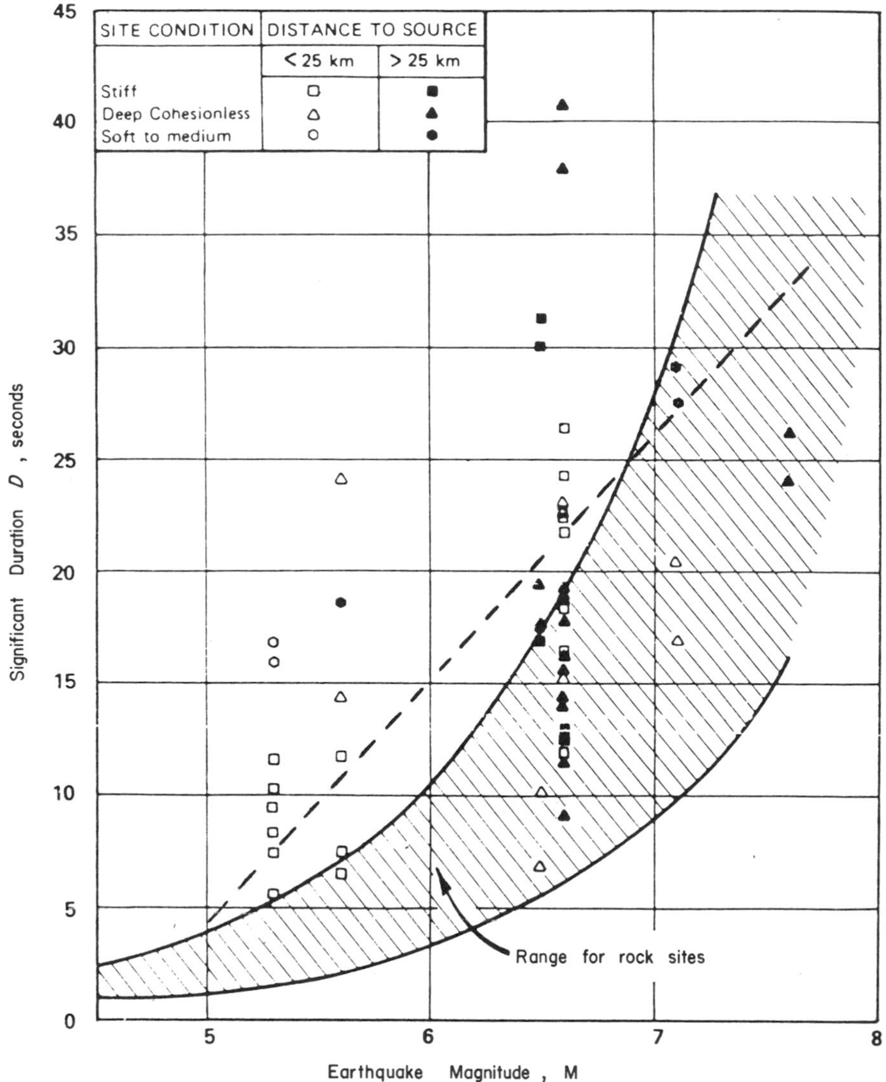

Figure 2.29 Duration of significant ground shaking (after Dobry *et al.*,[82] reproduced with permission from The Seism. Soc. of America). The straight line is Donovan's expression from Figure 2.28

was assumed the $T_d = f_o^{-1}$, which is an approximation that was observed to be accurate to ± 50 percent, and T_d varied from about 2 to 20 s for earthquakes in the magnitude range $M = 5$ to 7.7. The use of $T_1 + T_d$ has the merit that it was found that within this window T_d the earthquake records were essentially stationary processes, thus improving the stability of the a_{rms} values calculated. Obviously, with duration being variably defined, care is necessary in comparing a_{rms} values to ensure that they have been calculated on the same basis.

A further variable arises in comparing a_{rms} values because other means of defining a_{rms} exist, such as that of Hanks and McGuire:[62]

$$a_{rms} = 0.85 \frac{(2\pi)^2}{106} \frac{\Delta\sigma}{\rho R} \sqrt{\left(\frac{f_{max}}{f_o} \right)} \tag{2.38}$$

which utilizes a number of the source parameters previously discussed (Section 2.4.1).

The root-mean-square acceleration has long been of interest to earthquake engineers as a measure of the strength of ground motion, partly because the averaging involved could be expected to lead to a more stable parameter than peak ground acceleration. However, both the above studies[63,85] found, contrary to expectation, that a_{rms} is not less variable than peak ground acceleration.

2.4.2.7 Significant acceleration

Another approach to measuring acceleration strength of an earthquake lies qualitatively between a_{max} and a_{rms} described above. Significant acceleration or effective peak acceleration as defined by Bolt and Abrahamson[86] is the 90 percentile (or similar statistic) of all the peaks above some minimum value (0.005g) within a window in the record, where the ends of the window are marked by the first and last points above some given threshold, such as 0.02*g*. This procedure produces an acceleration value which is more stable than a_{max} by excluding scattered outliers of high-amplitude peaks which may not be representative of the general distribution of ground-motion amplitudes. The significant acceleration is obviously analogous to the characteristic loads used in probabilistic design elsewhere.

Using the above definition the significant acceleration for the El Centro 1940 earthquake was found[86] to be 0.127*g* compared with $a_{max} = 0.349g$, while the San Fernando 1971 Pacoima Dam record yielded a significant acceleration of 0.246*g*, showing how uncharacteristic was the a_{max} value of 1.172*g* for this accelerogram.

2.4.2.8 Frequency Content

The frequency of vibration associated with the amplitudes of motion is, of course, an intrinsic characteristic of earthquake ground motion, and frequency content is commonly studied in spectral form. As an example, Figure 2.21 shows a Fourier amplitude spectrum for the Pacoima Dam recording of the San Fernando 1971 earthquake, where the Fourier amplitude spectrum is given by

$$|F(\omega)| = \{ [\int_o^t \ddot{u}_g(\tau)\cos \omega\tau \, d\tau]^2 + [\int_o^t \ddot{u}_g(\tau)\sin \omega\tau \, d\tau]^2 \}^{1/2} \tag{2.39}$$

where the symbols are as defined Section 1.4.

Earthquake Fourier spectra are mainly used in seismology, while another spectral method of examining frequency content, namely the response spectrum, is generally used by engineers. Response spectra are referred to throughout this book and their theoretical basis is described in Section 1.4.

The frequency content of ground motions is a function of a number of phenomena, notably source mechanism, focal depth, distance from the epicentre, nature of the travel path and site soil, and the magnitude of the event. Considering source mechanisms, basic physics tells us that a rapid rupture in strong rock will produce more high-frequency vibration than gentler rupture mechanisms. In a study involving the use of source models of mid-plate earthquakes, which may have much smaller fault rupture lengths than interplate earthquakes (Section 2.2.4), Hermann *et al.*[71] have argued that mid-plate events have much more high-frequency content than interplate events. If this is substantiated, it demonstrates the regional dependence of frequency content.

Because higher frequencies attenuate more rapidly than low ones there is a tendency for the predominant period to increase with distance from the source, as found by Benioff[87] and Hermann *et al.*[71] This effect is illustrated in Figure 2.30, which compares the spectral accelerations for two focal distances ($R = 125$ km and $R = 50$ km) using McGuire's data[66] for the western USA.

The longer duration of larger earthquakes allows time for the generation of more cycles of longer period vibration. Thus the predominant period tends to increase with magnitude, as shown in Figure 4.4.

The effect of local soil conditions on frequency content is dramatic, as shown in the response spectra of Figure 3.3, and as discussed in Section 6.2.2.

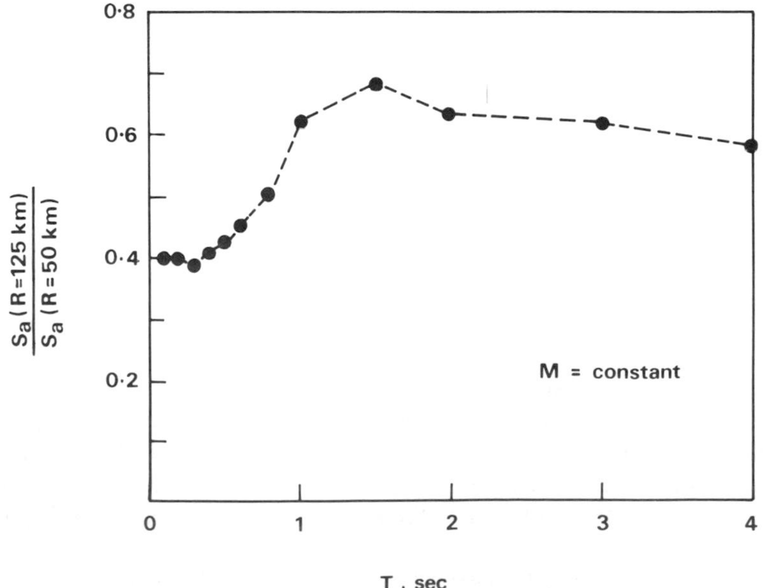

Figure 2.30 Comparison of spectral accelerations for two focal distances, showing how the predominant period lengthens with increasing R (based on McGuire's data set for the western USA[66])

2.5 DEFINING DESIGN EVENTS

Referring to the EERI Committee on Seismic Risk,[88] the *design event* is defined as *a specification of one or more earthquake source parameters, and of the location of energy with respect to the site of interest; used for the earthquake resistant design of a structure*. Thus a design event specification may consist simply of a magnitude M and a focal distance R.

Design events are required when normal code loadings are inappropriate or unavailable, as often arises in the case of very large, or critical, or novel structures such as tall buildings, liquefied petroleum gas storage depots, cooling towers, nuclear facilities, or offshore oil platforms. The size and location of design events will depend on the establishment of the design levels of seismic hazard, which may be related to some code or regulatory requirement or may need to be agreed with the owner.

In some cases more than one level of hazard may be required, say one for operational and one for survival performance, the equivalents in nuclear facility terminology being the operational basis earthquake (OBE) and the safe-shutdown earthquake (SSE). Also, two design events may need to be considered for a particular hazard level. For example, tall buildings with long natural periods will be more sensitive to larger, more distant events because of the greater long-period content of such events, while short-period structures designed to the same hazard level may be more sensitive to smaller, closer events, as can be seen from Figure 4.5. To further illustrate the point, Kuala Lumpur in Malaysia is located at the rather remote distance of about 400 km from the Alpide earthquake belt, and historically has therefore had little concern for earthquakes. However, in the 1970s, with the advent of taller buildings some instances of alarming swaying and cracking have occurred, but *only* in tall buildings. The author has been involved in investigations which causally linked these phenomena with large-magnitude events occurring 400 km away. Many other sites around the world located at similar distances from large earthquake sources, while safe for most traditional construction, may merit seismic design checks for longer period construction.

Studies of the seismic activity of a region as discussed in the previous sections of this chapter supply the material for defining the design earthquake for a given project. The characteristics of the design earthquake may be used in conjunction with the dynamic characteristics of the site to determine the dynamic design criteria for the project as discussed in Chapter 4.

An adequate definition of a design earthquake is very elusive, even prior to consideration of site conditions, because of difficulties in defining past earthquake behaviour and difficulties in predicting future seismic events. The main variables derived or implied in this chapter for use in defining the design event are: magnitude, return period, epicentral distance, focal depth, fault positions, fault types, and rupture length, while the associated dependent variables such as peak ground acceleration, peak ground velocity, peak ground displacement, duration of strong shaking, dominant period of shaking and attenuation relations are used for ground motion specification (Chapter 4).

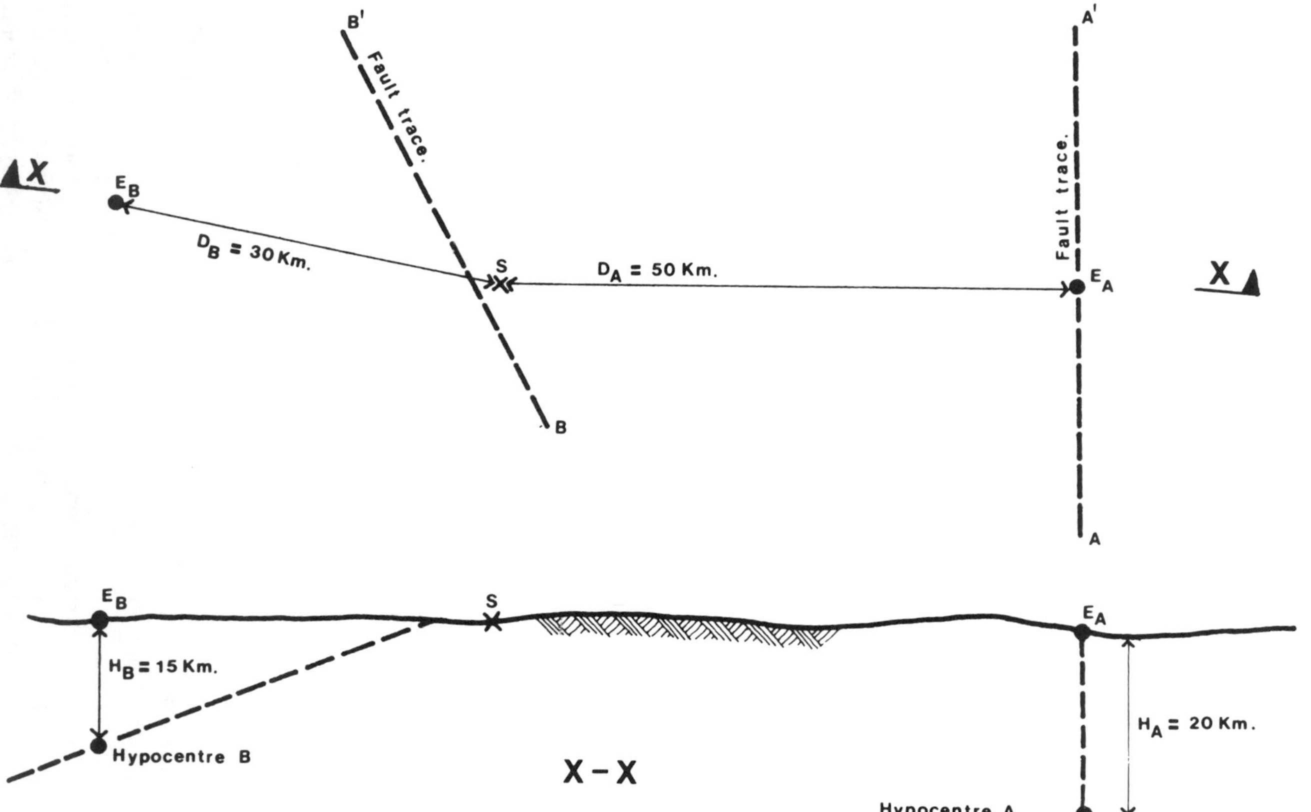

Figure 2.31 Hypothetical example of site relationship to two design earthquakes A and B with epicentres E_A and E_B, respectively

Data on the above aspects of earthquakes are variable, often inaccurate, and scarce. This means that the interpretation of the data must be highly subjective, and the use of mean values or some other value such as the 90 percent confidence level may be open to argument. A considerable amount of idealization is necessary.

In order to illustrate the definition of design earthquakes for a given site, reference will be made to Figure 2.31. Assume that studies of the earthquake history of the region have suggested the use of two design earthquakes, A and B, with the characteristics tabulated in Figure 2.31. It is quite common practice to consider two different design earthquakes with magnitudes and return periods as suggested above; normally the larger, less frequent, earthquake would be considered the worst design condition for use as ultimate loading, while the smaller, more frequent, earthquake would be used as a criterion for control of non-structural damage. However, in the situation illustrated in Figure 2.31, the associated fault types would render this use of the design earthquakes inappropriate.

Because the sloping plane for earthquake B outcrops near the site, the intensity of ground shaking at the site due to earthquake B may be as intense as at positions closer to the epicentre E_B. If the fault trace BB′ had been undetected or not allowed for at the time of the design, the intensity of ground motion at the site would be underestimated assuming normal attenuation from an epicentre 30 km away.

As an example of the type of information obtained by the methods of this chapter, consider the seismicity of Djakarta as illustrated in Figure 2.13. A probability study of the events within 300 km of Djakarta recorded between 1900 and 1972 indicated that for a 50-year return period there was a 95 percent probability that the greatest magnitude of event would be $M = 8.0$, and that for a 25-year return period the maximum magnitude of event would be $M = 7.6$. But Figure 2.13 indicates no larger shallow events recorded near to Djakarta. As can be seen from Figures 2.1 and 2.2 the tectonic discontinuity of the Sunda Arc lies adjacent to Java, the main concentration of events being near the South coast and many events being fairly deep seated. Without a detailed knowledge of the local fault lines, a reasonable 50-year design earthquake for Djakarta might therefore be a shallow event of magnitude 7.5 with an epicentral distance of 100 km, particularly for longer-period structures.

In regions where the locations of active faults are insufficiently defined the locations of the design events may have to be assumed to be uniformly distributed in plan and thus could be anywhere on a circle defined by the required epicentral distance, as assumed in the consideration of Djakarta above. A discussion of probabilistic assessment of seismic hazard is given in Section 4.2.

In conclusion, it is recommended that the proposed design events are carefully reviewed prior to use. All of the assumptions used in deriving them should be listed and their degrees of conservatism or non-conservatism should be noted, so that a rational attempt at a balanced assessment can be made.

REFERENCES

1. Smith, W. D., 'Pitfalls in the estimation of seismic hazard', *Bull. NZ Nat. Soc. for Earthq. Eng.*, **15**, No. 2, 77–81 (1982).
2. Vere-Jones, D., 'What are the main uncertainties in estimating earthquake risk?' *Bull. NZ Nat. Soc. for Earthq. Eng.*, **16**, No. 1, 39–44 (1983).
3. Barazangi, M., and Dorman, J., 'World seismicity map of ESSA coast and geodetic survey epicentre data for 1961–67', *Bull. Seism. Soc. Amer.*, **59**, 369–380 (1969).
4. Stevens, G. R., *New Zealand Adrift*, A. H. & A. W. Reed, Wellington (1980).
5. Walcott, R. I., 'The gates of stress and strain', in *Large earthquakes in New Zealand*, The Royal Society of New Zealand, Miscellaneous Series No. 5 (1981).
6. Mogi, K., 'Relationship between the occurrence of great earthquakes and tectonic structures', *Bull. Earthq. Res. Institute*, University of Tokyo, 47, Part 3, 429–51 (May, 1969).
7. Gubin, I. E., 'Earthquakes and seismic zoning', *Bull. Intl Institute of Seismology and Earthquake Eng.*, **4**, 107–26 (1967).
8. Panel on Earthquake Problems, Earthquake research for the safer siting of critical facilities, National Academy of Sciences, Washington, DC (1980).
9. Allen, C. R., 'Geological criteria for evaluating seismicity', in *Seismic Risk and Engineering Decisions*, (Eds C. Lomnitz and E. Rosenbleuth), Elsevier, Amsterdam (1976).
10. Dowrick, D. J., 'Earthquake risk and design ground motions in the U.K. offshore area', *Proc. Inst. Civ. Engrs*, **71**, 305–21 (1981).
11. Ambraseys, N. N., Imperial College, London, Personal communication (1985).
12. Klimkiewicz, G. C., Leblanc, G., Holt, R. J., and Thiruvengodam, T. R., 'Relative seismic hazard assessment for the north central United States', *Proc. 8th World Conf. on Earthq. Eng.*, San Francisco, **I**, 149–56, (1984).
13. Waldron, H. H., and Galster, R. W., 'Comparative seismic hazards study of western Montana', *Proc. 8th World Conf. on Earthq. Eng.*, San Francisco, **I**, 31–8 (1984).
14. Clark, R. H., Dibble, R. R., Fyfe, H. E., Lensen, G. J., and Suggate, R. P., 'Tectonic and earthquake risk zoning in New Zealand', *Proc. 3rd World Conf. on Earthq. Eng.*, New Zealand, **1**, I–107 to I–124 (1985).
15. Bolt, B. A., 'Causes of earthquakes', in *Earthquake Engineering*', (Ed. R. L. Wiegel), Prentice-Hall, Englewood Cliffs, NJ (1970).
16. Housner, G. W., 'Important features of earthquake ground motions', *Proc. 5th World Conf. on Earthq. Eng., Rome*, **1**, CLIX–CLXVIII (1973).
17. Bonilla, M. G., 'Surface faulting and related effects', in *Earthquake Engineering* (Ed. R. L. Wiegel), Prentice-Hall, Englewood Cliffs, NJ (1970).
18. Stevens, G. R., *Rugged Landscape*, A. H. & A. W. Reed, Wellington, (1974).
19. Eiby, G. A., *Earthquakes*, Heinemann Educational Books, Auckland, (1980).
20. Officers of the NZ Geological Survey, 'Active earth deformation', *Reprint NZGS 89*, Dept of Scientific and Industrial Research, New Zealand (1979).
21. Cluff, L. S., and Cluff, J. L., 'Importance of assessing degrees of fault activity for engineering decisions', *Proc. 8th World Conf. on Earthq. Eng.*, San Francisco, **II**, 629–36 (1984).
22. Anderson, J. G., 'Estimating the seismicity from geological structure for seismic risk studies', *Bull. Seism. Soc. Amer.*, **69**, No. 1, 135–58 (1979).
23. Beanland, S., and Fellows, D. L., 'Late quarternary tectonic deformation in the Kawarau River area, Central Otago', *NZ Geological Survey/EDS Immediate Report 84/019*, Dept of Scientific and Industrial Research, New Zealand (1984).
24. Officers of the NZ Geological Survey, 'Seismotectonic hazard evaluation for Upper Clutha Power Development', *NZGS Report EG 377*, Dept of Scientific and Industrial Research, New Zealand (1984).

25. Schwartz, D. P., Coppersmith, K. J., and Swan, F. H., 'Methods of estimating earthquake magnitudes', *Proc. 8th World Conf. on Earthq. Eng., San Francisco*, **I**, 279–85 (1984).
26. Slemmons, D. B., 'State-of-the-art for assessing earthquake hazards in the United States; Report 6, faults and earthquake magnitude', US Army Corps of Engineers, Waterways Experiment Station, Vicksburg, Mississippi, *Misc. Paper S-173-1* (1977).
27. Bonilla, M. G., Mark, R. K., and Lienkaemper, J. J., 'Statistical relations among earthquake magnitude, surface rupture length, and surface fault displacement', *Bull. Seism. Soc. Amer.*, **76**, No. 6, 2379–411 (1984).
28. Wyss, M., 'Estimating maximum expectable magnitude of earthquakes from fault dimensions', *Geology*, **7**, No. 7, 336–40 (1979).
29. Nuttli, O. W., 'Average seismic source–parameter relations for mid-plate earthquakes', *Bull. Seism. Soc. Amer.*, **73**, No. 2, 519–35 (1983).
30. Gutenburg, B., and Richter, C. F., *Seismicity of the Earth*, Hafner, New York (1965).
31. Karnik, V., *Seismicity of the European area*, Reidel, Dordrecht (Part 1, 1969; Part 2, 1971).
32. Lomnitz, C., *Global Tectonics and Earthquake Risk*, Elsevier, Amsterdam (1974).
33. Carmona, J. S., and Castano, J. C., 'Seismic risk in South America to the south of 20 degrees', *Proc. 5th World Conf. Earthq. Eng., Rome*, **2**, 1644–53 (1973).
34. Richter, C. F., *Elementary Seismology*, Freeman, San Francisco (1958).
35. Algermissen, S. T., 'Seismic risk studies in the United States', *Proc. 4th World Conf. on Earthq. Eng., Santiago*, **1**, Part A-1, 14–27 (1969).
36. Brooks, J. A., 'Seismicity of the Territory of Papua and New Guinea', *Proc. 3rd World Conf. on Earthq. Eng., Auckland*, **1**, III–15 to III–26 (1965).
37. Ambraseys, N. N., 'Value of historical records of earthquakes', *Nature*, **232**, No. 5310, 375–9 (1971).
38. Mei, Shi-yun, 'Characteristics of earthquake activity in China', *Acta Geophys. Sin.*, **9**, 1–9 (1960) (In Chinese).
39. Kaila, K. L., and Narain, H., 'A new approach for the preparation of quantitative seismicity maps', *Bull. Seism. Soc. Amer.*, **61**, No. 1275–91 (1971).
40. Evernden, J. F., 'Study of regional seismicity and associated problems', *Bull. Seism. Soc. Amer.*, **60**, No. 2, 393–446 (1970).
41. Scholz, C. H., 'The frequency–magnitude relationship of microfracturing in rock and its relationship to earthquakes', *Bull. Seism. Soc. Amer.*, **58**, No. 2, 399–415 (1968).
42. Chinnery, M. A., and North, R. G., 'The frequency of very large earthquakes', *Science*, **190**, 1197–8 (1975).
43. Anderson, J. G., and Luco, J. E., 'Consequences of slip rate constraints on earthquake occurrence relations', *Bull. Seism. Soc. Amer.*, **73**, No. 2, 471–96 (1983).
44. Schwartz, D. P., and Coppersmith, K. J., 'Fault behaviour and earthquake characteristics: Examples from the Wasatch and San Andreas fault zones', *J. Geophys. Res.*, **89**, 5681–98 (1984).
45. Lomnitz, C. 'What is a gap?', *Bull. Seism. Soc. Amer.*, **72**, No. 4, 1411–13 (1982).
46. Habermann, R. E., McCann, W. R., and Nishenko, S. P., 'A gap is. . . .', *Bull. Seism. Soc. Amer.* **73**, No. 5, 1485–6 (1983).
47. Adams, J., 'Paleoseismicity of the Alpine fault seismic gap, New Zealand', *Geology*, **8**, 72–6 (1980).
48. Kelleher, J. A., Sykes, L. R., and Oliver, J., 'Possible criteria for predicting earthquake locations and their applications to major plate boundaries of the Pacific and the Caribbean', *J. Geophys. Res.*, **78**, 2547–85 (1973).
49. Lomnitz, C., and Nava, F. A., 'The predictive value of seismic gaps', *Bull. Seism. Soc. Amer.*, **73**, No. 6, Part A, 1815–24 (1983).
50. Shah, H. C., and Dong, W. M., 'A re-evaluation of the current seismic hazard assessment methodologies', *Proc. 8th World Conf. on Earthq. Eng., San Francisco*, **I**, 247–54 (1984).

51. Dong, W. M., *et al.*, 'Utilization of energy flux, seismic moment and geological information in Bayesian seismic hazard models', *Proc. US–Japan Cooperative Research on Generalised Seismic Risk Analysis and Development of a Model Seismic Format* (1983).
52. Esteva, L., 'Seismicity', in Seismic Risk and Engineering Decisions, (Eds C. Lomnitz and E. Rosenblueth), Elsevier, Amsterdam (1976).
53. Vere-Jones, D., 'Stochastic models for earthquake occurrence', *J. R. Stat. Soc.*, **32**, No. 1, 1–45 (1970).
54. Schlien, S., and Toksöz, M. N., 'A clustering model for earthquake occurrence', *Bull. Seism. Soc. Amer.*, **60**, No. 6, 1765–87 (1970).
55. Savy, J. B., and Shah, H. C., and Boore, D., 'Nonstationary risk model with geophysical input', *J. Struct. Divn, ASCE*, **106**, 145–63 (1980).
56. Anagnos, T., and Kiremidjian, A. S., 'Temporal dependence in earthquake occurrence', *Proc. 8th World Conf. on Earthq. Eng., San Francisco*, **I**, 255–62 (1984).
57. Rhoades, D. A., and Millar, R. B., 'Estimating the hazard of surface faulting in a single fault zone', Appendix 1 in *Seismotectonic Hazard Evaluation for the Kawarau River Power Development* by G. T. Hancox *et al.*, New Zealand Geological Survey Report EG 384, Department of Scientific and Industrial Research, Lower Hutt, July (1985).
58. Kasahara, K., *Earthquake Mechanics*, Cambridge University Press, Cambridge (1981).
59. Aki, K., and Richards, R. G., *Quantitative Seismology*, W. H. Freeman and Co., San Francisco (1980).
60. Brune, J. N., 'The physics of earthquake strong motion', in *Seismic Risk and Engineering Decisions* (Eds C. Lomnitz and E. Rosenblueth), Elsevier, Amsterdam (1976).
61. Aki, K., 'Prediction of strong motion using physical models of earthquake faulting', *Proc. 8th World. Conf. on Earthq. Eng.*, **II**, 433–40 (1984).
62. Hanks, T. C., and McGuire, R. K., 'The character of high-frequency strong ground motion', *Bull. Seism. Soc. Amer.*, **71**, No. 6, 2071–95 (1981).
63. Brune, J. N., 'Tectonic stress and spectra of seismic shear waves from earthquakes', *J. Geoph. Res.*, **75**, No. 26, 4997–5009 (1970).
64. Hanks, T. C., 'f_{max}', *Bull. Seism. Soc. Amer.*, **72**, No. 6, Part A, 1867–79 (1982).
65. Esteva, L., and Villaverde, R., 'Seismic risk, design spectra and structural reliability', *Proc. 5th World Conf. on Earthq. Eng., Rome*, **2**, 2586–96 (1973).
66. McGuire, R. K., 'Seismic structural response risk analysis, incorporating peak response regressions on earthquake magnitude and distance', *Research Report R74–51*, Dept of Civil Engineering, Massachusetts Institute of Technology (1974).
67. Donovan, N. C., 'A statistical evaluation of strong motion data including the February 9, 1971 San Fernando earthquake', *Proc. 5th World Conf. on Earthq. Eng., Rome*, **1**, 1252–61 (1973).
68. Joyner, W. B., and Boore, D. M., 'Peak horizontal acceleration and velocity from strong-motion records including records from the 1979 Imperial Valley, California, earthquake', *Bull. Seism. Soc. Amer.*, **71**, No. 6, 2011–38 (1981).
69. Chiaruttini, C., and Siro, L., 'The correlation of peak ground horizontal acceleration with magnitude, distance and seismic intensity for Friuli and Ancona, Italy and the Alpide belt', *Bull. Seism. Soc. Amer.*, **71**, 1993–2009 (1981).
70. Ambraseys, N. N., 'Preliminary analysis of European strong-motion data 1965–1978', *Proc. 6th European Conf. on Earthq. Eng., Dubrovnik* (1978).
71. Hermann, R. B., and Nuttli, O. W., 'Scaling and attenuation relations for strong ground motion in eastern North America', *Proc. 8th World Conf. on Earthq. Eng., San Francisco*, **II**, 305–9 (1984).
72. Smith, W. D., 'Statistical estimates of the likelihood of earthquake shaking throughout New Zealand, *Bull. NZ Nat. Soc. for Earthq. Eng.*, **9**, No. 4, 213–21 (1976).

73. Katayama, T., Iwasaki, T., and Seaiki, M., 'Statistical analysis of earthquake response spectra', *Trans. Japanese Soc. Civ. Eng.*, **10**, 311–13 (1978).
74. Murphy, J. R., and O'Brien, J. L., 'The correlation of peak ground acceleration amplitude with seismic intensity and other physical parameters', *Bull. Seism. Soc. Amer.*, **67**, No. 3, 877–915 (1977).
75. Housner, G. W., 'Engineering estimates of ground shaking and maximum earthquake magnitude', *Proc. 4th World Conf. on Earthq. Eng., Santiago.*, **1**, A-1, 1–13 (1969).
76. Boatwright, J., and Boore, D. M., 'Analysis of the ground accelerations radiated by the 1980 Livermore Valley earthquakes for directivity and dynamic source characteristics', *Bull. Seism. Soc. Amer.*, **72**, No. 6, Part A, 1843–65 (1982).
77. McGarr, A., 'Upper bounds on near-source peak ground motion based on a model of inhomogeneous faulting', *Bull. Seism. Soc. Amer.*, **72**, No. 6, Part A, 1825–41 (1982).
78. Ambraseys, N. N., 'Dynamics and response of foundation materials in epicentral regions of strong earthquakes, *Proc. 5th World Conf. on Earthq. Eng., Rome*, **1**, CXXVI-CXLVIII (1973).
79. Ambraseys, N. N., and Hendron, A. J., 'Dynamic behaviour of rock masses', in *Rock Mechanics in Engineering Practice* (Eds K. G. Stagg and O. C. Zienciewicz), Wiley, London (1968).
80. Newmark, N. M., and Rosenblueth, E., *Fundamentals of Earthquake Engineering*, Prentice-Hall, Englewood Cliffs, NJ (1971).
81. Bolt, B. A., 'Duration of strong ground motion', *5th World Conf. on Earthq. Eng., Rome*, **1**, 1304–13 (1973).
82. Dobry, R., Idriss, I. M., and Ng, E., 'Duration characteristics of horizontal components of strong motion earthquake records', *Bull. Seism. Soc. Amer.*, **68**, No. 5, 1487–1520 (1978).
83. Donovan, N. C., 'Earthquake hazards for buildings', *Engineering Bulletin, No. 46*, Dames and Moore, Los Angeles, 3–20 (1974).
84. Trifunac, M. D., and Brady, A. G., 'A study on the duration of strong earthquake ground motion', *Bull. Seism. Soc. Amer.*, **65**, No. 3, 581–626 (1975).
85. McCann, M. W., and Boore, D. M., 'Variability in ground motions: root mean square acceleration and peak acceleration for the 1971 San Fernando, California, earthquake, *Bull. Seism. Soc. Amer.*, **73**, No. 2, 615–32 (1983).
86. Bolt, B. A., and Abrahamson, N. A., 'New attenuation for peak and expected accelerations of strong ground motion, *Bull. Seism. Soc. Amer.*, **72**, No. 6, 2307–22 (1982).
87. Benioff, H., Unpublished report to A. R. Golze, Chief Engineer, Department of Water Resources Consulting Board for Earthquake Analysis, USA, November 1962.
88. EERI Committee on Seismic Risk, Haresh C. Shah, Chairman, 'Glossary of terms for probabilistic seismic-risk and hazard analysis', *Earthquake Spectra*, **1**, No. 1, 33–40 (1984).

Chapter 3

Determination of site characteristics

3.1 INTRODUCTION

In seismic regions geotechnical site investigations obviously should include the gathering of information about the physical nature of the site and its environs that will allow an adequate evaluation of seismic hazard to be made. The scope of the investigation will be a matter of professional judgement, depending on the seismicity of the area and the nature of the site as well as of the proposed or existing construction. In addition to the effects of local soil conditions upon the severity of ground motion, the investigation should cover possible earthquake danger from geological or other consequential hazards such as:

(1) Fault displacement;
(2) Subsidence (flooding);
(3) Liquefaction of cohesionless soils;
(4) Failure of sensitive or quick clays;
(5) Landslides;
(6) Mudflows;
(7) Dam failures;
(8) Water waves (tsunamis, seiches);
(9) Groundwater discharge changes.

The seismic characteristics of local geology and soil conditions described briefly in the following section provides an introduction to the site investigations (Section 3.3), and to the determination of design ground motions and soil response analyses described in Chapters 4 and 6.

3.2 LOCAL GEOLOGY AND SOIL CONDITIONS

In many earthquakes the local geology and soil conditions have had a profound influence on site response. The term 'local' is a somewhat vague one, generally meaning local compared to the total terrain transversed between the earthquake focus and the site. On the assumption that the gross bedrock vibration will be similar at two adjacent sites, local differences in geology and soil produce different surface ground motions at the two sites. Factors influencing the local modifications to the underlying motion are the topography and nature of the bedrock and the nature and geometry of the depositional soils. Thus the term 'local' may involve a depth of a kilometre or more, and an area within a horizontal distance of several kilometres from the site.

Soil conditions and local geological features affecting site response are numerous and some of the more important are now discussed with reference to Figure 3.1.

(1) The greater the *horizontal extent* (L_1 or L_2) of the softer soils, the less the boundary effects of the bedrock on the site response. Mathematical modelling is influenced by this, as discussed in Section 6.2.2.2.
(2) The *depth* (H_1 or H_2) *of soil overlying bedrock* affects the dynamic response, the natural period of vibration of the ground increasing with increasing depth. This helps to determine the frequency of the waves amplified or filtered out by the soils and is also related to the amount of soil–structure interaction that will occur in an earthquake (Sections 6.2 and 6.3). The Mexico City earthquake of 1957 witnessed extensive damage to long-period structures in the area of the city sited on deep (>1000 m) compressible alluvium.[1] The natural tendency for long-period ground motions to be amplified in the structural response was intensified in this earthquake because the epicentral distance was quite large at 230 km. Another notable example of an earthquake where the fundamental period of structures which were damaged appeared closely related to depth of alluvium, was that in 1967 at Caracas.[2] Again long-period structures were damaged in areas of greater depth of alluvium.
(3) The *slope of the bedding planes* (valleys 2 and 3 in Figure 3.1) of the soils overlying bedrock obviously affects the dynamic response; but it is less easy to deal rigorously with non-horizontal strata.
(4) *Changes of soil types horizontally* across a site (sites F and G in Figure 3.1) affect the response locally within that site, and may profoundly affect the safety of a building straddling the two soil types.
(5) The *topography of both the bedrock and the deposited soils* has various effects on the incoming seismic waves such as reflection, refraction, focusing, and scattering. Unfortunately many of these effects will always remain suppositional; for instance, while focusing effects in bedrock (valleys 1 and 2 in Figure 3.1) may be amenable to calculation, how are the response modifications at sites G and J to be reliably predicted due to these effects in valley 3?

    It may well be that geological features such as hidden irregularities in the bedrock topography explain the otherwise unexplained differences of response observed at two nearby sites in the 1971 San Fernando earthquake.[3] At this time at two locations on the campus of the California Institute of Technology, the peak acceleration recorded at one site was 21 percent g while only 11 percent g was recorded at the other; whereas the local soil profiles at both locations were considered identical.
(6) Another topographical feature affecting response is that of *ridges* (Site B in Figure 3.1) where magnification of basic motion by factors as high as about two may occur (Section 6.2.2.3).
(7) *Slopes of sedimentary deposits* may, of course, completely fail in earthquakes. In steep terrain (Site H in Figure 3.1) failure may be in the

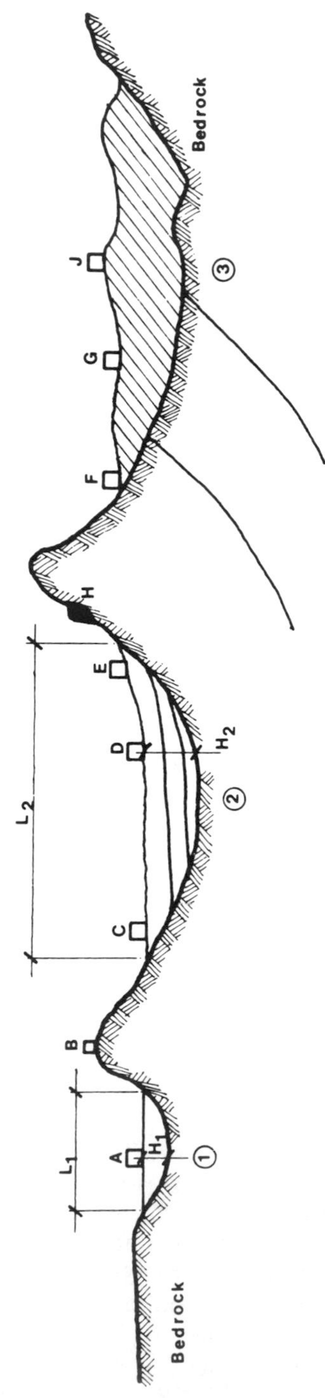

Figure 3.1 Schematic diagram illustrating local geology and soil features

form of avalanches. This occurred in the Northern Peru earthquakes of 31 May, 1970, in which whole towns were buried and about 20 000 people were killed[4] by one particular avalanche which travelled 18 km at speeds of 200–400 km/h.

(8) Spectacular soil failures can also occur in *gentle slopes*, as seen in the 1964 Alaskan earthquake[5] and again in the 1968 Tokachi–Oki earthquake.[6] The slope failures in the Alaskan earthquake were mostly related to liquefaction of layers of soil. For instance, landslides occurred in basically clay deposits (Figure 3.2) where liquefaction occurred in thin lenses of sand contained in the clay. In the Tokachi–Oki earthquake, some of the slope failures resulted from upper soil layers sliding on a slippery (wet) supporting layer of clay. This 'greasy back' situation could occur as illustrated in Figure 3.1, Site *E*.

Similar phenomena are known to occur on land in highly sensitive (i.e. quick) clays (Section 6.4.1), and on the sea floor, where normally consolidated clays with slopes of less than 1 degree can fail if subjected to external forces such as earthquake or waves.[7] During the development of the North Sea oil and gas fields the author was involved in a study[8] in which it was shown that slopes of less than 1 degree would fail under a ground acceleration of about $0.1g$.

(9) The *water content* of the soil is an important factor in site response. This applies not only to sloping soils as mentioned above, but liquefaction may also occur in flat terrain composed of saturated cohesionless soils (Section 6.2.2.5). Classical examples of failures of this type occurred in the Alaskan and Tokachi–Oki earthquakes referred to above, and in the much-studied 1964 Niigata earthquake.

(10) *Faults* of varying degrees of potential activity sometimes cross the site of proposed or existing construction and cases of damage have been recorded. The recurrence intervals of given levels of fault displacement both horizontal and vertical, and the structure's ability to tolerate the design displacement, sometimes need to be evaluated (Section 4.4).

(11) *Water waves* are sometimes generated by earthquakes. Those occurring in the sea, called *tsunami*, are caused by vertical displacements of blocks of sea bed. Where the resulting high-velocity, low-amplitude surface wave in the sea reaches the shore, waves of considerable height (10 m) may surge well beyond the normal high tide limit, hundreds of metres in flat terrain. These extreme effects only occur where the topography of the coastline focuses the wave energy, such as the narrow inlets of the southern Alaskan coast, where a disastrous tsunami struck in the 1964 Great Alaska earthquake. Various other coastlines are susceptible to damaging tsunami, particularly the Pacific Ocean, such as in Hawaii and Japan.

Water waves called *seiches* may also occur in the enclosed waters of lakes and harbours due to resonance effects or landslides, and, while not as large as tsunami, seiches have caused considerable damage.

More information on seismic water waves should be sought in the specialist literature.[10]

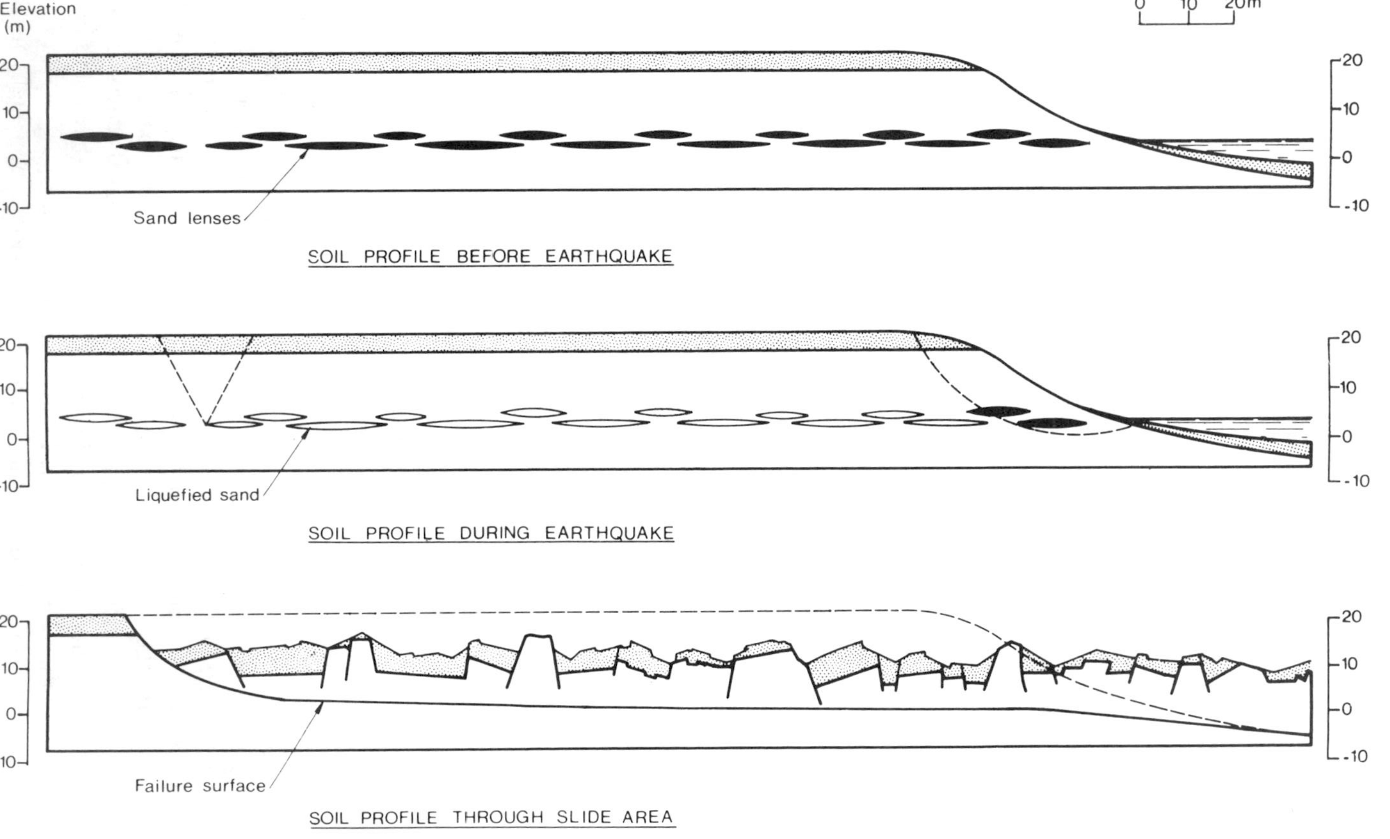

Figure 3.2 Conceptual development of Turnagain Heights landslide, Anchorage, Alaska, due to liquefaction of sand lenses (after Seed[5])

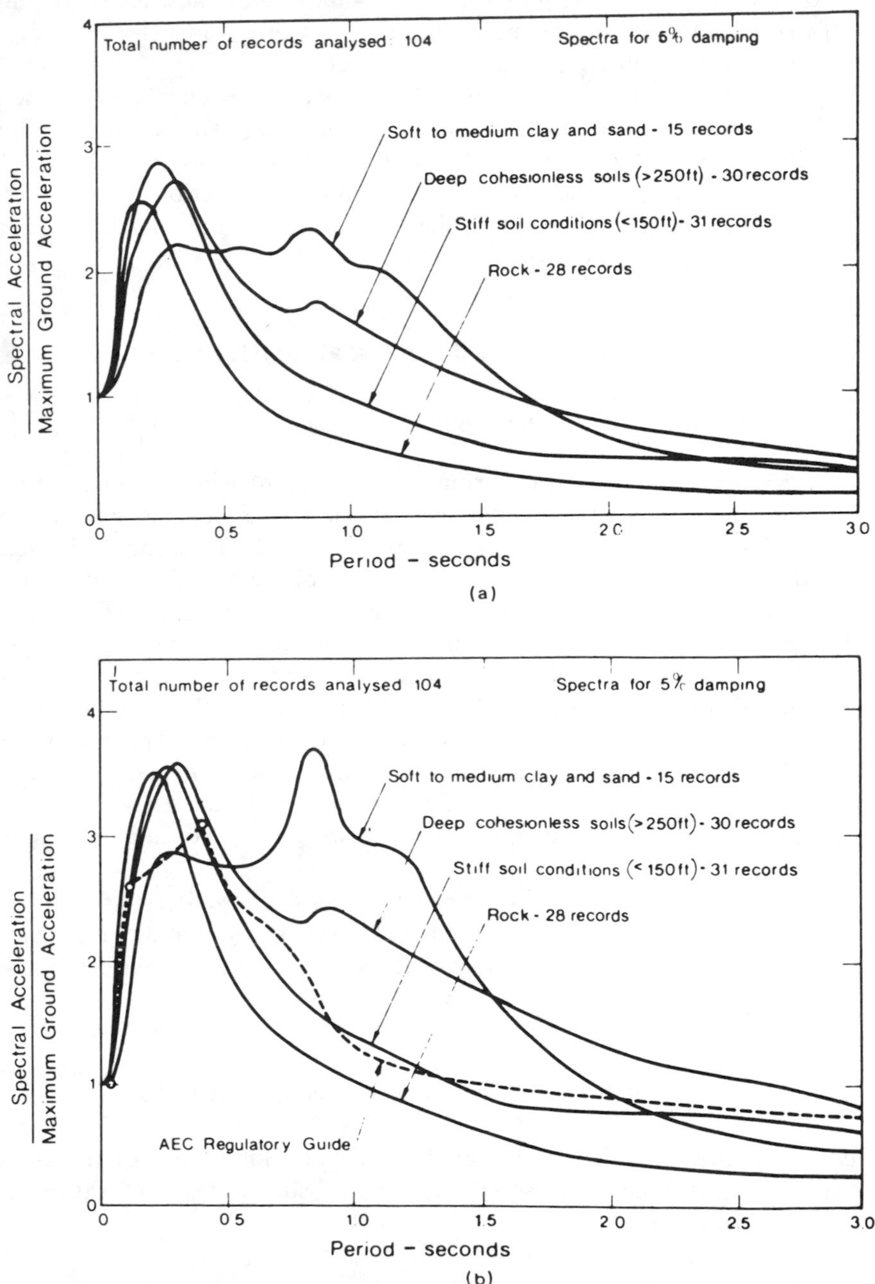

Figure 3.3 (a) Mean acceleration spectra for different site conditions (after Seed *et al.*[11]); (b) Mean plus one standard deviation (84 percentile) acceleration spectra for different site conditions (after Seed *et al.*[11])

(12) *Changes in groundwater discharge* occur after earthquakes, apparently due to changes in porewater pressure. The discharge may cause local flooding or streams to dry up, extensive sand boils, or erosion, such as observed in the 1983 Borah Peak, Idaho earthquake.[9]

(13) Finally, the seismic response of a site and structures on it is of course a function of the local *soil types* and their condition. This is illustrated by the very different response spectra for different soils shown in Figure 3.3. The dynamic properties of individual soils are described in terms of mechanical properties such as shear modulus, damping, density, and compactability as discussed in Section 6.2.

3.3 SITE INVESTIGATIONS AND SOIL TESTS

3.3.1 Introduction

For any construction project it is normal to carry out some investigations of the site, generally using fairly standardized operations in the field and in the laboratory such as drilling boreholes and carrying out triaxial tests. In this section only those investigating techniques related to the seismic response of soils are discussed.

The scope of the site investigations will depend on the site and on the budget and importance of the project, but in general it will be desirable to examine *to some degree* the factors relating to local geology and soil conditions discussed in Section 3.2. In Tables 3.1 and 3.2 the main variables is seismic site response have been related to some means for evaluating them. It is not proposed that these tables are exhaustive, but the field and laboratory test methods listed have been chosen because of their availability, reliability, or economy. For some parameters such as radiation damping and Poisson's ratio, no suitable tests for their evaluation exist.

For the description of the dynamic behaviour of soils see Section 6.2, where the main dynamic design parameters such as shear modulus and damping are defined. The application of the results of the site investigation to soil response and design problems may be found in various parts of Chapters 4–6.

3.3.2 Field determination and tests of soil characteristics

A brief description of the nature, applications and limitations of those site investigations pertaining to seismic behaviour of soils as listed in Tables 3.1 and 3.2 now follows.

3.3.2.1 Soil distribution and layer depth

Standard borehole drilling and sampling procedures are satisfactory for determining layer thicknesses for most seismic response analysis purposes as

Table 3.1. List of the main seismic soil factors with the most suitable tests used in their evaluation

		Field tests	Laboratory tests
Settlement of dry sands		Penetration resistance	Relative density
Liquefaction		Penetration resistance;	Relative density;
		Groundwater conditions	Particle size
Dynamic response parameters	Shear modulus	Shear wave velocity	Resonant column or cyclic triaxial
	Damping		Resonant column or cyclic triaxial
	Mass Density		Density
	Fundamental soil period	Vibration test	

Table 3.2. List of the best field and laboratory tests related to the evaluation of the seismic response of soils

Field determinations and tests	Related to
Soil distribution and layer depth	Response calculations
Depth to bedrock	Response calculations
Groundwater conditions	Response calculations and liquefaction
Penetration resistance	Settlement and liquefaction
Shear wave velocity	Shear modulus
Fundamental period of soil	Response calculations
Laboratory tests	
Particle size distribution	Liquefaction
Relative density	Liquefaction and settlement
Cyclic triaxial	Shear modulus and damping
Resonant column	Shear modulus
Unit mass	Response calculations

well as for normal foundation design. In the upper 15 m of soil, sampling is usually carried out at about 0.75 or 1.5 m intervals; from 15–30 m depth, a 1.5 m interval may be desirable; while below 30 m depth, 1.5 or 3.0 m may be adequate, depending on the soil complexity. If the site may be prone to liquefaction or slope instabilities, thin layers of weak materials enclosed in more reliable material may need to be identified, requiring more frequent or continuous sampling in some cases.

The depth to which the deepest boreholes are taken will depend, as usual, on the nature of the soils and of the proposed construction. For instance, for the design of a nuclear power plant on deep alluvium, detailed knowledge of the soil is required to a depth of perhaps 200 m, while general knowledge of the nature of subsoil will be necessary down to bedrock or rock-like material.

3.3.2.2 Depth to bedrock

For use in response calculation a knowledge of the depth to bedrock or rock-like material is essential. Beyond the ordinary borehole depth of 50–100 m, bedrock may be determined from geophysical refraction surveys, preferably checked by reference to information from geological records, artesian water or oil boreholes where available. In areas of deep overburden, for seismic response purposes the depth at which bedrock or equivalent bedrock is reached may have to be defined fairly arbitrarily. For example, on some sites it may be reasonable to say that equivalent bedrock is material for which the shear wave velocity at low strains (0.0001 percent) is $v_s \geqslant 760$ m/s, where such material is not underlain by materials having significantly lower shear wave velocities.[12] In California on a typical site, effective bedrock would be found within 30 m of the surface, while on virtually all Californian sites it would be found within 150 m depth. The order of accuracy of bedrock depth determination as currently required for seismic response calculations is as follows:

Bedrock depth (m)	Approximate accuracy (m)
0–30	1.5
30–60	1.5–3.0
60–150	6–15
150–300	15–30
>300	60

The large errors permissible in the measurement of deeper bedrock reflects the great approximations made in soil response analyses at the present time.

3.3.2.3 Groundwater conditions

Adequate standard borehole installations are available for accurately measuring groundwater conditions at any site. For response calculations this information is used indirectly through effective confining pressures as they affect both shear modulus and damping of the soil. Those sites which are most susceptible to liquefaction have their water table within 3 m of the surface, while sites with water tables within about 8 m of ground level may also be potentially liquefiable, depending on other soil parameters.

3.3.2.4 Penetration resistance tests

The penetration resistance test is really an indirect means of determining the relative density or degree of compaction of granular deposits. It is therefore an important factor in the study of settlement and liquefaction of soils in earthquakes. It may also be used to estimate shear modulus of the soil, as described in Section 6.2.2.1. Because it can be carried out simply, frequently, and cheaply as part of routine subsoil investigations, it is probably preferable to the direct laboratory test for determining relative density.

Two basic types of penetrometer are in common use for penetration tests, namely hollow tube samplers and cone penetrometers. Both types may be either driven by a falling weight (dynamic method) or by a static load into the undisturbed soil at the bottom of the borehole as drilling proceeds. In America and some other countries the preferred method is the standard penetration resistance test (SPT), which is a dynamic method having the advantage of sample recovery. The static cone tests, particularly the Dutch cone, have found favour in some countries because of the greater consistency of results deriving from the simple static load application. This advantage is offset, however, because the cone test does not recover samples, so that no visual examination of the material being tested is possible.

When using the results of penetration tests for assessing the condition of granular soils they may in some cases be used directly (for example, in Figure 6.11) or else indirectly, i.e. after conversion to relative density (Table 6.8). As the various penetrometer tests yield different numerical results for the same soils, the exact type of equipment used in each case must be known and appropriate conversions made where necessary for assessing results. For example Schmertmann[13] related the static cone penetration resistance, Q_c (kg/cm^2), to standard penetration resistance (blows/foot) for fine sands, but the relationship has been found to vary with grain size.

It is particularly important to bear in mind the large scatter of results obtained using all penetration tests; therefore penetrometer readings should be used to establish trends of soil compaction rather than be considered as absolute values.

For conversion of results of the American standard penetration test (ASTM designation: D1586–67) to relative density values, Figure 3.4 shows the correlation given by Holtz and Gibbs[14] and Bazaraa.[15] The Holtz and Gibbs criteria appear to be more widely accepted.[16]

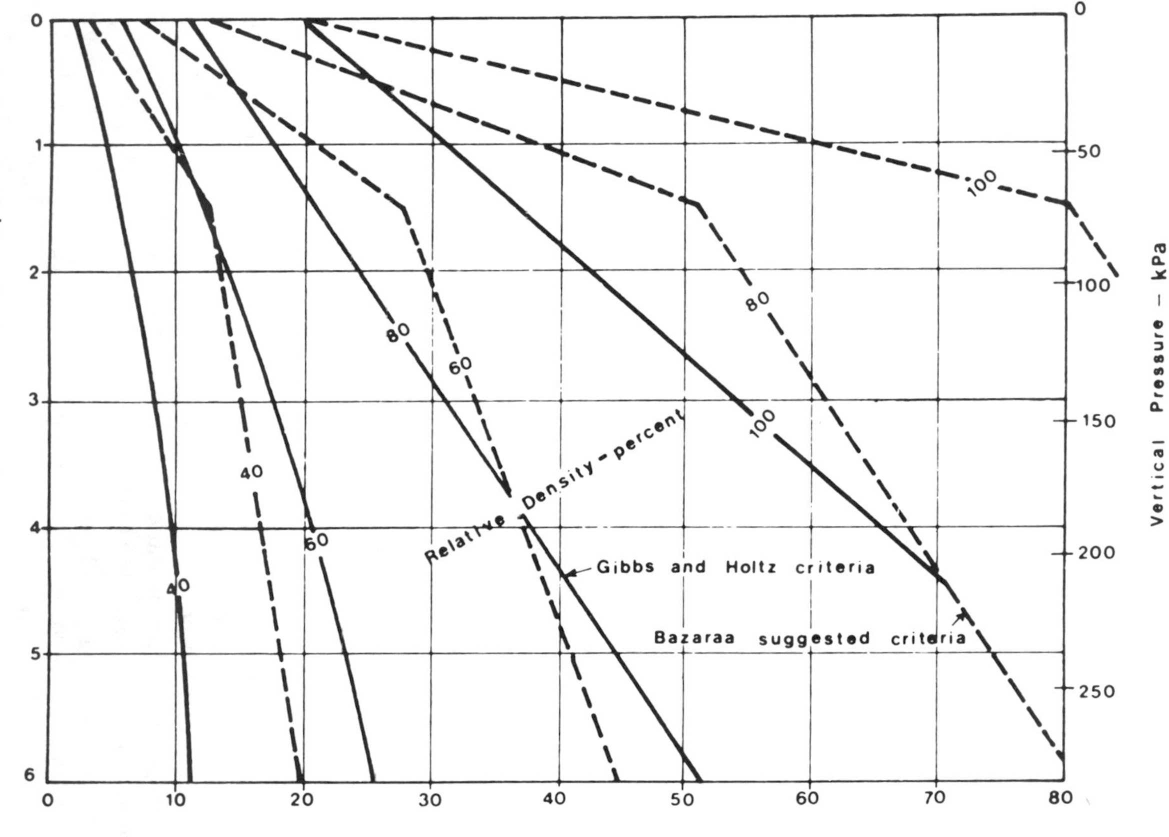

Figure 3.4 Relative densities derived from standard penetration resistances (after Holtz and Gibbs[14])

3.3.2.5 *Field determination of shear wave velocity*

Although the shear wave velocity is often used directly in response analyses (Sections 6.2 and 6.3), it may be thought of mainly as a means of determining the shear modulus G of a soil (Section 6.2.1.1) from the relationship

$$G = \rho v_s^2 \tag{3.1}$$

where ρ is the mass density of the soil. In the geophysical method of determining v_s low energy waves are propagated through the soil deposit, and the shear wave velocity is measured directly. Three techniques using boreholes are illustrated in Figure 3.5.

In each case waves are generated by an explosive charge or a hammer and the time of first arrival of the shear wave travelling from energy source to geophone is recorded. Difficulties in interpreting results arise from uncertainties in separating the first arrival of shear waves from the faster travelling longitudinal waves. Unfortunately these latter P-waves are not suitable for shear modulus calculations as they are greatly influenced by the presence of groundwater, whereas shear waves are not.

The cross-hole technique shown in Figure 3.5 measures shear wave velocities horizontally between adjacent boreholes, and is clearly well suited to response calculations of reasonably homogeneous or thick strata. With thinly bedded deposits, various routes may be taken by waves between source and geophone and the interpretation of arrival times is more problematical and should be viewed with caution. When using the up-hole and down-hole techniques of Figure 3.5 the different wave types can be distinguished more easily, but care must be taken to deal with misleading local borehole effects. For example, where casing has to be used in a borehole, the waves transmitted by the casing may disguise the slower and weaker signals in the soil and experienced resolution of the results is required.

The above geophysical methods of determining v_s are the most applicable field procedures because they involve a large mass of soil, they can be carried out in most soil types, and they permit v_s to be determined as a function of depth. Furthermore their cost is reasonable and in many countries the necessary equipment is available. Because these tests are only feasible at low levels of soil strain of $10^{-5} - 10^{-3}$ percent, compared with design earthquake strains of about $10^{-3} - 10^{-1}$ percent, values of shear modulus calculated from these values of v_s will be scaled down for seismic response purposes (Table 6.6). It is also wise to compare values of G computed in this manner with values determined from laboratory tests as discussed in Section 3.3.3.

3.3.2.6 *Field determination of fundamental period of soil*

A knowledge of the predominant period of vibration of a given site is helpful in assessing a design earthquake motion (Sections 4.3.3.3(i) and 6.2.2.1) and the vulnerability of the proposed construction to earthquakes (Sections 5.3.6 and 6.3).

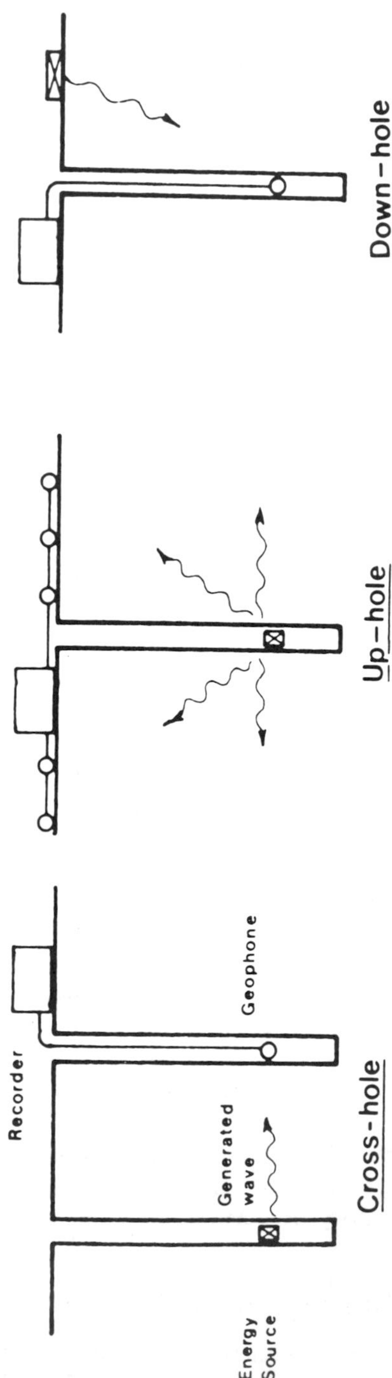

Figure 3.5 Geophysical methods of evaluating shear wave velocity

Many attempts have been made to measure the natural period of vibration of different sites; the vibrations measured have generally been microtremors, some arising from small earthquakes[17] or those induced artifically such as by explosive charges, pile driving, passing trains or nuclear test explosions.[18,19]

For an important or seismically vulnerable project, a vibration test may well be warranted, but problems of interpretation of results arise as such tests involve much lower magnitudes of soil strain than occur in design earthquakes. If a local correlation between soil periods in strong motion earthquakes and periods recorded during microtremors does not exist, cautious comparisons with strong motion results on similar soils in different areas will have to be made. In the case of vibration tests carried out for the Parque Central Development in Caracas,[18] the measured periods were increased by 50 percent in order to convert the microtremor behaviour into strong ground motion. This adjustment factor was derived through comparison of studies of the 1967 Caracas earthquake with the site tests.

It should be noted that the fundamental period of the soil will generally be between about 0.2 and 4.0 s, depending on the stiffness and depth of the soils overlying bedrock (Section 6.2.2.1).

A review of various microtremor recording techniques, and a detailed discussion of a particular method used in New Zealand are given in two papers by Parton and Taylor.[19,20]

3.3.3 Laboratory tests relating to dynamic behaviour of soils

A brief description of the nature, applications, and limitations of the laboratory tests relating to the dynamic behaviour of soils, as summarized in Tables 3.1 and 3.2, is set out below.

3.3.3.1 Particle size distribution

This soil property is related to the liquefaction of saturated cohesionless soils as discussed in Section 6.2.2.5. As the test for its determination is a standard laboratory procedure, it will not be described here. Although a number of classifications of grain size and standard sieves exist, correlations are straightforward, so that use of any scale of sizes can easily be applied to the liquefaction potential graph shown in Figure 6.12 which incorporates an American sieve grading.

3.3.3.2 Relative density test

The *in situ* relative density or degree of compaction is helpful in determining the likely settlement of dry sands and the liquefaction potential of saturated cohesionless soils is earthquakes (Sections 6.2.2.4 and 6.2.2.5). As this property has a significant influence on the dynamic modulus,[16] it indirectly relates to response analyses. Relative density for the void ratio must also be assessed in order to reproduce field conditions in samples which are recompacted in the laboratory for cyclic loading tests. As is well known by soils engineers, larger scatter occurs in the results of relative density tests, the chief reason being the virtual impossibility of retrieving reliable undisturbed samples of granular deposits.

The relative density may be found from either

$$D_r = \frac{e_{max} - e}{e_{max} - e_{min}} = \frac{\rho_{max}(\rho - \rho_{min})}{\rho(\rho_{max} - \rho_{min})} \tag{3.2}$$

where e_{max} and e_{min} are the maximum and minimum void ratios, e and ρ are the natural (*in situ*) void ratio and unit mass, respectively, and ρ_{max} and ρ_{min} are the maximum and minimum unit mass.

In the laboratory e, the void ratio of the undisturbed sample, is first determined by measuring the appropriate quantities in

$$e = \frac{G\rho_w}{\rho_d} - 1 \tag{3.3}$$

where G is the specific gravity of the solids, ρ_w is the unit mass of water, and ρ_d is the dry unit mass of the sample.

The minimum mass density may be found by pouring oven-dry material gently through a funnel into a mould, using a method such as the American one designated ASTM:D2049–69. For reasonably clean sands this method is reliable.

More difficulty is experienced in determining the maximum density ρ_{max} with equal consistency, different methods of compaction giving modestly different results. Vibratory compaction techniques seem better for uniform sands with few fines, while impact methods seem better for sands with more fines. Vibration and impact techniques generally used in America comply with ASTM tests designated D2049–69 and D1557–70, respectively.

If the percentage passing the 200-mesh sieve exceeds approximately 15 percent, laboratory determination of relative density is of doubtful validity. In this case more reliance will have to be made upon the penetration resistance tests as a measure of relative density as discussed in Section 3.3.2.

3.3.3.3 Cyclic triaxial test

This test is one of the best laboratory methods at present available for determining the shear modulus and damping of cohesive and cohesionless soils for use in dynamic response analyses (Sections 6.2 and 6.3) In this test cyclically varying axial compression stress–strain characteristics are measured directly.[21,22] The compressive modulus E so obtained is converted to the shear modulus G using the relationship.

$$G = \frac{E}{2(1+\nu)}$$

where ν is Poisson's ratio. The damping ratio may also be obtained from this test from the resulting hysteresis diagram as illustrated in Figure 6.1. Depending on the range of strains produced in the test, any desired level of strain may be chosen for plotting the hysteresis loops.

As well as having the facility for applying a variety of stress conditions, the cyclic triaxial test has the advantages that it can be applied to all types of soils except gravel, that the test equipment is widely available and precise in its control, and that testing is comparatively cheap. The disadvantages of this test are related to its inability to reproduce the stress conditions found in the field, i.e. that the cyclic shear stresses are not applied symmetrically in the test, that zero shear stresses are applied in the laboratory with isotropic rather than anisotropic consolidation, and also that the test involves deformations in the three principal stress directions, whereas in earthquakes the soil in many cases is thought to be deformed mainly unidirectionally in simple shear.

Cyclic shear tests are carried out at high strains ($10^{-2}-5$ percent) equal to and larger than the strains occurring in strong earthquakes; since geophysical test involve low strains, values of G at intermediate strains may be determined by interpolating between G values found from these different methods, but as there is no overlap between the strains occurring in these two tests cross-checking

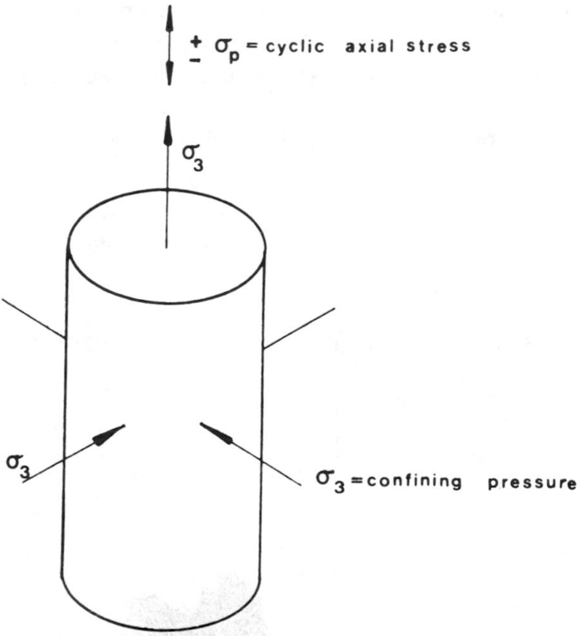

Figure 3.6 Cyclic triaxial test

between the field and laboratory methods is not possible. It is also to be noted that in the use of this test to determine soil damping characteristics, no field method of evaluating damping is as yet available for comparison, and hence any values of damping coefficient obtained should be treated with appropriate caution.

3.3.3.4 Resonant column test

This test provides a good alternative to the cyclic triaxial test for the laboratory determination of shear modulus of most soils. A cylindrical column of soil is vibrated at small amplitudes on one end, either torsionally or longitudinally (Figure 3.7), varying the frequency until resonance occurs. Wilson and Dietrich[23] proposed that the shear or compression modulus for a solid cylinder may be found from

$$G \text{ or } E = 1.59 \times 10^{-8} f^2 h^2 \rho \,(\mathrm{MN/mm^2}) \tag{3.4}$$

where h is the height of the soil cyclinder (mm), ρ is the unit mass of soil ($\mathrm{Mg/m^3}$), and f is the resonant frequency of torsional vibration in cycles per second when determining G, or the resonant frequency of longitudinal vibration in cycles per second when determining E.

It will be seen that by determining E and G separately from these tests, a value of Poisson's ratio ν may be determined, but as this test involves low strain and no suitable extrapolation method exists, such values of ν are not suitable for most earthquake engineering purposes. Although this test has the disadvantage of being carried out at low strains ($10^{-2}-10^{-4}$ percent), it has the advantages of simplicity, cheapness of equipment, and applicability to most soil types.

Details of the equipment used in this test may be found elsewhere.[23,24]

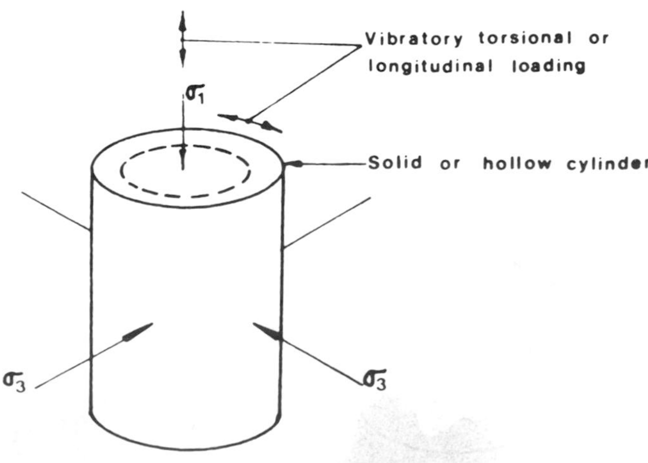

Figure 3.7 Resonant column test

REFERENCES

1. Rosenblueth, E., 'Earthquake of 28th July, 1957 in Mexico City', *Proc. 2nd World Conf. Earthq. Eng., Japan*, **1**, 359–79 (1960).
2. Seed, H. B., Whitman, R. V., Dezfulian, H., Dobry, R., and Idriss, I. M., 'Soil conditions and building damage in 1967 Caracas earthquake', *J. Soil Mechanics and Foundations Division, ASCE*, **98**, No. SM8, Aug. 1972, 787–806 (Aug. 1972).
3. Housner, G. W., and Jennings, P. C., 'The San Fernando California earthquake', *Int. J. of Earthquake Engineering and Structural Dynamics*, **1**, 5–32 (1972).
4. Cluff, L. S., 'Peru earthquake of May 31, 1970; engineering geology observations', *Bull. Seism. Soc. Amer.*, **61**, No. 3, 5111–533 (June 1971).
5. Seed, H. B., 'Landslides during earthquake due to soil liquefaction', *J. of the Soil Mechanics and Foundations Division*, *ASCE*, **94**, No. SM5, 1053–1122 (Sept. 1968).
6. Suzuki, Z., (Ed), *General report on the Tokachi-Oki earthquake of 1968*, Keigaku Publishing Co., Ltd., Tokyo (1971).
7. Henkel, D. J., 'The role of waves in causing submarine landslides', *Geotechnique*, **20**, 75–80 (1970).
8. Ove Arup and Partners, 'Earthquake effects on platforms and pipelines in the UK offshore area', *Report to the UK Dept of Energy, OT-R 7950* (1980).
9. Wood, S. H., *et al.* 'The Borah Peak, Idaho earthquake of October 28, 1983–Hydrologic effects', *Earthquake Spectra*, **2**, No. 1, 127–50 (1985).
10. Wiegel, R. L., 'Tsunamis', in *Earthquake Engineering*, (Ed. R. L. Wiegel), Prentice-Hall, Englewood Cliffs, NJ (1970).
11. Seed, H. B., Ugas, C., and Lysmer, J., 'Site dependent spectra for earthquake resistant design', *Report No. EERC 74-12, Earthquake Engineering Research Center*, Univerity of California, Berkeley, Nov. (1974).
12. Applied Technology Council, 'Tentative provisions for the development of seismic regulations for buildings', *ATC-06, NBS SP-510, NSF 78-8* (1982).
13. Schmertmann, J. H., 'Static cone to compute static settlement over sand', *J. of the Soil Mechanics and Foundations Division, ASCE*, **96**, No. SM3, 1011–43 (May, 1970).
14. Holtz, W. G., and Gibbs, H. J., Discussion of 'Settlement of spread footings on sand', (by D. J. D'Appolonia, E. D'Appolonia, and R. F. Brissette, *J. of the Soil Mechanics and Foundations Division, ASCE*, May, 1968), *J. of the Soil Mechanics and Foundations Division, ASCE*, **95**, No. SM3, 900–05 (May 1969).
15. Bazaraa, A. R. S. S., 'Use of the standard penetration test for estimating settlements of shallow foundations on sand', *Ph.D. thesis*, University of Illinois, 1967.
16. Shannon and Wilson, Inc., and Agbabian-Jacobsen Associates, *Soil behaviour under earthquake loading conditions—state of the art evaluation of soil characteristics for seismic response analyses*, prepared for the U.S. Atomic Energy Commission by Shannon and Wilson, Inc., Seattle, and Agbabian-Jacobsen Associates, Los Angeles (1972).
17. Espinosa, A. F., and Algermissen, S. T., 'Soil amplification studies in areas damaged by the Caracas earthquake of July 29, 1967', *Proc. Microzonation Conf., Seattle, Washington*, **II**, 455–64 (1972).
18. Ravara, A., Pereira, J., Oliveira, C., and Lourtie, P., 'Estudos estruturais dos edificios de Parque Central—2°- Relatónò: Análise dinâmica dos edifícios de apartamentos', *LNEC Report*, Lisbon (1971).
19. Parton, I. M., and Taylor, P. W., 'Analysis of microtremor recordings', *Bull. NZ Soc. Earthq. Eng.*, **6**, No. 3, 96–109 (Sept. 1973).
20. Parton, I. M., and Taylor, P. W., 'Microtremor recording techniques', *Bull. NZ Soc. Earthq. Eng.*, **6**, No. 2, 87–92 (June 1973).

21. Seed, H. B., and Lee, K. L., 'Pore-water pressures in earth slopes under seismic loading conditions', *Proc. Fourth World Conf. on Earthq. Eng., Chile*, **3**, A5, 1–11 (1969).
22. Parton, I. M. and Smith, R. W. M., 'Effect of soil properties on earthquake response', *Bull. NZ Soc. Earthq. Eng.*, **4**, No. 1, 73–93 (March 1971).
23. Wilson, S. D., and Deitrich, R. J., 'Effect of consolidation pressure on elastic and strength properties of clay', *Proc. ASCE Research Conf. on Shear Strength of Cohesive Soils, University of Colorado*, 1960, pp. 419–35, discussion pp. 1086–92.
24. Drnevich, V. P., Hall, J. R., and Richart, F. E., 'Effects of amplitude of vibration on the shear modulus of sand', *Proc. International Symposium on Wave Propagation and Dynamic Properties of Earth Materials, New Mexico*, 1967, pp. 189–92.

Chapter 4

Design earthquakes

4.1 INTRODUCTION

As defined by the EERI Committee on Seismic Risk,[1] a *design earthquake* is *a specification of the seismic ground motion at a site, used for the earthquake-resistant design of a structure.* The ground motions may be specified in a number of ways, i.e. by peak accelerations, velocities, and displacements, by accelerograms, and by response spectra. In some cases, the differential surface displacement of the ground due to fault displacement may be important, such as the design uplift or downthrow of the area, or in occasional instances where an active fault crosses the construction site, the design relative displacement across the fault will need to be specified.

As set out in the design flow chart, Figure (i), page 2, specifying design earthquakes requires information on seismic activity (Chapter 2) and on the site (Chapter 3). It is then necessary to establish the acceptable risk so that the appropriate rarity of event may be chosen. Again quoting the EERI Committee on Seismic Risk,[1] *acceptable risk* is *a probability of social or economic consequences due to earthquakes that is low enough to be judged by appropriate authorities to represent a realistic basis for determining design requirements for engineered structures.*

In general, establishing design earthquakes involves both deterministic and probabilistic considerations of various aspects contributing to the hazard assessment. The ratio between the two types of argument varies widely, but obviously there should always be some probability content, implicit or explicit in a rational assessment of seismic hazard. An introduction to methods of determining the probabilities of ground motion is therefore given below.

4.2 PROBABILITY DISTRIBUTIONS OF GROUND MOTION CRITERIA

The determination of design earthquakes ground motion criteria from seismic hazard analyses on a probabilistic basis was formulated by Cornell.[2] The method involves two separate models: a *seismicity model* describing the geographical distribution of event sources and the distribution of magnitudes; and an *attenuation model* describing the effect at any given site as a function of magnitude and source-to-site distance.

The *seismicity model* may comprise a number of source regions the seismicity of which may be expressed (Section 2.3.3) in the form

$$\log N = A - bm \tag{4.1}$$

where N is the number of earthquakes of magnitude exceeding m per year. The source regions may be described as lines representing known faults or areas of diffuse seismicity, so that N relates respectively to a unit length or a unit area. The value of N will also generally be found assuming that m has upper and lower bounds m_1 and m_o.

Attenuation models relate the effect i at a site to magnitude and distance, so that in general

$$i = i(m,r) \tag{4.2}$$

from which we have the inverted expression for magnitude

$$m = m(i,r) \tag{4.3}$$

More specifically the attenuation of peak ground motion amplitudes (a,v,d) and also response spectrum ordinates $(S_a,S_v,S_d,)$, are commonly expressed in the form of equation (2.16), discussed in Section 2.4.2.1 such that

$$\log i = b_1 + b_2 m - b_3 \log(r + b_4) \tag{4.4}$$

where b_1 to b_4 are empirically derived constants, such as those of Esteva and Villaverde,[3] McGuire,[4] or Katayama *et al.*[5] (Section 2.4.2.1 and Table 4.2).

Rewriting equation (4.4.) in the form of equation (4.3) we have

$$m = \frac{1}{b_2}\left\{ \log\ [i(r + b_4)^{b3}] - b_1 \right\} \tag{4.5}$$

Combining the above two models leads to the probability p_i that any earthquake occurring at random in the source region will produce an effect with strength exceeding i at the site:

$$p_i = P\ [I > i\,] = \int_{\text{source}} 10^{-bm(i,r)} f_R(r)\ \mathrm{d}r \tag{4.6}$$

where $f_R(r)$ is the probability density function of distance r.

As there are on average N earthquakes per year in the source region, the average annual probability of i being exceeded at the site is

$$p_D = p_i N \tag{4.7}$$

and hence the average return period of the effect exceeding i is

$$T_R = \frac{1}{p_D} = \frac{1}{p_i N} \tag{4.8}$$

A more detailed discussion of the above theory, including the proper treatment of scatter associated with the regression analysis for the attenuation model, is given in Appendix B.4.

As an example of the above method, consider a region of uniformly distributed seismicity of moderately high seismic activity as expressed by the recurrence relationship $\log N = 2.25 - 0.75M$, truncated by two different upper bounds on magnitude. Making the assumption that the earthquake foci are all located on a plane at a depth $H = 20$ km, the annual probability that peak ground acceleration will exceed any given value, as found from equation (4.7), is as shown by the curves on Figure 4.1. The curves represent the difference between assuming $M_{max} = 8$ and 9 in the recurrence relationship used for the seismicity model. At a return period typically used for the design of ordinary structures i.e. $T_r = 100$ years, the peak ground acceleration is about $0.24g$ for $M_{max} = 8$, and about $0.28g$ for $M_{max} = 9$, a difference of about 15 percent. At the return period $T_R = 10\,000$ years, which might be used for the Safe-Shut-Down design criteria of critical facilities, the peak ground acceleration using $M_{max} = 9$ is about 1.16 times the value obtained in $M_{max} = 8$. Obviously the use of different values of M_{max} can significantly affect the calculated design earthquake parameter values and therefore M_{max} should be assigned with care, particularly for long return period design criteria.

The curves on Figure 4.1 represent mean expected values as they are based on Esteva and Villeverde's[3] attenuation expression (equation (2.17)) which is a mean value relationship. The $M_{max} = 8$ curve on Figure 4.1 asymptotes to the maximum value of $a = 0.95g$. The uncertainty in the data is illustrated by comparison with peak ground accelerations obtained using the attenuation expression equation (2.20). For a position on the causative fault, so that $d = 0$, and assuming that the moment magnitude $M_m = 7.6$ which corresponds approximately to magnitude $M = 8.0$, this yields

$$a(\text{mean}) = 0.98g$$
$$a(84\ \text{percentile}) = 1.78g$$

Obviously, allowance for scatter should be made in selecting design values. The close correspondence between the two mean values of $0.95g$ and $0.98g$ is coincidental, considering that the two equations use different magnitude scales, and also that the two values correspond to different focal depths (i.e. $H = 20$ and 7.3 km, respectively).

The curves in Figure 4.1 are based on an attenuation expression which is an empirical relationship based on firm ground data, and thus they do not include the limits that less competent site soils may place on the ground motions, such as those noted at the bottom-left corner of the figure. Means of dealing with this situation include various levels of site response analysis (Section 6.2) which, of course, introduce their own sets of uncertainties into the analysis.

Another important source of uncertainty in calculating probabilities of ground motion arises from the basic earthquake recurrence relationship (equation (4.l)).

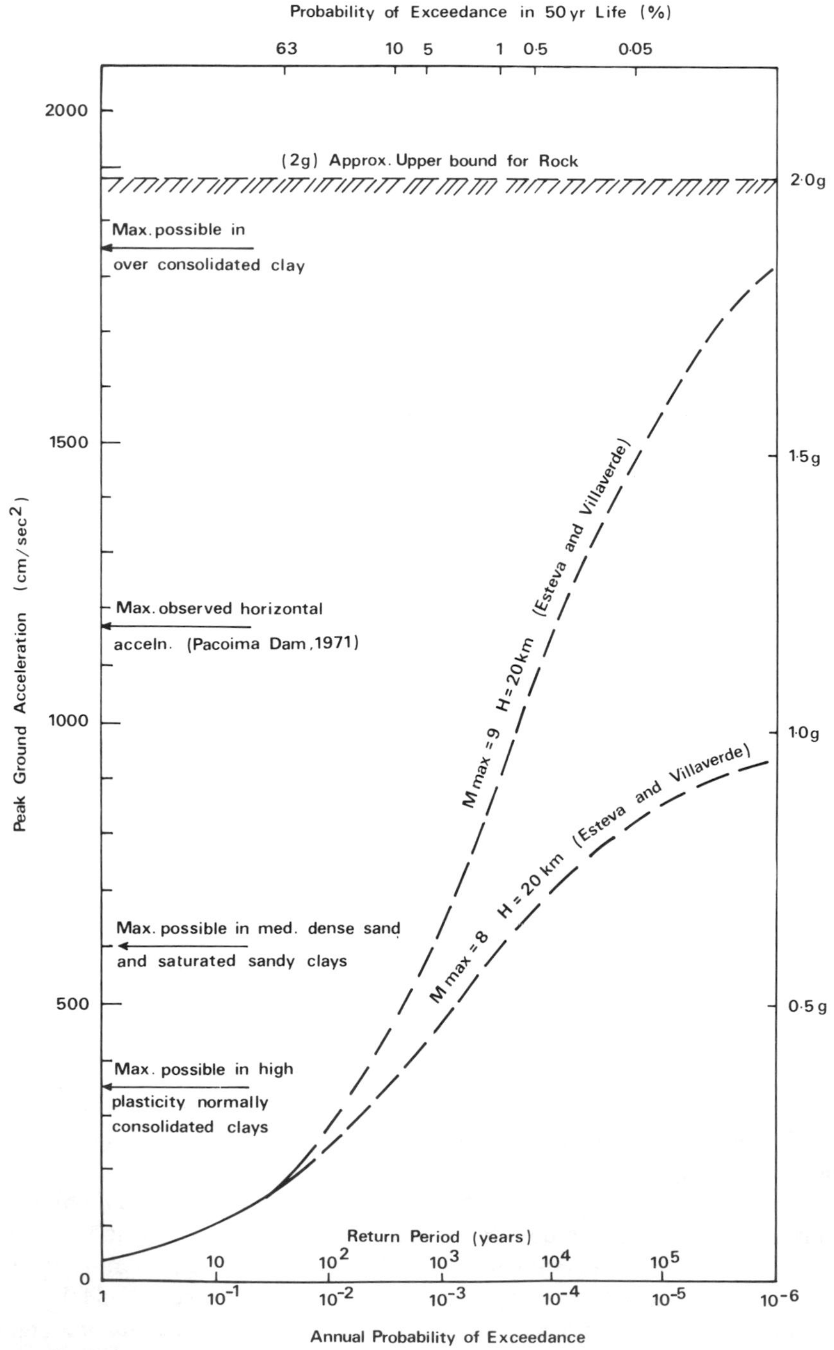

Figure 4.1 Peak horizontal acceleration probabilities as a function of M_{max} for a moderately high hazard zone and some physical bounds from soil strength

As discussed in Section 2.3.3, the rate of occurrence of earthquakes varies considerably over different time periods (Figure 2.16), and the great uncertainty involved in extrapolating the curve in this figure into the future is very apparent. Is a quieter or more active phase approaching?

In a study of the *c.* 3000-year long Chinese earthquake catalogue, McGuire[6] found that 50- and 100-year data intervals provided better estimates of probabilities of felt shaking in the 50 year period following each time segment than 200-year data intervals. McGuire argued that 'at a specific time, the most recent seismic activity is therefore the best data base to use for calculation of probabilities of shaking in the near future'. However, these results concern the global reliability for an ensemble of data sets for 62 cities in China, and the reliability of such projections for individual cities would presumably be less. The above conclusion also defies the seismic gap theory (Section 2.3.4). In contrast, based purely on the statistics, it would not be wise to argue that the 50-year long quiescent phase in New Zealand (from *c.* 1940 to 1986+) is likely to be a more reliable basis for estimating seismic activity in the coming 50-years than taking into account the known higher seismic activity of the century prior to 1940. As discussed in Chapter 2, geological data should be used to adjust deterministically the constants A and b of equation (4.1) to obtain greatest forecasting reliability.

In the preceeding paragraphs we have discussed three sources of uncertainty in the probabilistic evaluation of design ground motion criteria, i.e.:

(1) The earthquake recurrence relationship;
(2) The attenuation expression;
(3) Site response.

For completeness we should note that a fourth set of uncertainties is introduced by the remaining main component of the analysis, the spatial distribution of the events, i.e. whether they are uniformly distributed in plan or clustered on known faults or particular segments of those faults. Some of the problems involved in putting these components together and controlling the uncertainties through uncertainty analyses have been usefully discussed by Shah and Dong.[7]

Design events, discussed in Section 2.5, may be determined from probabilities of peak ground accelerations or velocities at the site, such as those of Figure 4.1. If the design event is defined as that giving rise to a peak ground acceleration, a, equal to or exceeding that of a given probability level, then the design event will be an (M,R) pair that give this value a. For example, if the design probability is $P=0.01$ per annum, and $M_{max}=8$, then from Figure 4.1 the peak ground acceleration $a=240\ \text{cm/s}^2$. Using the attenuation model used in deriving Figure 4.1, in this case equation (2.17), various M,R pairs may be found, corresponding to larger or smaller magnitude events; e.g. ($M=8.5$, $R=53$ km) and ($M=6$, $R=13$ km) both give $a=240\ \text{cm/s}^2$.

4.3 DETERMINING DESIGN EARTHQUAKES

4.3.1 Introduction

Several widely differing ways exist for specifying the design earthquake, the principal ones being:

(1) Equivalent-static loadings in codes;
(2) Response spectra
  —from the Design Event (various methods);
  —uniform risk spectra;
(3) Accelerograms
  —from records of real earthquakes;
  —from theoretical simulation.

Equivalent-static loadings are discussed in Section 6.6.7 and are of no particular interest here, so only response spectra and accelerograms will be considered below. In establishing design earthquakes, two major factors which need early consideration are

(1) The nature of the site;
(2) The type(s) of seismic response analysis to be carried out.

As discussed in Chapter 3, the site investigation should have determined the nature of the soil and the topography at the site. Regarding soil conditions, the main issue is the location of bedrock; does it occur at the surface, or is it overlain by sedimentary soils? As shown on the flow chart (Figure 4.2), sites having surface bedrock will not require a site response analysis, while sites having subsurface bedrock may or may not require a site response analysis. The need to carry out a site response analysis will increase with increasing thickness and softness of the overburden, and with the size and sensitivity of the structures under consideration.

The types of response analysis which are to be used for both the soil and the structure dictate whether design earthquakes are specified as accelerograms or as response spectra, and whether they are to be located at or below the ground surface (Figure 4.2). While response spectra are satisfactory for the majority of projects, more sophisticated analyses requiring the use of accelerograms, such as studies of non-linear material behaviour, are sometimes required, depending on the nature of the site and the size and sensitivity of the structure.

Dynamic response analysis of soils or any type of structure may be carried out using accelerograms or response spectra as input (Chapters 1 and 6). Whichever form of dynamic input is used a number of earthquakes should be used or implied. Because of the random nature of earthquakes it seems unlikely that any single seismic event can be shown to be safely representative of the design risk, without choosing an uneconomically powerful ground motion.

The use of accelerograms in a time-dependent analysis is analytically more powerful than response spectrum analysis and may be significantly more

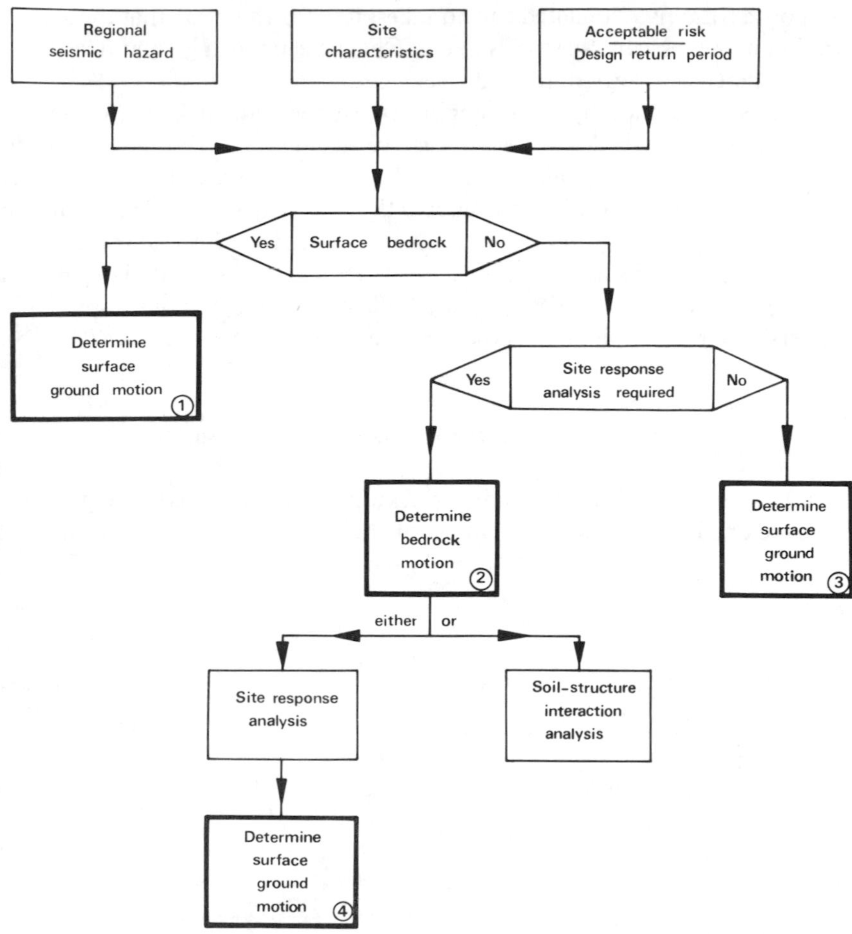

Figure 4.2 Flow chart showing four different procedures for determining design earthquakes (boxes (1) to (4)), depending on the situation

informative about the dynamic response of the structure. Individual accelerograms may induce local response peaks (Figure 1.10) in elastic analyses which may be difficult to interpret or justify. In non-elastic analyses this difficulty is partly overcome, but a number of accelerograms should be taken in all cases. It is common practice to take three or four accelerograms in a given study, these often being a mixture of real events (sometimes scaled) and simulated records.

When using response spectra as input, either several response spectra from individual events (e.g. Figure 1.10) or a single spectrum which is the average of several events (Figures 1.11 or 4.3) should be taken. This will help to allow for the randomness of earthquakes, and smoothed average spectra will eliminate undue influence of local peaks in response. Figure 4.3 indicates the scatter of response from five simulated earthquakes; it is worth noting that the standard deviation in the response spectra is likely to have been much greater if real rather than simulated accelerograms had been used.

It is strongly argued by many engineers[8] that where surface motions have been computed from bedrock motions as in Section 6.2.2, these surface motions should be used for structural analysis in the form of averaged response spectra rather than accelerograms. It is considered that so many simplifying assumptions are made in site response analyses, that very sophisticated use of the computed surface accelerograms can scarcely be justified for practical design purposes.

Further discussions on different methods of seismic response are given in Chapters 1 and 6, including a comparative review in Section 6.6.7.

4.3.2 Sources of accelerograms and response spectra

Earthquake engineers experienced at working outside basic code requirements have developed sources of information of their own, through government and

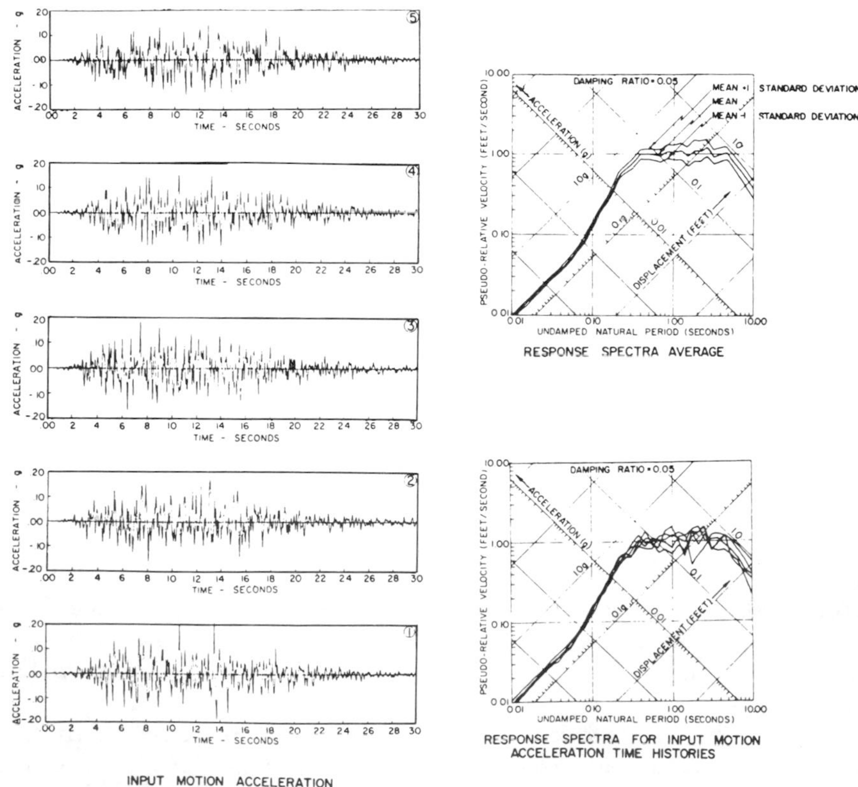

Figure 4.3 Set of simulated input accelerograms and response spectra for a magnitude 7.0 earthquake with a closest distance to the causative fault of 50 km (after Valera and Donovan[8])

university organizations specializing in seismology and earthquake engineering. As the problem of availability of information varies so widely from place to place and as the situation is changing so rapidly, this section will simply discuss a few of the chief sources of data existing, at present.

(i) *Accelerograms of real earthquakes.*

The major source of accelerograms is a worldwide collection of strong-motion records for dissemination (in various forms) to the scientific and engineering community, which is available from the World Data Center for Solid Earth Geophysics. A list of their available data is obtainable from National Geophysical Center (D622) NOAA, Code E/GC11, 325 Broadway Street, Boulder, CO 80303, USA. Countries contibuting to the strong-motion data base include Australia, Italy, Japan, New Zealand, Rumania, the USA, the USSR, and Yugoslavia. The US Geological Survey has furnished records from its network of co-operative strong-motion stations, including those in Central and South America. Perhaps the best-known national collection of earthquake records is that of events with epicentres within the USA, started by the California Institute of Technology[9,10] and being continued by the US Geological Survey.

(ii) *Accelerograms of simulated earthquakes.*

Many earthquake engineering research organizations throughout the world have computer programs for generating artificial earthquakes. Eight simulated earthquakes have been described by Jennnings *et al.*,[13] and the tape for these was available from the California Institute of Technology. Two computer programs called PSEQGN AND SIMEAR for the generation of simulated earthquakes are available as listings and as tapes, from the National Information Service for Earthquake Engineering, address NISEE, 47th Street and Hoffman Boulevard, Richmond, California 94804. White noise segments can readily be generated at most computer centres. When filtering white noise, reference may be made to Fourier amplitude spectra of actual earthquakes.[12]

(iii) *Response spectra of real earthquakes.*

Response spectra are more readily available than accelerograms as they are easily described in diagram form in the literature. The classical averaged response spectra (Figure 1.11) developed by Housner,[14] are widely referred to elsewhere. Response spectra can be computed from earthquake accelerograms by computer programs such as SPECEQ, which is available through NISEE at address given in part (ii) above. A considerable early collection of American response spectra was prepared by Alford *et al.*,[15] but this has been largely superseded by the series being produced by the California Institute of Technology.[11]

(iv) *Response spectra of simulated earthquakes.*

As mentioned in (ii) above, Jennings *et al.*[13] have described eight simulated earthquakes, giving response spectra as well as accelerograms. Response spectra may readily be computed from simulated accelerograms from computer programs such as SPECEQ, which are available through NISEE as described in part (ii) above.

4.3.3 Response spectra as design earthquakes

As noted in Section 4.3.1, response spectra used as design earthquakes may be derived in a number of ways, all of which have considerable uncertainties and need subjective input. Some of these methods are described below.

4.3.3.1 Elastic response spectra derived from Design Events

Design Events (Sections 2.5 and 4.2), provide a base from which response spectra may be readily determined in a number of ways.

(i) *Elastic response spectra from selected records of real earthquakes*

Having determined the magnitude and focal distance of the Design Event, ideally it may be possible to select a number of records of earthquakes with similar M and R values, and with appropriate source mechanisms and similar site soil conditions. These spectra may be applied to the analytical model individually, or the average or an envelope may be determined creating a smoothed design spectrum, as with the Housner[14] approach (Figure 1.11). As well as trying to match the M,R values it is often considered appropriate to scale the individual events to have either the same peak ground acceleration or the same peak ground velocity. While acceleration is probably more commonly used as the scaling criterion, velocity may often be the better arbiter of damage to the structure. For example, in a study of offshore concrete oil platforms for survivability in extreme magnitude earthquakes, Watt *et al.*[16] scaled three earthquakes (El Centro, 1940; Borrego Mountain, 1968; and Taft, 1952) to have the same peak ground velocity of 100 cm/s.

Scaling of earthquake accelerograms (or spectra) should be done with caution if the change in amplitudes is large (changes of more than 50 percent, say). Large changes imply either that the scaled event is much larger or smaller than the original event or that the focal distance is different. These conditions imply a different source-controlled or attenuation-controlled frequency content (Section 2.4), and also imply possible non-linearities regarding soil behaviour.

While peak ground motions are commonly used for scaling response spectra, they are not very satisfactory for this purpose. Hall *et al.*[17] report that a three-parameter system, using response spectrum intensities, may offer a better means of scaling response spectra.

(ii) *Elastic response spectra obtained by factoring ground motions*

In Newmark and Hall's method[18] the spectral ordinates S_a,S_v,S_d, are obtained by multiplying, respectively, the peak ground motion amplitudes a,v,d (corresponding to the Design Event), by simple factors which vary with the degree of damping, as shown in Table 4.1. At low periods ($T<0.1$ s) S_a asymptotes to a, and at long periods $S_a = d$. When these values of S_a, S_v, and S_d are plotted on tripartite log paper of the form shown in Figure 6.28, the period ranges controlled by each are easily defined. Depending on the design ground motions and the degree of damping, velocity governs the spectral values in the approximate period range 0.3–0.6 to 3–5s.

The factors in Table 4.1 correspond to averages of many earthquakes, presumably mostly from the western USA, but, while appealingly simple, the source mechanisms, soil conditions, and the statistical properties represented by these factors are obscure. In an effort to improve the method, Blume *et al.*[19] produced a revised system of amplification factors representing the 84 percentile (mean plus one standard deviation) spectral values for a set of earthquakes mostly recorded in the western USA (plus one each from Japan and Lima). The spectra are constructed on tripartite log paper by drawing straight lines between four control points A,B,C,D, at specific frequencies, defined as follows:

Point A, 33 Hz

$$S_a(A) = 1.0a \tag{4.9}$$

Point B, 9 Hz

$$S_a(B) = (4.25 - 1.02 \ln\xi)a \tag{4.10}$$

Point C, 2.5 Hz

$$S_a(C) = 1.2S_a(B) = (5.1 - 1.224 \ln\xi)a \tag{4.12}$$

Point D, 0.25 Hz

$$S_d(D) = (2.85 - 0.5 \ln\xi)d \tag{4.12}$$

in which S_a = spectral acceleration; S_d = spectral displacement; a = design peak ground acceleration; d = design peaking ground displacement; and ξ = damping factor, as a percentage of critical value. The peak ground displacement was taken as proportional to the peak ground acceleration, a value of 910 mm (36 in) for a ground acceleration of 1.0g being proposed by Blume *et al.*[19] In a subsequent study McGuire[4] found that for western USA earthquakes the average d/a ratio was 665 mm per 1.0g acceleration. The difference between this and Newmark's result was attributed[4] to different correction and integration procedures used on the accelerograms processed by the EERL.[10] However, the differences in computed displacement produces no effect on the design spectra, as long as the same displacements which are used to develop amplification factors for the low-frequency range are also used to predict design ground displacements.

The spectra resulting from the use of equations (4.9) to (4.12) are for firm ground sites, and Blume *et al.*[19] recommend that for sites which are significantly responsive to ground motion components with periods longer than 0.5s that the spectral shapes should be modified appropriately.

(iii) *Elastic response spectra from McGuire's western USA data set*

This method has some advantages over those given in (ii) above. If the Design Event is in a plate boundary area, it may be appropriate to use the data from

Table 4.1. Response spectrum amplification factors (after Newmark and Hall[18])

Damping	Amplification factor		
ξ(%)	Acceleration	Velocity	Displacement
0	6.4	4.0	2.5
1	5.2	3.2	2.0
2	4.3	2.8	1.8
5	2.6	1.9	1.4
7	1.9	1.5	1.2
10	1.5	1.3	1.1
20	1.2	1.1	1.0

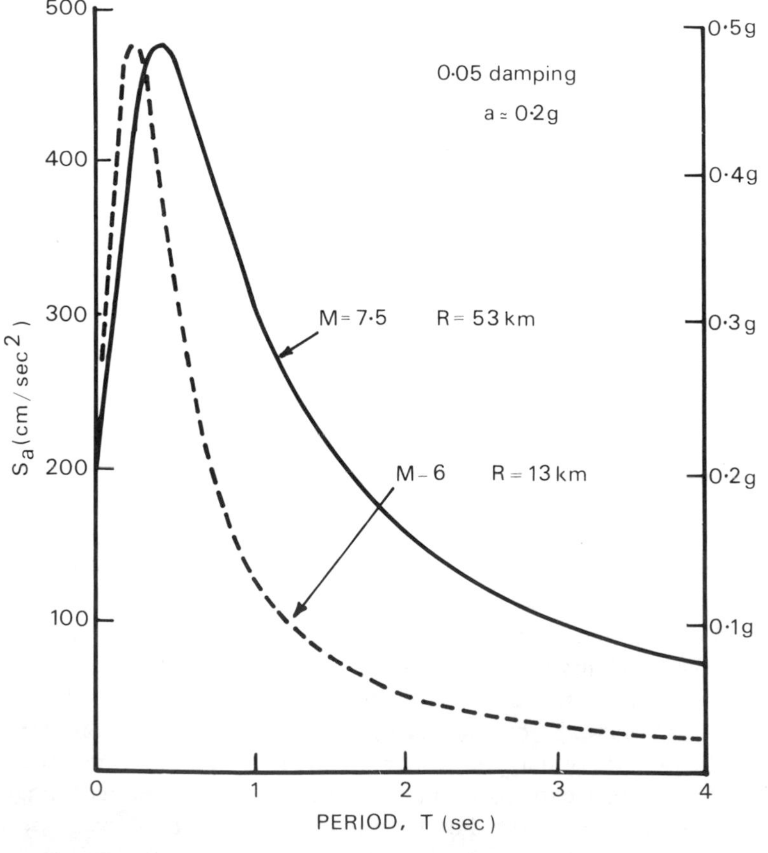

Figure 4.4 Two design earthquakes based on two different (M,R) pairs giving the same peak ground acceleration at the site, using the mean spectral ordinates from McGuire's data[4]

Table 4.2. McGuire's attenuation expressions[4] for spectral acceleration with 5 percent damping

$$S_a = b_1' \, 10^{b_2 M} \, (R+25)^{-b_3} \quad \text{cm/s}^2$$

$$\log S_a = b_1 + b_2 M - b_3 \log (R+25)$$

Period (s)	b_1'	b_1	b_2	b_3	Coeff. of var. of S_a
0.1	1610	3.173	0.233	1.341	0.651
0.2	2510	3.373	0.226	1.323	0.577
0.3	1478	3.144	0.290	1.416	0.560
0.5	183.2	2.234	0.356	1.197	0.591
1.0	6.894	0.801	0.399	0.704	0.703
2.0	0.974	−0.071	0.466	0.675	0.941
3.0	0.497	−0.370	0.485	0.709	1.007
4.0	0.291	−0.620	0.520	0.788	1.191

McGuire,[4] based on regression analyses of 68 earthquake records from sites in western USA, with $5.3 \leqslant M \leqslant 7.6$ and $15 \leqslant R \leqslant 125$ km. Supposing that a probability analysis similar to that described in Section 4.2 has given a design peak ground acceleration (PGA) $a = 0.2g$, then using McGuire's attenuation equation for PGA (Table 2.7), two Design Events would be ($M = 7.5$, $R = 53$ km) and ($M = 6$, $R = 13$ km). The spectral ordinates for 5 percent damped acceleration response for these two design events calculated from the equation and constants given in Table 4.2 are shown on Figure 4.4. For structures with periods of vibration greater than $T = 0.3$ s the $M = 7.5$ event is obviously much more demanding than the $M = 6$ event.

(iv) *Elastic response spectra from source and attenuation data of shear waves*

This method of obtaining response spectra demonstrates the growing maturity of knowledge of source behaviour (Section 2.4.1).

McGuire *et al.*[20] developed expressions for finding the spectral velocity S_v as a function of frequency f, and damping ξ, which reduce to

$$S_v(f, \xi) = q\sigma_v \tag{4.13}$$

where q is the ratio of peak to rms amplitude of a stationary process:

$$q = \{2 \ln(2fS/-\ln p)\}^{1/2} \tag{4.14}$$

where S is the duration of shear wave motion taken as equal to the duration of faulting and p is the probability of exceedance. The second part of equation (4.13), σ_v, is defined by a complex expression involving f, ξ, focal distance R, shear wave velocity v_s, soil density ρ, corner frequency f_o, fault stress drop $\Delta\sigma$, and quality factor Q.

The response spectra obtained by this method compared well with those obtained from records of a wide range of California earthquakes (local magnitudes 4.0 to 7.2, distances 9 to 130 km). This method offers the advantage over purely empirical methods of a physical basis for extrapolating to predict ground motions for magnitudes and distances which are poorly documented with data.

4.3.3.2 Uniform risk response spectra

Using the techniques noted in Section 4.2 and Appendix B.4, response spectra may be generated for a given site such that the spectral ordinates for all of the periods of vibration have the same probability of occurrence. Such a response does not represent just one design even (M,R) as given by either of the curves on Figure 4.4, but represents all of the M,R pairs contributing to the distribution of spectral values at each period and damping value for which they are calculated. Such design spectra are therefore sometimes referred to as *uniform risk spectra* or *consistent risk spectra*,[4] and may be used for specific sites or for codes.

As an example, the spectra in Figure 4.5 relate to code design requirements for bridges in New Zealand.[21] The two curves for ductility factor $\mu = 1$ are elastic risk spectra for 5 percent of damping. They are based on two different evaluations of the limited New Zealand strong-motion data supplemented by Japanese attenuation data from Katayama *et al.*,[5] because New Zealand and Japan are tectonically similar. The large difference between the two $\mu = 1$ curves illustrates the difficulty of establishing reliable results from probability analysis, as discussed by Shah and Dong,[7] especially if the data are sparse.

4.3.3.3 Special features of design earthquake response spectra

In establishing design earthquake response spectra by the methods outlined in Sections 4.3.3.1 and 4.3.3.2 it will be important to ensure that certain features specific to the proposed use of the spectra are incorporated, as discussed below.

(i) *Site soil conditions*

As shown by Figure 3.3, the soil conditions at the site have a profound influence on the shape of the response spectrum; the longer the predominant period of vibration of the site, the greater will be the period at which the peak in the response spectrum occurs, although the values of these two parameters will not necessarily be identical. Ideally this effect should be allowed for by using a data set derived from earthquakes recorded on sites with similar soil conditions to that of the sites in question. Much of readily available data, such as that of the western USA, is from relatively firm ground sites. Thus Blume *et al.*[19] and McGuire[4] both give data specific to firm ground. A useful feature of the spectral attenuation data of Katayama *et al.*[5] is its division into three different soil categories.

When using the ground motion factoring methods of Section 4.3.3.1(ii), some allowance for soil conditions may be made by estimating the peak ground

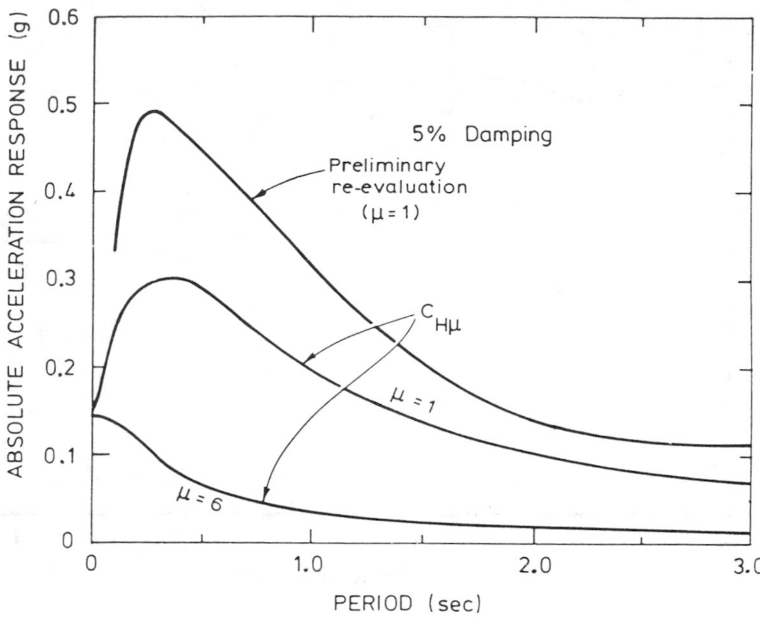

Figure 4.5 Uniform risk response spectra under development for seismic zone C in New Zealand (after Berrill *et al.*[21])

motions for the appropriate soil conditions. For example, peak ground velocity attenuation is given as a function of rock or soil sites for the western USA by Joyner *et al.* (equation 2.21) in Section 2.4.2.1. This would allow ground motion spectra of Newmark and Hall[18] and Blume *et al.*[19] to be constructed appropriately, but unfortunately amplifications appropriate to soft soil sites, instead of the firm ground ones in Table 4.1, are not at present available.

In a study of the subduction zone earthquakes of Taiwan, Singh[22] concludes that the spectral shapes are different from those of California earthquake for the same soil conditions. There was more amplification in the mean Taiwanese spectra at periods less than about 1.5 s than in the American Petroleum Institute[23] spectra for deep deposits or the Applied Technology Council[24] spectra for deep sediments. Also these Taiwanese mean spectra envelope the 84 percentile spectra of Seed *et al.*[25] given in Figure 3.3.

(ii) *Foundation size effects*

Ground motions will not be identical throughout space and time under the whole of the foundation of a structure during any given earthquake, so that the effective amplitude in any given direction at any given moment during an event tends to be less than the relevant peak amplitude given by a single accelerogram when averaging across a finite width of foundation. Arrays of closely spaced instruments have been providing the data to verify what theoretical analyses predict. In a study of recordings made in the SMART 1 array in Taiwan, at

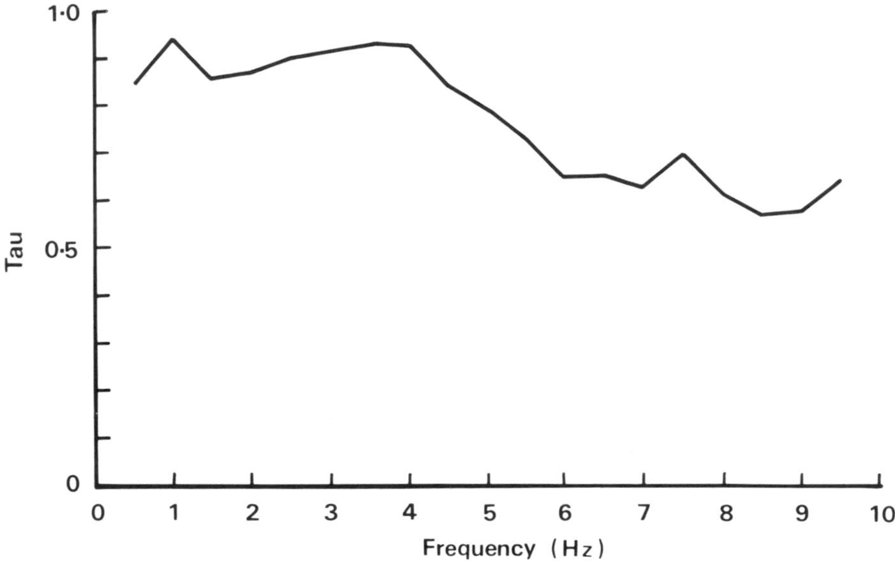

Figure 4.6 Response spectral ratio (tau) for the averaged response from two stations 200 m apart, using the transverse motion with 5 percent damping (from Bolt *et al.*[26] Reproduced by permission of Earthquake Engineering Research Institute (EERI))

points 200 m apart, Bolt *et al.*[26] found that effective spectral amplitudes for horizontal excitation at 5 percent of damping ranged between 60 and 100 percent of those for a single accelerogram (Figure 4.6), the effect being significant for frequency of vibration greater than about 4 Hz.

From studies of sinusoidal waves[27] it is clear that this averaging effect will be significant for foundations with dimensions approaching the wave length of the excitation. For buildings of average plan size or smaller this reduction effect is therefore likely to be small, but for very wide structures the economies offered by allowing for this effect make it worthy of investigation. For structures of 100 m or more in diameter much of the purely translational energy associated with periods less than about 1.0 s may not be transmitted above the foundation slab. However, the averaging of horizontal ground motions applied across the width of a foundation implied corresponding horizontal tensions and compressions in the foundation structure, and such tensions should be estimated for design purposes. Also the passage of the waves across the structure implies torsional excitations that may peak at a period of about 0.5 s.

Finally it is noted that more field data and theoretical insight into the effects of foundation size are needed to allow them to be confidently predicted. A step in this direction has been made with a theoretical model proposed by Vanmarcke and Harichandran.[28]

(iii) *Near-field versus far-field effects*

Studies of earthquake ground motions have produced a growing awareness of significant differences between the spectra of near-field and far-field shaking.[22] Those that are most evident at present are the difference in frequency content through differential attenuation (Section 2.4.2.8) and the greater effects of directivity (Section 2.4.2.3) closer to the source.

Attention has been drawn to this phenomenon in California by the establishment of an extra seismic zone about 8 km wide on each side of major faults capable of generating earthquakes of magnitude 7 or greater. In this zone it is required to study special effects on important structures caused by near-source events.

4.3.3.4 *Inelastic response spectra as design earthquakes*

In the foregoing discussion the response spectra presented have been those derived assuming linear elastic structural behaviour according to the analysis set out in Section 1.4.2.9. In design practice, in some cases the structure will be required to remain elastic in the design earthquake, but more commonly some degree of inelastic behaviour will be assumed. As discussed in Chapter 6, inelastic behaviour is often expressed in terms of the ductility factor μ, where $\mu=1$ represents elastic behaviour and $\mu=6$ is about the greatest degree of inelastic deformation that can usefully be achieved in most structures.

In order to arrive at a response spectrum corresponding to the desired degree of inelasticity, i.e. the design ductility level μ, the usual (least effort) technique is simply to divide the elastic response spectrum for the design earthquake by factors which allow for μ. As an example, Figure 4.5 shows the elastic response spectrum ($\mu=1$) and the corresponding $\mu=6$ response spectrum proposed for bridge design in New Zealand's seismic zone C, based on an alluvial site located in Auckland. Here the ordinates of the elastic spectrum were multiplied by a factor R for any given level of ductility μ, as a function of the period of vibration T, as follows:

$$\text{for } T=0\text{:} \qquad R = 1.0 \tag{4.15}$$

$$\text{for } 0<T<0.7\text{ s:} \qquad R = \frac{0.7}{(\mu-1)(T+0.7)} \tag{4.16}$$

$$\text{for } T\geqslant 0.7\text{ s:} \qquad R = \frac{1}{\mu} \tag{4.17}$$

A discussion of the reasons for using the above values of R is given in Section 6.6.7.3(i).

4.3.4. Accelerograms as design earthquakes

Accelerograms used as design earthquakes may be derived using the parameter values (M,R) or (a,v,d) representing the Design Event (Sections 2.5 and 4.2), or may be made to match a target design response spectrum. Using such criteria the accelerograms are obtained either from records of real events or by simulation techniques as discussed below.

4.3.4.1 Accelerograms of selected real earthquakes

In choosing from amongst real earthquake records (Section 4.3.2) it will be desirable to match as nearly as possible the design conditions of magnitude, epicentral distance, focal depth, source mechanism, and soil profile with those of the real earthquakes. Close matching of magnitude and distance is desirable for minimising scaling errors. Unfortunately not all of the above factors may be known or be readily available for the real events, and a second-best criterion is that the records should come from a geologically similar area. Unfortunately it will not be possible to fulfil this latter condition in seismic areas which have few, if any, strong-motion recordings of their own and little in common geologically with regions rich in accelerograms such as Japan and California. However, even when soil conditions are reasonably matched, it is well known that each individual earthquake record has strong features characteristic only of that particular earthquake and site. To rely *only* on records of similar earthquakes in some cases may be the least satisfactory method of choosing dynamic input for structural analysis.

Major difficulties arise over the choice of a real earthquake, with suitable peak accelerations, as some small earthquakes have much greater peak accelerations than larger earthquakes.

Hence rather than scale accelerograms according to peak ground acceleration (or velocity) it is sometimes better to scale them to have the same maximum spectral ordinate of a target design response spectrum.

The above remarks should be read in conjunction with the discussion on response spectra from real earthquakes given in Section 4.3.3.1(i).

For sites with surface bedrock there is an added difficulty, because very few strong ground motions have as yet been recorded on rock sites. Accelerograms of this type have been recorded in the USA as follows:

Helena, Montana, 31 Oct. 1935

$$M = 6.0, \quad I_{max} = \text{VIII}$$

Golden Gate Park, San Francisco, 22 Mar. 1957

$$M = 5.3, \quad I_{max} = \text{VII}$$

Pacoima Dam, California, 9 Feb. 1971

$$M = 6.6, \quad I_{max} = \text{XI}$$

Although the above is not a complete list, insufficient recordings of bedrock motion have been made for a definitive study to be possible. Care is required in the application of individual existing recordings to other sites; for example, at Pacoima Dam, peak horizontal acceleration was the very large value of 1.17*g*. Part of this enormous surface acceleration may be explained by the amplification occurring due to the cracking of the rock below the instrument station, and there may also have been magnification of base bedrock motion arising from the location of the recording instrument on a steep ridge of the valley.[29,30] This particular ground motion record should be directly used elsewhere for design *only* if the site characteristics are similar.

As so few bedrock strong ground motions have been recorded to date, it may be necessary to derive dynamic design criteria from random vibration theory as discussed in the next subsection.

4.3.4.2 Simulated accelerograms for surface bedrock

For most design purposes it can be assumed that ground motion is a random vibratory process, and that accelerograms can be mathematically simulated with random vibration theory. This will be most true at distances from the causative fault sufficient to ensure that the details of the fault displacement are not significant in the ground shaking. Because of the scarcity of actual bedrock recordings, at present the modelling of simulated earthquake is necessarily based on the more numerous accelerograms recorded on softer soils. This is considered reasonable, as there is much to suggest that the main difference between bedrock and soft-soil motions is one of frequency content; this difference can be dealt with in the simulation process.

By far the most common pattern of ground motion is one of an abrupt transition from zero to maximum shaking, followed by a portion of more or less uniformly intense vibration, and finally a rather gradual attenuation (Figure 4.3). In the terminology of random vibration theory, the middle portion may be considered as a stationary random process, whereas the initial and final phases, being transitional, are non-stationary. At present no other type of ground motion is seriously considered as a model for deriving earthquake loads. A short sharp jolt such as experienced at Agadir (1960) and Skopje (1964) is the only type of earthquake, essentially different from the random process described above, that has so far been identified.

(1) Non-stationary processes of various types[13,31-4]
(2) Stationary processes such as the white noise type[35-7] and the Gaussian white noise type.[38-40]

(i) *Non-stationary random processes.*

As only the middle portion of strong ground motion can be reasonably described as a stationary process, it has been argued that the shaking of smaller earthquakes and the tail of larger shocks are most accurately modelled by non-stationary processes. As some of the latter become mathematically very complex,

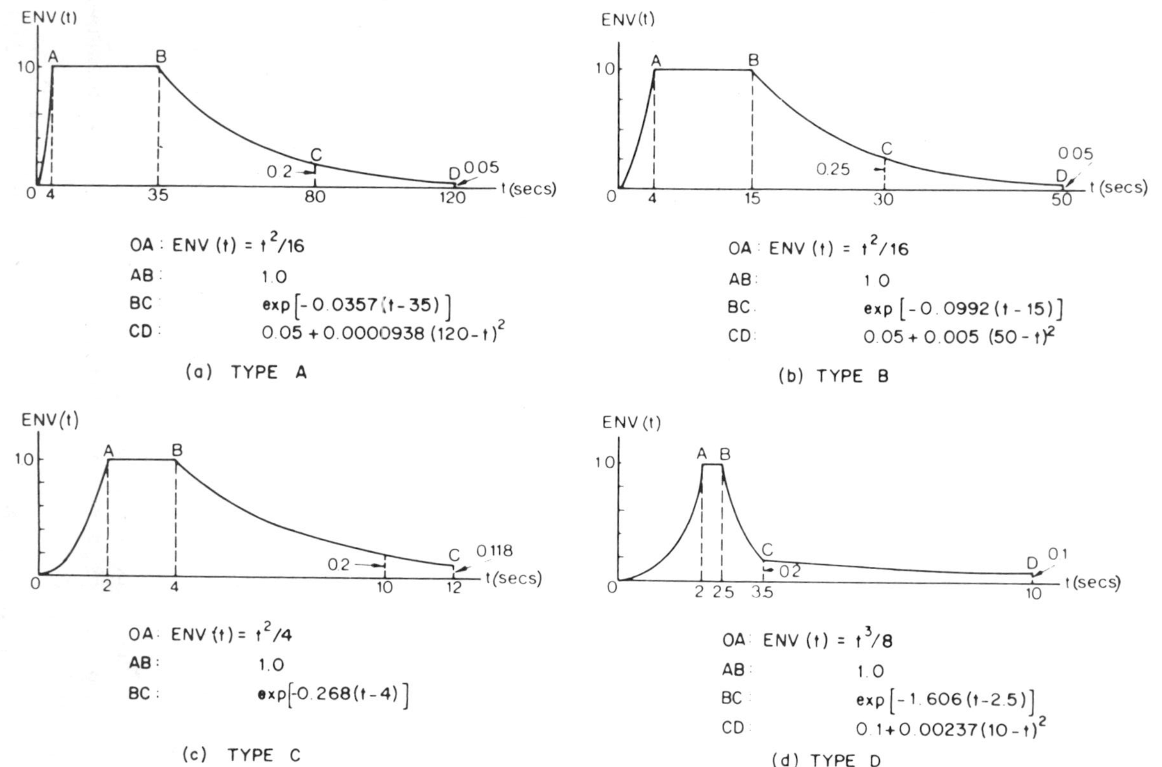

Figure 4.7 Envelope functions for non-stationarity of simulated earthquakes based on typical Californian earthquakes (after Jennings *et al.*[13] by permission of the California Institute of Techology)

it is fortunate that one of the simplest types of non-stationary models seems satisfactory. This consists of a stationary process multiplied by a non-stationary envelope function. The stationary process is usually derived from a segment of band-limited or filtered white noise;[41] this means that the frequency content has been limited to a prescribed band, or that the ensemble average Fourier spectrum has a prescribed shape which may be deduced from actual records.[12]

Various forms of the stationary process have been taken such as the Gaussian random process used by Jennings *et al.*[13] and Ruiz and Penzien,[42] or the filtered Poisson process used by Amin and Ang.[32] In these cases the non-stationarity was achieved with am amplitude envelope as mentioned in the preceding paragraph. Figure 4.7 illustrates the amplitude envelopes used by Jennings *et al.* in modelling their four earthquakes, which ranged in magnitude from about 5.0 to 8.0.

Various other envelope functions have been used, which have generally been assumed to have simple parametric forms, such as that using the shape of a γ function

$$\sigma(t) = At^{\gamma}e^{\alpha t} \tag{4.18}$$

which was used by Saragoni and Hart.[33] Such envelopes, while having the merit of simplicity,[34] have been based on matching a few observed average characteristics of real earthquakes rather than from theory or estimated in a systematic way from data. In seeking to improve on this situation, Nau *et al.*[34] have developed a non-parametric method for estimating an envelope.

For modelling ground motion *in the vicinity of the causative faults* Housner[41] suggests characteristics related to the types of faulting discussed above in Section 2.2.

(ii) *Stationary random processes*

Some workers consider that as so few records of bedrock motion exist, it is reasonable to carry out strong-motion studies using simple stationary processes in the form of banded white noise segments to simulate accelerograms.[35,37,43] Parton[43] found good agreement when comparing the velocity response spectra for three white noise segments with Housner's averaged response spectra (Figure 4.8). Any inaccuracy involved in not modelling the tail of the earthquake may be considered tolerable in relation to other important simplifications which are usually made in analysis, such as the assumption of elastic material behaviour. In this way the use of segments of white noise for the motion of bedrock overlain by softer deposits is arguably even more justifiable, as the simplifying assumptions in dynamic soil analyses are gross compared with those in structures. Band-limited white noise has the practical advantages that it is simple to generate and use in analysis, and programs for its generation are readily available in computer centres.

The three main characteristics of an accelerogram requiring consideration in the simulation process are peak acceleration, duration, and frequency content. These are functions of the design earthquake, particularly its magnitude and

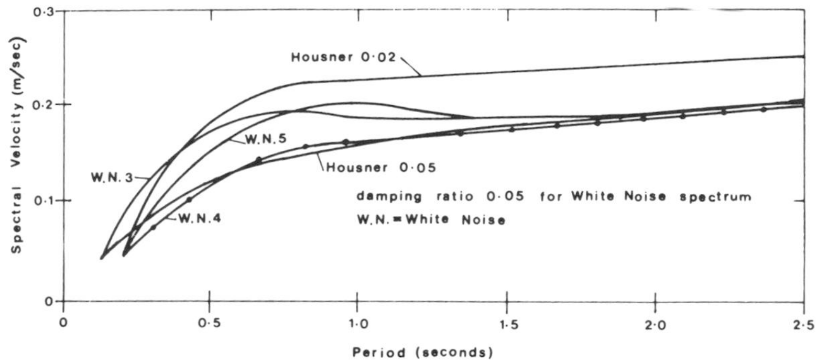

Figure 4.8 Comparison of Housner's averaged curves with response spectra from three white noise segments (after Parton[43])

distance as discussed in Chapter 2. As well as taking account of the preceding factors, simulated accelerograms should be integrated to check that realistic peak velocities and displacements occur.

In determining the duration of the simulated earthquake a white noise segment is clearly equivalent to the region of strongest shaking of a real earthquake. As the tails of most earthquakes subside gradually the cut-off point of strong shaking is clearly fairly arbitrary, depending on the definition of 'strong'. Various methods of estimating the duration of strong shaking are noted in Section 2.4.2.5, but care should be taken to apply them in a manner consistent with their definitions. In elastic analyses the duration of excitation is relatively unimportant, except that in order to determine peak response the duration should be several times the fundamental period of the system.

When considering relatively short epicentral distances, the nominally uniform frequency content of white noise seems at least as appropriate as more refined models of bedrock motion. At great epicentral distance the attenuation of high frequency vibration may be worth consideration. It is recommended that a number of different white noise accelerograms be used in any response analysis.

An interesting variant on the use of filtered Gaussian white noise is that of Boore,[39] who combined this method with seismological models of radiated spectra. The amplitude spectrum of the stochastic time series is equal, on average, to a theoretical spectrum with a sound physical basis. Boore uses the ω-squared spectrum with a high-frequency cut-off f_{max} and a constant stress parameter $\Delta\sigma$, as used by Hanks and McGuire,[44] which is scaled for source size using only one parameter, i.e. seismic moment or moment magnitude. The method permits the generation of accelerograms as a function of magnitude and distance.

4.3.4.3 Simulated accelerograms for soil sites

When simulating surface ground motion accelerograms at sites where softer soils overlie bedrock, two main procedures may be followed, i.e.:

(1) A site response analysis may be carried out, as discussed in Section 6.2.2.2; or
(2) Simulation may be effected using an approach similar to that described above (Section 4.3.4.2).

In the latter case allowance needs to be made for the softer soils. When filtering the white noise, deducing the prescribed shape of the average Fourier spectrum from actual records may have more relevance than for surface bedrock, as most surface records have been made on material other than bedrock. For the same reason non-stationary envelope functions such as shown in Figure 4.7 are also likely to be more relevant to simulated accelerograms for softer surface layers. Because of the filtering effects of softer layers non-stationary processes are even more appropriate to the surface of softer layers than to bedrock motion.

4.4 FAULTS—RISK AND DESIGN CONSIDERATIONS

4.4.1 Introduction

Intuitively the thought of building across an active fault is alarming, and obviously in general it is best avoided. However, in some circumstances structures can safely ride a fault rupture. For example, in the 1972 Managua earthquake the Banco Central de Nicaragua was astride a fault which moved 17 cm, and was strong enough to deflect the rupture around itself and survive intact.[45,46] Indeed the situation not infrequently arises when it is highly desirable to build across, or immediately beside, an active fault. Typically this happens when the location of a structure is conditioned by factors such as

(1) Topography, e.g. with dams;
(2) The structure's function, e.g. tunnels or pipelines; or
(3) Where land is valuable, e.g. in city centres.

In such circumstances seismic risk evaluation of alternative designs may be necessary.

In evaluating the seismic hazard of the displacement of a particular fault it will be important to know not only the probability of fault rupture during the lifetime of the structure, but also to differentiate between the likely amounts of vertical and horizontal displacement. Obviously some structures may be much more severely affected by vertical than by horizontal displacements, and there may be no practicable way of designing around such displacements. In such a case the possibility of building across a fault may have to be abandoned, unless the implied risk is accepted.

4.4.2 Probability of occurrence of fault displacements

In order to carry out a hazard analysis of fault displacements an investigation of the degree of activity of the fault in question, along the lines discussed in

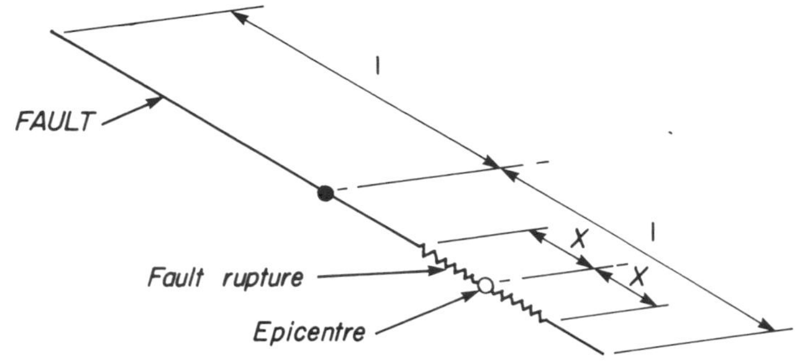

Figure 4.9 Model of fault used for estimating displacement probabilities (after Berrill[47])

Section 2.2.4, may be required. The more significant the structure, the greater will be the effort that will be appropriate in the hazard study. For major structures the study will hopefully result in recurrence intervals being associated with fault displacements of given magnitudes with an appropriate level of confidence.

Considering that epicentres are uniformly distributed along the length of a fault of length $2l$, as illustrated in Figure 4.9, Berrill[47] used the theory of probability of earthquake occurrence of Cornell[2] (Section 4.2) to find the probability of fault displacement u exceeding some value $\mathbf{U}$. Summarizing Berrill's work, we first assume that the rupture extends a length X in each direction from the epicentre (Figure 4.9). Then, provided that the fault is long compared with the length of rupture, we find the probability

$$P\,[\,S|X] = \frac{x}{l} \tag{4.19}$$

where S denotes the occurrence of rupture at a specified site on the fault.

Making the assumption that the relationship between earthquake magnitude and fault displacement is of log-normal form (equations (2.7) to (2.9) in Section 2.2.4.4) we write

$$\ln u = a_1 m + a_2 \tag{4.20}$$

where a_1 and a_2 are constants, leads to a power relationship between rupture length and displacement of the form

$$x = c_1 u c_2 \tag{4.21}$$

where c_1 and c_2 are constants obtained by regression analysis. These assumptions lead to the probability p_u that the displacement will exceed some value u at a particular site along a given fault:

$$p_u = \frac{hc_1\beta \exp\,(\beta a_2/a_1)}{la_1\alpha}(u^{-\alpha} - u_1{}^{-\alpha}), \qquad u_{\text{o}} < u < u_1 \tag{4.22}$$

where $h=(e^{-\beta m_o}-e^{-\beta m_1})^{-1}$ (4.23)

$\beta=b \ln 10$, where b is the Gutenberg and Richter seismicity parameter; m_o and m_1 are the minimum and maximum magnitudes considered; u_1 is the displacement corresponding to m_1;

$$\alpha=u_1/a_1-c_2 \tag{4.24}$$

Letting ν denote the average annual rate of occurrence of earthquakes with magnitudes greater than m_o, per unit length of fault, then the average rate of occurence λ_u of displacements exceeding u at a site is

$$\lambda_u=2l\nu p_u \tag{4.25}$$

Finally, making the assumption that earthquakes on the fault occur as a Poisson process, the return period T_R for displacements exceeding u is

$$T_R=(1-e^{\lambda_u})^{-1} \tag{4.26}$$

which for larger (rarer) values of u reduces to

$$T_R=\frac{1}{\lambda_u} \tag{4.27}$$

Using the above theory, Berrill[47] considered the moderately active Wellington fault which passes through the built-up area of the capital city of New Zealand, and found that horizontal displacements with return periods of 100 to 200 years were about 100–150 mm. This analysis was based on a maximum displacement of $u_1=6m$ at a maximum magnitude $m_1=8.6$, but because of the rarity of these extreme events, the results are insensitive to differences in u_1 (ranging from $4m$ to $10m$) at return periods of less than *c.* 5000 years.

An example of an extensive investigation of fault displacement hazard is that carried out for the site of a very large water-filtration plant at Sylmar near Los Angeles, California.[48] Site exploration consisted of logging 550 m of trench, and four faults were observed and dated by age of soil profile development. One of these faults was judged to be an active reverse fault. Its displacement increased with depth, indicating that more than one fault had occurred within the age span of the soil profile, the maximum single displacement being *c.* 0.2m and of a total observed cumulative vertical displacement of $0.8m$. This maximum displacement event of $0.2m$ occurred about 5000 years ago, and the probability of a recurrence of this size of displacement during a period of 100 years was estimated as being less than 2 percent, but the method of obtaining this estimate was not stated.

A further example of fault displacement hazard analysis comes from part of the seismotectonic hazard evaluation for a major hydroelectric power development in New Zealand.[49] Here a theoretical feature of particular interest is that it was *not* assumed that events occur as a Poisson process; this is in contrast to the assumption of Cornell[2] and others (Section 4.2) and by Berrill[47] noted above. Instead, the probability of fault movements occurring was estimated assuming that the recurrence intervals between events are log-normally distributed using the method described in Section 2.3.5. Using the Hazard Refinement Factor (HRF) in Figure 2.19, the probability of fault movements occurring during a given period

$$P(\text{faulting}) = \frac{\text{HRF}}{(\text{Mean Recurrence Interval})} \times (\text{Time period of interest}) \quad (4.28)$$

Because the data on the movement of the faults were poorly defined probability ranges were calculated, with the lower and upper bounds corresponding to HRF = 1 and 5, respectively. These values of HRF were chosen on the evidence of the faulting hazard functions for Nankai Trough and Pallet Creek given in Figure 2.19.

4.4.3 Designing for fault movements

The largest observed surface fault movements are a vertical displacement of 10.7*m* (India, 1877) and 8.9*m* vertically (Mongolia, 1957). While designing against such extreme displacements would seldom be contemplated, the feasibility of designing against more modest movements has been demonstrated by the Banco Central (Section 4.4.1), which survived 17 cm of horizontal rupture. As well as sudden rupture, allowance for slow creep movements is also sometimes desirable.

As occurred for the Banco Central in Managua, the possibility exists that the fault rupture may be diverted by the structure, and that this situation may be predicted at the design stage. For rupture diversion to occur the structure must be sufficiently strong and heavy in relation to the underlying soil, and in general diversion is likely to be realized only for soil (and not rock) sites and for horizontal (strike-slip) fault displacements. Berrill[50] studied this situation using a two-dimensional analysis for buildings of shallow foundations on both cohesionless and cohesive soil, finding limiting conditions for rupture diversion in terms of dimensionless parameters such as $B = b/h$, where $2b$ is the foundation width and h is the depth of the soil layer, and $Q = q'/\rho gh$, where q' is the uniform bearing stress under the foundation, and ρ is the density of the soil.

For cohesionless soils the analysis predicted rupture diversion if the dimensionless bearing stress Q exceeds about 0.2 to 0.4, depending on the ratio of base friction to internal friction in the soil, and on the value B. This bearing stress would usually not be met with buildings of less than four or five storeys in height.[50]

In conjunction with the above theoretical study, tests of models on sand were also carried out, which verified that rupture diversion can take place, and showed that end conditions are significant, making the two-dimensional analysis unconservative by a factor of perhaps 1.5 to 2.

As noted previously, large dams are sometimes constructed in location by topography and suitability of foundation conditions to sites traversed by active faults. This was the case with the Clyde Dam[51] in New Zealand. The minor fault in the river channel under the dam site was estimated to be capable of 0.2*m* horizontal displacement with a probability of occurrence of 1 in 100 to 1 in 1000 in a period of 100 years. This local fault displacement would occur as a consequence of an event of $M_{max} = 7.5$ occurring on a major fault lying within about 3 km of the site. The dam, of concrete gravity construction, was provided with a vertical construction joint along the line of the river channel fault, which could slip sympathetically to the design displacement of the fault.

Long *pipelines* for water supply, sewerage, or oil and gas supplies quite often have to cross active fault zones, e.g. in Alaska, California, and the Himalayas. Fortunately, because of their configuration, pipes are relatively amenable to design solutions for fault displacements. This may be achieved by providing loops and/or flexible connections, such as is done for pipework in buildings as illustrated in Section 11.3.7, or flexible joints and pipework on offshore oil and gas platforms and single-buoy moorings.

REFERENCES

1. EERI Committee on Seismic Risk, Haresh C. Shah, Chairman, 'Glossary of terms for probabilistic seismic-risk and hazard analysis', *Earthquake Spectra*, **1**, No. 1, 33–40 (1984).
2. Cornell, C. A., 'Engineering seismic risk analysis', *Bull. Seism. Soc. Amer.*, **58**, No. 5, 1583–1606 (1968).
3. Esteva, L., and Villaverde, R., 'Seismic risk, design spectra and structure reliability', *Proc. 5th World Conf. Earthq. Eng., Rome*, **2**, 2586–97 (1973).
4. McGuire, R. K., 'Seismic structural response risk analysis, incorporating peak response regressions on earthquake magnitude and distance', *Res. Report R74–51*, Dept of Civil Engineering, Massachusetts Institute of Technology (1974).
5. Katayama, I., Iwasaki, T., and Seaiki, M., 'Statistical analysis of earthquake acceleration response spectra', *Tran. Japanese Soc. Civ. Eng.*, **10**, 311–13 (1978).
6. McGuire, R. K., 'Adequacy of simple probabilistic models for calculating felt-shaking hazard, using the Chinese earthquake catalog', *Bull. Seism. Soc. Amer.*, **69**, No. 3, 877–92 (1979).
7. Shah, H. C., and Dong, W. M., 'A re-evaluation of the current seismic hazard assessment methodologies', *Proc. 8th World Conf. Earthq. Eng., San Francisco*, **II**, 247–54 (1984).
8. Valera, J. E., and Donovan, N. C., 'Incorporation of uncertainties in the seismic response of soils', *Proc. 5th World Conf Earthq. Eng., Rome*, **1**, 370–79 (1973).
9. California Institute of Technology, *Strong motion accelerograms, Digitized and plotted data, Vol. I—Uncorrected accelerograms*, Earthq. Eng. Res. Lab., Calif. Inst. Tech., Pasadena (Issued serially).
10. California Institute of Technology, *Strong motion accelerograms, Digitized and plotted data, Vol II—Corrected accelerograms and integrated ground velocity and displacement curves*, Earthq. Eng. Res. Lab., Calif. Inst. Tech., Pasedena (Issued serially).
11. California Institute of Technology, *Analysis of strong motion earthquake accelerograms, Vol. III—Response spectra*, Earthq. Eng. Res. Lab., Calif. Inst. Tech., Pasadena (Issued serially).

12. California Institute of Technology, *Analyses of strong motion earthquake accelerograms, Vol. IV—Fourier amplitude spectra*, Earthq. Eng. Res. Lab., Calif. Inst. Tech., Pasadena (Issued serially).
13. Jennings, P. C., Housner, G. W., and Tsai, N. C., *Simulated earthquake motions*, Earthq. Eng. Res. Lab., Calif. Inst. Tech., Pasadena, April, 1968.
14. Housner, G. W., 'Behaviour of structures during earthquakes', *J. Engineering Mechanics Division, ASCE*, **85**, No. EM4, 109–29 (1959).
15. Alford, J. L., Housner, G. W., and Martell, R. R., *Spectrum analyses of strong-motion earthquakes*, California Institute of Technology, Pasedena, 1951 (Revised 1964).
16. Watt, B. J., Boaz, I. B., Ruhl, A. J., Shipley, S. A., Dowrick, D. J., and Ghose, A., 'Earthquake survivability of concrete platforms', Paper 3159, *Proc. Offshore Technology Conference*, Houston, Texas (1978).
17. Hall, W. J., Nau, J. M., and Zahrah, F. T., 'Scaling of response spectra and energy dissipation in SDOF Systems', *Proc. 8th World Conf. Earthq. Eng., San Francisco*, **IV**, 7–14 (1984).
18. Newmark, N. M., and Hall, W. J., 'Seismic design criteria for nuclear reactor facilities', *Proc. 4th World Conf. Earthq. Eng., Santiago, Chile*, **2**, B-4, 37–50 (1969).
19. Blume, J. A., Newmark, N. M., and Kapur, K. K., 'Seismic design spectra for nuclear power plants', *J. Power Division, ASCE*, **99**, No. PO2, 287–303 (1973).
20. McGuire, R. K., Becker, A. M., and Donovan, N. C., 'Spectral estimates of seismic shear waves', *Bull. Seism. Soc. Amer.*, **74**, No. 4, 1427–40 (1984).
21. Berrill, J. B., Priestley, M. J. N., and Peek, R., 'Further comments on seismic design loads for bridges', *Bull. NZ Nat. Soc. for Earthq. Eng.*, **14**, 1, 3–11 (1981).
22. Singh, J. P., 'Earthquake ground motions: Implications for designing structures and reconciling structural damage', *Earthquake Spectra*, **1**, No. 2, 239–70 (1985).
23. American Petroleum Institute, *Recommended practice for planning, designing, and construction of fixed offshore platforms*, API-RP2A, 12th edn (1981).
24. Applied Technology Council, 'Tentative provisions for the development of seismic regulations for buildings', *ATC 3-06*, California (1978).
25. Seed, H. B., Ugas, C., and Lysmer, J., 'Site dependent spectra for earthquake resistant design', *Report No. EERC 74-12*, Earthquake Engineering Research Center, University of California, Berkeley (1974).
26. Bolt, B. A., Abrahamson, N., and Yeh, Y. T., 'The variation of strong ground motion over short distances', *Proc. 8th World Conf. Earthq. Eng., San Francisco*, **II**, 183–9 (1984).
27. Wolf, J. P., 'Seismic response to travelling shear wave including soil-structure interaction with base-mat uplift, *Proc. of Specialist Meeting*, 'The Anti-seismic Design of Nuclear Installations', Nuclear Energy Agency, OECD, Paris (1975).
28. Vanmarcke, E. H., and Harichandran, R. S., 'Models of the spatial variation of ground motion for seismic analysis of structures', *Proc. 8th World Conf. on Earthq. Eng., San Francisco*, **II**, 597–604 (1984).
29. Housner, G. W., and Jennings, P. C., 'The San Fernando California earthquake', *Int. J. Earthq. Eng. and Struct. Dynamics*, **1**, 5–32 (1972).
30. Reimer, R. B., Clough, R. W., and Raphael, J. M., 'Evaluation of the Pacoima dam accelerogram', *Proc. 5th World Conf. on Earthq. Eng., Rome*, **2**, 2328–37 (1973).
31. Bolotin, V. V., 'Statistical theory of a seismic design of structures' *Proc. 2nd World Conf. on Earthq. Eng., Tokyo*, **2**, 1365–74 (1960).
32. Amin, M., and Ang, A. H. S., 'A nonstationary stochastic model for strong-motion earthquakes', *Structural Research Series No. 306*, University of Illinois, Department of Civil Engineering, April, 1966.
33. Saragoni, G. R., and Hart, G. C., 'Nonstationary analysis and simulation of earthquake ground motions', *UCLA-ENG-7238*, Earthq. Eng. and Structures Lab., University of California, Los Angeles (1972).

34. Nau, R. F., Oliver, R. M., and Pister, K. S., 'Simulating and analyzing artificial nonstationary earthquake ground motions', *Bull. Seism. Soc. Amer.*, **72**, No. 2, 615–36 (1982).
35. Bycroft, G. N., 'White noise representation of earthquakes', *J. of the Engineering Mechanics Division, ASCE*, **86**, No. EM2, 1–16 (April 1960).
36. Housner, G. W., 'Characteristics of strong-motion earthquakes', *Bull. Seism. Soc. Amer.*, **37**, No. 1, 19–31 (Jan. 1947).
37. Werner, S. D., *A study of earthquake input motions for seismic design*, prepared for US Atomic Energy Commission by Agbabian-Jacobsen Associates, Los Angeles, June (1970).
38. Tajimi, H., 'A statistical method of determining the maximum response of a building structure during an earthquake', *Proc. 2nd World Conf. on Earthq. Eng., Tokyo*, **2**, 781–797 (July 1960).
39. Housner, G. W., and Jennings, P. C., 'Generation of artificial earthquakes', *J. of the Engineering Mechanics Division, ASCE*, **90**, No. EM1, 113–150 (Feb. 1964).
40. Boore, D. M., 'Stochastic simulation of high-frequency ground motions based on seismological models of the radiated spectra, *Bull. Seism. Soc. Amer.*, **73**, No. 6, Part A, 1865–94 (1983).
41. Housner, G. W., 'Important features of earthquake ground motion', *Proc. 5th World Conf. on Earthq. Eng., Rome*, **1**, CLIX-CLXVIII (1973).
42. Ruiz, P., and Penzien, J., 'Stochastic seismic response of structures, *J. of the Engineering Mechanics Division, ASCE*, **97**, No. EM2, 441–456 (April 1971).
43. Parton, I. M., 'Site response to earthquakes, with reference to application of microtremor measurements,'*Report No. 80*, School of Engineering, University of Auckland, (May 1972).
44. Hanks, T. C., and McGuire, R. K., 'The character of high frequency strong ground motion', *Bull. Seism. Soc. Amer.*, **71**, 2071–95 (1981).
45. Wyllie, L. A., Chamorro, F., Cluff, L. S., and Niccum, M. R., 'Performance of Banco Central related to faulting', *Proc. 6th World Conf. Earthq. Eng., New Delhi*, 2417–22 (1977).
46. Niccum, M. R., Cluff, L. S., Chamorro, F., and Wyllie, L. A., 'Banco Central de Nicaragua: a case history of a building that survived surface fault rupture', *Proc. 6th World Conf. Earthq. Eng., New Delhi*, 2423–8 (1977).
47. Berrill, J. B., 'Building over faults: A procedure for evaluating risk', *Earthq. Eng. and Struct. Dynamics*, **11**, 427–36 (1983).
48. Spellman, H. A., Stellar, J. R., and Shlemon, R. J., 'Risk evaluation for construction over an active fault, Sylmar, California', *Proc. 8th World Conf. on Earthq. Eng., San Francisco*, **I**, 15–22 (1984).
49. Hancox, G. T., Beanland, S., and Brown, I. R., 'Seismotectonic hazard evaluation for the Kawarau River Power Development', *NZ Geological Survey Report EG 384*, Dept of Scientific and Industrial Research, Lower Hutt, New Zealand, (July 1985).
50. Berrill, J. B., 'Two dimensional analysis of the effect of fault rupture on buildings with shallow foundations', *Int. J. Soil Dynamics and Earthquake Engineering*, **2**, No. 3, 156–60 (1983).
51. Williams, M. J., 'MWD confident Clyde Dam is "safe and economical"', *New Zealand Engineering*, **40**, No. 3, 23–8 (March 1985).

Chapter 5

Earthquake resistant design philosophy—choice of form and materials

5.1 INTRODUCTION

In this chapter, *which is the most important chapter in this book*, we have now reached the beginning of the design proper. The preceding four chapters are concerned with the gathering of information essential to the design as described in the flow chart of Figure (i) on page 2. This chapter discusses the broader aspects of design, to do with the basic concepts which govern the overall behaviour of the structure and which respond to the general demands of the owner's design brief.

Various aspects of this chapter apply to structures of all types: buildings, bridges, or industrial structures. It is emphasized that *poor design concepts cannot be made to perform well in strong earthquakes*, whereas good design concepts often perform well despite major shortcomings in analysis and detailing. This observation is addressed to the whole design team, which may include engineers of all disciplines and architects, depending on the nature of the project.

The need for co-operation between the various members of a design team is, of course, not restricted to earthquake resistant design, but the need for good conceptual design is especially important in this context. As well as its need for earthquake resistance *per se*, good design philosophy has become increasingly important in recent years for countering the dangers of errors and loss of design direction arising from rapidly growing complexities in analytical techniques and detailing requirements. It is thus apparent that decisions made by the design team at this conceptual stage will generally be more important than any other aspect of the design process. Hence this chapter is the most important in this book.

5.2 CRITERIA FOR EARTHQUAKE RESISTANT DESIGN

5.2.1 Function, cost, and reliability

The basic principle of any design is that the product should meet the owner's requirements, which may be reduced to just three criteria, i.e.

(1) Function;
(2) Cost; and
(3) Reliability.

While the terms *function* and *cost* are simple in principle, *reliability* concerns various technical factors relating to serviceability and safety. As the above three criteria are interrelated, and because of the normal constraints on cost, compromises with function and reliability generally have to be made. In considering the means of achieving the above requirements it is necessary to take into account both the limitations and the opportunities arising from the availability of construction materials and components and of construction skills. These considerations, of course, apply to any project regardless of earthquake design content, and, although noted because they are fundamental, further discussion of function and cost is outside the chosen subject area of this book. The criteria governing reliability in earthquake resistance are discussed below.

5.2.2 Reliability Criteria

5.2.2.1 General Serviceability and Safety Criteria

The term *reliability* is used here in its normal language qualitative sense and in its technical sense, where it is a quantitative measure of performance stated in terms of probabilities (of failure or survival). Aspects of the probabilistic ingredients of reliability control are discussed in other parts of this book, notably the evaluation of seismic hazards and the question of acceptable risks. The required reliability is achieved if enough of the elements of the design behave satisfactorily under the design earthquake (ground shaking and other geological hazards, listed in Section 3.1). The elements that may be required to behave in agreed ways during earthquakes include structure, architectural elements, equipment, and contents.

The design criteria governing the satisfactory behaviour or reliability of the above elements relate to one or more levels of loading which in some codes are referred to as *limit state design criteria*. These criteria vary widely for different elements. For structures or equipment a typical hierarchy of earthquake limit states is shown in Table 5.1, where column (A) is for *Serviceability* criteria and columns (B) and (C) are principally concerned with *Safety* through damage control. The choice of the terminology used to identify the hazard levels (B) and (C) is problematical, as usage varies in the literature. Referring to column B it is noted that the British concrete code[1] and the proposed revisions to the New Zealand loadings code[2] both use the term 'ultimate limit state' for the design loading and strengths used to determine the safety or strength of the structure. The word 'ultimate' could, of course, equally well apply to the more extreme loading state of Column (C).

Up to the mid-1980s it was common practice to design normal structures or equipment to meet two criteria:

Table 5.1. Hierarchy of limit state design criteria for different levels of earthquake hazard

		(A) Serviceability limit state Serviceability earthquake (or OBE)	(B) Ultimate limit state 'Usual' design earthquake	(C) Survivability limit state Survivability earthquake (or SSE)
Normal structures or equipment	Response condition	(1) Undamaged Elastic	(2a) No collapse Post-yield cycling Limited deformation (Repairable) (2b) pre-yield	(3) Pre-collapse
	Typical return period (yr)	5–10 or $T_A \approx T_B/5$ (?)	50–100	500–1000
Critical structures or equipment	Response condition	(4) Pre-yield		(5) As (2a) or Pre-collapse
	Typical return period (yr)	500–1000		5000–10 000

(1) In moderate, frequent earthquakes the structure or equipment should be undamaged (Serviceability limit state, Box 1, Table 5.1);
(2) In strong, rare earthquakes the structure or equipment could be damaged but should not collapse (usual design earthquake limit state, Box 2a, Table 5.1).

The main intention of the second of these criteria was to save human lives, while the definition of the terms 'strong', 'rare', 'moderate', and 'frequent' have varied from place to place, and have tended to be rather imprecise because of the uncertainties in the state-of-the-art. Indeed, design has generally only been carried out explicitly for criterion (2), the assumption being made that, in so doing, it could be deemed that criterion (1) would automatically be satisfied. As it was not the practice to specify the strength of the serviceability earthquake, criterion (1) was of no quantitative use, although it has been an important feature of earthquake resistant design philosophy. However, with the growing concern about the cost of repairs for earthquake damage, and the improvement in our ability to specify loadings and carry out stress analyses, more specific attention to the serviceability limit state has become appropriate. This has been recognized in New Zealand, where the revised earthquake loadings code[2] includes the Serviceability Limit State as a design condition.

Table 5.1 gives the hierarchy of design criteria for two different classes of construction with acceptable risks near the two ends of the risk spectrum for engineered structures or equipment, i.e. what we have called *normal* and *critical*, respectively. For normal structures designed for the ultimate limit state with what we have called the 'usual' design earthquake (e.g. usual code design loads) there is a range of response condition criteria which need to be satisfied as represented by Boxes (2a) and (2b), depending on the degree of post-yield behaviour (i.e. ductility) that may be called upon. Box (2a), where some degree of ductility can be demanded, is the most common condition, while Box (2b), where the structural materials must stay in the elastic state, is generally reserved for brittle materials. However, this latter design criterion is sometimes invoked for construction in an intermediate risk class between normal and critical. As examples, some offshore oil platforms are designed to remain elastic in the 100-year return period earthquake,[3] and a multi-storey bank building in Los Angeles also was designed to this criterion at the owner's request.[4]

5.2.2.2 Criteria for post-yield behaviour

The following remarks refer particularly to Box (2a) of Table 5.1, where normal risk structures are being designed to the normal degree of seismic hazard and where the post-yield behaviour of the construction materials is being utilized, and some (unspecified) strength capacity is in reserve for larger earthquake loading. Traditionally, design criteria have concentrated on the important strength-related properties of ductility and energy absorption, which comprise a subject requiring further elucidation, as discussed in Section 6.6.2. This creates a complex physical situation where damage is, by definition, occurring, and limits on deformation and hence damage are both very important and difficult to ensure. He Guangquian *et al.*[5] note that deformations often control the design of multi-storey buildings, and have therefore proposed a design method based on deformation criteria rather than the traditional strength approach.

Conflicts between requirements of strength and deformation sometimes arise, because increasing the stiffness to reduce deformations often increases the earthquake response of the system, and may involve the use of more brittle components which reduces the amount of post-yield capacity available. A discussion of stiff versus flexible structures is given in Section 5.3.6. Some means of reducing the strength demands on structures or equipment, such as base isolation and rocking foundations (Section 5.4), also introduce special deformation considerations.

In Box (2a) of Table 5.1 is the criterion *repairable*, implying that normal risk construction should be repairable after the occurrence of the design earthquake. It is placed inside brackets in the table, because it was not a widespread requirement of earthquake codes at the time of writing. However, for bridge design in New Zealand, if the potential plastic hinge zones in piers are expected to form in inaccessible locations for repairs (e.g. underground), the bridge has to be designed to withstand higher loads. Also some code requirements implicitly reduce damage, e.g. limitations on drift. However, as noted above regarding the serviceability limit state, the growing concern over the costs of earthquake damage and the difficulty of repairing much post-yield damage, especially to non-structural elements, suggests that more attention should be given to repairability at the design stage. This, of course, could also help reduce the danger of life, which is the traditional fundamental design criterion, as noted above. However, the subject of repairability is a very difficult one, requiring consideration of the interplay between all components, structure, architecture, and equipment, and much research is required to provide more definitive design criteria.

It will be seen from the above discussion that the principles for reliable design for post-yield behaviour require that the system conforms to criteria for strength, deformation, and repairability, which in some cases are complex, conflicting, and/or ill-defined.

5.2.2.3 Survivability in extreme events

In addition to the two earthquake limit states discussed above we sometimes need to consider that of *survivability* in more extreme events. While this has long been a required design condition for critical facilities (notably nuclear power plants) there has recently been a tendency to consider the survivability of other types of construction with acceptable risk levels much nearer to the norm, e.g. offshore platforms[3,6] and a 30-storey building near San Francisco.[7] This practice has arisen because of the desire of some owners to improve the quantification of the reliability of their construction. This desire has arisen because (1) earthquake loads can be much stronger in rarer events than the usual design earthquake of typically 100-year return period, and (2) there is considerable uncertainty involved in predicting behaviour of structures in the post-yield condition.

In studying the behaviour of concrete offshore platforms for the Gulf of Alaska Watt *et al.*[6] designed their structure to remain elastic in the 'usual' design earthquake (return period of 100 years), and then found that moderate demands would be made on the ductile capacity of the structure in the survivability event. While some permanent residual displacements would have occurred, with some spalling of concrete cover, the platform would not only survive but would probably be repairable. Although the return period of the survivability event was not evident in this study, it was estimated by Dowrick[8] that the return period was in the range 1000–3000 years.

In designing the above-mentioned 30-storey reinforced concrete framed building Tai *et al.*[7] required the structure to remain elastic in the 'usual' design earthquake which had a return period of 73 years (50 percent exceedance in 50 years), and required that no major structural damage would occur in the survivability event which had a return period of 950 years (10 percent exceedance in 100 years).

It is of interest that in both the above structures the design was essentially governed by the 'usual' design earthquake rather than the survivability event.

As a further note on terminology it is pointed out that for the 30-storey building cited above, Tai *et al.*[7] use the term *maximum credible earthquake* rather than *survivability earthquake*. Because of its vagueness, the former term is one of a number of terms that are discouraged by the EERI Committee on Seismic Risk.[9]

5.2.2.4 Criteria for Critical Structures or Equipment

The design criteria for critical risk construction vary widely, depending on the nature of the facility concerned. In general they are similar in principle to those discussed above for normal risk construction, the main difference being that the various response conditions are required to occur at much lower probabilities, as indicated by the typical criteria given in Table 5.1.

Here again, terminology is varied. For instance, in the nuclear power industry the Survivability event is generally referred to as the Safe Shutdown Earthquake (SSE), and the Operating Basis Earthquake (OBE) may perhaps be best related to the Serviceability Limit State despite the nuclear industry's preoccupation with safety.

5.2.2.5 Effect of workmanship and buildability on reliability

Good workmanship, complying with design requirements, obviously is fundamental to reliability. Designs which are easy to build are more likely to conform to specification than less buildable construction. While these factors are difficult to quantify, the importance of buildable design in creating reliable earthquake resistance should be borne in mind throughout the design process.

5.3 PRINCIPLES OF RELIABLE SEISMIC BEHAVIOUR—FORM, MATERIAL, AND FAILURE MODES

5.3.1 Introduction

In seeking the optimum of the proposed construction designers should choose forms and materials that give the best failure modes in earthquakes with functional and cost requirements. The form or configuration of the construction is the geometrical arrangement of all of the elements, i.e. structure, architecture, equipment, and contents. The importance of form was first highlighted by the author[10] and the baton has been taken up from the architectural point of view by Arnold.[11,12] Following studies of the performance of buildings in earthquakes the Applied Technology Council[13] concluded that configuration and detailing may play the key roles in providing earthquake resistance, while further confirmation was provided by the 1985 Chilean earthquake.[14] Obviously similar principles apply not only to buildings but to other forms of construction as well.

In order to achieve reliable earthquake resistance the form of construction should be decided from consideration of the following factors:

(1) Simplicity and symmetry;
(2) Length in plan;
(3) Shape in elevation;
(4) Uniformity and continuity;
(5) Stiffness;
(6) Failure modes;
(7) Foundation conditions.

These topics are discussed below, together with the influence of construction materials and failure mode control on reliability. Foundation conditions are discussed in relation to stiffness and failure modes.

5.3.2 Simplicity and symmetry

Earthquakes repeatedly demonstrate that the simplest structures have the greatest chance of survival. There are three main reasons for this. First, our ability to understand the overall behaviour of a simple structure is markedly greater than it is for a complex one, e.g. torsional effects are particularly hard to predict on an irregular structure. Second, our ability to understand simple structural details is considerably greater than it is for complicated ones. Third, simple structures are likely to be more buildable than complex ones.

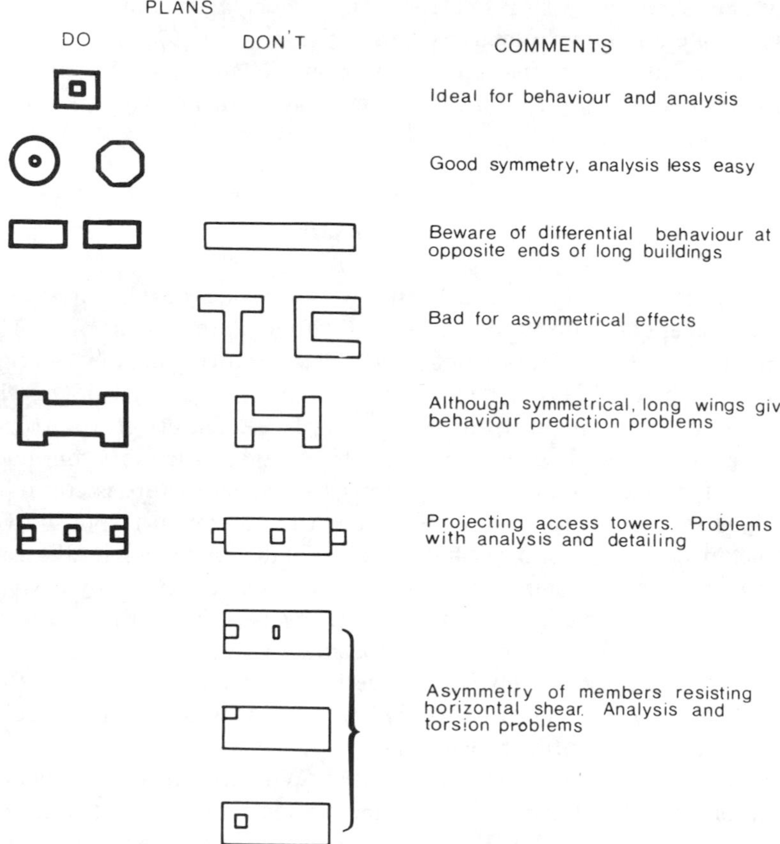

Figure 5.1 Simple rules for plan layouts of aseismic buildings. (*Only* with dynamic analysis and careful detailing should these rules be broken)

Symmetry is desirable for much the same reasons. It is worth pointing out that symmetry is important in both directions in plan (Figure 5.1) and helps in elevation as well. Lack of symmetry produces torsional effects which are sometimes difficult to assess and can be very destructive. This has been recognized in New Zealand by the code[2] requiring a three-dimensional analysis for buildings having more than a given degree of torsional eccentricity.

The introduction of deep re-entrant angles into the facades of buildings introduces complexities into the analysis which makes them potentially less reliable than simple forms. Buildings of H-, L-, T-, and Y-shape in plan have often been severely damaged in earthquakes, such as the Hanga Roa Building in Viña del Mar in the 1985 San Antonio Chile earthquake. This 1970 15-storey Y-shaped reinforced concrete building failed at the junction between one of the wings and central core area. Such plan forms should only be adopted if an appropriate three-dimensional earthquake analysis is used in the design.

External lifts and stairwells provide similar dangers, and should be uscd with the appropriate attention to analysis and design. In the 1971 San Fernando, California, earthquake external access towers at the Olive View Hospital were tied into the buildings they were meant to serve, and either collapsed or rotated so far as to be useless.

5.3.3 Length in plan

Structures which are long in plan naturally experience greater variations in ground movement and soil conditions over their length than short ones. These variations may be due to out-of-phase effects or to differences in geological conditions, which are likely to be most pronounced along long bridges where depth to bedrock may change from zero to very large. The effects on structure will differ greatly, depending on whether the foundation structure is continuous, or a series of isolated footings, and whether the superstructure is continuous or not. Continuous foundations may reduce the horizontal response of the superstructure at the expense of push–pull forces in the foundation itself (Section 4.3.3.3(ii)). Such effects should be allowed for in design, either by designing for the stresses induced in the structure or by permitting the differential movements to occur by incorporating movement gaps.

Movement gaps are relatively easy to design in bridge structures, but tend to be unreliable in buildings because of design, workmanship or cost difficulties. Insufficient gap width is often provided, perhaps because the true deformations in the post-elastic state were underestimated. Where adequate gap width is provided, in practice the gaps often become ineffective because of solids such as dirt or builder's rubble blocking them, and hammering between adjacent structures occurs. For example, there were many examples of damage from inproper articulation in buildings in the 1985 San Antonio, Chile, earthquake.[14] Also Wada *et al.*[15] have studied examples of collapse of buildings due to battering across movement gaps, analysing the dynamics of battering.

5.3.4 Shape in elevation

As indicated in Figure 5.2, very slender structures and those with sudden changes in width should be avoided in strong earthquake areas. Very slender buildings have high column forces and foundation stability may be difficult to achieve. Also higher mode contributions may add significantly to the seismic response of the superstructure. Height/width ratios in excess of about 4 lead to increasingly uneconomical structures and require dynamic analysis for proper evaluation of seismic responses. For comparison, in the design of latticed towers for wind loadings, aspect ratios in excess of about 6 become uneconomical.

Sudden changes in width of a structure, such as setbacks in the facades of buildings, generally imply a step in the dynamic response characteristics of the

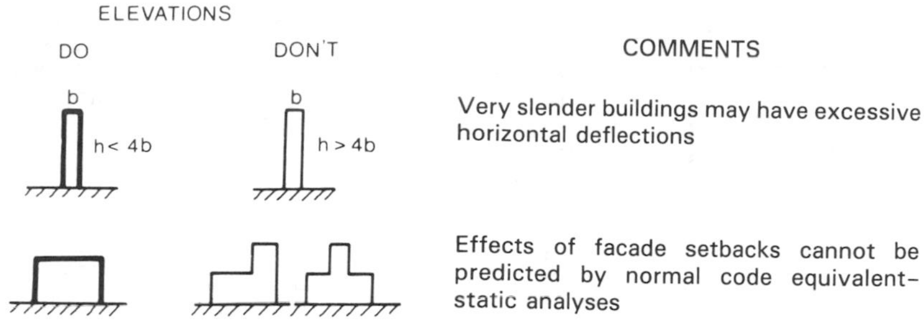

Figure 5.2 Simple rules for elevation shapes of aseismic buildings. (*Only* with dynamic analysis and careful detailing should these rules be broken)

structure at that height, and modern earthquake codes have special requirements for them. If such a shape is required in a structure it is best designed using dynamic earthquake analysis, in order to determine the stress concentrations at the notch and the shear transfer through the horizontal diaphragm below the notch.

5.3.5 Uniform and continuous distribution of strength and stiffness

This concept is closely related to that of simplicity and symmetry. The structure will have the maximum chance of surviving an earthquake if:

(1) The load bearing members are uniformly distributed;
(2) All columns and walls are continuous and without offsets from roof to foundation;
(3) All beams are free of offsets;
(4) Columns and beams are coaxial;
(5) Reinforced concrete columns and beams are nearly the same width;
(6) No principal members change section suddenly;
(7) The structure is as continuous (redundant) and monolithic as possible.

In qualification of the above recommendations it can be said that while they are not mandatory they are well proven, and the less they are followed the more vulnerable and expensive the structure will become.

While it can readily be seen how these recommendations make structures more easily analysed and avoid undesirable stress concentrations and torsions, some further explanation may be warranted. The restrictions to architectural freedom implied by the above sometimes make their acceptance difficult. Perhaps the most contentious is that of uninterrupted vertical structure, especially where cantilevered facades and columns supporting shear walls are fashionable. But sudden changes in lateral stiffness up a building are *not* wise (Figure 5.3), first because even with the most sophisticated and expensive computerized analysis the earthquake stresses cannot be determined adequately, and second, in the

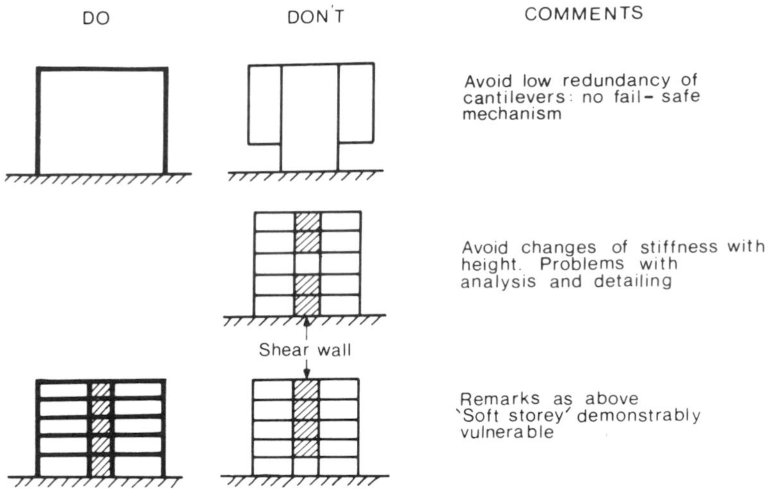

Figure 5.3 Simple rules for vertical frames in aseismic buildings

present state of knowledge we probably could not detail the structure adequately at the sensitive spots even if we knew the forces involved. The damage to the Sheraton–Macuto Hotel in the 1967 Caracas earthquake[16] illustrates this point, which is further discussed in Section 6.6.6.2.

This leads naturally into a discussion of the so-called 'soft storey' concept. In principle it is advantageous to isolate a structure from excessive ground movements by some sort of spongy layer. It has been proposed that a basically stiff structure could be protected from short-period vibrations by making the bottom storey columns relatively flexible (Figure 5.3). Unfortunately, many modern buildings of this type have not performed well in earthquakes. Studies have shown the soft storey concept to have theoretical as well as practical problems. Chopra *et al.*[17] found that a very low yield force level and an essentially perfectly plastic yielding mechanism are required in the first storey, and that the required displacement capacity of the first storey mechanism is very large. Consequently the soft storey concept is best avoided in moderate to strong earthquake areas, if obtained by the above means. It should be noted, however, that *base isolation* techniques may be described as a variation on the soft storey principle and provide reliable means of achieving some of the desired attributes of the conventional soft storey (Section 5.5). The above views have been supported by the architect Arnold,[12] who, however, notes that some apparent soft storeys may not be such, implying that they may not be dangerous—this claim needs substantiation.

Item (5) above recommends that in reinforced concrete structures, contiguous beams and columns should be of similar width. This promotes good detailing and aids the transfer of moments and shears through the junctions of the members concerned. Very wide, shallow beams have been found to fail near junctions with normal sized columns (Figure 5.4).

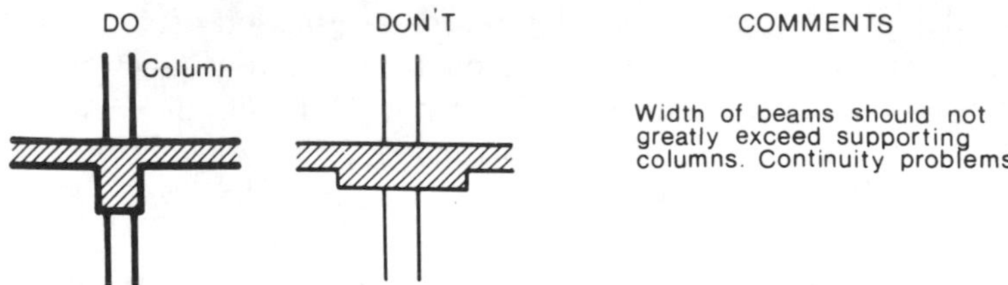

Figure 5.4 Simple rules for widths of beams and columns in aseismic reinforced concrete moment-resisting frames

The remaining main point worth elaborating is item (7) above, which says that a structure should be as redundant as possible. The earthquake resistance of an economically designed structure depends on its capacity to absorb apparently excessive energy input, mainly in repeated plastic deformations of its members. Hence the more continuous and monolithic a structure is made, the more plastic hinges and shear and thrust routes are available for energy absorption. This is why it is so difficult to make precast concrete structures work for strong earthquake motions.

Making joints monolithic and fully continuous is not only important for energy absorption; it also eliminates a frequent source of serious local failure due to high local stresses engendered solely by the large movements and rotations caused by earthquakes. This problem can arise in such places as the connection of major beams to slabs or minor beams, and beams to columns or corbels.

5.3.6 Appropriate stiffness

In designing construction to have reliable seismic behaviour the design of structures to have appropriate stiffness is an important task which is often made difficult because so many criteria, often conflicting, may need to be satisfied. The criteria for the stiffness of a structure fall into three categories, i.e. the stiffness is required:

(1) To create desired vibrational characteristics of the structure (to reduce seismic response, or to suit equipment or function);
(2) To control deformations (to protect structure, cladding, partitions, services);
(3) To influence failure modes.

5.3.6.1 Stiffness to suit required vibrational characteristics

Discussing item (1) above, first, we note that it would be desirable in general to avoid resonance of the structure with the dominant period of the site as indicated by the peak in the response spectrum (Figure 3.3). For example, short-period (stiff, low-rise) structures are good for long-period sites, i.e. those sites where the local soil is soft and deep enough to filter out much of the high-frequency ground motion, as in Mexico City. Similarly taller, more flexible structures will suit rock sites. Unfortunately, in terms of conventional construction, often it will not be possible to arrange the structure to benefit in this respect.

In industrial installations it may be necessary to have very stiff structures for functional reasons or to suit the equipment mounted thereon, and this will of course override any preference for seismic performance.

However, if we turn to unconventional techniques, notably the use of base isolation (Section 5.5), it is often possible to greatly modify the horizontal vibrational characteristics of a structure whether it is inherently stiff or flexible above the isolating layer. This not only allows the horizontal seismic responses to be greatly reduced but does not conflict with some functional needs for high stiffness (e.g. nuclear reactors or containment structures are inherently stiff).

5.3.6.2 Stiffness to control deformation

Referring to item (2) above, the importance of deformation control in enhancing safety and reducing damage and thus improving the reliability of construction in earthquakes is now well recognized (Section 5.2). The stiffness levels required to control damaging interaction between structure, cladding, partitions, and equipment vary widely, depending on the nature of components and the function of the construction, but stiff construction is obviously better than flexible in this regard. The seismic deformations of conventional construction can be greatly reduced by the use of base isolation (Section 5.5) or the Muto slitted wall (Section 5.5.7.1), so that relatively flexible moment-resisting frames may be able to satisfy the design deformation criteria, and P-delta column moments will be greatly reduced.

5.3.6.3 Stiffness affects failure modes

Different levels of stiffness can be created by such widely differing structural configurations that wide differences in potential failure modes arise. In general, stiffer construction implies the existence of less favourable failure modes from an earthquake design point of view, and this needs special design attention, as discussed in Section 5.3.8.

5.3.6.4 Stiff structures versus flexible

The terms 'stiff' and 'flexible' are relative ones, and must be interpreted with care. Some of their effects depend in part on the height of structure concerned. Table 5.2 summarizes some of the comparative merits of stiff and flexible construction, some of which have been discussed in the earlier parts of this section. A few further points of comparison are highlighted by discussing fully flexible structures.

Table 5.2. Comparative merits of stiff and flexible construction (which is not base isolated)

	Advantages	Disadvantages
Flexible Structures	(1) Specially suitable for short period sites, for buildings with long periods (2) Ductility arguably easier to achieve (3) More amenable to analysis	(1) Higher response on long period sites (2) Flexible framed reinforced concrete is difficult to reinforce (3) Non-structure may invalidate analysis (4) Non-structure difficult to detail
Stiff Structures	(1) Suitable for long period sites (2) Easier to reinforce stiff reinforced concrete (i.e., with shear wall) (3) Non-structure easier to detail	(1) Higher response on short period sites (2) Appropriate ductility not easy to knowingly achieve (3) Less amenable to analysis

Fully flexible structures may be exemplified by many modern beam and column buildings, where non-structure has been carefully separated from the frame. No significant shear elements exist, actual or potential: all partitioning and infill walls are isolated from frame movements, even the lift and stair shaft walls are completely separated. The cladding is mounted on rocker and roller brackets (of non-corrosive material). This type of completely ductile frame is currently fairly popular in Japan and California. Apart from the points listed in Table 5.2 it has further disadvantages. Floor-to-floor lateral drift and permanent set may be excessive after a moderate earthquake. In reinforced concrete the joint detailing is very difficult. There is no hidden redundancy (extra safety margin) provided by non-structure as in traditional construction.

In order to overcome the difficulties imposed by the deformability of more flexible construction in recent years there has been a trend to avoid using traditional moment-resisting frames by various means such as shear walls (various forms), bracing (various forms), base isolation, and energy absorbing devices. These will:

(1) Reduce lateral drift;
(2) Reduce reinforced concrete joint detailing problems;
(3) Help to ensure that plasticity develops uniformly over the structure;
(4) Prevent column failure in sway due to the $P \times \Delta$ effect (i.e. secondary bending resulting from the product of the vertical load and the lateral deflection).

In conclusion it can be said that in many situations either a stiff or a flexible structure can be made to work, but the advantages and disadvantages of the two forms need careful consideration when choosing between them.

5.3.7 Choice of construction materials

Reliability of construction in earthquakes is greatly affected by the materials used for the constituent elements of structure, architecture, and equipment. It is seldom possible to use the ideal materials for all elements, as the choice may be dictated by local availability or local construction skill, cost constraints, or political decisions.

Purely in terms of earthquake resistance the best materials have the following properties:

(1) High ductility;
(2) High strength/weight ratio;
(3) Homogeneity;
(4) Orthotropy;
(5) Ease in making full strength connections.

Generally the larger the structure, the more important the above properties are. By way of illustration the applicability of the major structural materials to buildings is given in Table 5.3. The term 'good reinforced masonry' refers to properly detailed hollow concrete block as discussed in Section 8.4.

Table 5.3. Suitability of construction materials for moderate to high earthquake loading

	Type of building		
	High-rise	Medium-rise	Low-rise
Best Structural materials in approximate order of suitability Worst	(1) Steel (2) *In situ* reinforced concrete	(1) Steel (2) *In situ* reinforced concrete (3) Good precast concrete* (4) Prestressed concrete (5) Good reinforced masonry*	(1) Timber (2) *In situ* reinforced concrete (3) Steel (4) Prestressed concrete (5) Good reinforced masonry* (6) Precast concrete (7) Primitive reinforced masonry

*These two materials only just qualify for inclusion in the medium-rise bracket. Indeed many earthquake engineers would not use either material. In Japan masonry is not permitted for buildings of more than three storeys.

Most fully precast concrete systems are *not* suitable for highly ductile earthquake resistance, because of the difficulty of achieving a monolithic and continuous structure.

The order of suitability shown in Table 5.3 is, of course, far from fixed, as it will depend on many things such as the qualities of materials as locally available, the type of structure, and the skill of the local labour in using them.

All these factors being equal, there is arguably little to choose between steel and *in situ* reinforced concrete for medium-rise buildings, as long as they are both well designed and detailed. For tall buildings steelwork is generally preferable, though each case must be considered on its merits. Timber performs well in low-rise buildings partly because of its high strength/weight ratio, but must be detailed with great care. Further discussion of the use of different materials is given in Chapters 7 to 10. Underdeveloped countries have special problems in selecting building materials, from the points of view of cost, availability, and technology. Further discussion of these factors has been made by Flores.[18]

The choice of construction material is important in relation to the desirable stiffness (Section 5.3.6). It is worth bearing in mind while choosing materials that if a flexible structure is required then some materials, such as masonry, are not suitable. On the other hand, steelwork is used essentially to obtain flexible structures, although if greater stiffness is desired diagonal bracing or reinforced concrete shear panels may sometimes be incorporated into steel frames. Concrete, of course, can readily be used to achieve almost any degree of stiffness.

A word of warning should be given here about the effect of non-structural materials on the structural response of buildings. The non-structure, mainly in the form of partitions, may enormously stiffen an otherwise flexible structure and hence must be allowed for in the structural analysis. This subject is discussed in more detail in Section 5.3.8.2.

5.3.8 Failure mode control

5.3.8.1 Failure modes of a complete system

Underlying the principle of failure mode control is the assumption that structural elements of a certain minimum strength will be provided, as required by the strength limit states of codes of practice (Section 5.2). This means that some overall probability of failure should not be exceeded. For structures which are required to be stiff, or for those which are inherently brittle, it may suffice simply to design the structure to remain elastic in the design earthquake, i.e. to conform with Box (2b) in Table 5.1.

However, in general, good design not only seeks to keep the overall probability of failure below a given level but it arranges the system such that less desirable modes of failure are less likely to happen than others. This increases the reliability of the design by decreasing the potential for damage and increasing the overall safety. The less desirable modes of failure for structures are:

(1) Those resulting in total collapse of the structure (notably through failure of vertical load-carrying members); and
(2) Those involving sudden failure (e.g. brittle or buckling modes).

While the above principle is good practice for *any* type of loading, it is particularly important for moderate to strong earthquake loading, because such loading is so much more demanding on structures than other environmental loadings, and generally involves stress incursions well into the post-elastic range in the parts of the structure. It is therefore highly desirable to control both the location and the manner of the post-elastic behaviour, i.e. to design for failure mode control.

In order to reduce the probability of occurrence of modes of failure (1) and (2) above, earthquake codes commonly have requirements that give added strength (i) to vertical load-carrying elements and (ii) to members carrying significant shear or compressive loads. Figure 5.5 illustrates alternative failure modes for a multi-storey moment-resisting frame. Clearly, the column sidesway mechanism is less desirable than the beam sidesway mechanism, as the former will lead to earlier total collapse than the latter. However, while it is possible and desirable to design so that plastic hinges form in beams rather than columns, it is not possible to eliminate plastic hinges from vertical structure completely. A number of potential plastic hinge zones are generally required in the lowest level of columns or walls even in the preferred failure mode, as in Figure 5.5(b).

While beam-hinging failure mechanisms are obviously preferable, the desired configuration for a structure sometimes dictates that a column-hinging failure mode cannot be avoided. In this case, in line with the above philosophy, some earthquake codes require that the structure be designed for a higher level of loading.

In the general case, the number of possible failure modes increases with increasing number of elements, and plastic hinges are likely to form at different locations in different earthquakes.[19] Detailing for plastic hinge control may not be sufficient based on solely linear frame analysis, because hinge positions do not necessarily occur only at the locations of maximum moment indicated in a linear analysis.

The number of possible failure modes is significantly reduced by suppressing the chances of occurrence of undesirable failure mechanisms, as discussed above, but some uncertainty over the manner of overall failure remains unless failure mode control is systematically carried out for all elements of the construction. With this objective in mind, Mueller[20] has studied the means and advantages of enforcing a preselected sequence of plastification in multi-storey frames, recommending that the number of plastic hinges of the desired failure mode are restricted to as few as possible. This should help to improve reliability by decreasing the possible range of variability of plastic hinge patterns.

In New Zealand failure mode control has been formalized in the earthquake code,[2,21] where it is called *capacity design*. This in effect requires the designer to impose a mode of overall failure on the structure, which demands that the

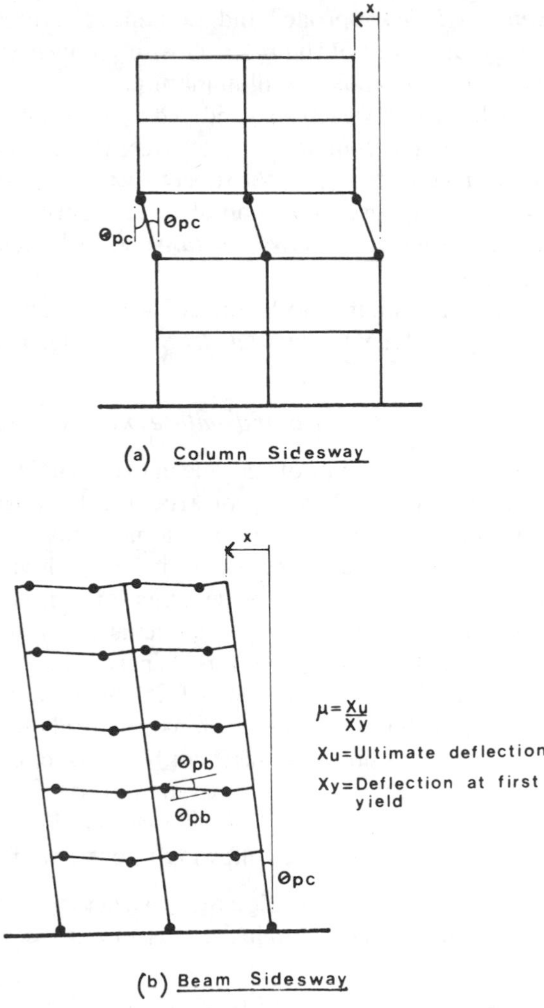

Figure 5.5 Alternative plastic hinge mechanisms for a typical multi-storey frame

parts of the structure that yield in the chosen failure mode are detailed for high energy absorption, and that the remainder of the structure has the strength capacity to ensure that no other yielding zones are likely to occur. This principle is straightforward to apply to most structures with very few members, but otherwise may be problematical (Section 7.3.3.1). In theory it not only helps to maximize safety but, by dictating where the damage will occur, it enables designers to improve the repairability of the structure and interacting elements.

While the merits of failure mode control are not in dispute, the manner in which it is achieved deserves further examination. For example, no comparison has been made of structures of any type designed by the full rigour of the

New Zealand *capacity design* approach and the same structures designed to the less global *toughness* approach of the USA. Cost-effectiveness comparisons in terms of relative reliability would be illuminating.

In the foregoing discussion we have considered how to control failure modes by structuring a system in certain ways. However, these good intentions are often frustrated if elements other than the superstructure, i.e. the part normally analysed for seismic response, are not also appropriately designed and constructed. Thus it is essential that *non-structure* and *substructure* have suitable forms, as discussed below.

Finally it is noted that failure mode control will be implemented through effective *workmanship*, and that the *buildability* of the design plays a crucial role.

5.3.8.2 Non-structure and failure mode control

Non-structural elements have an important role in the reliability or predictability of seismic response of any given type of construction. In considering the form of a structure it is important to be aware that some items which are normally non-structural become structurally very responsive in earthquakes. This means anything which will interfere with the free deformations of the structure during an earthquake. In buildings the principal elements concerned are cladding, perimeter infill walls, and internal partitions. Where these elements are made of very flexible materials, they will not affect the structure significantly. However, very often it will be desirable for non-structural reasons to construct them of stiff materials such as precast concrete or blocks or bricks. Such elements can have a significant effect on the behaviour and safety of the structure. Although these elements may be carrying little vertical load, they can act as shear walls in an earthquake with the following important effects. They may;

(1) Reduce the natural period of vibration of the structure, hence changing the intake of seismic energy and changing the seismic stresses of the 'official' structure;
(2) Redistribute the lateral stiffness of the structure, hence changing the stress distribution;
(3) Cause premature failure of the structure usually in shear or by pounding;
(4) Suffer excessive damage themselves, due to shear forces or pounding.

The more flexible the basic structure is, the worse the effects will be; and they will be particularly dangerous when the distribution of such 'non-structural' elements is asymmetric or not the same on successive floors. Stratta and Feldman[22] have discussed some of the effects of infill walls during the Peruvian earthquake of May 1970.

In attempting to deal with above problems either of two opposite approaches may be adopted. The first is knowingly to include those extra shear elements into the official structure as analysed, and to detail accordingly. This method is appropriate if the building is essentially stiff anyway, or if a stiff structure is desirable for low seismic response on the site concerned. It means that the shear elements themselves will probably require aseismic reinforcement. Thus 'non-structure' is made into real structure. For notes on the analysis of such composite structures, see Section 6.6.6.

The second approach is to prevent the non-structural elements from contributing their shear stiffness to the structure. This method is appropriate particularly when a flexible structure is required for low seismic response. It can be effected by making a gap against the structure, up the sides and along the top of the element. The non-structural element will need restraint at the top (with dowels, say) against overturning by out-of-plane forces. If the gap has to be filled, a really flexible material must be used. Some advice on the detailing of infill walls is given in Sections 8.5 and 12.2.

Unfortunately, neither of the above solutions is very satisfactory, as the fixing of the necessary ties, reinforcement, dowels, or gap treatments is time-consuming, expensive, and hard to supervise properly. Also, flexible gap fillers will not be good for sound insulation.

Finally the client should be warned not to permit construction of solid infill walls without taking structural advice about the earthquake effects.

5.3.8.3 Substructure and failure mode control

Although the form of the substructure must have a strong influence upon the seismic response of structures, little comparative work has been done on this subject. The following notes briefly summarize what appears to be good practice at the present time.

The basic rule regarding the earthquake resistance of substructure is that *integral action* in earthquakes should be obtained. This requires adequate consideration of the dynamic response characteristics of the superstructure and of the subsoil. If a good seismic-resistant form has been chosen for the superstructure (Sections 5.3.1 to 5.3.6) then at least the plan form of the substructure is likely to be sound, i.e.;

(1) Vertical loading will be symmetrical;
(2) Overturning effects will not be too large;
(3) The structure will not be too long in plan.

As with non-seismic design, the nature of the subsoil will determine the minimum depth of foundations. In earthquake areas this will involve consideration of the following factors;

(a) Transmission of horizontal base shears from the structure to the soil;
(b) Provision for earthquake overturning moments (e.g. tension piles);
(c) Differential settlements (Figure 5.6);
(d) Liquefaction of the subsoil;
(e) The effects of embedment on seismic response.

The effects of depth of embedment are not fully understood at present (Section 6.3.4), but some allowance for this effect can be made in soil–structure

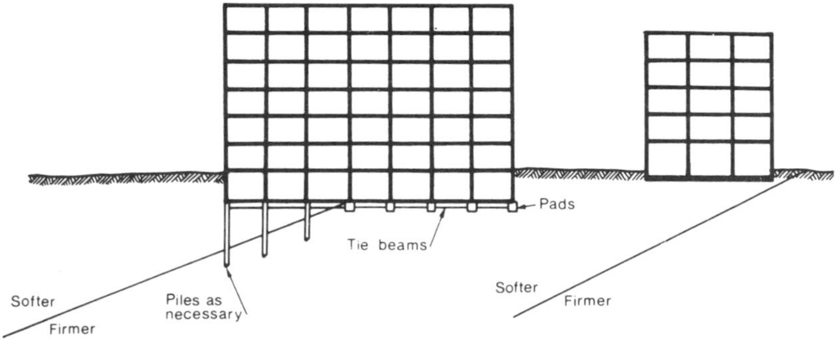

Figure 5.6 Typical structures founded on two types of soil, requiring precautions against differential seismic movements

interaction analyses (Section 6.3), or when determining at what level to apply the earthquake loading input for the superstructure analysis.

Three basic types of foundation may be listed as;

(1) Discrete pads;
(2) Continuous rafts;
(3) Piled foundations.

Piles, of course, may be used in conjunction with either pads or rafts. Continuous rafts or box foundations are good aseismic forms only requiring adequate depth and stiffness. Piles and discrete pads require more detailed consideration in order to ensure satisfactory integral action which deals with so many of the structural requirements implied in (1) to (3) and (a) to (e) above. Integral action should provide sufficient reserves of strength to deal with some of the differential ground movements which are not explicitly designed for at present. Where a change of soil type occurs under a structure (Figure 5.6), particular care may be necessary to ensure integral substructure action.

This discussion of substructure form is applicable to structures on softer soils only, as structures on rock are naturally integral per media of the rock itself. For a more detailed discussion of foundation design see Section 6.4.

Finally it is noted that piled foundations offer a special opportunity for failure mode control through base isolation, as discussed in Section 5.5.4.

5.4 SPECIFIC STRUCTURAL FORMS FOR EARTHQUAKE RESISTANCE

In the preceding sections of this chapter we have considered the principles underlying good earthquake resistant design, which should be applied to the specific structural forms utilized for a given project. The various structural forms in use around the world all have their strong and weak points, conforming better to some of the above principles than others. The main structural forms suitable for earthquake resistance are:

(1) Moment-resisting frames;
(2) Framed tube structures;
(3) Structural walls (shear walls);
(4) Concentrically braced frames;
(5) Eccentrically braced frames;
(6) Hybrid structural systems.

Design details of various aspects of the above forms are discussed under the relevant construction materials in Chapters 7 to 10, while a general overview of the seismic resistant attributes of these forms is given below.

5.4.1 Moment-resisting frames

Moment-resisting frames comprise one of the commonest forms of modern structure, in widespread use in building and industrial structures. Their great advantage for seismic resistance is that, by definition, they avoid potentially brittle shear failure modes, but they tend to sway excessively. They are made from steelwork, concrete, and timber.

5.4.2 Framed tube structures

The framed tube system is a special case of the moment-resisting frame, which usually consists of closely spaced wide steel columns combined with relatively deep beams. These frames are usually, but not only, located on the perimeter of the structure, and introduce more stiffness to overcome the problems of excessive horizontal deflection of orthodox moment-resisting frames, at the expense of a reduction in ductility. They have been widely used for tall buildings in high wind regions since the 1960s and more recently have been used in earthquake zones.[23] The framed tube system may be seen as a compromise between 'pure' moment-resisting and shear structures.

5.4.3 Structural walls (shear walls)

The terms *structural walls* or *shear walls* refer to structures in which the resistance to horizontal forces is principally provided by walls. These walls are usually constructed of concrete, masonry, timber, or steel, while other lesser structural materials such as gypsum, or composites are also encountered.

As mentioned earlier in this chapter, the great advantage of structural walls is the protection their natural stiffness offers to non-structure through limiting interstorey deflections. Earlier designs of concrete structural walls exhibited classical brittle shear failure modes in some earthquakes, particularly the

1964 Alaska event. Subsequent research has shown how these walls should be designed to overcome this problem, through appropriate reinforcing of ordinary *cantilever walls* or through the use of concrete *coupled walls* (Section 7.2.4.5).

Coupled walls make special use of lintel beams between adjacent walls (Figure 7.14) such that these coupling beams have ductility and energy-dissipating characteristics, which help to protect the walls from excessive damage.

A further variant of walls specifically developed for earthquake resistance is the Muto slitted wall, which is discussed in Section 5.5.7.1.

5.4.4 Concentrically braced frames

Concentrically braced frames are here defined as those where the centre lines of all intersecting members meet at a point (Figure 9.7). This traditional form of bracing is, of course, widely used for all kinds of construction such as towers, bridges, and buildings, creating stiffness with great economy of materials in two-dimensional trusses or three-dimensional space frames. Concentrically braced frames are constructed from steel, timber, and concrete, and composite forms are frequently met such as timber beams and columns with steel diagonals (Figure 10.11).

The bracing may take the form of either a single diagonal in a bay or a double bracing in an X shape (Figure 9.7). Braced frames have the advantage over moment-resisting frames, of having smaller horizontal deflections in moderate earthquakes, but are more inclined to undesirable buckling modes and have less reliable ductility.

If the diagonals are very slender and hence capable of tensile resistance only, as is often the case in steel construction, the seismic resistance is not as good as when the bracing is capable of compressive as well as tensile resistance. This is partly because in tension-only bracing there is a greater tendency for incremental permanent deflections to occur in one direction only. Also shock loadings tend to occur as bracings straighten from the buckled zero-load state to the tensile load-carrying state.

5.4.5 Eccentrically braced frames

Traditional design of trussed structures lays great importance on keeping the forces in the structure to axial only, avoiding moments by ensuring that the centre-lines of all intersecting members meet at a point, i.e. concentrically (Section 5.4.4). However, starting in the late 1970s, the concept of using deliberately eccentric bracing for earthquake resistance purposes has been found to have certain advantages, so far principally for steel structures, with major structures being designed this way (Figure 9.8).

In eccentrically braced frames the axial forces in the braces are transmitted to the columns through bending and shear in the beams, and, if designed correctly, the system possesses more ductility than concentrically braced frames while retaining the advantage of reduced horizontal deflections which braced systems have over moment-resisting frames. This system conforms to the requirement for good earthquake design of *failure mode control* (Section 5.3.8), with post-elastic behaviour being largely confined to selected portions of the beams and sudden failure modes being suppressed. However, the questions of the degree of damage that will be incurred in the floors, and the repairability of the floors and the beams, need examination.

5.4.6 Hybrid structural systems

Structures are often built in which the lateral resistance is provided by more than one of the above methods. The most common of these hybrid systems are those in which moment-resisting frames are combined with either structural walls or diagonally braced frames.

While hybrid systems are often unavoidable and can provide good seismic resistance, care must be taken to ensure that the structural behaviour is correctly modelled in the analysis. Interaction between the different components can be large and is not necessarily obvious, and many papers have been written on this subject. For example, for low-rise buildings it may be reasonable in many cases to assume that the walls or the braced bays resist the entire horizontal earthquake load, and the moment-resisting frame is not required to resist horizontal earthquake forces. However, deformations are still imposed on the moment-resisting members which require some seismic design consideration such as detailing for ductility.[24]

For a given plan layout, the contribution of the moment-resisting frame to lateral load resistance increases with height of structure, so that while the walls may take most of the horizontal shear in low-rise building, the moment-resisting frame becomes the dominant partner for very tall buildings.[25]

5.5 ENERGY ISOLATING AND DISSIPATING DEVICES

5.5.1 Introduction

Earthquake ground motions impart kinetic energy into structures, and the principles outlined above seek to control the location and extent of the damage caused by this energy. This philosophy can be extended beyond the structural forms described in Section 5.4 to any means which may further protect the structure by reducing the amount of energy which enters it. This may be done by:

(1) Limiting the energy entry at source (using energy avoiding devices at foundation level i.e. base isolation), and/or
(2) Providing energy-dissipating devices within the structure, from foundation level upwards.

Such devices have proved very effective in similar non-seismic applications for many years. Flexible bearings, for example, have been used to protect bridges from temperature movements, or to isolate machinery from structures, while damping devices have many energy-absorbing functions in mechanical engineering such as shock absorbers on motor vehicles. A very close parallel to seismic base isolation is found in the isolation of buildings from vibrations caused by underground trains. In 1985 there were over a hundred such buildings in Europe and Australia, a notable example being the vibration-sensitive concert hall, the Royal Festival Hall, built in London in 1951 on rubber bearings.

Suggestions of applying the same principles to earthquake engineering have been made for many years, a patent for such a system being taken out in 1909, but the complexities of verifying designs using any specific device have required considerable non-linear analysis skills and shake-table technology which only became available in the 1970s. The innovative research work was done in the New Zealand Physics and Engineering Laboratory,[26-30] with great interest subsequently being shown in the USA[31,32] and many other countries.[33] By 1985 a number of major structures had been built incorporating various devices, with design methods being well established,[29,32] and extensive further developments in this area may be anticipated.

5.5.2 Isolation from seismic motion

The principle of isolation is simply to provide a discontinuity between two bodies in contact so that the motion of either body, in the direction of the discontinuity, cannot be fully transmitted. The discontinuity consists of a layer between the bodies which has low resistance to shear compared with the bodies themselves. Such discontinuities may be used for isolation from horizontal seismic motions of whole structures, parts of structures, or items of equipment mounted on structures. Because they are generally located at or near the base of the item concerned, such systems are commonly referred to as *base isolation* (Figure 5.7).

The layer providing the discontinuity may take various forms, ranging from infinitely thin sliding surfaces (e.g. PTFE bearings), through rubber bearings a few centimetres thick, to flexible or lifting structural members of any height. To control the seismic deformations which occur at the discontinuity, and to provide a reasonable minimum level of damping to the structure as a whole, the discontinuity must be associated with energy-dissipating devices. The latter are also usually used for providing the required rigidity under serviceability loads, such as wind or minor earthquakes. Because vertical stiffness is generally required for gravity loads, seismic isolation is only appropriate for horizontal motions.

The soft layer providing discontinuity against horizontal motions cannot completely isolate the structures. Its effect is to increase the natural periods of vibration of the structure, and to be effective the periods must be shifted so as to reduce substantially the response of the structure. For example, a

BEARINGS LOCATED AT BOTTOM OF FIRST STORY COLUMNS

ADVANTAGES
- Minimal added structural costs
- Separation at level of base isolation is simple to incorporate
- Base of columns may be connected by diaphragm
- Easy to incorporate back-up system for vertical loads

DISADVANTAGES
- May require cantilever elevator pit

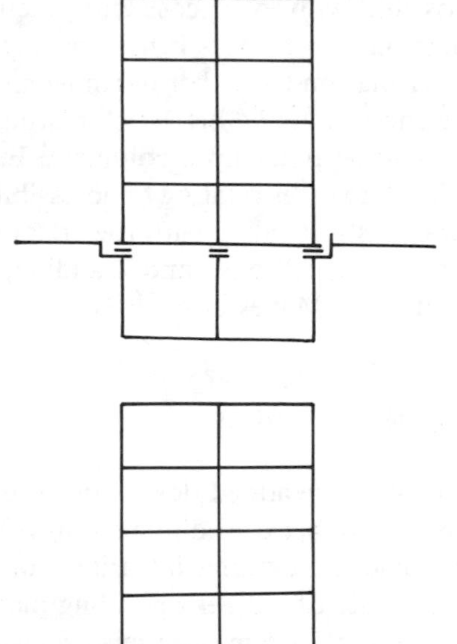

BEARINGS LOCATED AT TOP OF BASEMENT COLUMNS

ADVANTAGES
- No sub-basement requirement
- Minimal added structural costs
- Base of columns connected by a diaphragm at isolation level
- Backup system for vertical loads provided by columns

DISADVANTAGES
- May require cantilevered elevator shaft below first floor level
- Special treatment required for internal stairways below first floor level

BEARINGS LOCATED AT MID-HEIGHT OF BASEMENT COLUMNS

ADVANTAGES
- No sub-basement required
- Basement columns flex in double curvature and therefore may not be required to be as stiff as for bearings located at top or bottom

DISADVANTAGES
- Special consideration required for elevators and stairways to accommodate displacements at mid-story
- No diaphragm provided at isolation level
- Difficult to incorporate back-up system for vertical loads

BEARINGS LOCATED IN SUB-BASEMENT

ADVANTAGES
- No special detailing required for separation of internal services such as elevators and stairways
- No special cladding separation details
- Base of columns connected by diaphragm at isolatio level
- Simple to incorporate back-up system for vertical loads

DISADVANTAGES
- Added structural costs unless sub-basement required for other purposes
- Requires a separate (independent) retaining wall

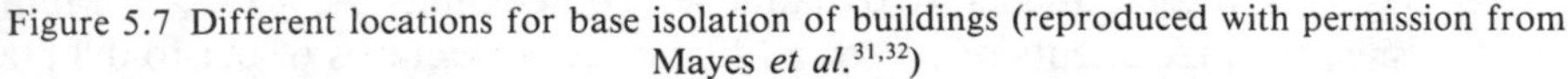

Figure 5.7 Different locations for base isolation of buildings (reproduced with permission from Mayes *et al.*[31,32])

three-storey building might typically have its fundamental period shifted from 0.3 to 1.0 s by being changed from fixed base to isolated. If the structure were located on a rock site, and had a design response spectrum as for rock in Figure 3.3(a), this would reduce the elastic response of the structure by a factor of about 5. However, a similar period shift for a structure on softer soil might not achieve any reduction in response, or could even result in an increased response, as may be inferred from the spectra for softer soil sites in Figure 3.3. Clearly, the shape of the design spectrum, the fixed base period, and the period shift are the three factors which determine whether base isolation has any force-reducing effect (or, indeed, the opposite!).

The *location of the isolating devices* obviously should be as low as possible to protect as much of the structure as possible. However, cost and practical considerations influence the choice of location. On bridges it may mostly be convenient to isolate only the deck, because isolation from thermal movements is required there anyway. In buildings the choice may lie between isolating at ground level, or below the basement, or at some point up a column.[31] Each of these locations has its advantages and disadvantages relating to accessibility and to the very important design considerations of dealing with the effects of the shear displacements on building services, partitions, and cladding, as described in Figure 5.7, which was derived from Mayes *et al.*[31,32]

5.5.3 Base isolation using flexible bearings

The most commonly used method of introducing the added flexibility for base isolation is to seat the item concerned on either rubber or sliding bearings. The energy dissipators (dampers) that must be provided may come in various forms. For use with standard bridge-type bearings made of rubber or sliding plates, any of the energy dissipators mentioned in Section 5.5.6 may be suitable, while the more recent invention of lead–rubber bearings is discussed below.

Lead–rubber bearings

The lead–rubber bearing[34,35] is conceptually and practically a very attractive device for base isolation, as it combines all of the required design features of flexibility and deflection control into a single component. As shown in Figure 5.8, it is similar to the laminated steel and rubber bearings used for temperature effects on bridges, but with the addition of a lead plug energy dissipator. Under cyclic shear loading the lead plug causes the bearing to have high hysteretic damping behaviour, of almost pure bilinear form (Figure 5.9). The high initial stiffness is likely to satisfy the deflection criteria for serviceability limit state loadings, while the low post-elastic stiffness gives the potential for a large increase in period of vibration desired for the ultimate limit state design earthquake.

As shown by Tyler and Robinson,[35] hysteretic behaviour is very stable under increasing cyclic displacements. In dynamic tests on bearings $280 \times 230 \times 113$ mm in size, shear displacements of up to ± 140 mm at frequencies of 0.1 to 0.3 Hz

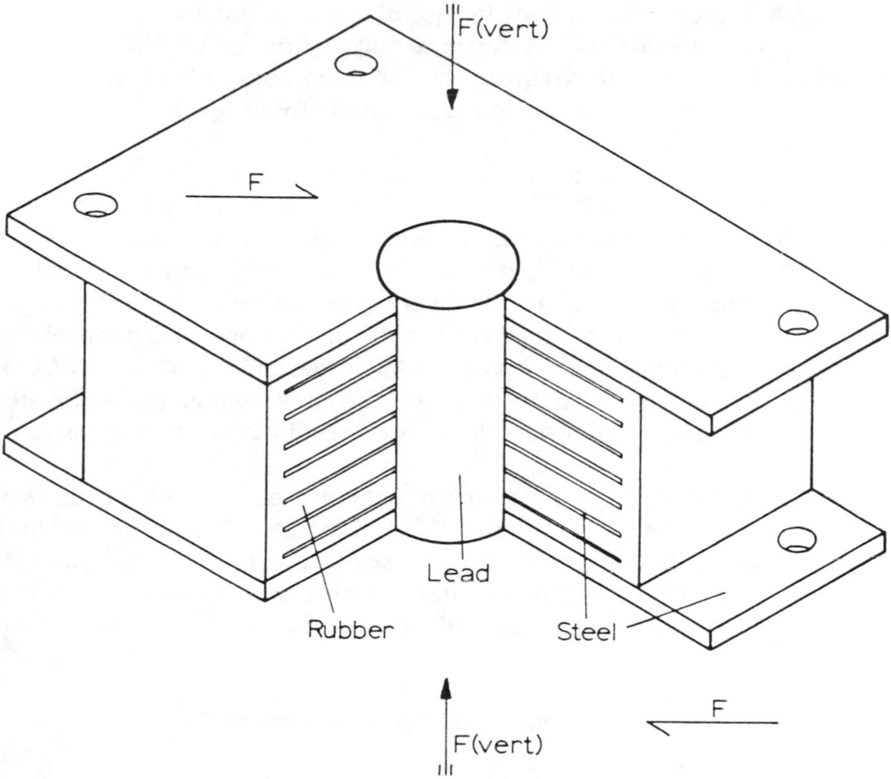

Figure 5.8 Construction of a patented lead–rubber bearing (after Robinson and Tucker[34])

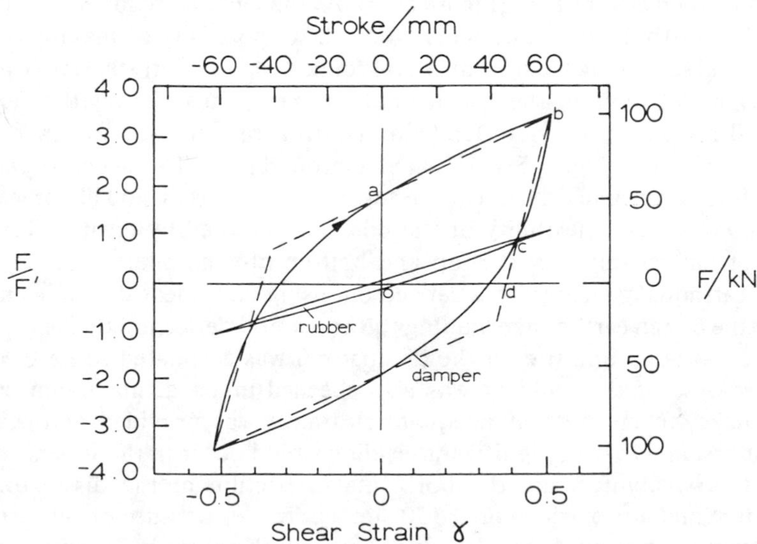

Figure 5.9 Typical hysteretic behaviour of a lead–rubber bearing (after Robinson and Tucker[34])

were applied giving shear strains in the rubber of up to ±200 percent. The weight of the structure on the bearings ranged from 35 to 455 kN. It was concluded that with peak strains in the rubber in excess of 100 percent, the bearings would continue to function satisfactorily for a sequence of very large earthquakes.

The first building to be built using lead–rubber bearings for seismic isolation was the William Clayton Building[36,37] in Wellington, New Zealand, designed *c*. 1978. It has a four-storey ductile moment-resisting frame, a section through the building being shown in Figure 5.10. The interstorey drifts calculated for the isolated building were about 10 mm and were uniform over the building's height. For comparison, the maximum drift for the non-isolated model was 52 mm per storey for the top two storeys. An overall deflection ductility factor of only $\mu = 1.6$ (μ is defined in Section 6.6.7.3(i)) was required for the isolated building, whereas $\mu = 7.6$ would have been required for the non-isolated condition.

Lead–rubber bearings have also been used in a rapidly growing number of bridges in New Zealand (over 30 in 1985) and the USA. These bearings have a wide range of applications where they are likely to lead not only to less damaged structures in earthquakes but also to cheaper construction.[31] Design procedures are well established, including methods using design graphs.[29,31,32]

5.5.4 Isolation using flexible piles and energy dissipators

An interesting alternative to the use of lead–rubber bearings is the isolation system used for Union House,[38] a 12-storey office block in Auckland, New Zealand, completed in 1983 (Figure 5.11). As the building required end-bearing piles about 10 m long, the designers took the opportunity of making the piles flexible and separating them from lateral contact with the soft soil layer overlying bedrock by surrounding them with a hollow sleeve, thus creating the flexibility required for base isolation. Deflection control was imposed by tapered steel energy dissipators (Figure 5.12) located at ground level. The structure was built of reinforced concrete except that the superstructure was diagonally braced with steel tubes. Lateral flexibility of the piles was attained by creating hinges of low moment resistance at the top and bottom of each pile.

The earthquake analysis was carried out using non-linear dynamic analysis. Under the design earthquake loading the horizontal deflection of the first floor relative to the ground (i.e. at the dissipators) was calculated to be ±60 mm. The response of the building was also checked under a 'maximum credible earthquake' to ensure that adequate clearance was provided at the energy dissipators, and that no significant yielding would occur in the superstructure. In this survivability state the horizontal deflection at the dissipators was ±130 mm and a provision for ±150 mm was made. Because of the structural discontinuity at ground level, the lift shaft and the bottom storey facade had to be supported from the first floor above ground level.

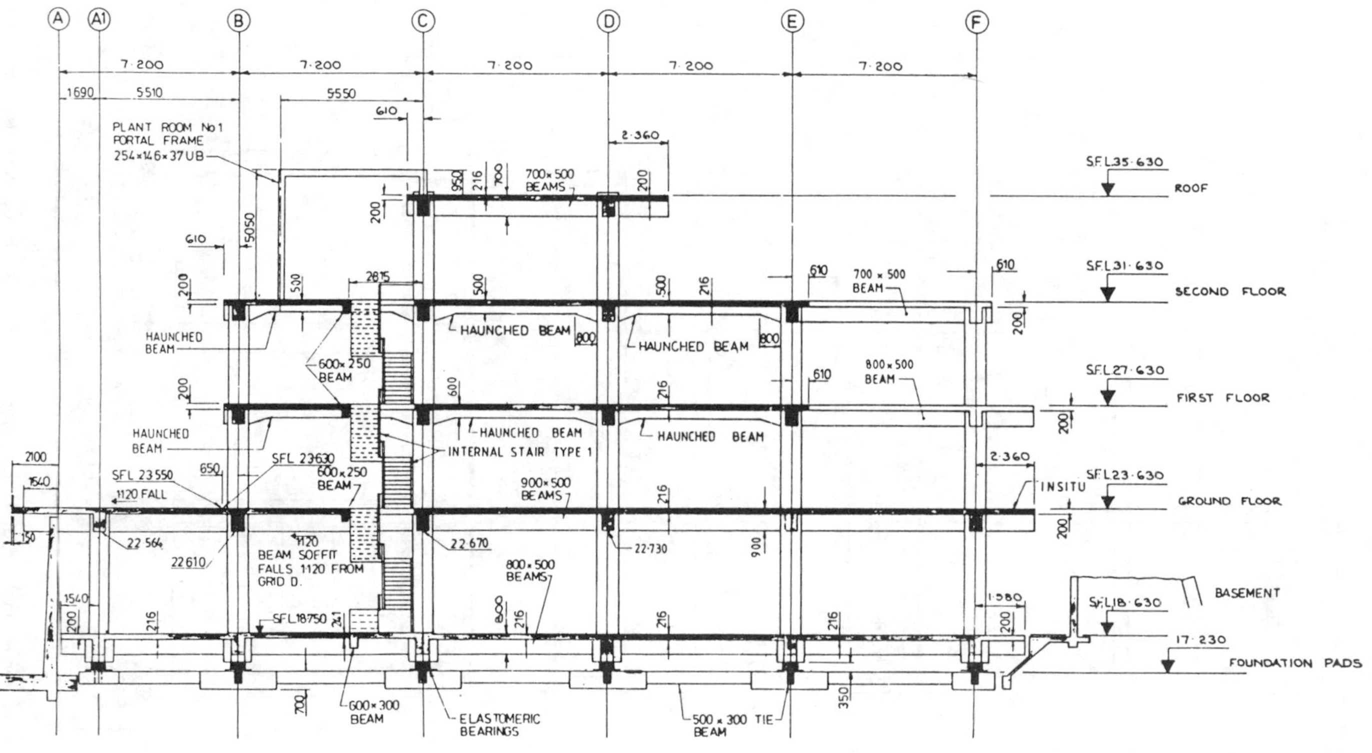

Figure 5.10 Section through the William Clayton Building, Wellington, New Zealand, the first building to be seismically isolated. The lead–rubber bearings are shown beneath the basement (after Megget[36,37])

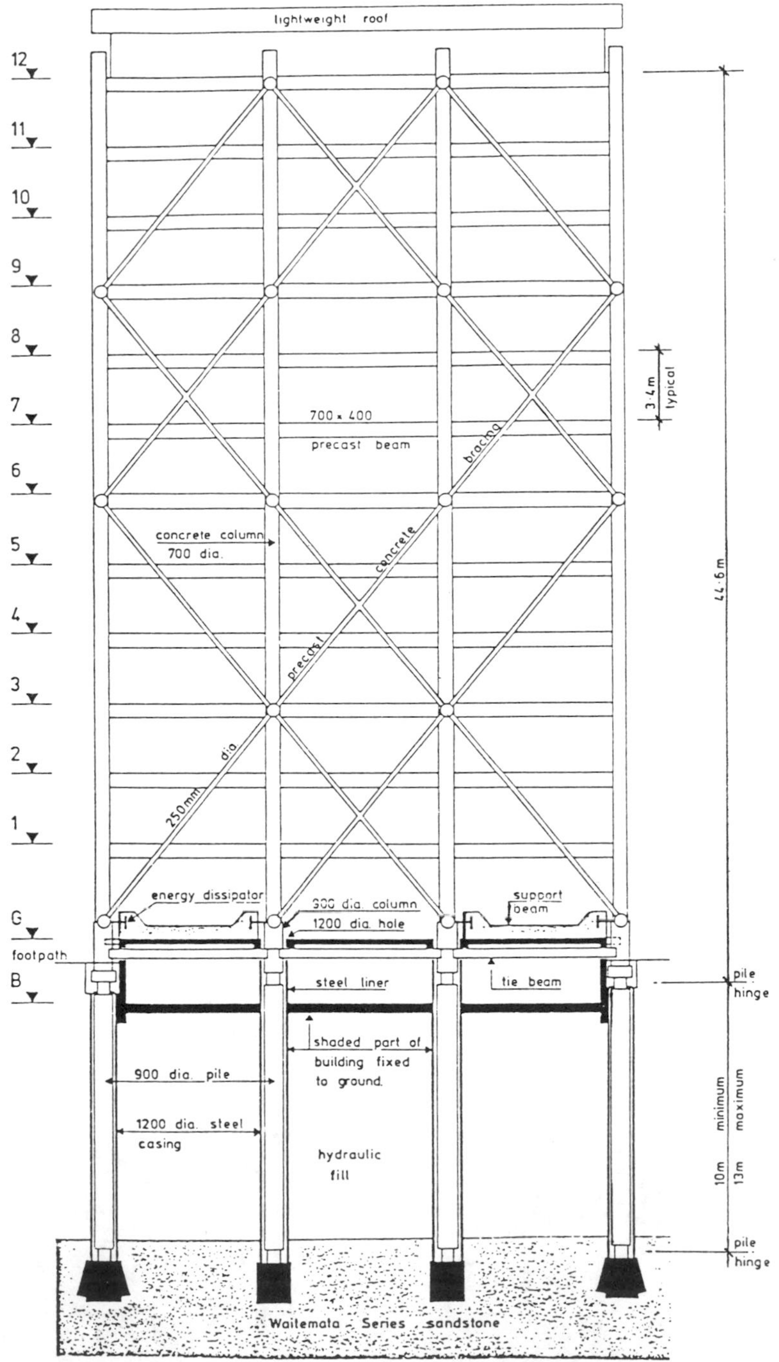

Figure 5.11 Section through Union House, Auckland, New Zealand, showing isolating piles and energy dissipators (after Boardman *et al.*[38])

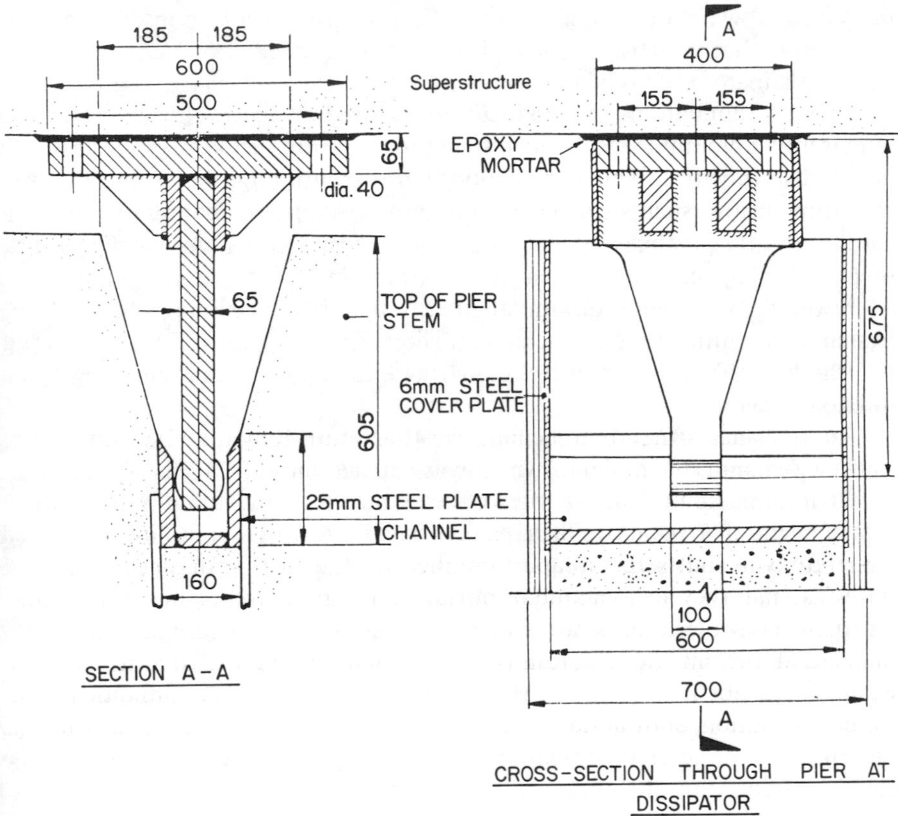

Figure 5.12 Cantilever steel plate dampers of the type used in Union House (Figure 5.11) and Dunedin Motorway Bridge[46]

Cost and time comparisons of the isolated and non-isolated equivalent structure estimated a capital cost saving of $300 000 and a construction time saving of 3 months, representing $150 000. Together these equal a substantial saving of nearly 7 percent in the total construction cost of $NZ 6.6 million.

5.5.5 Isolation using uplift

As well as the methods described in the preceding sections, the flexibility required to reduce seismic response in isolation systems may be obtained by allowing part of the structure to lift during large horizontal motions. This mechanism is referred to variously as uplift, rocking, or stepping, and involves a discontinuity of contact between part of the foundations and the soil beneath, or between a vertical member and its base.

The response of structures experiencing uplift has been a subject of increasing interest in recent years and a variety of systems have been studied,[39–44] generally showing a considerable reduction in structural responses compared with non-uplift alternatives. Some of the systems studied have not incorporated energy dissipators,[41-4] relying solely on uplifting columns or rocking of raft or local pad foundations to produce the desired effects. However, despite apparently favourable results such structures have not yet been enthusiastically adopted in practice. This is probably due to continuing design uncertainties regarding factors such as soil behaviour under rocking foundations in the design earthquake, the possible overturning of slender structures in survivability events, or possible impact effects when the separated interfaces slam together.[41]

However, with the addition of energy absorbers the above hazards are lessened, and utilization of the advantageous flexibility of uplift has been put to practical effect in completed constructions, a bridge and chimney being discussed below.

The first such structure to be built was the South Rangitikei Railway Bridge in New Zealand, the design of which was carried out *c.* 1971. The bridge deck is 320 m long, comprising six prestressed concrete spans, about half of which is at a height of 70 m above the riverbed. The piers consist of hollow reinforced concrete twin shafts 10.7 m apart coupled together with cross beams at three levels, so that they act as a kind of portal frame lateral to the line of the bridge. At their base, these shafts are seated on an elastomeric bearing (Figure 5.13) and lateral rocking of the portals is possible under the control of a steel torsion-beam energy dissipator[45] of the type shown in Figure 5.14. As with other forms of base isolation, substantial reductions in earthquake stresses are possible, as described for an early investigation of this bridge by Beck and Skinner,[39] the final configuration differing in detail but not in principle.[46]

The second example of the use of uplift is a free-standing industrial chimney 37.5 m, high in Christchurch, New Zealand, designed *c.* 1977.[40] Its base was 7.5 m wide, giving an aspect ratio of 5.0. Rocking of the base was allowed under the control of hysteretic dampers of the tapered steel cantilever type, similar to those used for the Union House building discussed in Section 5.5.4. The fundamental period of vibration prior to the onset of uplift was $T_o = 0.43$ s, and thereafter increases as a function of displacement, which in this case may be approximated to a single-mass elastoplastic resonator. For such a system it has been shown[47] that

$$T_1 = T_o \frac{1}{3\mu} \left[1 + 2\mu^{3/2} \right] \tag{5.1}$$

where $\mu = \left[\dfrac{\text{max. lateral displ. of single mass}}{\text{lateral displ. at start of rocking}} \right]$

This equation was used in the hand spectral analysis as part of the design. A non-linear dynamic analysis was also carried out using El Centro 1940 N–S component as the design earthquake, and it was found that the period T_1 lengthened to about 1 s during the strong shaking, which implies a much-reduced

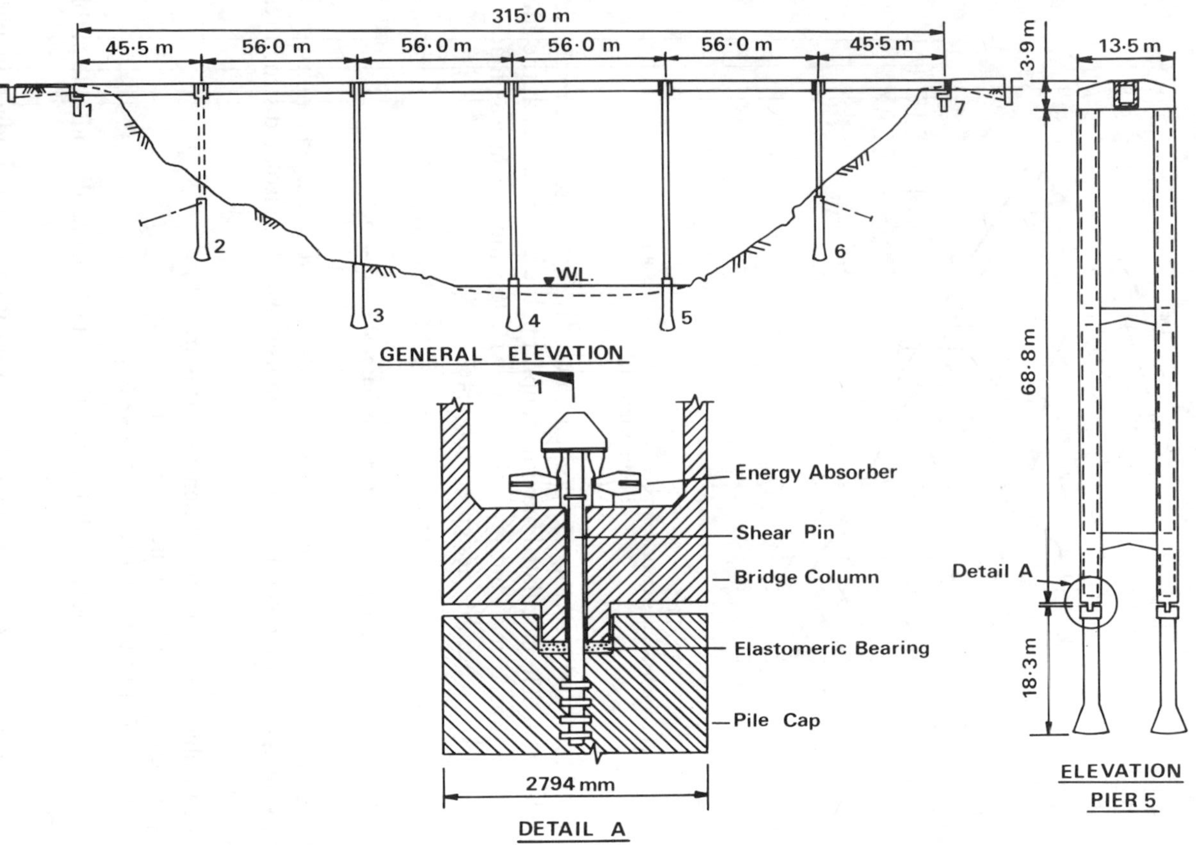

Figure 5.13 South Rangitikei railway bridge, New Zealand, showing locations of bearings and torsion beam energy dissipators[46]

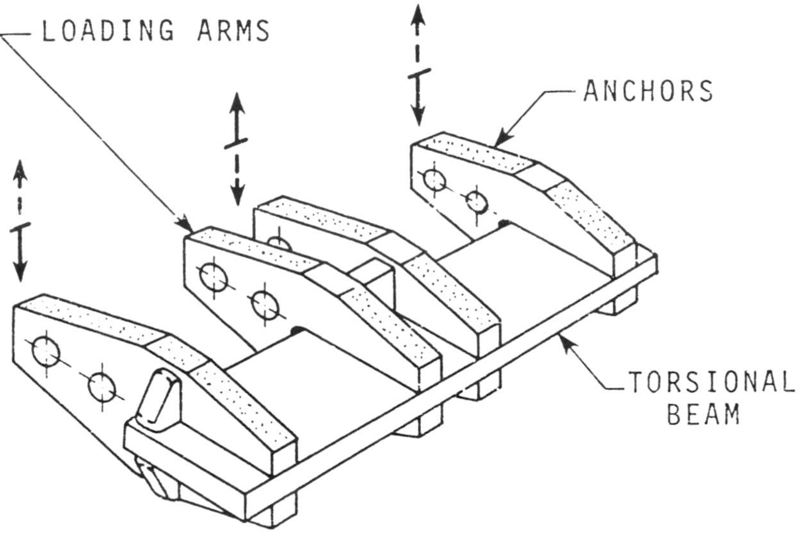

Figure 5.14 A torsion beam hysteretic damper. The arrows show the opposing actions of the structure and its support[45]

force response to this earthquake compared with the response in the restrained base condition. The calculated maximum uplift at the dampers (i.e. at the edge of the base) was 25 mm, but provision for twice this movement was made to allow for less favourable design earthquakes.

5.5.6 Energy dissipators for base-isolated structures

In the preceding sections on isolation methods we have discussed a number of energy dissipators (dampers) that have been used with base isolated structures, namely:

(1) Lead plugs, in lead-rubber bearings (Figure 5.8);
(2) Tapered steel plate cantilevers (Figure 5.12);
(3) Steel torsion-beam (Figure 5.14).

A variety of other devices have been investigated which are also suitable in this situation:

(4) Lead extrusion devices[48] which have been used in two bridges in New Zealand;[45]
(5) Flexural beam dampers,[46] which have been used in a bridge in New Zealand;
(6) Curved steel bars[49] or plates.[50]

A general overview of some of these energy dissipators has been made by Skinner *et al.*[46]

5.5.7 Energy dissipators for non-isolated structures

As discussed above, energy-dissipating devices are an essential component of base isolation systems, and they also may be used to reduce seismic stresses in non-isolated structures. Various forms of energy-dissipating devices have been developed for such structures, some of which are discussed below.

5.5.7.1 Reinforced concrete energy dissipators

A notable first entry to this field is the *Muto slitted wall.* Developed by Muto[51] in the 1960s, this has been used effectively in a number of tall buildings in Japan. It consists of a precast panel designed to fit between adjacent pairs of columns and beams of moment-resisting steel frames. The panel is divided by slits into a group of vertical ductile beam elements connected by horizontal ductile beams at the top and bottom, thus suppressing shear failure modes and creating a stiff energy-dissipating device. It is connected to the beams of the steel frame and effectively stiffens the building against wind load while providing high energy dissipation in larger earthquakes.

5.5.7.2 Energy dissipators in diagonal bracing

Diagonal bracings incorporating energy dissipators provide a structurally comparable alternative to the Muto slitted wall panel in that they control the horizontal deflections of the frame and also the locations of damage, thus protecting both the main structure and non-structure. A practical example is provided by a six-storey government office building[52] constructed in Wanganui, New Zealand, in 1980. This building obtains its lateral load resistance from diagonally braced precast concrete cladding panels (Figure 5.15), thus minimizing the amount of internal structure to suit architectural planning. The steel insert in the figure consists of a sleeve housing a specially fabricated steel tube 90 mm diameter and 1.4 m long, which was designed to yield axially at a given load level. A movement gap was provided through the surrounding structure, and buckling was prevented by the surrounding sleeve and concrete.

A number of devices to be connected to diagonal steel bracing show high energy-absorbing capabilities, such as the lead extrusion damper[48] also used in base isolated structures (Section 5.5.6). Tyler has tested[53] and produced design data[54] for steel rings which yield in bending and which are located at the intersection of the diagonals, while Pall and Marsh[55,56] have developed friction damped devices to suit both X- and K-bracing (but the reliability of the friction forces needs to be established).

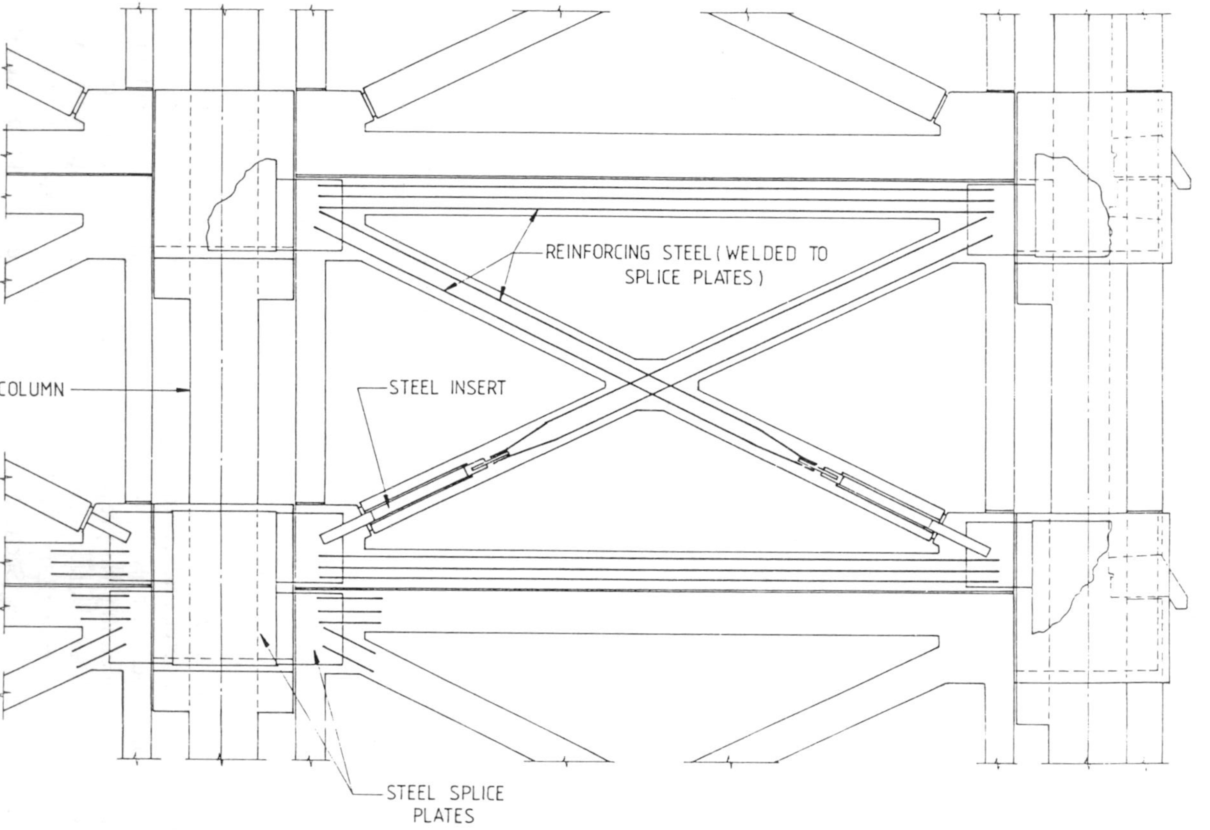

Figure 5.15 Energy dissipators housed in the bracing of precast cladding in a government building in Wanganui, New Zealand[52]

REFERENCES

1. British Standards Institution. *The structural use of concrete*, British Standard Code of Practice, CP 110:1972.
2. Standards Association of New Zealand. *Draft revisions for code of practice for general structural design and for design loadings*, DZ 4203 (1986).
3. American Petroleum Institute. *Recommended practice for planning, designing, and construction of fixed offshore platforms*, API-RP2A (1982).
4. Lord, James, Consulting Engineer, Los Angeles, Personal communication.
5. He Guangquian, Wei Lian, Zhong Yicun, and Dai Guoying, 'New procedure for aseismic design of multi-storey shear type building based upon deformation checking', *Proc. 8th World Conf. on Earthq. Eng., San Francisco*, **V**, 483–90 (1984).
6. Watt, B. J., Boaz, I. B., Ruhl, J. A., Shipley, S. A., Dowrick, D. J., and Ghose, A., 'Earthquake survivability of concrete platforms', *Proc. Offshore Technology Conference*, OTC 3159, Houston, Texas (1978).
7. Tai, J. C., Yang, Y. C., and Lin, T. Y., 'Design and construction of a thirty-storey concrete ductile framed structure Emeryville-East, San Francisco Bay', *Proc. 8th World Conf. on Earthq. Eng., San Francisco*, **V**, 371–8 (1984).
8. Dowrick, D. J., 'Survivability of structures in extreme magnitude earthquakes', in *Large Earthquakes in New Zealand*, Misc. Series No. 5, The Royal Society of New Zealand, 77–81 (1981).
9. EERI Committee on Seismic Risk, Haresh C. Shah, Chairman, 'Glossary of terms for probabilistic seismic-risk and hazard analysis', *Earthquake Spectra*, **1**, No. 1, 33–40 (1984).
10. Dowrick, D. J., 'Structural form for earthquake resistance', *Proc. 6th World Conf. on Earthq. Eng., New Delhi*, **2**, 1826–33 (1977).
11. Arnold, C., and Reitherman, R., *Building Configuration and Seismic Design*', John Wiley and Sons, New York (1982).
12. Arnold, C., 'Soft first stories: Truths and myths', *Proc. 8th World Conf. on Earthq. Eng., San Francisco*, **V**, 943–9 (1984).
13. Applied Technology Council, *An investigation of the correlation between earthquake ground motion and building performance*, ATC-10, Berkeley, California (Nov 1982).
14. Dowrick, D. J., 'Preliminary field observations of the Chilean earthquake of 3 March 1985', *Bull. NZ Nat. Soc. for Earthq. Eng.*, **18**, No. 2, 119–27 (1985).
15. Wada, A., Shinozaki, Y., and Nakamura, N., 'Collapse of building with expansion joints through collision caused by earthquake motion', *Proc. 8th World Conf. on Earthq. Eng., San Francisco*, **IV**, 855–62 (1984).
16. Sozen, M. A., Newmark, N. M., and Housner, G. W., 'Implications on seismic structural design of the evaluation of damage to the Sheraton–Macuto', *Proc. 4th World Conf. on Earthq. Eng., Chile*, **III**, J-2, 137–50 (1969).
17. Chopra, A. K., Clough, D. P., and Clough, R. W., 'Earthquake resistance of buildings with "soft" first storey', *Earthquake Engineering and Structural Dynamics*, **1**, No. 4, 347–55 (1973).
18. Flores, R., 'An outline of earthquake protection criteria for a developing country', *Proc. 4th World Conf. on Earthq. Eng., Chile*, **III**, J4, 1–14 (1969).
19. Sharpe, R. D., and Carr, A. J., 'The seismic response of inelastic structures', *Bull. NZ Nat. Soc. for Earthq. Eng.*, **8**, No. 3, 192–203 (1975).
20. Mueller, P., 'On aseismic design', *Proc. 8th World Conf. on Earthq. Eng., San Francisco*, **V**, 411–18 (1984).
21. Standards Association of New Zealand, *Code of practice for general structural design and design loadings for buildings*, NZS 4203:1984.
22. Stratta, J. L., and Feldman, J., 'Interaction of infill walls and concrete frames during earthquakes', *Bull. Seism. Soc. Amer.*, **61**, No. 3, 609–12 (1971).

23. Amin, N. R., and Louie, J. J. C., 'Design of multiple framed tube high rise steel structures in seismic regions', *Proc. 8th World Conf. on Earthq. Eng., San Francisco*, **V**, 347–54 (1984).
24. Selna, L., Martin, I., Park, R., and Wyllie, L., 'Strong and tough concrete columns for seismic forces', *J. Structural Division, ASCE*, **106**, No. ST8, 1717–34 (1980).
25. Ghoubhir, M. L., 'Earthquake resistance of structural systems for tall buildings', *Proc. 8th World Conf. on Earthq. Eng., San Francisco*, **V**, 491–8 (1984).
26. Skinner, R. I., Beck, J. L., and Bycroft, G. N., 'A practical system for isolating structures from earthquake attack', *Earthquake Engineering and Structural Dynamics*, **3**, No. 3, 297–309 (1975).
27. Skinner, R. I., Kelly, J. M., and Heine, A. J., 'Hysteretic dampers for earthquake-resistant structures', *Earthquake Engineering and Structural Dynamics*, **3**, No. 3, 287–96 (1975).
28. Skinner, R. I., and McVerry, G. H., 'Base isolation for increased earthquake resistance of buildings', *Bull. NZ Nat. Soc. for Earthq. Eng.*, **8**, No. 2, 93–101 (1975).
29. Blakeley, R. W. G., *et al.* (Working Group NZNSEE), 'Recommendations for the design and construction of base isolated structures', *Bull. NZ Nat. Soc. for Earthq. Eng.*, **12**, No. 2, 136–157 (1979).
30. Skinner, R. I., 'Base isolated structures in New Zealand', *Proc. 8th World Conf. on Earthq. Eng., San Francisco*, **V**, 927–34 (1984).
31. Mayes, R. L., Jones, L. R., Kelly, T. E., and Button, M. R., 'Design guidelines for base-isolated buildings with energy dissipators', *Earthquake Spectra*, **1**, No. 1, 41–74 (1984).
32. Dynamic Isolation Systems, *Seismic base isolation using lead-rubber bearings—Design procedures for buildings*, Dynamic Isolation Systems, Berkeley, California (1984).
33. Various authors, *Proc. 8th World Conf. on Earthq. Eng., San Francisco*, **V**, Pt 7.5 (1984).
34. Robinson, W. H., and Tucker, A. G., 'A lead-rubber shear damper', *Bull. NZ Nat. Soc. for Earthq. Eng.*, **10**, No. 3, 151–3 (1977).
35. Tyler, R. G., and Robinson, W. H., 'High strain tests on lead-rubber bearings for earthquake loadings', *Bull. NZ Nat. Soc. for Earthq. Eng.*, **17**, No. 2, 90–105 (1984).
36. Megget, L. M., 'Analysis and design of a base-isolated reinforced concrete frame building', *Bull. NZ Nat. Soc. for Earthq. Eng.*, **11**, No. 4, 245–54 (1978).
37. Megget, L. M., 'The design and construction of a base-isolated concrete frame building in Wellington, New Zealand, *Proc. 8th World Conf. on Earthq. Eng., San Francisco*, **V**, 935–42 (1984).
38. Boardman, P. R., Wood, B. J., and Carr, A. J., 'Union House—cross braced structure with energy dissipators', *Bull. NZ Nat. Soc. for Earthq. Eng.*, **16**, No. 2, 83–97 (1983).
39. Beck. J. L., and Skinner, R. I., 'The seismic response of a reinforced concrete bridge pier designed to step', *Earthq. Eng., Struct. Dyn.*, **2**, 343–58 (1974).
40. Sharpe, R. D., and Skinner, R. I., 'The seismic design of an industrial chimney with a rocking base', *Bull. NZ Nat. Soc. for Earthq. Eng.*, **16**, No. 2, 98–106 (1983).
41. Meek, J. L., 'Effects of foundation tipping on dynamic response', *J. Struct. Divn, ASCE*, **101**, No. ST7, 1297–1311 (1975).
42. Huckelbridge, A. A., and Clough, R. W., 'Seismic response of uplifting building frame', *J. Struct. Divn, ASCE*, **104**, No. ST8, 1211–29 (1978).
43. Huckelbridge, A. A., and Ferencz, R. M., 'Overturning effects in stiffened building frames', *Earthq. Eng. Struct. Dyn.*, **9**, No. 1, 69–83 (1981).
44. Yim, S. C. S. and Chopra, A. K., 'Earthquake response of buildings on Winkler foundation allowed to uplift', *Proc. 8th World Conf. on Earthq. Eng., San Francisco*, **V**, 275–82 (1984).
45. Kelly, J. M., Skinner, R. I., and Heine A. J., 'Mechanisms of energy absorption

in special devices for use in earthquake resistant structures', *Bull. NZ Soc. for Earthq. Eng.* **5**, No. 3, 63–88 (1972).
46. Skinner, R. I., Taylor, R. G., and Robinson, W. H., 'Hysteretic dampers for protection of structures from earthquakes', *Bull. NZ Nat. Soc. for Earthq. Eng.*, **13**, No. 1, 22–36 (1980).
47. Newmark, N. M., and Rosenblueth, E., *Fundamentals of Earthquake Engineering*, Prentice-Hall, Englewood Cliffs, NJ (1971).
48. Robinson, W. H., and Greenbank, L. R., 'Properties of an extrusion energy absorber', *Bull. NZ Nat. Soc. for Earthq. Eng.*, **8**, No. 3, 187–91 (1975).
49. Tyler, R. G., 'A tenacious base isolation system using round steel bars', *Bull. NZ Nat. Soc. for Earthq. Eng.*, **11**, No. 4, 273–81 (1978).
50. Steimer, S. F., and Chow, F. L., 'Curved plate energy absorbers for earthquake resistant structures', *Proc. 8th World Conf. on Earthq. Eng., San Francisco*, **V**, 967–74 (1984).
51. Muto, K., 'Earthquake resistant design of 36-storied Kasumigaseki building', *Proc. 4th World Conf. on Earthq. Eng., Chile*, **III**, J4, 15–33 (1969).
52. Matthewson, C. D., and Davey, R. A., 'Design of an earthquake resisting building using precast concrete cross-braced panels and incorporating energy-absorbing devices', *Bull. NZ Nat. Soc. for Earthq. Eng.*, **12**, No. 4, 340–45 (1979).
53. Tyler, R. G., 'Preliminary tests on an energy absorbing element for braced structures under earthquake loading', *Proc. 3rd Sth Pacific Regional Conf. on Earthq. Eng., Wellington, New Zealand*, **3**, 545–63 (1983).
54. Tyler, R. G., 'Further notes on a steel braced energy-absorbing element for braced frameworks', *Bull. NZ Nat. Soc. for Earthq. Eng.*, **18**, No. 3, 270–79 (1985).
55. Pall, A. S., and Marsh, C., 'Response of friction damped braced frames', *J. Struct. Divn, ASCE*, **108**, No. ST6, 1313–23 (1982).
56. Pall, A. S., 'Response of friction damped buildings', *Proc. 8th World Conf. on Earthq. Eng., San Francisco*, **V**, 1007–14 (1984).

Chapter 6

Seismic response of soils and structures

6.1 INTRODUCTION

This chapter is principally concerned with the determination of seismic motions, stresses, and deformations necessary for detailed design. The design earthquake (Chapter 4) is applied to the soil and/or the proposed form and materials of the structure (Chapter 5).

In earthquake conditions the relationship

'Subsoil — Substructure — Superstructure — Non-structure'

ideally should be analysed as a structural continuum. Although in practice this is seldom feasible, each of the parts should be seen as part of the whole when considering boundary conditions.

The problems involved in adequately representing seismic behaviour in theoretical analysis are numerous, and many compromises have to be made. In order to obtain the maximum benefit from any method of seismic analysis, an understanding of the dynamic response characteristics of materials is essential. For the adequate earthquake resistance of most structures, satisfactory post-elastic performance as well as elastic performance must occur.

6.2 SEISMIC RESPONSE OF SOILS

6.2.1 Dynamic properties of soils

Soil behaviour under dynamic loading depends on many factors, including:

(1) The nature of the soil;
(2) The environment of the soil (static stress state and water content); and
(3) The nature of the dynamic loading (strain magnitude, strain rate, and number of cycles of loading).

Some soils increase in strength under rapid cyclic loading, while others such as saturated sands or sensitive clays may lose strength with vibration.

This section provides background information on soil and rock properties required for dynamic response analysis of soil or soil–structure systems. Ways of estimating the basic parameters of shear modulus, damping, and shear wave velocity are suggested, and typical values of these and other parameters are given. A more detailed discussion of these properties may be found elsewhere.[1] In order to obtain appropriate design values of these parameters for a given site, suitable field and laboratory tests as discussed in Section 3.3 may be necessary.

6.2.1.1 Shear modulus

For soils the stress–strain behaviour of most interest in earthquakes is that involving shear, and, except for competent rock, engineering soils behave in a markedly non-linear fashion in the stress range of interest.

For small strains the shear modulus of a soil can be taken as the mean slope of the stress–strain curve. At large strains the stress–strain curve becomes markedly non-linear so that the shear modulus is far from constant but is dependent on the magnitude of the shear strain (Figure 6.1).

There are various field and laboratory methods available for finding the shear modulus G of soils. Field tests may be used for finding the shear wave velocity, v_s and calculating the shear modulus from the relationship

$$G = \rho v_s^2 \tag{6.1}$$

where ρ is the mass density of the soil.

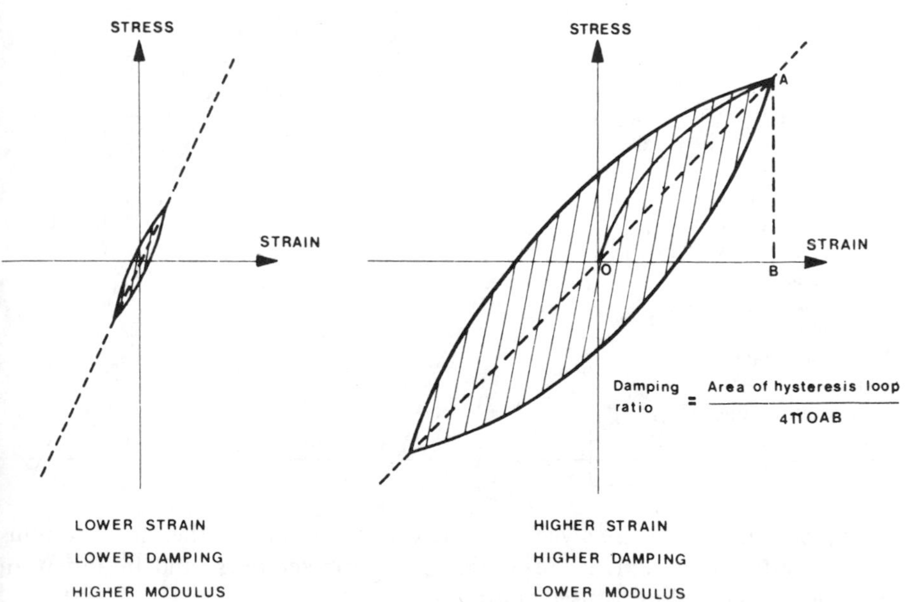

Figure 6.1 Illustration defining the effect of shear strain on damping and shear modulus of soils (after Seed and Idriss[2])

Table 6.1. Typical shear wave velocities (m/s) in foundation and building materials

	Depth of deposit		
Material	3–20 m	21–50 m	>50 m
Loose sand saturated	60		—
Fluvial sand	60	100	125
Clay	60	200	300
Silt	60	—	—
Silty clay	60	240	—
Marshland	80	—	—
Reclaimed land, recent	50	100	—
Sandy clay	100	250	—
Gravel, loose	100	300	600
Fine sand, saturated	110	—	—
Medium sand, uniform grading	100	140	—
Tertiary moist clay	130	—	—
Clay mixed with sand	140	—	—
Loam	150	200	—
Dense sand	160	—	—
Saturated medium sand	160	—	—
Argillaceous sand	170	—	—
Gravel with stones	180	—	—
Clay, saturated	190	—	—
Medium sand with fines	190	—	—
Clayey sand with gravel	200	—	—
Medium sand *in situ*	220	220	—
Marl	220	—	—
Dry clay	220	—	—
Compacted clay fill	240	—	—
Dry loess	260	—	—
Puddled clay heavily compacted	—	320	—
Coarse gravel tightly packed	420	—	—
Medium gravel	—	330	—
Quartz sandstone	—	—	780
Atlantic muck, ooze	—	—	1000–1500
Hard sandstones (mesozoic)	—	—	1200
Ice, glaciers	—	—	1600
Tuffaceous sandstone	—	—	2000
Concrete	—	—	2200
Mesozoic shales	—	—	2350
Granite (intact)	—	—	2700
Limestone (palaeozoic)	—	—	3340
Clay-slate (palaeozoic)	—	—	3610

Typical values of v_s are given in Table 6.1. Alternatively, the shear modulus may be estimated from the results of penetrometer tests, and in Japan an empirical relationship[3] is used, i.e.:

$$G = 1200N^{0.8} \tag{6.2}$$

where N is the blow count appropriate to the Japanese methods of testing.

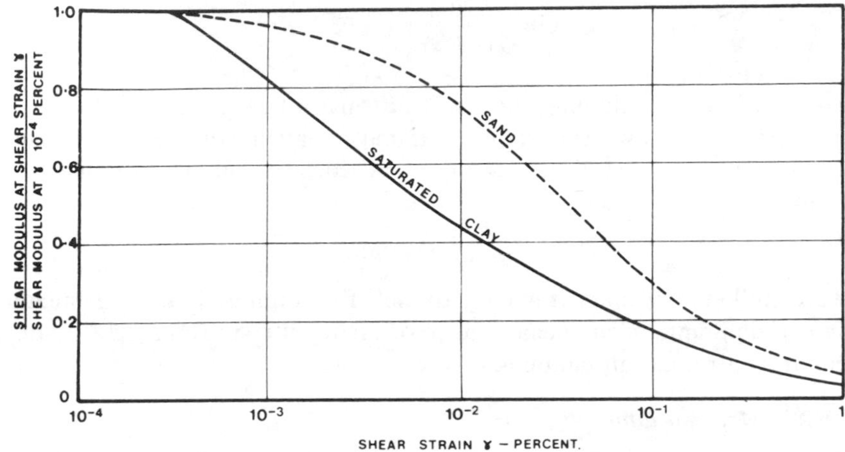

Figure 6.2 Average relationships of shear modulus to shear strain for sand and saturated clays (after Seed *et al.*[4,5])

Laboratory methods generally measure G more directly from stress–strain tests. It is clear from Figure 6.1 that the level of strain at which G is measured must be known. Average relationships of shear modulus to strain are shown for clay and sand in Figure 6.2 as produced by Seed *et al.*[4,5] The shear modulus for clays, while always having the general form shown in Figure 6.2, appears to vary as a function of the plasticity index.[6]

An alternative description of the non-linear stress–strain relationship for clays and sands is that used by Sugimura and Ohkawa.[3]

$$\gamma = \frac{\sigma}{G_o}\left\{ 1 + a \left| \frac{\sigma}{C_u} \right|^b \right\} \tag{6.3}$$

where γ = shear strain;
σ = shear stress;
G_o = initial shear modulus;
c_u = shear strength;
$a = 0.01(G_o/c_u) - 1.0$; and
$b = 1.4$ for clay, $b = 1.6$ for sand.

Shear strains developed during earthquakes may increase from about 10^{-3} percent in small earthquakes to 10^{-1} percent for large motions, and the maximum strain in each cycle will be different. Whitman[7] suggests that for earthquake design purposes a value of two thirds G measured at the maximum strain developed may be used. Alternatively, an appropriate value of G can be calculated from the relationship

$$G = \frac{E}{2(1+\nu)} \tag{6.4}$$

where E is Young's modulus and ν is Poisson's ratio. In the absence of any more specific data, low strain values of E may be taken from Table 6.3. Values of Poisson's ratio from Table 6.4 may be used in the above formula.

6.2.1.2 *Damping*

The second key dynamic parameter for soils is damping. Two fundamentally different damping phenomena are associated with soils, namely material damping and radiation damping.

6.2.1.2(i) *Material damping.*

Material damping (or internal damping) in a soil occurs when any vibration wave passed through the soil. It can be thought of as a measure of the loss of vibration energy resulting primarily from *hysteresis* in the soil. Damping is conveniently expressed as a fraction of critical damping, in which form it is refereed to as the damping ratio.

Considering the hysteresis loop on the right-hand side of Figure 6.1 it can be shown the equivalent viscous damping ratio may be expressed as

$$\xi = \frac{W}{4\pi\ \Delta W} \tag{6.5}$$

where W = energy loss per cycle (area of hysteresis loop);
ΔW = strain energy stored in equivalent perfectly elastic material (area OAB).

Published data on damping ratios are sparse, and consist only of values deduced from tests on small samples, or theoretical estimates. It should be appreciated that to date no *in situ* determinations of material damping have been made, and that damping ratios may only be used in analyses in a comparative sense. As dynamic soils analyses are required for some projects, at least for its qualitative information, a means of choosing values of material damping is required. Some material damping values are therefore given in Figure 6.3. These represent average values of laboratory test results on sands and saturated clays as presented elsewhere.[4,5] In the absense of any other information it may be reasonable to take the damping of gravels as for sand.

6.2.1.2(ii) *Radiation damping.*

In considering the vibration of foundations radiation damping is present as well as material damping. Radiation damping is a measure of the energy loss from the structure through radiation of waves away from the footing, i.e. it is a purely geometrical effect. Like material damping, it is very difficult to measure in the field. The theory for the elastic half-space has been used to provide estimates

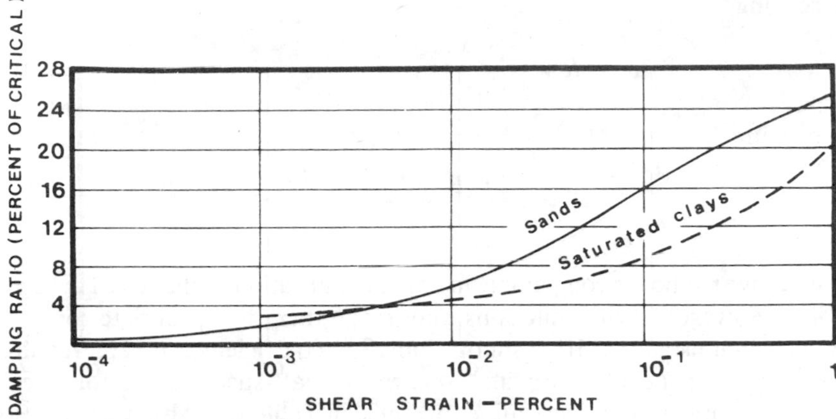

Figure 6.3 Average relationship of internal damping to shear strain for sands and saturated clays (after Seed *et al.*[4,5])

for the magnitude of radiation damping. Whitman and Richart[8] have calculated approximate values of radiation damping for circular footings for machines by this method and their results are reproduced in Figure 6.4

As with the values for material damping, the limitations of the values in Figure 6.4 must be emphasized. First, they involve the approximation that radiation damping is frequency independent, a reasonable assumption in some cases; second, because they are only theoretical values for a particular type of footing, they should be applied with circumspection. In the analysis of foundations of buildings the usefulness of Figure 6.4 may be for qualitative rather than quantitative assessments, but the following generalizations may be helpful. For horizontal and vertical translations, radiation damping may be quite large (>10 percent of critical), while for rocking or twisting it is quite small (about 2 percent of critical) and may be ignored in most practical design problems.

A further limitation of the half-space theory is that it takes no account of the reflective boundaries provided by harder soil layers or by bedrock at some distance vertically or horizontally from the structure. Any such reflection of radiating waves will naturally reduce the beneficial radiation damping effect. Various aspects of radiation damping are discussed in Section 6.3 below.

In Figure 6.4 m is the mass of the foundation block plus machinery, I_m is the mass moment of inertia of the foundation block plus machinery, R is the radius (or equivalent radius) of the soil contact area at the foundation base, ϱ is the mass density of the soil and ν is Poisson's ratio for the soil. For rectangular bases of plan size B×L the equivalent radius is given by the following;

for translation

$$R = \left(\frac{BL}{\pi} \right)^{1/2} \tag{6.6}$$

for rocking

$$R = \left(\frac{BL^3}{3\pi}\right)^{1/4} \tag{6.7}$$

for twisting

$$R = \left\{\frac{BL(B^2 + L^2)}{6\pi}\right\}^{1/4} \tag{6.8}$$

The above method is comparable in ease of application to that given in Section 6.3.4.3. However, their limitations may make them inappropriate for use in various circumstances. If dashpot damping coefficients, c, are required, frequency independent approximations to the half-space values for circular foundations are given by the simple formulae in Table 6.9. Alternatively, more widely applicable means of allowing for radiation damping, and a method of combining material and radiation damping, are given in Section 6.3.3.3.

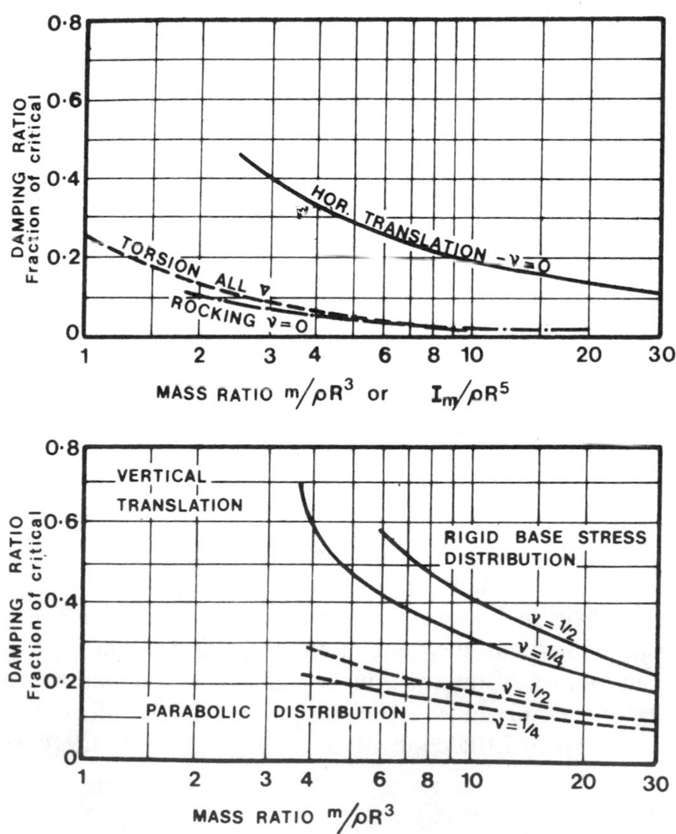

Figure 6.4 Values of equivalent damping ratio for radiation damping, of machines, derived from the theory of circular footings on elastic half-space (after Whitman and Richart[8])

6.2.1.3 Other basic soil properties

Typical values of shear wave velocity (v_s), soil density (ρ), modulus of elasticity (E), and Poisson's ratio (ν) are given below in Tables 6,1, 6,2, 6.3 and 6.4, respectively.

Table 6.2. Typical mass densities of basic soil types

Soil type	Mass density ρ (Mg/m^3)*			
	Poorly graded soil		Well-graded soil	
	Range	Typical value	Range	Typical value
Loose sand	1.70–1.90	1.75	1.75–2.00	1.85
Dense sand	1.90–2.10	2.00	2.00–2.20	2.10
Soft clay	1.60–1.90	1.75	1.60–1.90	1.75
Stiff clay	1.90–2.25	2.07	1.90–2.25	2.07
Silty soils	1.60–2.00	1.75	1.60–2.00	1.75
Gravelly soils	1.90–2.25	2.07	2.00–2.30	2.15

*Values are representative of moist sands and gravels and saturated silts and clays.

Table 6.3. Typical modulus of elasticity values for soils and rocks

Soil type	E (MN/m^2	E/c_u
Soft clay	up to 15	300
Firm, stiff clay	10 to 50	300
Very stiff, hard clay	25 to 200	300
Silty sand	7 to 70	
Loose sand	15 to 50	
Dense sand	50 to 120	
Dense sand and gravel	90 to 200	
Sandstone	up to 50,000	400
Chalk	5,000 to 20,000	2000
Limestone	25,000 to 100,000	600
Basalt	15,000 to 100,000	600

Note the values of E vary greatly for each soil type depending on the chemical and physical condition of the soil in question. Hence the above wide ranges of E valuc provide only vague guidance prior to tcst results being available. The ratio E/c_u may be helpful, if the undrained shear strength c_u is known, although the value of this ratio also varies for a given soil type. See elsewhere for information E values for Clays,[9] Sands,[10] and Rocks.[11]

Table 6.4. Typical values of Poisson's ratio for soils[13]

Soil type	Poisson's ratio, ν
Clean sands and gravels	0.33
Stiff clay	0.40
Soft clay	0.45

A value of 0.4 will be adequate for most practical purposes

6.2.2 Site response to earthquakes

6.2.2.1 Introduction

As outlined in Chapter 3, there is a great variety of possible geological and soil conditions at construction sites, which give rise to a variety of responses in earthquakes. The basic response phenomena which will be considered below are:

(1) Modification of bedrock excitation during transmission through the overlying soils (amplification or attenuation);
(2) Topographical effects;
(3) Settlement of dry sands;
(4) Liquefaction of saturated cohesionless soils.

The methods of analysing these responses vary in complexity, from simple empirical criteria to highly sophisticated analytical techniques. Regardless of the resources available, it should be borne in mind that knowledge of the real dynamical characteristics of the underlying soils is always incomplete, and the sophistication of the analyses used should not exceed the quality of the available data.

In the following discussion emphasis will be placed more on practical design procedures than on research methods. As has been stressed by Ambraseys,[12] there is a great need for simple methods correlated to field experience in the subject of soil dynamics.

6.2.2.2 Effect of soil layers on bedrock excitation

The presence of soil layers overlying bedrock modifies the excitation in a complex manner, with conflicting effects dependent on dynamic characteristics of the soil layers and the strength of the excitation. In many earthquakes the degree of damage to structures situated on soils has been reported as worse than that occurring on adjacent bedrock sites. Measured on the subjective intensity scales, the intensity may increase by 2 or 3 units compared with bedrock, depending on the soil type. Such measures of soil effects are very crude but give a broad indication of the effect of soil layers when amplification occurs, and typical figures are those given in Table 6.5.

In addition to changes in strength of shaking, dramatic changes in frequency content are caused by different soil layers, as shown by Figures 3.3 and 6.5. The deeper and softer the soil, the more the frequency content is modified. It should be noted that the spectra in Figure 3.3 are normalized to the same peak ground acceleration, so that this figure does not show the relative effect on maximum spectral acceleration of different soils for the same bedrock motion. Relative spectral effects are more apparent in Figure 6.5 (Section 6.2.2.2(iii).

It follows from the above that it is important to understand the dynamical properties of soils as structures in order to predict their response.

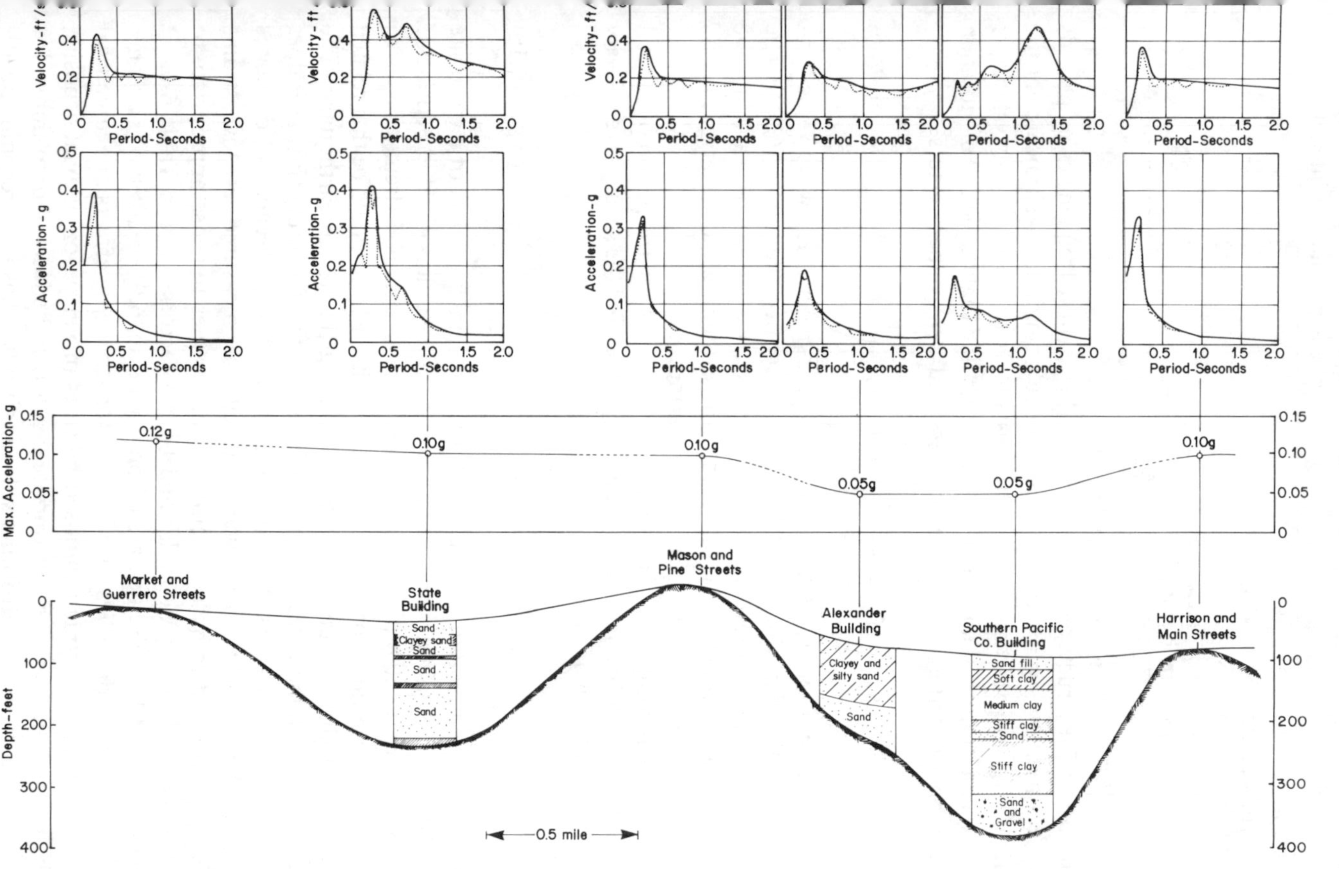

Figure 6.5 Soil conditions and response spectral characteristics of ground motions of six sites, San Francisco earthquake, 1957 (after Seed,[23] reproduced with permission from Van Nostrand Reinhold)

Table 6.5. Typical change in intensity due to soil layers (when amplification occurs)

Soil type	Average change in intensity
Bedrock	0
Firm sediments	+1
Medium sediments	+2
Saturated sediments, reclamations	+2½
Loose or recent sediments	+3

6.2.2.2(i) Period of vibration of soil sites

As discussed for structures in Section 1.4, the natural frequencies or periods of vibration of any dynamical system comprise a fundamental indicator of the dynamic response characteristics of the system. In the case of soil systems, if we consider a stratum of uniform thickness H we find that the period of vibration T is a simple function of stratum stiffness and density parameters

$$T_n = \frac{4H}{(2n-1)v_s} \tag{6.9}$$

where N is an integer, 1,2,3, . . . , and v_s is the mean shear wave velocity in the layer and a function of stiffness and density (equation (6.1)). The fundamental period, corresponding to $N=1$, occurs when a shear wave of wave length $4H$ passes through and is reflected in the stratum, while the larger integers $N=2,3,$ . . . , correspond to the higher harmonics.

Where a site is composed of more than one layer of soil the period of the soil may be estimated by using a weighted average value for the shear wave velocity in equation (6.9) such that

$$\bar{V}_s = \frac{\sum_1^n v_{si} H_i}{H} \tag{6.10}$$

In practice, in attempting to assess the fundamental period T_1 of a given site it is difficult to obtain a value from equation (6.9) unless reliable periods of similar sites are available for tuning purposes. The chief difficulty arises in deriving a suitable value for the shear wave velocity, which should be that related to the level of shear strain in the soil, G, during the design earthquake. The value of T for soil increases with increasing strength of shaking (just as it does for structures when stressed beyond the elastic state), because G decreases (Section 6.2.1.1). The shear wave velocity is measured at low strains (0.0001 percent) and in order to convert such values to those appropriate to strong shaking they may be multiplied by the factors given in Table 6.6.

The values of T for soil calculated from equation (6.9) are likely to be higher than reality, unless due allowance is made for stiffening effects of geometrical features such as the restraint imposed by sides of valleys on alluvial deposits, and by properly judging the appropriate depth H to bedrock. For example, for using equation (6.9) in Californian soil conditions it has been recommended[13] that the maximum depth to bedrock should be limited to 183 m, and bedrock is defined as having a low strain shear wave velocity of 760 m/s.

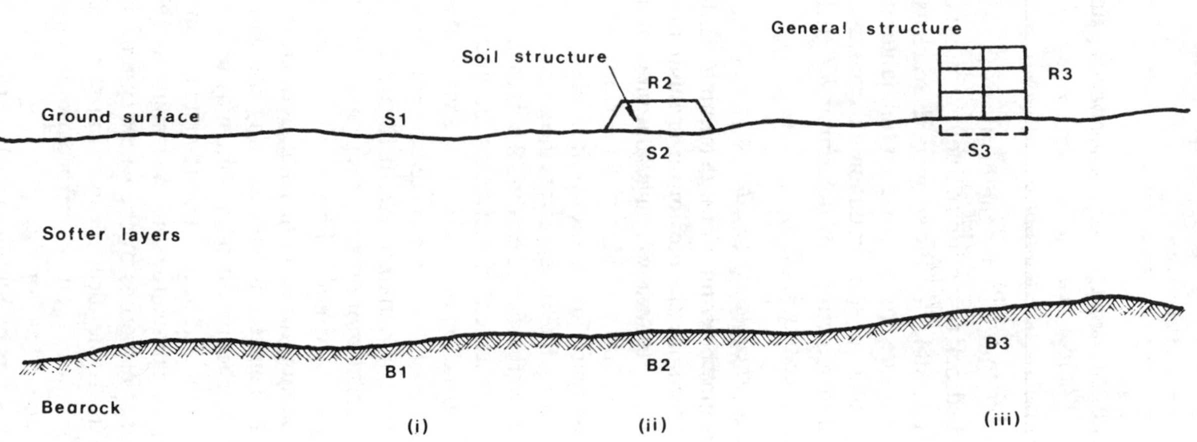

Figure 6.6 Schematic diagram of sites with subsurface rock

Table 6.6. Factors[13] for reducing shear wave velocity measured at low shear strain ($\leqslant 0.001\%$)

Effective peak ground acceleration	$\dfrac{v_s \text{ (high strain)}}{v_s \text{ (low strain)}}$
$a_{max} \leqslant 0.1g$	0.9
$a_{max} = 0.15g$	0.8
$a_{max} = 0.2g$	0.7
$a_{max} \geqslant 0.3g$	0.65

In the Lake Zone of Mexico City the depth to the stiff soil layer which constitutes effective bedrock ranges up to about 60 m, and at this location the superficial clays are so flexible that the site periods reach as high as $T = 5$ s. This may be taken as a worldwide upper bound for buildable sites. Values of fundamental periods representative of some common soil deposits are given in Table 6.7.

In addition to the above methods of determining T, field measurements are sometimes also made (Section 3.3.2.6).

6.2.2.2.(ii) The vertical shear beam model

Site response analysis may be required for determining either what will happen within the body of the soil or the motions at ground surface. At the time of writing, practical design procedures were available only for horizontal layers, and hence dipping strata are not considered here. Consider the three analytical situations indicated in Figure 6.6 for sites with regular geometry (i.e. horizontal layers and flat topography). Site (i) represents the general site evaluation problem; here the stability of the overburden in earthquakes is to be determined with regard to phenomena such as settlement, liquefaction, or landslides, in relation to the feasibility of future construction on this site or the safety of adjacent sites. For the dynamic response analysis of this site, an accelerogram or response spectrum must be applied at B1 to the soil system between B1 and S1. This necessitates the choice of a suitable bedrock motion, and it is proposed that this should be done as described in Section 4.3. Attempts have also been made to compute bedrock motion from surface motion from another site, as discussed later in this section.

Sites (ii) and (iii) in Figure 6.6 represent any site with any structure. The dynamic analysis of a structure on such a site may be carried out in either of two ways. First, the total soil and structure system from bedrock to the top of the structure may be analysed together with applying bedrock motion at B2, B3 and determining the responses of the whole system, including that of the structure R2, R3. This is the ideal means of analysis, as full allowance for interaction between *in situ* soils and constructed soils is included. The dynamic input at bedrock is chosen as for Site (i). Or second, the structure may be analysed by applying a dynamic input at its base (*S*2, *S*3 or at some arbitrary distance below ground surface. The dynamic input appropriate for application at *S*2, *S*3 may be either: (1) surface motion accelerograms or spectra derived specially for the site by computing the modifications caused by the overlying soils on the bedrock motions input at *B*2, *B*3, or (2) surface motion accelerograms

Table 6.7. Typical values of fundamental period for soil deposits (for rock motions with $a_{max} = 0.4g$)[14]

Soil depth (m)	Dense sand (s)	5m of fill over normally consolidated clay* (s)
10	0.3–0.5	0.5–1.0
30	0.6–1.2	1.5–2.3
60	1.0–1.8	1.8–2.8
90	1.5–2.3	2.0–3.0
150	2.0–3.5	

*Representative of San Francisco Bay area.

or spectra derived without specific dynamic analysis of the soil layers as described in Section 4.3.

At present the most common and probably most practical technique for modelling the dynamic behaviour of the soil above bedrock is that of the *vertical shear beam model*, which is so called because of the use of shear wave theory. Several types of errors or limitations apply to the shear beam model as discussed below.

(1) Errors arise in representing a three-dimensional problem by a one-dimensional model.[15]
(2) Nearly all shear beam models assume linear material behaviour as a crude approximation to the real non-elastic behaviour.
(3) Errors arise from the use of viscous rather than hysteretic damping.
(4) Errors from the use of approximate mathematical solutions.
(5) The shear beam model is valid only for sites where ground motion is dominated by shear waves propagating vertically through the soil. It is argued[16] that this is reasonably true at many sites related to the earthquake focus as illustrated in Figure 6.7, but is more likely to be valid at special sites like Mexico City rather than at locations where normal thicknesses of overburden exist. The shear waves will be approximately vertical for deep-focus earthquakes, but this will not be true at sites near the source of shallow earthquakes. At some distancc from shallow-focus earthquakes the significant seismic waves may approach local soils horizontally; for this case it has been suggested that the shear beam model should be used if the half-wave length $\lambda/2$ of incoming waves is large compared to the lateral extent of the soil layers (Figure 6.8).
(6) For the shear beam model to be applicable the boundaries of the site must be essentially horizontal, allowing the soil profile to be treated as a series of semi-infinite layers.
(7) Finally the effect of the presence of the proposed structure (or other structures) is not readily included in the computation of surface motion. For further discussion of the soil–structure interaction problem see Section 6.3.

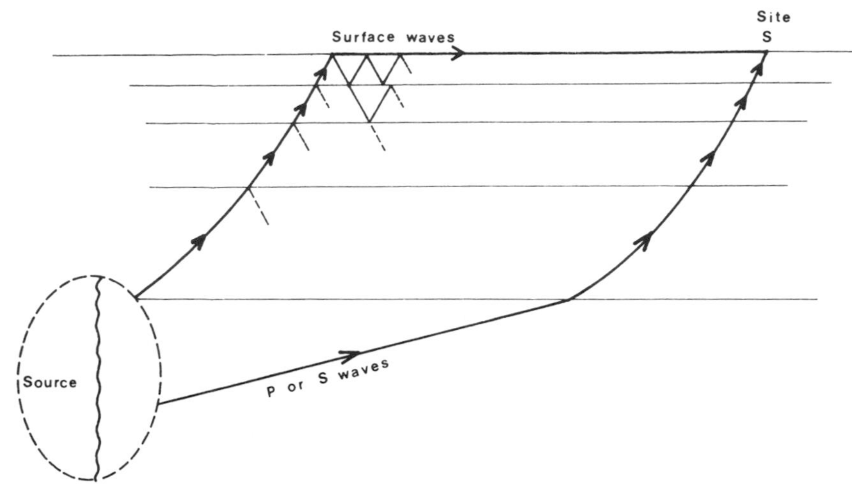

Figure 6.7 Schematic relationship of source, travel paths, and site as assumed in one-dimensional shear wave studies

Two main types of vertical shear beam model are in use, first, the lumped mass methods and second, continuous solutions in the frequency domain. The chief characteristics of each of these methods are now discussed briefly.

In the *lumped mass model* the soil profile is idealized with discrete mass concentrations interconnected by stiffness elements which represent the structural properties of the soil. A time-step modal analysis is commonly used because of the consequent computer economies and because of the familiarity of modal superposition to earthquake engineers. In modal analysis it is necessary to assume linear material behaviour and viscous damping. As the damping of soils is more nearly hysteretic it is common to use an *equivalent* viscous damping which assumes constant damping in all modes, rather than true viscous damping in which the critical damping ratio would increase in proportion to the natural frequency of each mode.

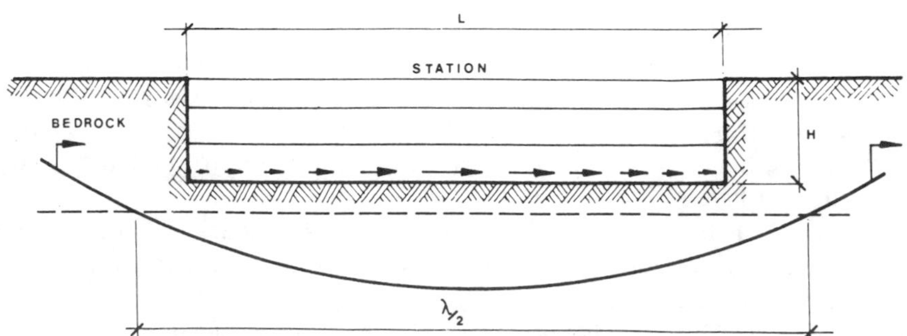

Figure 6.8 Waves arriving horizontally from shallow-focus earthquake. Shear-beam site response theory applicable only if $\lambda/2 \geqslant L$ (after Tsai[16])

Further allowances for the non-linearity of soil behaviour and hysteretic damping have been made by Seed and Idriss,[2] who used an iterative procedure to adjust the soil properties according to the level of strain. Even so they make the considerable simplification of averaging the properties of all layers in the soil profile. Further discussion on the problems involved in determining suitable soil properties for use in modal solutions may be found in Section 6.2.1. and in papers by Whitman *et al.*[17] and Ambraseys.[12]

A well-known computer program using this iterative linear lumped mass shear beam model is SHAKE.[18] More recently, Kausel and Roesset[19] have proposed some simple refinements which can be made to programs like SHAKE to improve their efficiency and to extend their capability to handling non-vertically incident SH waves.

Using data from two sites in Japan, Tazoh *et al.*[20] have compared computer ground motions with those recorded in earthquakes at ground surface and at two levels below surface. In order to obtain good correlation, when the peak shear strain in the soil exceeds about 10^{-3} it was found necessary to consider non-linearity, and a modified Ramberg–Osgood stress–strain model (Figure 6.34) found to be effective.

A *continuous solution in the frequency domain* provides an 'exact' alternative to the lumped mass treatment of the vertical shear beam model. In this method the transfer of the bedrock motion to the surface is derived by consideration of the equation of motion of one-dimensional wave propagation in a continuous medium:

$$\rho \frac{\partial^2 u}{\partial t^2} = G \frac{\partial^2 u}{\partial x^2} + \eta \frac{\partial^3 u}{\partial t\, \partial x^2} \tag{6.11}$$

where ρ is the density of a semi-infinite soil layer, G = shear modulus, η the viscosity constant, and $u(x, t)$ the displacement of a point in the soil layer.

Transfer functions may be derived which modify input bedrock harmonic motions into corresponding surface motions in terms of the elastic properties of the intervening layers and of the bedrock layer itself. By multiplying the Fourier spectrum corresponding to the time-dependent bedrock motion by the transfer function, the surface Fourier spectrum is found. This Fourier spectrum may then be converted into the surface accelerogram. Fuller discussion of the continuous solution to the shear beam problem has been given by Roesset[21] and Schnabel *et al.*[22] Computer programs involving Fourier analysis and transfer functions are simple and may be more economical than those using the lumped mass solution if output at only a few points is desired. The continuous solution has the advantages that it readily handles many soil layers with different properties including the bedrock layer, and any linear damping may be used, but it has the disadvantage of handling linear properties only.

An interesting feature of the transfer function technique is the facility with which bedrock motions can be estimated from surface motion recorded at a given site. This is a useful source of input bedrock data at another site. The main problem with transferring surface motion at one site to surface motion at another site lies in the incorporation of *two* sets of errors implicit in modelling ground motion transfer downwards through one soil profile as well as upwards through another.

6.2.2.2(iii) Response studies of regular soil sites

As noted earlier (Section 6.2.2.1), soil layers sometimes cause *amplification* of bedrock motions. For example, in the 1980 magnitude 6.1 Chiba-ken Chubu earthquake in Japan,[20] at one site the peak ground acceleration increased from 31 cm/s^2 at a depth of 60 m to 104 cm/s^2 at the surface vertically above, while at another site a_{max} increased from 64 cm/s^2 at 42 m depth to 194 cm/s^2 at the surface. As a further example, amplification was particularly strong in part of Mexico City in the September 1985 earthquakes. At the one instrumented site in the zone of interest, the 30 m deep soft clay remained essentially elastic (and hence had low damping) throughout the long excitation, such that the peak bedrock acceleration was amplified about five times. The spectral accelerations at the site period of 2 s was amplified even more.[156]

The opposite effect, *attenuation*, appears to have occurred in the 1957 magnitude 5.3 San Francisco earthquake, as shown in Figure 6.5. Here there were several sites all about equidistant from the earthquake focus, and the peak ground acceleration at two of the soil sites was only half those at the adjacent rock sites. The increase in the response ordinates at longer periods on the soil sites is also evident in Figure 6.5 and has been further illustrated by Seed.[23]

An example of a site response study[24] in which surface ground response spectra were computed from bedrock motion is illustrated in Figure 6.9. An artificial accelerogram for a magnitude 7.0 earthquake at an epicentral distance of 50 km was generated using a non-stationary process, and applied to each of the soil profiles shown. The surface accelerograms (not shown) were computed using the continuous solution mentioned above, and the corresponding response spectra are shown at the top of Figure 6.9 for two values of damping. An indication of the effect of the soil layers on the site response can be obtained by comparing the output response spectra with the response spectra for the input motion shown at the bottom of Figure 6.9.

These show that peak ground acceleration decreases between bedrock and the surface for soil profile 2 but decreases for soil profiles 3 and 4. More revealingly, the spectral velocity increases from bedrock to surface for short periods (e.g. $T = 0.2$ s) for the shallower (stiffer) sites (1 and 2) and decreases for the more flexible sites (3 and 4). Similar differences can be seen between sites 1 and 2, site 2 with relative density $D_R = 0.8$ being stiffer than site 1 with $D_R = 0.4$. It is noted that the soil depth is 30.5 m at sites 1 and 2 and 91.5 m at sites 3 and 4.

According to the above study soil layers cause amplification or attenuation of bedrock responses, depending on soil stiffness and depth, i.e. depending on the site period and on the response period considered. Similar results have been obtained by Sugimura and Ohkawa,[25] who analysed many different soil profiles in a microzoning study of the Tokyo area using a direct integration non-linear analysis of a lumped mass model. They found the ratio of surface a_{max} depended on both site period T_G and on the level of bedrock a_{max}. Higher values of input acceleration cause more non-linearity in the soils, thus reducing the response. Their regression expressions for two input motions for all sites studied were

$$a_{max} = 78.2 T_G^{-0.767} \qquad \text{for } 100 \text{ cm/s}^2 \text{ input} \qquad (6.12)$$

$$a_{max} = 142.1 T_G^{-0.874} \qquad \text{for } 300 \text{ cm/s}^2 \text{ input} \qquad (6.13)$$

Hence for a stiff soil site with $T_G = 0.3$ s, the input bedrock a_{max} is amplified by factors of 2.0 and 1.4 at the surface for input $a_{max} = 100 \text{ cm/s}^2$ and 300 cm/s^2, respectively. In contrast, for a flexible soil site with $T_G = 1.5$ s, the input bedrock a_{max} is *attenuated* by factors of 0.57 and 0.33 at the surface for the above input motions, respectively.

The above results suggest that Table 6.5 does not tell the whole story, and further explanation of how ground accelerations are limited by weak soil layers is given in Section 2.4.2.4. These attenuation effects have been widely observed but are as yet imperfectly understood.

6.2.2.3 Topographical effects

When the surface topography is not flat, the hills or valleys constitute structures which obviously will have dynamical characteristics different from a flat plain. Such effects have been demonstrated in recent years by the evidence of amplification on ridges or hilltops when compared to adjacent flat ground. For example, the high ground motions recorded at the Pocoima Dam site in the 1971 San Fernando, California, earthquake were believed to be in part due to the ridge location.[26,27] However, in subsequent studies Brune[28] found that both amplification or attenuation was possible, depending on the angle of incident waves. This illustrates the significance, complexity, and immaturity of the subject.

An apparently unequivocal example of amplification is that given by several groups of similar houses on topographically different sites in the 1985 San Antonio, Chile, earthquake. The houses on hilltops or ridges were heavily damaged while those on nearly flat or valley sites were only slightly damaged.[29] A further example is given by accelerograph recordings of two earthquakes at three adjacent rock sites across a valley in New Zealand. It was found[30] that amplitudes on the two hilltop sites on opposite sides of the valley were much the same as each other, but they were about twice the size of those recorded for the third size, which was near the bottom of the valley.

Analytical techniques for studying ridge and valley topography have naturally started with simple geometrical forms, e.g. triangular, circular, and elliptical, and various studies[27,31–5] have suggested amplifications of about 2, or narrow

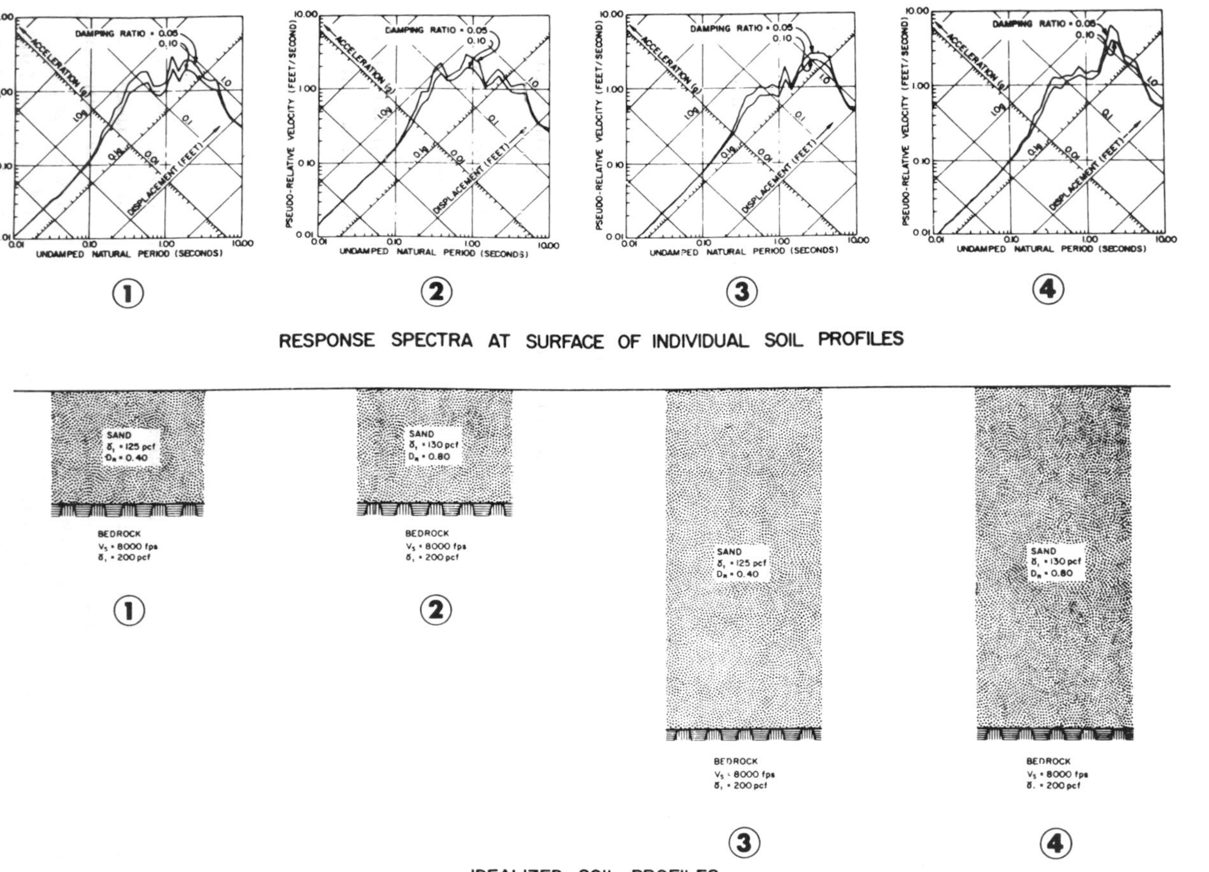

DAMPING RATIO = 0.05
0.10
ACCELERATION (g)
DISPLACEMENT (FEET)
PSEUDO-RELATIVE VELOCITY (FEET/SECOND)
UNDAMPED NATURAL PERIOD (SECONDS)
1
2
3
4
RESPONSE SPECTRA AT SURFACE OF INDIVIDUAL SOIL PROFILES
SAND
δ, = 125 pcf
D = 0.40
BEDROCK
V = 8000 fps
δ, = 200 pcf
SAND
δ, = 130 pcf
D = 0.80
IDEALIZED SOIL PROFILES

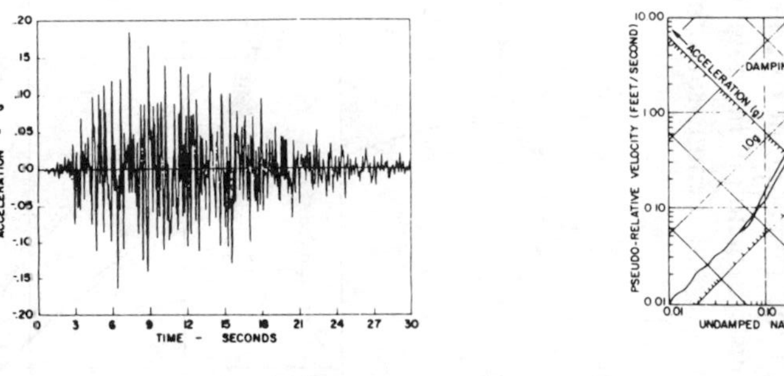

Figure 6.9 Site response analysis summary showing ground-motion criteria at bedrock and at surface after one-dimensional shear wave propagation through soil layer (after Valera and Donovan[24])

frequency band amplifications as high as 5–10. Unfortunately most of these studies have been done in terms of Fourier spectra, which are not directly applicable for engineering purposes. A promising method aimed at engineering design is that formulated by Ayala *et al.*[36] Obviously, with such large amplifications being possible in some topographical circumstances, this subject deserves to be thoroughly researched.

6.2.2.4 Settlement of dry sands

It is well known that loose sands can be compacted by vibration. In earthquakes such compaction causes settlements which may have serious effects on all types of construction. It is therefore important to be able to assess the degree of vulnerability to compaction of a given sand deposit. Unfortunately this is difficult to do with accuracy, but it appears that sands with relative density less than 60 percent or with standard penetration resistance less than fifteen are susceptible to significant settlement. The amount of compaction achieved by any given earthquake will obviously depend on the magnitude and duration of shaking as well as on relative density, as demonstrated by the laboratory test results plotted in Figure 6.10.

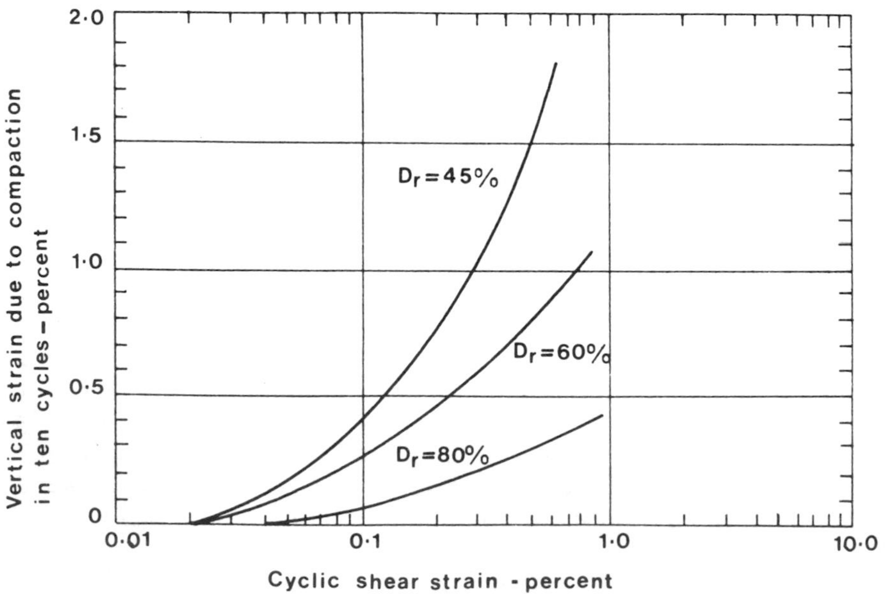

Figure 6.10 Effect of relative density on settlement in ten cycles (after Silver and Seed[37])

Attempts have been made to predict the settlement of sands during earthquakes and a simple method[38] is presented below. It should be noted that this ignores the effect of important factors such as confining pressure and number of cycles, but no fully satisfactory method of settlement prediction as yet exists.

There is a critical void ratio e_{cr} above which a granular deposit will compact when vibrated. If the void ratio of the stratum is $e > e_{cr}$ the maximum amount of settlement possible can be shown to be

$$\Delta H = \frac{e_{cr} - e}{1 - e} H \tag{6.14}$$

where H is the depth of the stratum.

The critical void ratio can be obtained from

$$e_{cr} = e_{min} + (e_{max} - e_{min}) \exp\left[-0.75a/g\right] \tag{6.15}$$

where e_{min} = minimum possible void ratio as determined by testing;
e_{max} = maximum possible ratio;
a = amplitude of applied acceleration;
g = acceleration due to gravity.

6.2.2.5 Liquefaction of saturated cohesionless soils

Under earthquake loading some soils may compact, increasing the pore water pressure and causing a loss in shear strength. This phenomenon is generally referred to as 'liquefaction'. Gravel or clay soils are not susceptible to liquefaction. Dense sands are less likely to liquefy than loose sands, while hydraulically deposited sands are particularly vulnerable due to their uniformity. Liquefaction can occur at some depth causing an upward flow of water. Although this flow may not cause liquefaction in the upper layers it is possible that the hydrodynamic pressure may reduce the allowable bearing pressures at the surface.

Extensive liquefaction at Niigata, Japan, during the 1964 earthquake[39] gave a great impetus to the search for methods of quantifying liquefaction potential. Attempts have been made to relate the latter individually to relative density[40] (Table 6.8), to standard penetration resistance N-values[40] (Figure 6.11) and to particle size distribution[1,39–42] (Figure 6.12). Unfortunately these criteria often give conflicting results for a given site, and a more reliable method has also since been sought.

One method promising better reliability is based on field data from earthquakes, which attempts to differentiate between sites which have liquefield and those which have not, in terms of the strength and duration of shaking and the soil penetration resistance. Using this approach, Ambraseys[43] has found an expression for the cyclic shear stress ratio, $Q = \tau/\sigma_o'$, which has become the principal parameter commonly used to assess soil resistance to liquefaction. For deposits at a critical depth to less than about 25 m it was found that

$$Q = 69.2 \exp(0.154 N^{60}) \exp(-1.15M) \tag{6.16}$$

Table 6.8. Liquefaction potential related to relative density D_r, of soil (after Seed and Idriss[40])

Maximum ground surface acceleration	Liquefaction very likely	Liquefaction depends on soil type and earthquake magnitude	Liquefaction very unlikely
$0.10g$	$D_r < 33\%$	$33\% < D_r < 54\%$	$D_r > 54\%$
$0.15g$	$D_r < 48\%$	$48\% < D_r < 73\%$	$D_r > 73\%$
$0.20g$	$D_r < 60\%$	$60\% < D_r < 85\%$	$D_r > 85\%$
$0.25g$	$D_r < 70\%$	$70\% < D_r < 92\%$	$D_r > 92\%$

where N^{60} is the normalized resistance, i.e. the blow count the deposit would have under an effective overburden pressure of 1 kg/cm^2, corrected to a standard testing procedure.[43] The field information came from data from many earthquakes,[44,45] which was then corrected in the Imperial College study.[43] Equation (6.16) may be rewritten[43] to yield the critical acceleration a_{max}, i.e. the minimum value of the ground acceleration required to cause liquefaction:

$$a_{max}/g = 106.5(\sigma_o/\sigma_o')r_d^{-1} \exp(0.154N^{60}) \exp(-1.15M) \qquad (6.17)$$

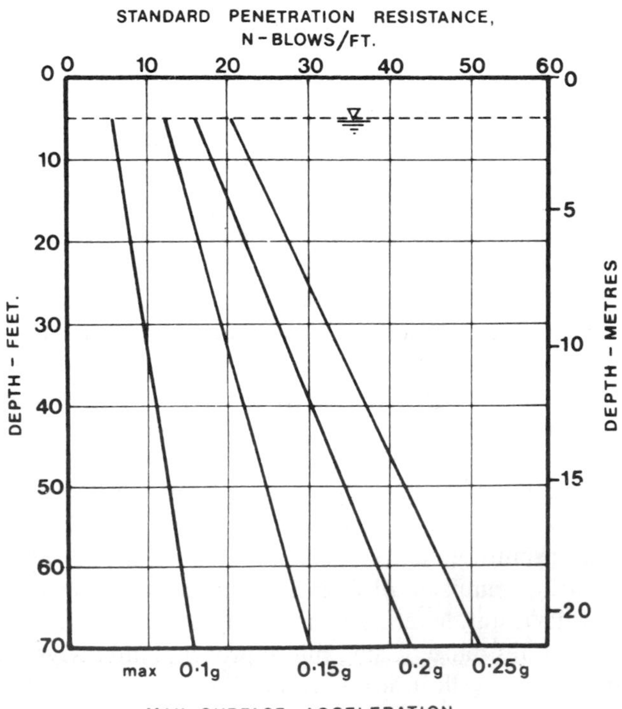

Figure 6.11 Standard penetration resistance values above which liquefaction is unlikely to occur under any conditions (after Seed and Idriss[40])

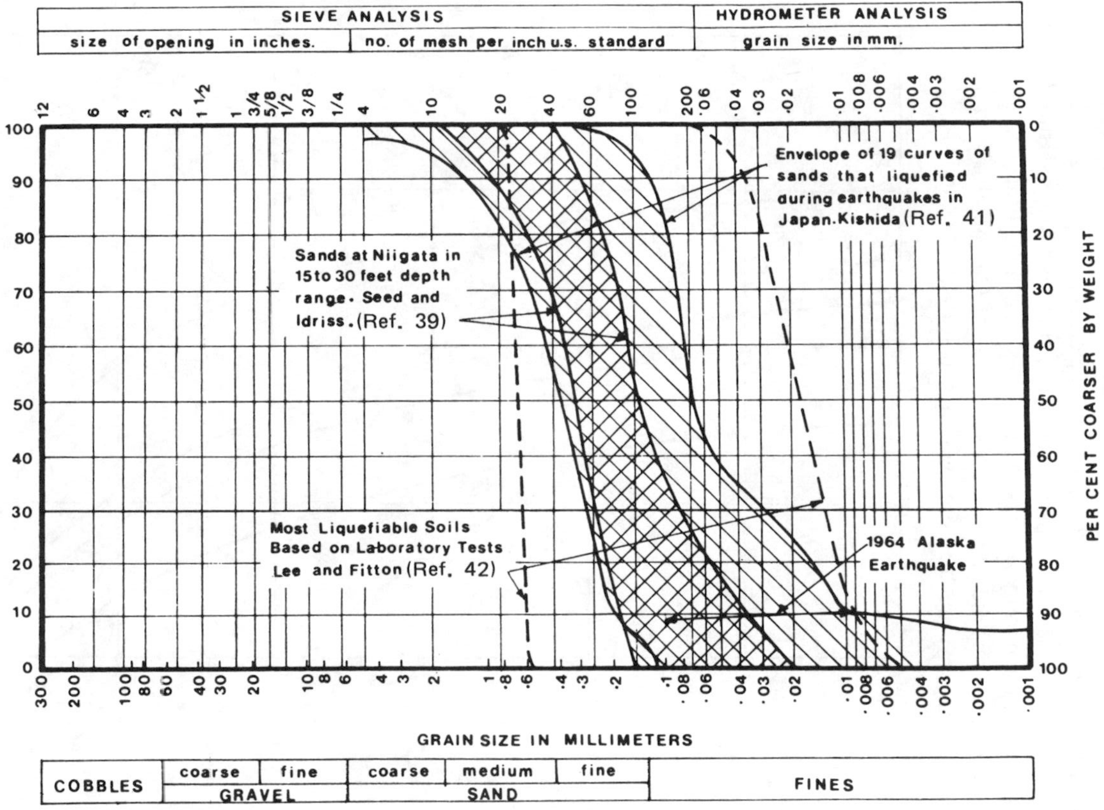

Figure 6.12 Liquefaction potential related to particle size (after Shannon *et al.*[1])

where σ_o and σ_o' are the total and effective overburden pressures at the critical depth; r_d = stress reduction factor[46] accounting for non-rigid response of the overburden, and varies from 1.0 at the surface to 0.9 at a depth of 10 m.[46]

Equation (6.17) is plotted in Figure 6.13 for three values of earthquake magnitude, a water table at ground surface, and critical depths of 5 and 15 m.

As a corollary, because the critical acceleration a_{max} is also the maximum acceleration that can be transmitted to the surface, a_{max} measured at ground surface would generally be smaller for sites that have liquefied than for sites that have not. Thus the attenuation equations of Chapter 2 would overestimate the surface a_{max} values for sites which will liquefy.

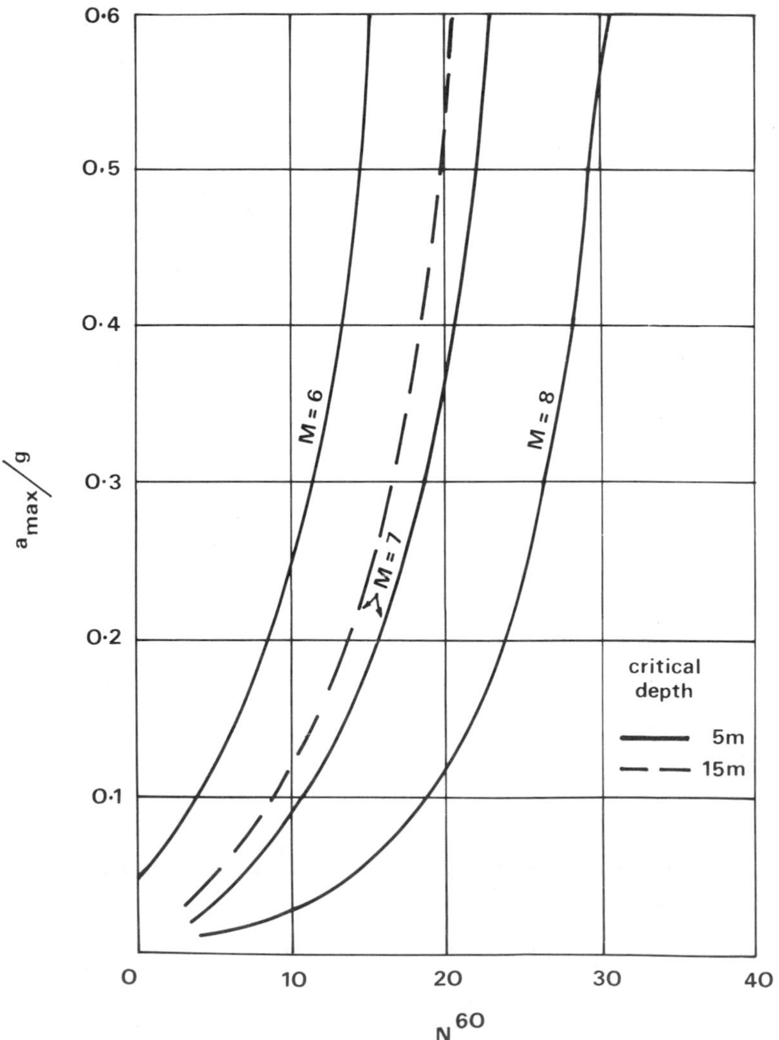

Figure 6.13 Conditions for liquefaction at depths 5 m and 15 m of sand deposits with water table at surface, in terms of a_{max}, M, and N^{60} (reproduced with permission from Ambraseys[43])

6.3 SEISMIC RESPONSE OF SOIL–STRUCTURE SYSTEMS

6.3.1 Introduction

The importance of the nature of the sub-soil for the seismic response of structures has been demonstrated in many earthquakes, but a reasonable understanding of the factors involved has only recently begun to emerge. For example, it seems clear from studies of earthquakes that the relationship between the periods of vibration of structures and the period of the supporting soil is profoundly important regarding the seismic response of the structure. An example from Mexico City is given in Section 3.2, item (2). In the case of the 1970 earthquake at Gediz, Turkey, part of a factory was demolished in a town 135 km from the epicentre while no other buildings in the town were damaged. Subsequent investigations revealed that the fundamental period of vibration of the factory was approximately equal to that of the underlying soil. Further evidence of the importance of periods of vibration was derived from the medium-sized earthquake of Caracas in 1967, which completely destroyed four buildings and caused extensive damage to many others. The pattern of structural damage has been directly related to the depth of soft alluvium overlying the bedrock.[154] Extensive damage to medium-rise buildings (5–9 storeys) was reported in areas where depth to bedrock was less than 100 m while in areas where the alluvium thickness exceeded 150 m the damage was greater in taller buildings (over 14 storeys). The depth of alluvium is, of course, directly related to the periods of vibration of the soil (equation (6.9)).

In order to evaluate the seismic response of a structure at a given site, the dynamic properties of the combined soil–structure system must be understood. The nature of the sub-soil may influence the response of the structure in four ways:

(1) The seismic excitation at bedrock is modified during transmission through the overlying soils to the foundation. This may cause *attenuation* or *amplification* effects (Figure 3.3 and Section 6.2.2.2).
(2) The fixed base dynamic properties of the structure may be significantly modified by the presence of soils overlying bedrock. This will include changes in the mode shapes and periods of vibration.
(3) A significant part of the vibrational energy of the flexibly supported structure may be dissipated by material damping and radiation damping in the supporting medium.
(4) Structures sited on soft alluvium may be damaged by differential vertical displacements occurring before and/or during earthquakes. Although this phenomenon is not properly understood it seems logical that structures with relatively low horizontal strength will suffer worst from this phenomenon, i.e. low-rise structures will be most vulnerable. This effect is in contrast to resonance which, in the case of soft ground, will, of course, occur for longer-period structures.

Items (2) and (3) above are investigated under the general title of *soil–structure interaction*, which may be defined as the interdependent response relationship between a structure and its supporting soil. The behaviour of the structure is dependent in part upon the nature of the supporting soil and similarly the behaviour of the stratum is modified by the presence of the structure.

It follows that *soil amplification* and *attenuation* (item (1) above) will also be influenced by the presence of the structure, as the effect of soil–structure interaction is to produce a difference between the motion at the base of the structure and the free-field motion which would have occurred at the same point in the absence of the structure. In practice, however, this refinement in determining the soil amplification is seldom taken into account, the free-field motion generally being that which is applied to the soil–structure model, as discussed in the following section. Because of the difficulties involved in making dynamic analytical models of soil systems, it has been common practice to ignore soil–structure interaction effects simply treating structures as if rigidly based regardless of the soil conditions. However, intensive study in recent years has produced considerable advances in our knowledge of soil–structure interaction effects and also in the analytical techniques available, as discussed below.

6.3.2 Dynamic analysis of soil–structure systems

Comprehensive dynamic analysis of soil–structure systems is the most demanding analytical task in earthquake engineering. The cost, complexity, and validity of such exercises are major considerations. There are two main problems to be overcome. First, the large computational effort which is generally required for the foundation analysis makes the choice of foundation model very important; five main methods of modelling the foundation are discussed in the next section. Second, there are great uncertainties in defining a design ground motion which not only represents the nature of earthquake shaking appropriate for the site but also represents a suitable level of risk.

Ideally the earthquake motion should be applied at bedrock to the complete soil–structure system. This is not a very realistic method at present because much less is known about bedrock motion than surface motion, and there is a great scatter in possible results for the soil amplification effect defined above. At present the most realistic methods of analysis seem to be those which apply the *free-field* motion to the base of the structure, the free-field motion being that which would occur at the surface in the absence of the structure. This may be done most simply using simple springs at the base of the structure (Figure 6.14), as described in Section 6.3.3.1, or using a sub-structuring technique in which the foundation dynamic characteristics are predetermined and superposition of soil and structure response is carried out. The latter technique has been described by Penzien and Tseng[47] using half-space modelling of the soil, and by Vaish and Chopra,[48] who illustrate their presentation with finite element modelling. These two types of soil model are discussed in the next section.

It should be noted that where the dynamic behaviour is expressed in frequency-dependent terms, the problem must be analysed in the *frequency domain* not the *time domain*. For this purpose acceleration-time records must be transformed into acceleration-frequency terms using Fourier transform methods before application to the system. An inverse transformation is required to obtain the response time record. These techniques are described in the above two papers.[47,48]

For projects in which soil–structure interaction effects are likely to be important, the choice of analytical method requires careful consideration. The reader will find useful extra guidance in Wolf's book[49] and in the state-of-the-art report by Seed *et al.*[50]

6.3.3 Soil models for dynamic analysis

A dynamic model of the soil which attempts to fully model reality requires the representation of soil stiffness, material damping, and radiation damping, allowing for strain-dependence (non-linearity) and variation of soil properties in three dimensions. While various analytical techniques exist for handling different aspects of the above soil behaviour they all suffer from varying combinations of expensiveness or inaccuracy. Therefore there is some difficulty for any given project in choosing an analytical model for the soil which will permit an appropriate level of understanding of the soil–structure system.

The methods of modelling the soil may be divided into five categories of varying complexity;

(1) Equivalent static springs and viscous damping located at the base of the structure only;
(2) Shear beam analogy using continua or lumped masses and springs distributed vertically through the soil profile;
(3) Elastic or viscoelastic half-space;
(4) Finite elements;
(5) Hybrid model of (3) and (4).

A brief discussion of each of the above modelling methods follows below.

6.3.3.1 Springs and dashpots at the base of the structure

The most rudimentary method of modelling the soil is to use only springs, located at the base of the structure, to represent the appropriate selection of horizontal, rocking, vertical, and torsional stiffnesses of the soil (Figure 6.14). An increase in the rigorousness of the model may be effected by adding dashpots at the same location. In the system shown in Figure 6.14(b), the stiffness of the individual vertical springs must be chosen to sum to either the global rocking stiffness or the global vertical stiffness, as used in Figure 6.14(a), as it is unusual to achieve both conditions simultaneously. The same is true for damping. This

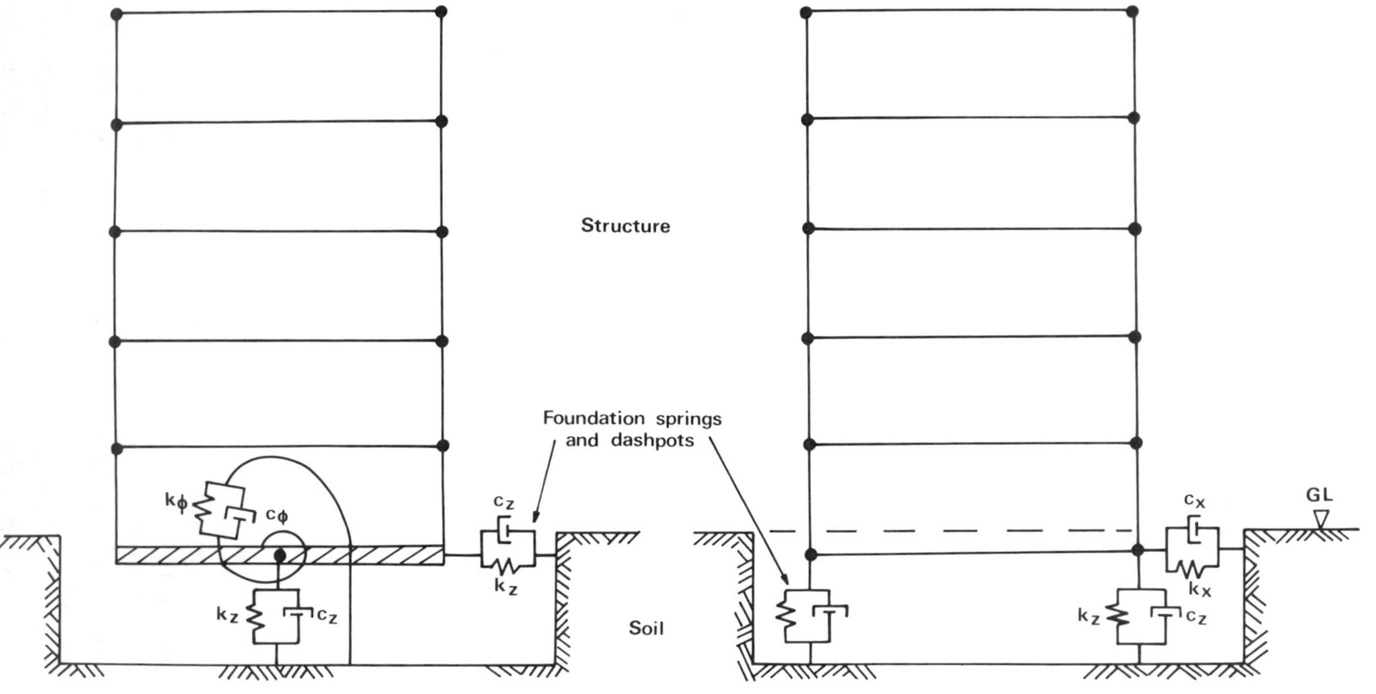

Figure 6.14 Rudimentary soil–structure analytical models representing soil properties by springs and dashpots

discrepancy may not matter in analyses in which horizontal and vertical excitations are not applied simultaneously, but generally a conflict arises.

As a simple illustration, consider modelling a circular disc foundation by 32 vertical springs located around its perimeter (Figure 6.17). The total vertical spring stiffness is

$$\mathrm{K}_{zo} = \frac{4GR}{(1-\nu)} \tag{6.18}$$

This stiffness would therefore be provided by the sum of 32 vertical springs of stiffness

$$k_{zi} = \frac{GR}{8(1-\nu)} \tag{6.19}$$

These springs give a rotational stiffness

$$k_{\phi} = 16R^2 k_{\nu i}$$

$$= \frac{2GR^3}{(1-\nu)} \tag{6.20}$$

However, this value is only three quarters of that given by the half-space rocking spring formula

$$k_{\phi} = \frac{8GR^3}{3(1-\nu)} \tag{6.21}$$

In these circumstances the stiffness value for the vertical springs will need to be chosen to give a conservative result depending on the nature of the loading. This will usually be done by increasing k_{zi} so that the value of k_{ϕ} equates to the half-space solution. In some cases it may be possible to equate the vertical and rotational stiffness criteria by locating the vertical springs on an increased radius, but this necessitates introducing very stiff dummy members into the foundation model, which may lead to numerical or local modelling problems.

A convenient method for determining the overall foundation spring stiffnesses is to use the zero-frequency (static) stiffnesses derived from elastic half-space theory as given in Table 6.9. It should be noted that the values in Table 6.9 are for a homogeneous elastic half-space, but need to be factored to give some equivalence to layered soils or to allow for a given degree of non-linearity in the soil behaviour. Solutions for the stiffness of various shapes of footings may be found elsewhere, conveniently collected by Poulos and Davis.[51]

Table 6.9. Discrete foundation properties for rigid plate on elastic half-space

Motion	Circular footings			Rectangular footings
	Spring stiffness k	Viscous damper*	Added mass*	Spring stiffness k
Vertical	$\frac{4GR}{1-\nu}$	$1.79\surd(k\rho R^3)$	$1.5\rho R^3$	$\frac{G}{1-\nu}\beta_z\surd(BL)$
Horizontal	$\frac{8GR}{2-\nu}$	$1.08\surd(k\rho R^3)$	$0.28\rho R^3$	$2G(1+\nu)\beta_x\surd(BL)$
Rocking	$\frac{8GR^3}{3(1-\nu)}$	$0.47\surd(k\rho R^5)$	$0.49\rho R^5$	$\frac{G\beta_\phi BL^2}{1-\nu}$
Torsion	$\frac{16GR^3}{3}$	$1.11\surd(k\rho R^5)$	$0.7\rho R^5$	†

G is the shear modulus for the soil, where $G=E/\{2(1+\nu)\}$, ν is Poisson's ratio for soil, ρ is mass density for soil, R is radius of footing, B,L, are the plan dimensions of rectangular pads, and β_x, β_z, β_ϕ are coefficients given in Figure 6.15.

* The properties come from Clough and Penzien, reference 53.

† For torsional spring stiffnesses of rectangular footings see Newmark and Rosenblueth, p. 98 reference 38.

As an example of layered soils, consider a circular disc footing of 70 m radius on soils consisting of a layer of depth $H = 32$ m overlying a half-space. The soil properties of these two elements are as follows:

Soil element	ν	G(MPa)
(1) Layer	0.4	430
(2) Half-space	0.275	3610

In order to find the approximate equivalent half-space spring stiffness of the foundation system by simple hand calculation, first determine the spring stiffness of the two elements. Considering vertical stiffness for a layer on a rigid base, Bycroft[52] gives

$$k_{z1} = \frac{4G_1R_1}{(1-\nu_1)}\left(1 + 1.4\frac{R_1}{H_1}\right) \tag{6.22}$$

$$= 315 \times 10^9 \quad \text{N/m}$$

For the lower half-space make the approximate assumption that the stress from the disc footing spreads out through the layer at an angle of 45 degrees, so that the stiffness of element (2) relates to an effective disc radius of

$$R_2 = 40 + 32 = 72 \text{ m}$$

Thus

$$k_{z2} = \frac{4G_2R_2}{(1-\nu_2)}$$

$$= 1434 \times 10^9 \quad \text{N/m}$$

The vertical spring stiffness for the combined soil system is obtained by adding the flexibilities of elements (1) and (2), so that

$$k_z(\text{system}) = \frac{k_{v1}\,k_{v2}}{k_{v1} + k_{v2}}$$

$$= 258 \times 10^9 \quad \text{N/m}$$

It can be seen that the presence below the layer of a half-space of moderate stiffness makes the foundation more flexible than if the layer had been underlain by effectively rigid rock.

The spring stiffnesses are dependent on the shear modulus, which in turn varies with the level of shear strain. Hence for linear elastic calculations, spring stiffnesses should be calculated corresponding to a value of shear strain which is less than the maximum expected shear strain. For instance, if the spring stiffness at low strain is k_o, then a value of k equal to 0.67 k_o may be used in the analysis. Alternatively a series of comparative analyses may be done using a range of values of k, particularly if *in situ* tests have not been made, in this case it may be appropriate to select values of k from the following ranges:

for translation

$$0.5\,k_o \leqslant k \leqslant k_o \tag{6.23}$$

for rocking

$$0.33\,k_o \leqslant k \leqslant k_o \tag{6.24}$$

In some computer programs spring supports may not be available and members with appropriate area and stiffness can be used instead.

Table 6.9 gives viscous damper values equivalent to radiation damping in a half-space foundation, where the degrees of freedom are represented by single discrete dashpots, as in Figure 6.14(a). These values will generally be reduced (often substantially) if layering exists in the upper regions of the soil, due to wave reflections at the interfaces (see Figure 6.18 and related text).

When the chosen method of analysis does not allow the use of foundation dashpots difficulties arise in accurately representing the effects of material damping and radiation damping in the foundation, as the total amount of equivalent viscous damping for the foundation in some cases exceeds considerably that for the superstructure. A conservative compromise between the structural and soil damping values will generally be necessary.

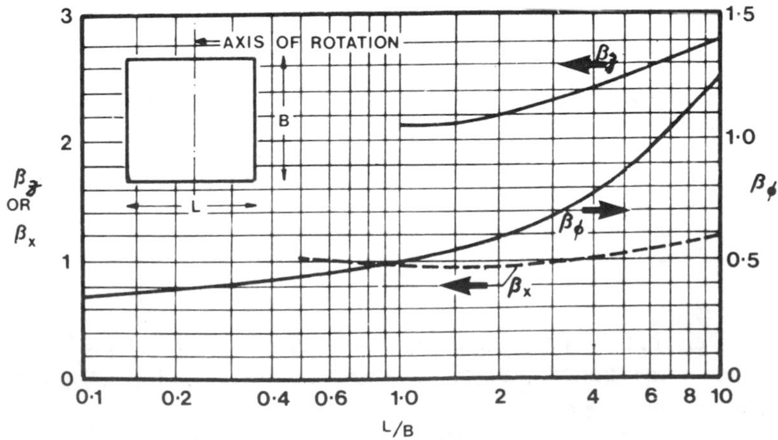

Figure 6.15 Coefficients β_x, β_z, and β_ϕ for rectangular footings (after Whitman and Richart[8])

Also the damping in the soil in different modes of vibration varies considerably. As most currently available dynamic analysis computer programs are written for equal damping in all modes, some intermediate value of damping has to be chosen which hopefully will lead to the most realistic result. The value of damping used should not vary too greatly from that of the mode in which most of the vibrational work is done. Hence a trial mode shape analysis may have to be done to determine which modes predominate. Use of too high or too low a value of damping will lead to unconservative or conservative results, respectively.[54]

6.3.3.2 Shear beam

The shear beam approach may be used to model the soil layers overlying bedrock (Figure 6.16), although difficulties arise in choosing appropriate stiffness and damping values for the soil. Non-linearity may be allowed for using iterative linear analyses such as those used in soil amplification studies (Section 6.2.2.2(ii)), or by non-linear foundation springs as in the research on piled bridge foundations reported by Penzien.[55]

6.3.3.3 Elastic or viscoelastic half-space

Modelling the foundation as a homogeneous linear elastic or viscoelastic half-space in which the stiffness and damping are treated as frequency-dependent provides a very useful means of allowing for the radiation damping effect. Various numerical and partly closed-form formulations of the theory have been made, such as those of Luco and Westmann[56] and Veletsos and Wei,[57] and others as noted in the following discussion.

Consider a rigid circular plate of radius R on the surface of an *elastic homogeneous half-space* of density ρ, Poisson's ratio ν, and shear wave velocity v_s. Let u_x, u_ϕ, and u_z be the amplitudes of the horizontal, rotational and vertical displacements of the plate. Neglecting the small coupling between the horizontal and rocking motions, the relationship between forces and displacements may be stated as

$$F_j = K_j u_j \tag{6.25}$$

where the subscript j denotes x, ϕ, or z, and K_j are complex-valued stiffness (impedance) functions of the form

$$K_j = k_j(\beta_{kj} + i a_o \beta_{cj}) \tag{6.26}$$

The symbol k_j in equation (6.26) is the zero-frequency stiffness of the foundation, as given by the expressions for spring stiffness k in Table 6.9, and a_o is a dimensionless frequency parameter $a_o = \bar{\omega} R / v_s$, where $\bar{\omega}$ is the forcing frequency. Veletsos and Verbic[58] have found analytical expressions

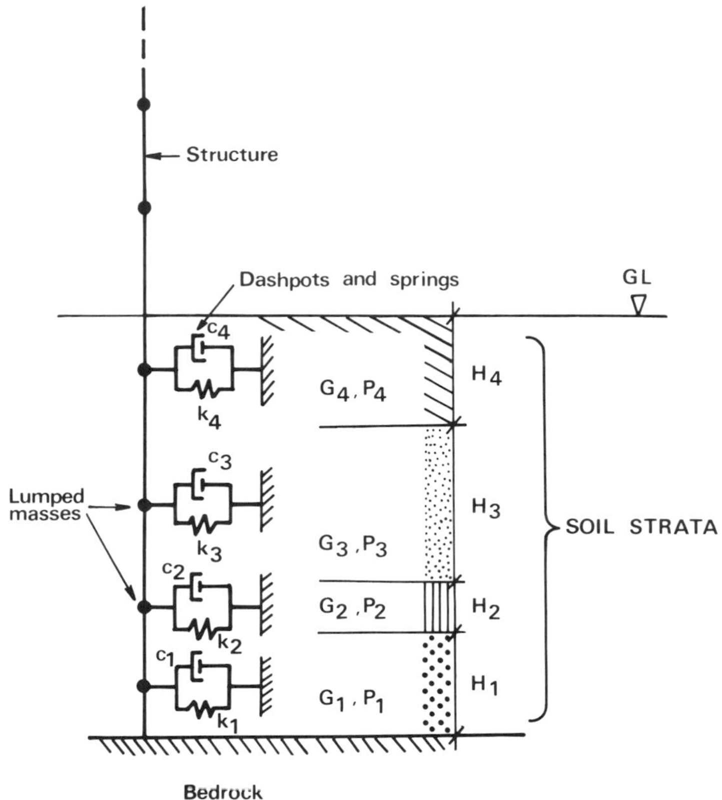

Figure 6.16 Soil–structure analytical model representing the soil vertical profile by a lumped parameter system of masses, springs, and dashpots

approximating to the 'exact' numerical solutions such that β_{kj} and β_{cj} in equation (6.26) are given by

$$\beta_{kj} = 1 - \left[\frac{b_1 b_2^2}{(1 + b_2^2 a_0^2)} + b_3 \right] a_0^2 \tag{6.27}$$

$$\beta_{cj} = b_4 + \frac{b_1 b_2^3 a_0^2}{(1 + b_2^2 a_0^2)} \tag{6.28}$$

where the parameters b_1 to b_4 are dimensionless functions of Poisson's ratio, and vary for horizontal, vertical and rocking motions, as given in Table 6.10. In this equivalent spring-dashpot representation of the supporting medium, i.e. the half-space, β_k is a measure of the dynamic stiffness of the spring and β_c is a measure of the damping coefficient of the dashpot. In this case the damping is solely due to radiation damping.

Although the general expressions for β_{kj} and β_{cj} are both functions of frequency (a_o in equations (6.27) and (6.28)), it should be noted that the horizontal motion the terms involving a_o are zero, so that $\beta_{kx}=1$ and $\beta_{cx}=b_4$.

The expression for β_{cj} given in equation (6.28) may be used for evaluating the foundation dashpots for a structure supported on an elastic half-space (e.g. Figure 6.14) using the dashpot coefficient c_j, obtained from

$$c_j = \frac{Rk_j\beta_{cj}}{V_s} \tag{6.29}$$

This dashpot coefficient is used for obtaining the dynamic damping forces F_{Dj} from

$$F_{Dj} = c_j\dot{u}_j$$

where $\dot{u}_j$ is the velocity experienced by the dashpot as decribed for equation (1.5) in Section 1.4.

A more widely applicable alternative to the above method of estimating radiation damping has been given by Gazetas and Dobry.[59] They found closed form expressions for determining the frequency dependent radiation damping coefficients for footings of various shapes and also for piles, and extended their method to deal with inhomogeneous soil conditions as well as the idealized half-space. For example, they found for a strip footing of width $2B$, on a uniform elastic half-space, that the radiation damping coefficient is given by

$$c_j = \varrho VA \;\Re\left[-i\,\frac{H_1^{(2)}(a)}{H_o^{(2)}(a)}\right] \tag{6.30}$$

Table 6.10. Coefficients b_1 to b_4 for use in equations (6.27) and (6.28), from Veletsos and Verbic[58]

Motion	Poisson's Ratio	b_1	b_2	b_3	b_4
Horizontal	0	0	0	0	0.775
	⅓	0	0	0	0.65
	½	0	0	0	0.60
Vertical	0	0.25	1.0	0	0.85
	⅓	0.35	0.8	0	0.75
	½	0	0	0.17	0.85
Rocking	0	0.8	0.525	0	0
	⅓	0.8	0.5	0	0
	½	0.8	0.4	0.027	0

$$with\ a = \frac{\omega B}{V} \tag{6.31}$$

where ϱ is the density of the soil;

$A = 2B$ = the area (per unit length) of the footing;

ω = frequency (rads/s);

$H_1^{(2)}$ = first-order Hankel function of second kind;

$H_o^{(2)}$ = second-order Hankel function of second kind;

$\Re$ denotes the real part of the complex quantity impled by $i = \sqrt{(-1)}$;

V = wave velocity, where for obtaining c_z for vertical motion, *Lysmer's analogue* V_{La} is appropriate, i.e.

$$V = V_{La} = \frac{3.4\ v_s}{\pi(1-\nu)} \tag{6.32}$$

while for finding c_x for horizontal motion, $V = v_s$.

Inhomogeneous soil.

As well as the above uniform half-space solution, Gazetas and Dobry[59] considered semi-infinite elastic soils having stiffness varying with depth z, from a value of G_o at the surface, in the form

$$G = G_o \left(\frac{z}{B}\right)^m \tag{6.33}$$

$$\text{with} \quad o \leqslant m \leqslant 1 \tag{6.34}$$

The resulting expressions[59] for c_j should be compared with those derived by Werkle and Waas,[155] who used a semi-analytical method to find stiffness and damping coefficients for the four modes of vibration for a half-space of linearly increasing stiffness with depth.

Non-linear soil behaviour cannot be explicitly modelled in the frequency domain solutions used for the above formulations, but the *viscoelastic* hysteretic model may be thought of as representing a limited degree of non-linearity.

In the above formulations for an elastic half-space the *material damping* is neglected. The viscoelastic formulation of foundation impedance improves on this by allowing for material damping through the parameter tan δ defined by

$$\tan \delta = \frac{\Delta W}{2\pi W} \tag{6.35}$$

ΔW and W are defined in Section 6.2.1.2(i), where it will be seen that tan δ is equal to twice the equivalent viscous damping for soil as defined by equation (6.35).

Veletsos and Verbic[58] modified the elastic parameters β_{kj} and β_{cj} into the viscoelastic terms β^{v}_{kj} and β^{v}_{cj}, which are given here in a rearranged form due to Danay,[60] such that

$$\beta^{v}_{kj} = \beta_{kj} - a_{o}\beta_{cj}\left[\frac{(1+\tan^{2}\delta)^{1/2}-1}{2}\right]^{1/2} \tag{6.36}$$

$$\beta^{v}_{cj} = \beta_{cj}\left[\frac{(1+\tan^{2}\delta)^{1/2}+1}{2}\right]^{1/2} + \beta_{kj}\frac{\tan\delta}{a_{o}} \tag{6.37}$$

As may be expected, the inclusion of hysteretic damping increases the overall damping of the system and reduces the deformations.[61] However, inspection of equation (6.37) shows that in cases where radiation damping is large, the effect of including the material damping will often be negligible.

The values obtained for c_x and c_z for circular footings on homogeneous half-spaces by the various methods outlined above are virtually *frequency independent*, and will usually be found to be in fair agreement with values obtained by the simple expressions given by Clough and Penzien as quoted here in Table 6.9. The dashpot coefficient for the *rocking mode*, however, is strongly *frequency dependent*, and an appropriate value of frequency has to be chosen when using equations (6.36) or (6.37). In some cases it may be deemed sufficient simply to take the frequency of the dominant mode of vibration, or perhaps a mean of the main modes weighted according to their participation factors. In a sophisticated analysis where foundation dashpots were required for non-linear analysis of an offshore oil platform, Watt *et al.*[62] carried out a series of constant dashpot analyses until a constant rocking dashpot value was found with which the peak response of the system was similar to that obtained from an analysis using the 'exact' frequency dependent impedance.

The effects of *soil layers* may be studied using the half-space model. The work of Luco[63] has demonstrated the importance of shallow reflective layer interfaces, i.e. low values of H/R, where H is the layer depth and R is the radius of the footing. This causes wave energy radiating away from the footing to be reflected back, thereby reducing radiation damping. In some cases the damping oscillates very rapidly with frequency, so that caution with selection of frequencies is required. As an example, consider a cooling tower structure with an annular footing supported on a layered half-space (Figure 6.17). For horizontal vibrations we may assume that the annular footing behaves like a solid circular plate, so that we use $H/R = 32/38 = 0.85$ for entering Luco's graph (Figure 6.18). From a modal response analysis it was found that the first three horizontal modes of vibration had periods of vibration $T_1 = 0.44$ s, $T_2 = 0.22$ s, $T_3 = 0.11$ s. From these values the dimensionless frequency parameter $a_o = \omega R/v_s$ has respective values 1.1, 2.2, and 3.3.

Entering Figure 6.18 using these values of H/R and a_o, we can determine that the damping in the first three modes is reduced compared to the unlayered case ($H/R = \infty$), by multiplying by the factors 0.3, 0.7 and 1.0. Finally, the

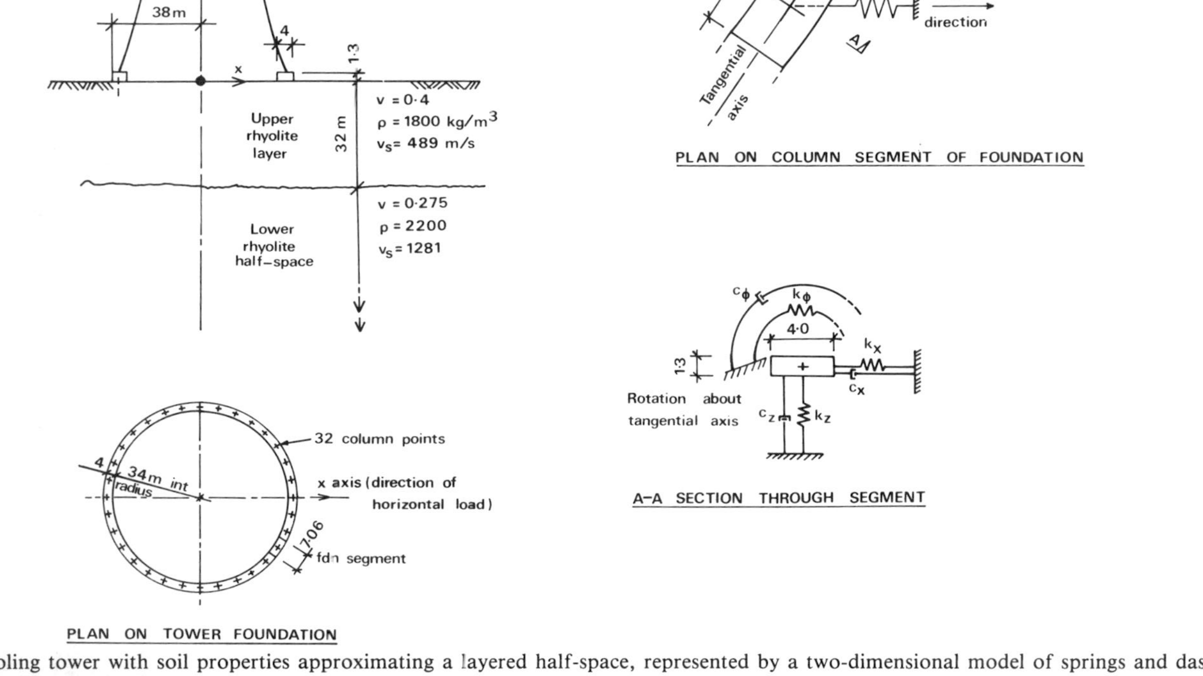

Figure 6.17 Cooling tower with soil properties approximating a layered half-space, represented by a two-dimensional model of springs and dashpots

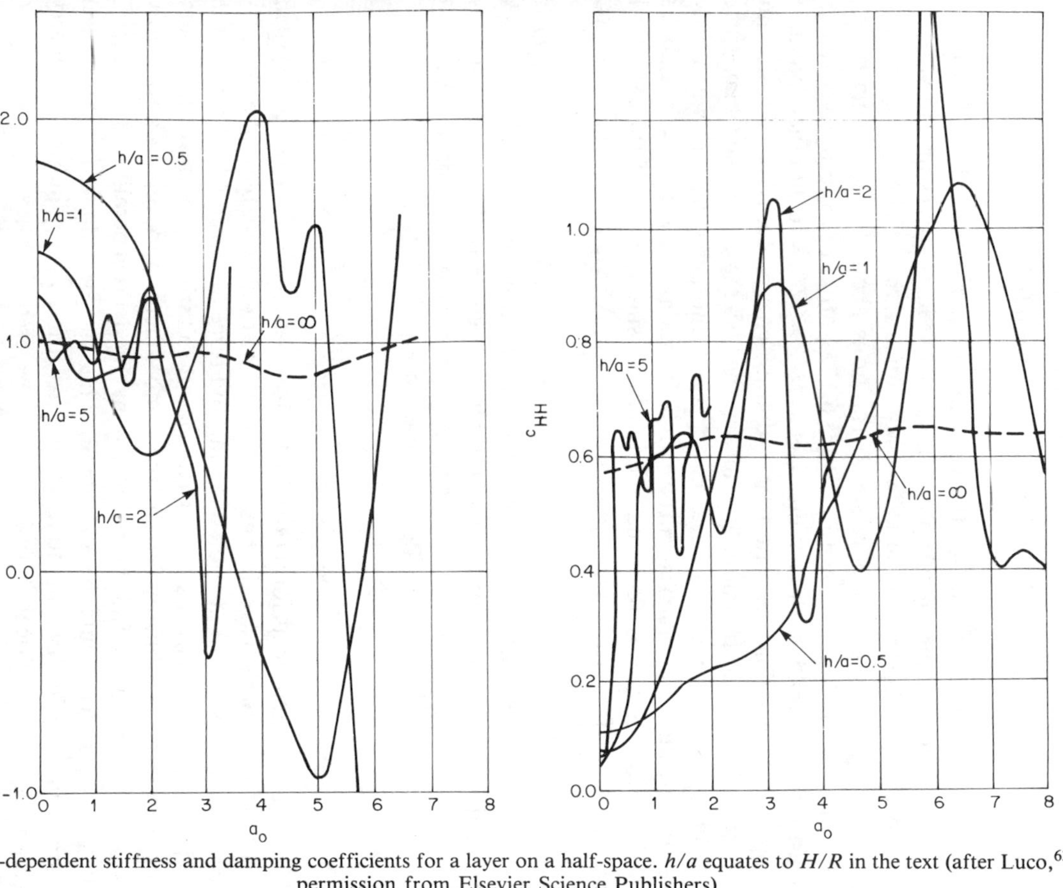

Figure 6.18 Frequency-dependent stiffness and damping coefficients for a layer on a half-space. h/a equates to H/R in the text (after Luco,[63] reproduced with permission from Elsevier Science Publishers)

effective reduction factor for horizontal motion is estimated by weighting the above reduction factors according to the contributions of each mode to the earthquake horizontal forces, which were 0.71, 0.28, and 0.01 for the first, second, and higher modes, respectively. Thus the effect of the layer is to reduce the radiation damping by multiplying it by the reduction factor

$$(0.3 \times 0.71) + (0.7 \times 0.28) + (1.0 \times 0.01) = 0.42$$

Using the same argument for the vertical mode, a reduction factor for the vertical radiation damping dashpot because of the layer was estimated at 0.51. For rocking, however, the situation was somewhat different, because the foundation rocking dashpots relate to rotations about the centreline of the annular strip footing as shown in Figure 6.17, so that the appropriate radius to use is the half width of the annulus. Hence $H/R = 32/2 = 16$. For these physical proportions negligible reflection of radiating energy occurs and hence no reduction is required in damping values in the rocking mode due to layering.

Embedment of footings into a half-space is another situation which may deserve attention. For practical purposes modifications to the results of the above methods for surface footings may be relatively simply made if the increased foundation stiffness caused by embedment is used. An example of this approach is given in Section 6.3.4.6.

Finally, in modelling soil–structure systems for dynamic analysis the *added mass of soil* which participates with the vibration of the footing may sometimes be significant. Estimates of this quantity have been derived in various studies of footings supported on a half-space, such as those of Veletsos and Verbic[58] or those due to Clough and Penzien given above in Table 6.9.

6.3.3.4 Finite elements

The use of finite elements for modelling the foundation of a soil–structure system is the most comprehensive (if most expensive) method available. Like the half-space model it permits radiation damping and three-dimensionality, but has the major advantage of easily allowing changes of soil stiffness both vertically and horizontally to be explicitly formulated. Embedment of footings is also readily dealt with. Although a full three-dimensional model is generally too expensive, three dimensions should be simulated. This can be achieved either by an equivalent two-dimensional model, or for structures with cylindrical symmetry an analysis in cylindrical co-ordinates can be used.[64]

In order to simulate radiation of energy through the boundaries of the element model three main methods are available.

(1) *Elementary boundaries* that do not absorb energy and rely on the distance to the boundary to minimize the effect of reflection waves.
(2) *Viscous boundaries* which attempt to absorb the radiating waves, modelling the far field by a series of dashpots and springs, as used by Lysmer and

Kuhlemeyer.[61] The accuracy of this method is not very good for thin surface layers or for horizontal excitation, although an improved version has been developed by Ang and Newmark.[66]

(3) *Consistent boundaries* are the best absorptive boundaries at present available, reproducing the far field in a way consistent with the finite element expansion used to model the core region. This method was developed by Lysmer and Waas[67] and generalized by Kausel.[64] The latter method, among other things, allows the lateral boundary to be placed directly at the side of the foundation, with a considerable reduction in the number of degrees of freedom.

Non-linearity of soil behaviour can be modelled with non-linear finite elements, but the necessary *time-domain* analysis, is very expensive with most methods. Alternatively, non-linearity could theoretically be simulated in repetitive linear model analyses with adjustment of modulus and damping in each cycle as a function of strain level. In *frequency-domain* solutions (for example, when using consistent boundaries) non-linearity can be approximately simulated again using an iterative approach. In a study of a nuclear containment structure, Kausel[68] showed that the iterative linear approach was adequate for structural response calculations, the costly full non-linear analysis only being warranted for detailed investigation of soil behaviour at or near failure.

As in the half-space solutions, material damping may be accounted for by using a viscoelastic finite element model as used by Kausel and Roesset,[64,69] or the Rayleigh damping model may be used.[70]

A recent major development by Bayo and Wilson[70,71] permits a time-domain solution with much greater computational efficiency than was previously possible, due to the use of Ritz vectors rather than exact eigenvalues for free-vibration mode shapes. Factors that may be incorporated include structural embedment, arbitrary soil profile, flexibility of the foundations, spatial variations of free field motions, interaction between two or more structures, and non-linearity of soil and structure.

6.3.3.5 Hybrid half-space/finite element model

This method combines the methods described in the two previous sections and claims[72] to take advantage of the desirable factors of the finite element and semi-infinite methods and minimizes their undesirable features. The modelling is achieved by partitioning the total soil–structure system into a near-field and a far-field with a hemispherical interface. The near-field, which consists of the structure to be analysed and a finite region of soil around it, is modelled by finite elements. The semi-infinite far-field is modelled by distributed impedance functions at the interface. According to Gupta *et al.*[72] this model is realistic and economical for three-dimensional soil–structure interaction analyses for both surface and embedded structures.

6.3.4 Useful results from soil–structure interaction studies

In recent years there have been intensive theoretical investigations of the dynamics of soil–structure systems using soil modelling techniques as described above. Although many of the conclusions of these studies are still tentative, requiring experimental or field verification, some of the results are physically or intuitively sound. A brief summary of the more important conclusions is therefore included here.

6.3.4.1

Perhaps the leading question to be answered about soil–structure interaction is: '*For what soil conditions will the rigid base assumption lead to significant errors in the response calculations*?' Veletsos and Meek[73] have suggested that consideration of soil–structure interaction is only warranted for values of the ratio

$$\frac{v_s}{fh} < 20 \tag{6.38}$$

where v_s is the shear wave velocity in the soil half-space, f is the fixed-base frequency of the single degree-of-freedom structure, and h is its height. Substituting $f \approx 30/h$ for framed buildings, and $f \approx 45/h$ for shear wall buildings in the above equation implies that soil–structure interaction effects may be important for framed buildings when $v_s \leqslant 600$ m/s, or for shear wall buildings when $v_s \leqslant 900$ m/s. It thus seems inappropriate that shear wave velocities of this order are used to define rock-like material, i.e. bedrock, for site response purposes in the USA,[13,14] as noted in Sections 3.3.2.2 and 6.2.2.2(i).

It is of interest that equation (6.38) correctly predicts that soil–structure interaction is important for the concrete gravity oil platforms studied by Watt *et al.*[74] Radiation damping effects were found to reduce the base shear of a platform on 'very hard' ground ($v_s = 480$ m/s) by about 50 percent (the relevant value of v_s/fh was 6.6), despite the fact that the foundation was effectively rigid regarding its effect on the mode shapes and periods. It is relevant that these offshore structures have a high mass density factor, $m/\rho\pi R^2 h$, where ρ is the density of the soil and m is the participating mass of the structure.

6.3.4.2

The *periods of vibration* of a given structure increase with decreasing stiffness of the sub-soil. This logical phenomenon has been widely noted such as by Veletsos and Meek[73] and Watt *et al.*[74] The latter found this effect to be very marked for a large offshore oil platform, where the fundamental period was 2.95 s for the rigid foundation condition and 5.9 s when allowance was made for a sub-stratum of 'firm' overconsolidated clay (Figure 6.19).

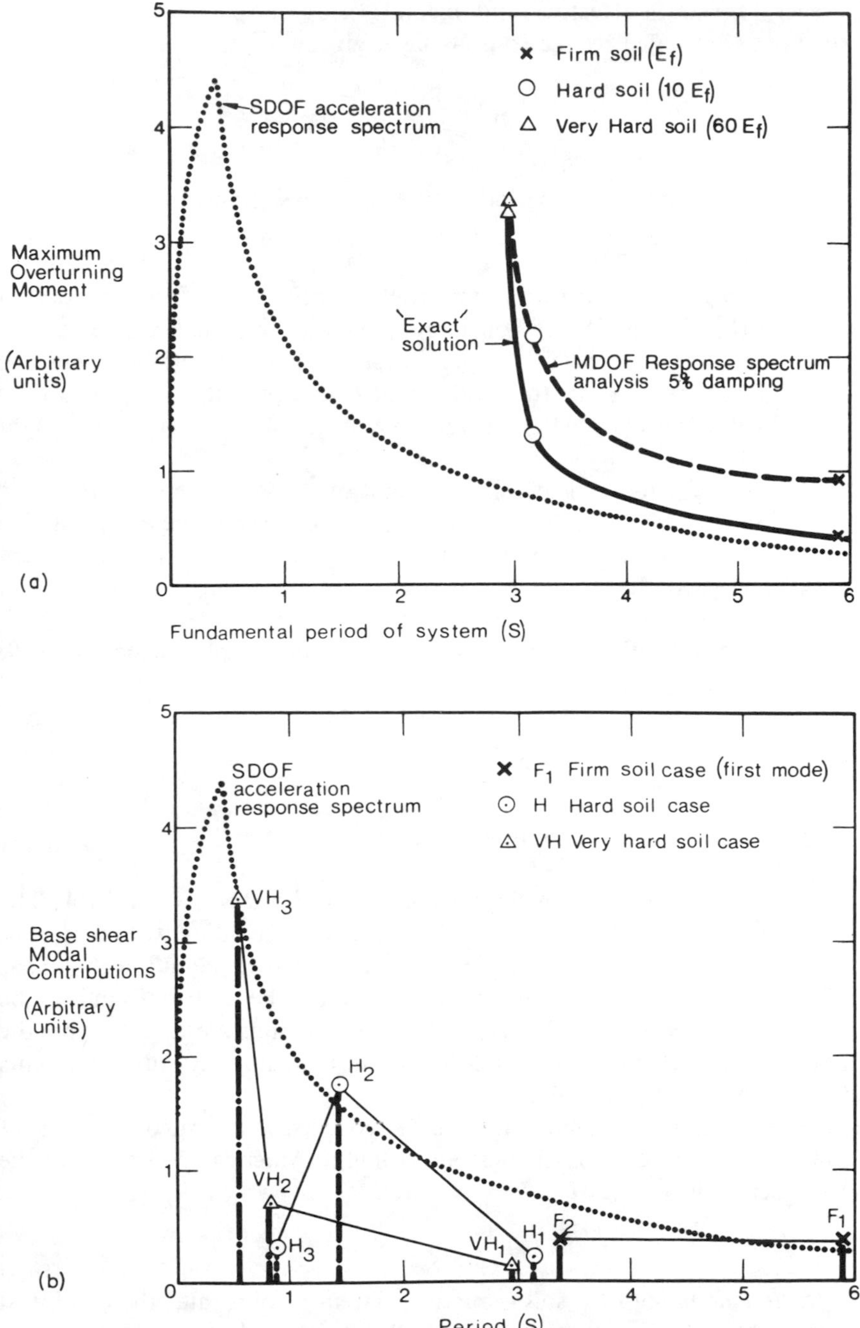

Figure 6.19 (a) Variation of peak seismic base moment of a concrete gravity oil platform, resulting from three different foundation stiffnesses; (b) base shear contributions of the significant modes for a concrete gravity oil platform with three different foundation stiffnesses

In general form, the effective fundamental period (horizontal translation) of a structure as modified by the soil has been given [73,75] as

$$\tilde{T} = T \sqrt{\left[1 + \frac{\bar{k}}{k_x}\left(1 + \frac{k_x \bar{h}^2}{k_\phi}\right)\right]} \tag{6.39}$$

where T is the fundamental period of the fixed base structure;

$\bar{k}$ is the stiffness of the structure when fixed at the base, i.e. $\bar{k} = 4\pi^2 \overline{W}/gT^2$;

k_x and k_ϕ are the horizontal and rocking stiffnesses of the foundation in the direction being considered, such as given in Table 6.9;

$\bar{h}$ is the effective height of the structure. For buildings this may be taken[13] as 0.7 times the total height h, except that where the gravity load is concentrated at a single level it should be taken as the height to that level;

$\overline{W}$ is the effective or generalized weight of the structure vibrating in its fundamental natural mode. For buildings this may be taken[13] as 0.7 times the total gravity load used in the earthquake analysis, except that where the gravity load is concentrated at a single level, the total gravity load should be used.

For simple consideration of buildings which are square in plan, equation (6.39) may be restated as

$$\frac{\tilde{T}}{T} = \sqrt{\left[1 + \frac{1.47Jb^2}{v_s^2 T^2}(1 + 1.65J^2)\right]} \tag{6.40}$$

where b is the width of the building; and

J is the aspect ratio h/b.

Considering a building of height 80 m and width 20 m, then $J = 4$. The fundamental period of the building is $T = 1.8$ s. If the building is sited on soils for which the shear wave velocity $v_s = 100$ m/s, then from equation (6.40) the effective period of the structure as modified by the soil is found to be $\tilde{T}/T = 1.73$, i.e. $T = 3.11$ s. Clearly, the soil has a substantial effect on the vibrational characteristics of the building, even though $v_s = 100$ m/s represents moderately firm soil (Table 6.1).

The same building is significantly affected even when sited on soils of $v_s = 200$ m/s, i.e. of bedrock stiffness according to American codes,[13,14] as the effective period is $\tilde{T} = 1.22T$.

6.3.4.3

The *effective damping* of a soil–structure system incorporates the combined material and *radiation damping* in the soil, the radiation damping in some cases leading to substantial reductions in response. For large concrete gravity platforms this reduction may be as much as 50 percent, as shown for the overturning moments in Figure 6.19(a).

The effective damping factor for structure-foundation systems has been proposed[61,76] as

$$\tilde{\beta} = \beta_o + \frac{\beta}{(\tilde{T}/T)^3} \tag{6.41}$$

where β is the damping ratio for the fixed base structure; and
β_o is the foundation damping factor given in Figure 6.20.

The quantity r in Figure 6.20 is a characteristic foundation length derived as follows:

$$\text{For } \bar{h}/b_o \leqslant 0.5, \quad r = \sqrt{\left(\frac{A_o}{\pi}\right)}$$

$$\text{For } \bar{h}/b_o \geqslant 1.0, \quad r = \left(\frac{4I_o}{\pi}\right)^{1/4}$$

where b_o is the length of the foundation in the direction being analysed;
A_o is the area of the foundation;
I_o is the static moment of inertia of the foundation about a horizontal axis normal to the direction being analysed.

The tentative American earthquake code for buildings[13] has adopted equation (6.41), and makes the structural damping constant by letting $\beta = 0.05$ and ruling that the effective damping is never less than this value, i.e. $\tilde{\beta} \geqslant 0.05$. For the example building discussed in Section 6.4.3.2 above, the effective damping is $\tilde{\beta} = 0.065$ when $v_s = 100$ m/s and $\tilde{\beta} = 0.052$ when $v_s = 200$ m/s.

6.3.4.4

The *mode shapes* of a given structure change as some function of the soil stiffness. Coupled with this effect there may be a corresponding change in the predominant mode; it can be seen from Figure 6.19(b) that for a given oil platform structure the dominant mode changed from third to first with decreasing soil stiffness.

6.3.4.5

Because of the complexity and expense of rigorously computing the effects of radiation damping in the foundation, an *equivalent viscously damped* response spectrum technique would be desirable. For estimating an equivalent viscous damping for a soil–structure system, the foundation damping (radiation plus hysteretic) is not directly additive to the structural damping, as described in Section 6.3.4.3. Watt *et al.*[74] found that the equivalent damping concept for use with response spectrum analyses of concrete gravity oil platforms was not satisfactory, as the amount of damping required depended strongly on the part of the structure under consideration (Figure 6.19). However, for structures with more uniform stiffness and mass distributions the equivalent viscous damping concept may be reliable.

Therefore a simple method of modifying the results of analyses of fixed-base buildings has been proposed by the Applied Technology Council,[13] such that the base shear may be reduced to the value $\tilde{V}$:

$$\frac{\tilde{V}}{V} = 1 - \frac{\Delta V}{V} \tag{6.42}$$

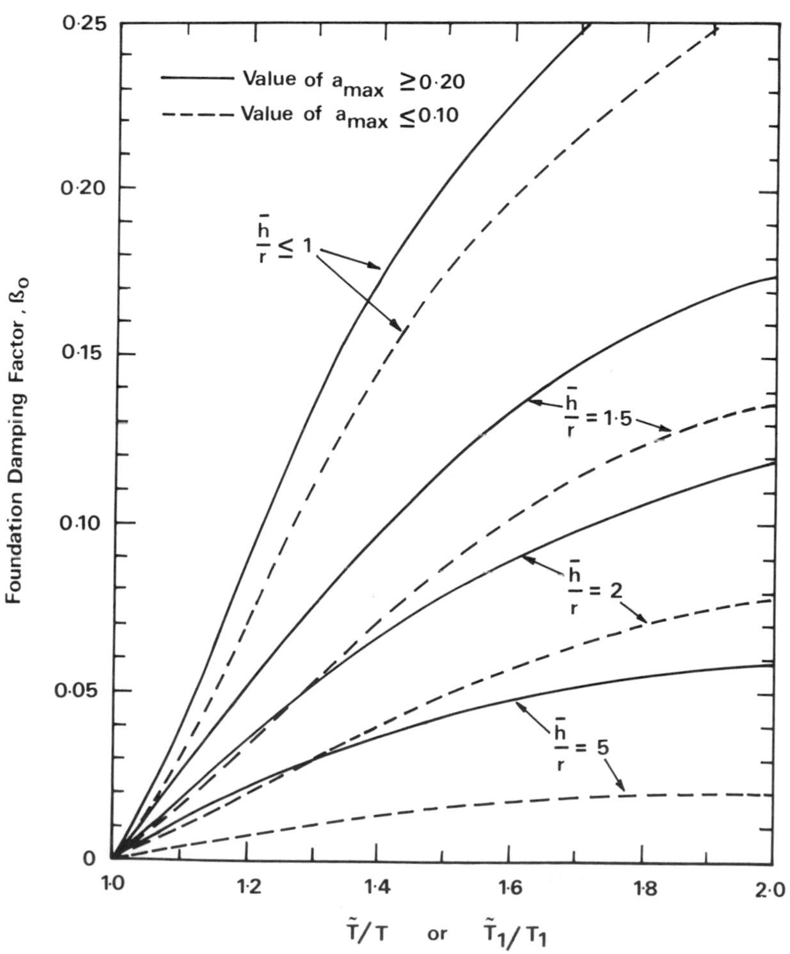

Figure 6.20 Foundation damping factor β_o (reproduced with permission from ATC[13]). a_{max} is the effective ground acceleration

where V is the base shear for a fixed base structure and ΔV is the reduction in base shear given by

$$\Delta V = \left[C_s - \tilde{C}_s \left(\frac{0.05}{\tilde{\beta}} \right)^{0.4} \right] \overline{W} \tag{6.43}$$

where $C_s = V/\overline{W}$ is the seismic design coefficient for the fixed base structure of period T; and

$\tilde{C}_s$ is the seismic design coefficient for the flexibly supported structure of period $\tilde{T}$.

These expressions may be used to modify the equivalent-static lateral forces derived from the code, or the moment and shears derived from a fixed base modal analysis (the ATC restricts the reduction to the first mode forces). The above expressions relate to conditions where the soil may be regarded as a homogeneous half-space, and will hence be unconservative when radiation damping is reduced by shallow reflective soil layer interfaces (see Figure 6.18 and the related text).

The above expressions may be used to arrive at

$$\frac{\tilde{V}}{V} = \frac{W - \overline{W}}{W} + \frac{\overline{W}S}{W} \left(\frac{1}{\tilde{T}/T} \right)^{2/3} \left(\frac{0.05}{\tilde{\beta}} \right)^{0.4} \tag{6.44}$$

where $\tilde{T}/T$ and $\tilde{\beta}$ are given in equations (6.39) or (6.40) and (6.41); and

S is the appropriate value of the soil profile coefficient as given in Table 6.11.

For a building of uniformly distributed mass and stiffness $\overline{W} \approx 0.7W$ (Section 6.3.4.2), so that equation (6.44) reduces to

$$\frac{\tilde{V}}{V} = 0.3 + 0.75 \left(\frac{1}{\tilde{T}/T} \right)^{2/3} \left(\frac{0.05}{\tilde{\beta}} \right)^{0.4} \tag{6.45}$$

In the above equations $\tilde{\beta}$ has a minimum value of 0.05, and the reduction in base shear due to soil–structure interaction is limited to 30 percent, giving the following range of values: $0.7 \leqslant \tilde{V}/V \leqslant 1.0$

Table 6.11. Soil profile coefficient, S, proposed by the ATC[13]

	Soil profile type*		
	S_1	S_2	S_3
S	1.0	1.2	1.5

*S_1 denotes rock of any characteristic, and having a (low strain) $v_s \geqslant 760$ m/s, or where sands, gravels, or stiff clay deposits less than 60 m thick overlie rock.

S_2 denotes profiles where sands, gravels and stiff clays exceeding 60 m thick overlie rock.

S_3 denotes soft to medium stiff clays and sands exceeding 9 m in thickness.

Referring again to the example building discussed above in Sections 6.3.4.2 and 6.3.4.3, and values of $\tilde{T}/T$ and $\tilde{\beta}$ derived there may be inserted into equation (6.44), and $\overline{W}$ may be taken as equal to 0.75. Taking $S=1$ for soils with $v_s=200$ m/s, and $S=1.2$ for $v_s=100$ m/s, the 80 m high by 20 m square building would have $\tilde{V}/V=0.90$ and $\tilde{V}/V=0.81$, respectively, for the two soil conditions, both representing significant favourable soil–structure interaction. It is interesting to observe that for a building of quite high flexibility (a fixed base period of $T_1=1.8$ s) sited on reasonably stiff soils with $v_s=200$ m/s, there is a predicted reduction of 10 percent in base shear compared with the fixed base condition.

6.3.4.6

The *effects of embedment* have been studied by various workers.[69,77–83] To date the findings are not very comprehensive but there is general agreement that increasing embedment increases the static stiffness of the system, decreases the periods of vibration, and decreases the displacement responses. These effects are evident in all four modes of vibration i.e. vertical, horizontal, rocking, and torsion. Where backfill is softer than the undisturbed soil, the effects of embedment are obviously reduced. In cases where theoretical results have been compared with experimental, agreement is qualitative rather than quantitative.

Approximate factors for estimating the increase in horizontal and rocking foundation stiffness have been proposed[13] as follows:

$$K_x \text{ (embedded)} = k_x\left(1+\tfrac{2}{3}\tfrac{d}{R}\right) \tag{6.46}$$

$$K_\phi \text{ (embedded)} = k_\phi\left(1+2\tfrac{d}{R}\right) \tag{6.47}$$

where k_x and k_ϕ are the horizontal and rocking stiffnesses for surface footings such as those given in Table 6.9, R is the equivalent radius for the footing, and d is the effective depth of embedment for the conditions that would prevail in the design earthquake. Because of the above-noted lack of good quantitative correlation between experiment and theory, the selection of values for d will be somewhat subjective.

6.3.4.7

The *uncertainty due to the use of different analytical techniques* is a subject of much concern, on account of the wide variety of techniques and the complexity involved which permit little useful closed-form checking. Some encouragement may be found from the results of a study by Then *et al.*,[84] who carried out comparative dynamic analyses on a nuclear reactor containment building on a particular site. They used (1) the substructure method with the program CLASSI, (2) the non-linear finite element procedure with DYNA3D, and (3) the equivalent linear finite element approach with ALUSH. They found reasonably good agreement of the responses calculated at some points on the structure, but a wide discrepancy in the spectrum shape was found at one location in the operating floor.

6.4 ASEISMIC DESIGN OF FOUNDATIONS

6.4.1 Introduction

Before completing the design of the foundations it is assumed that the dynamic characteristics of the sub-soil have been determined as discussed in Chapter 3 and Section 6.2, and a suitable form for the sub-structure should also have been chosen as suggested in Section 5.3.8.3.

It then remains to design the foundations for appropriate seismic forces which arise (1) directly from the deformation of the adjacent soil and (2) as a result of the earthquake forces acting in the superstructure. While our ability to estimate the seismic forces from (2) above is now quite advanced, there remains a great deal of uncertainty about the magnitude and effect of the forces induced directly by the ground. This is true despite the increasing attempts to elucidate the soil–structure interaction problem by sophisticated analytical and experimental techniques.

In current design practice it is often found convenient to consider two separate stress systems: (1) the seismic vertical stresses (e.g. due to overturning moments) and (2) the seismic horizontal stresses (e.g. due to the base shear on the structure). Overturning moments are not usually a problem for buildings as a whole, unless it is very slender, but can be difficult for individual footings such as column pads or shearwall strip footings. The foundation should, of course, be proportioned so as to keep the maximum bearing pressures due to the overturning moments and gravity loads within the allowable seismic value for the soil concerned. Unfortunately there is little agreement on what constitutes safe seismic bearing pressures on sedimentary soils. Most earthquake codes do not discuss the effect of soil type on bearing pressures. It appears that most soils are capable of sustaining higher short-term loads than long-term loads, with the exception of some sensitive clays which lose strength under dynamic loading.[85]

For preliminary design purposes only, the bearing pressures taken from out-of-print publication by the New Zealand Ministry of Works quoted in Table 6.12 may be helpful; here the bearing pressures are reduced by 25 percent for medium gravel and medium sand and increased by 50 percent for rock and very stiff or medium stiff clay. The latter values are given some support by Ishihara *et al.*[87] They found that the cohesion component c of partially saturated clays, i.e. a volcanic clay (PI = 30) and a sandy clay (PI = 18), was higher under dynamic loading (c_d) than under static loading (c) such that $c_d/c = 2.4$ and 1.86, respectively. The angle of internal friction was unchanged by the rate of loading. The values in Table 6.12 may in some cases be over-conservative, and well-informed geotechnical advice should in any case be taken for the actual soil conditions for final design of each project.

The horizontal interaction stresses between the soil and the foundation are arguably more problematical than the vertical stresses, as comparatively little is known about allowable seismic passive pressures and the effect of seismic active pressure in different foundation situations. Indeed it is customary to

Table 6.12. Allowable bearing pressures on soils[86] (for preliminary design only)

Soil types		Allowable bearing pressures (kN/m²): Long-term loads		Allowable bearing pressures (kN/m²): Total loads (including seismic loads)		Standard penetration blow count (N)	Apparent cohesion c_u (kN/m²)
Soft or broken rock		960		1440		30	
Gravel	Dense	285–570		285–570			
Gravel	Medium	96–285		72–215			
		Well graded	Uniform	Well graded	Uniform		
Sand*	Dense	240–525	120–265	240–525	120–265	30	
Sand*	Medium	96–240	48–120	72–180	40–90	15–30	
Clay†	Very stiff	190–380		285–570		15–30	100–200
Clay†	Medium stiff	48–190		72–285		4–15	25–100
Clay†	Soft	0–48		0–48		0–4	0–25
Peat, silts made ground		To be determined after investigation					

*Reduce bearing pressures by half below the water table.

†Alternatively: Allow 1.2 times c_u for round and square footings, and 1.0 times c_u for length/width ratios of more than 4.0. Interpolate for intermediate values.

assume even more arbitrary distributions for horizontal stress between foundations and soil than for vertical stress. The main problems (peculiar to earthquakes) of foundation design as presently understood occur in transferring the base shear of the structure to the ground, and in maintaining structural integrity of the foundation during differential soil deformations. Some design guidance on these problems now follows under the headings of;

(1) Shallow foundations;
(2) Deep box foundations;
(3) Caissons;
(4) Piled foundations.

In addition to the following discussion, further advice on aseismic foundation design is given by Zeevaert,[88] Allardice *et al.*,[89] and Taylor and Williams.[90]

6.4.2 Shallow foundations

The horizontal seismic shear force at the base of the structure must be transferred through the sub-structure to the soil. With shallow foundations it is normal to assume that most of the resistance to lateral load is provided by friction between the soil and the base of the members resisting horizontal load. Other footings and slabs in contact with the ground may also be assumed to provide shear resistance if they are suitably connected to the main resisting elements. The total available resistance to lateral movement of the structure may be taken to be equal to the product of the dead load carried by the elements considered and the coefficient of sliding friction between the soil and the sub-structure. Typical values of friction angles for foundations are given in Table 6.13.

In some cases further horizontal resistance will arise from the passive soil pressures developed against subsurface elements. If this resistance is taken into account it is often deemed wise to restrict the calculated total restraint by reducing either the frictional force or the passive resistance force by 50 percent. In order to ensure that the passive restraint can be developed, appropriate measures must be taken on site, as adequate compacting of backfill against sides of footings.

Shallow foundations are often of a form that is highly vulnerable to damage from differential horizontal and vertical ground movements during earthquakes. It is therefore good practice even in quite low structures, especially those founded on soft soils, to provide ties between column pads. In the absence of a more realistic method an arbitrary design criterion for such ties is to make them capable of carrying compression and tension loads equal to 10 percent of the maximum vertical load in adjacent columns (Section 7.2.6). However, it may be possible to resist some or all of these horizontal forces by passive action of the soil, particularly for light buildings. The designer may also have a choice between providing the tie action at the bottom floor level (in tie beams or in the slab), or at some other position in relation to the foundations.

Table 6.13. Typical friction angles and adhesion values for bases without keys[91]

Interface materials	Friction angle (degrees)	Adhesion (kN/m^2)
Mass concrete on the following foundation material:		
Clean sound rock	35–45	
Clean gravel, gravel–sand mixtures, coarse sand	29–31	
Clean fine to medium sand, silty medium to coarse sand, silty or clayey gravel	24–29	
Clean fine sand, silty or clayey fine to medium sand	19–24	
Fine sandy silt, non-plastic silt	17–19	
Very stiff and hard residual or preconsolidated clay	22–26	
Medium stiff and stiff clay and silty clay	17–19	
Formed concrete on the following foundation material:		
Clean gravel, gravel–sand mixtures, well graded rock fill with spalls	22–26	
Clean sand, silty sand–gravel mixture, single size hard rock fill	17–22	
Silty sand, gravel or sand mix with silt or clay	17	
Fine sandy silt, non-plastic silt	14	
Soft clay and clayey silt		10–35
Stiff and hard clay and clayey silt		35–60

6.4.3 Deep box foundations

Unfortunately at present an authoritative aseismic design rationale for deep box foundations does not exist, as discussed by Barnes.[92] Designers must rely mainly on normal structural and geotechnical static design techniques, supplemented where appropriate by consideration of known seismic phenomena such as seismically enhanced soil pressures. The natural stiffness and strength of box-shaped foundations should be utilized to advantage in distributing the seismic forces from the soil and the superstructure throughout the foundation with an adequate safety factor.

Although less susceptible to damage from ground motions than isolated pad footings, deep box foundations nevertheless require proper design to withstand strong earthquakes. This was exemplified by the virtual destruction of underground water tanks in the 1971 San Fernando earthquake.[93] This failure demonstated the importance of internal walls to provide an egg-crate type of stiffening; it also showed the valuable contribution that concrete keys could have provided to the strength of construction joints which moved 0.67 m in shear despite the presence of steel reinforcement normal to the joint.

6.4.4 Caissons

Caissons are similar to piles in that they are relatively slender at least in one direction and are used as isolated foundations spaced at intervals to support structures which may be long in plan, such as bridges or large buildings. In bridge construction the terms *caissons* and *piers* are sometimes used interchangeably. Where caissons penetrate the soil deeply, they need special consideration of soil–structure interaction effects, as discussed below for piles. Caissons may differ from piles in that they may often be treated as rigid rocking structures, rather than bending structures. This typically occurs for bridge piers in the direction lateral to the axis of the bridge.

Few studies of the seismic response of bridge caissons have been made, but Iwasaki and Hagiwara[94] report the results of a valuable soil–structure interaction study on deeply embedded bridge caissons in Japan. They found that resonance of the ground and the pier is an important indicator of seismic response. Using the ratio T_g/T_p, where T_g and T_p are the fundamental periods of the ground and the pier respectively, they found for a range of $T_g/T_p = 0.7$ to 1.25 that dynamic amplification of the Japanese seismic design shear occurred. For $T_g/T_p = 1.0$ this amplification factor was as high as 10 for hard ground and about 2.3 for soft ground.

6.4.5 Piled foundations

6.4.5.1 Introduction

The reliable design of *piles* for earthquake loads is difficult because of the uncertainties involved in determining the design deformation state of the piles. This is partly due to the uncertainties involved in assessing lateral soil–pile interaction and partly due to the complexity of behaviour of pile groups. As indicated in Figures 6.21 to 6.23, high bending moments may occur at various locations up the pile. In addition to the locations of high bending moments indicated by these idealized moment diagrams, high stresses may be induced at other depths due to local shear failure of weak layers of soil or due to liquefaction, or due to loss of lateral support from the soil because of scour in waterways or settlement of loose deposits.

The aseismic design of piled foundations will include consideration of the vertical and horizontal stresses and the structural integrity of the foundation. Vertical seismic loads in individual piles may vary greatly depending on their position in relation to the rest of the pile group and to the superstructure (Figure 6.21). Some piles, particularly those at the edges or corners of pile systems, may have to carry large tensile as well as compression forces during earthquakes.

Lack of structural integrity has caused failure of piled foundations in earthquakes, such as that of San Fernando, 1971. Sufficient continuity reinforcement must be provided between the piles and the pile cap, and the piles themselves must obviously be able to develop the required tensile, compression

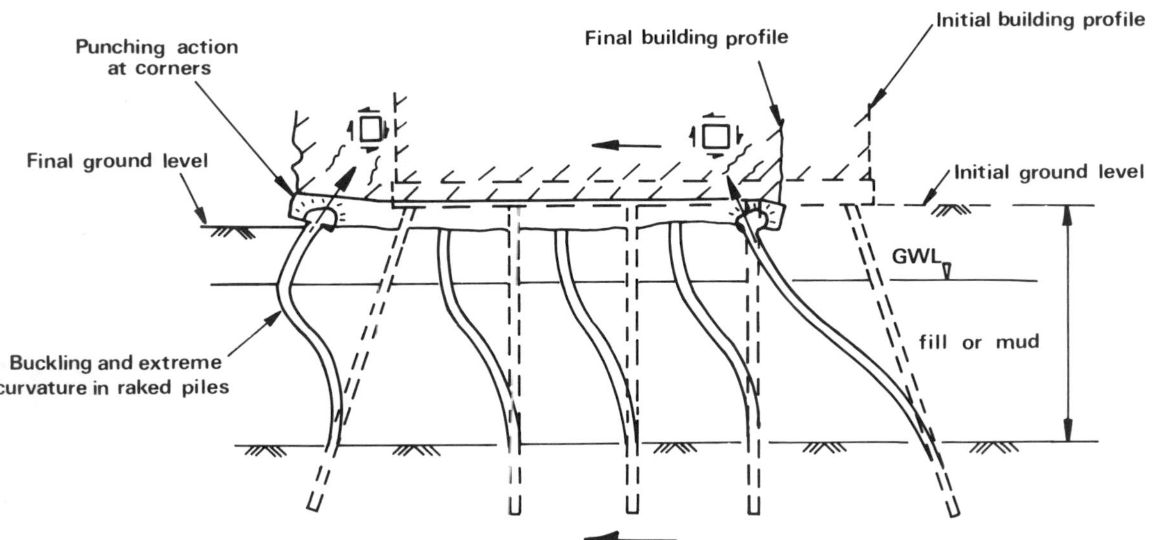

Figure 6.21 Interaction of raked piles and pilecap during an earthquake

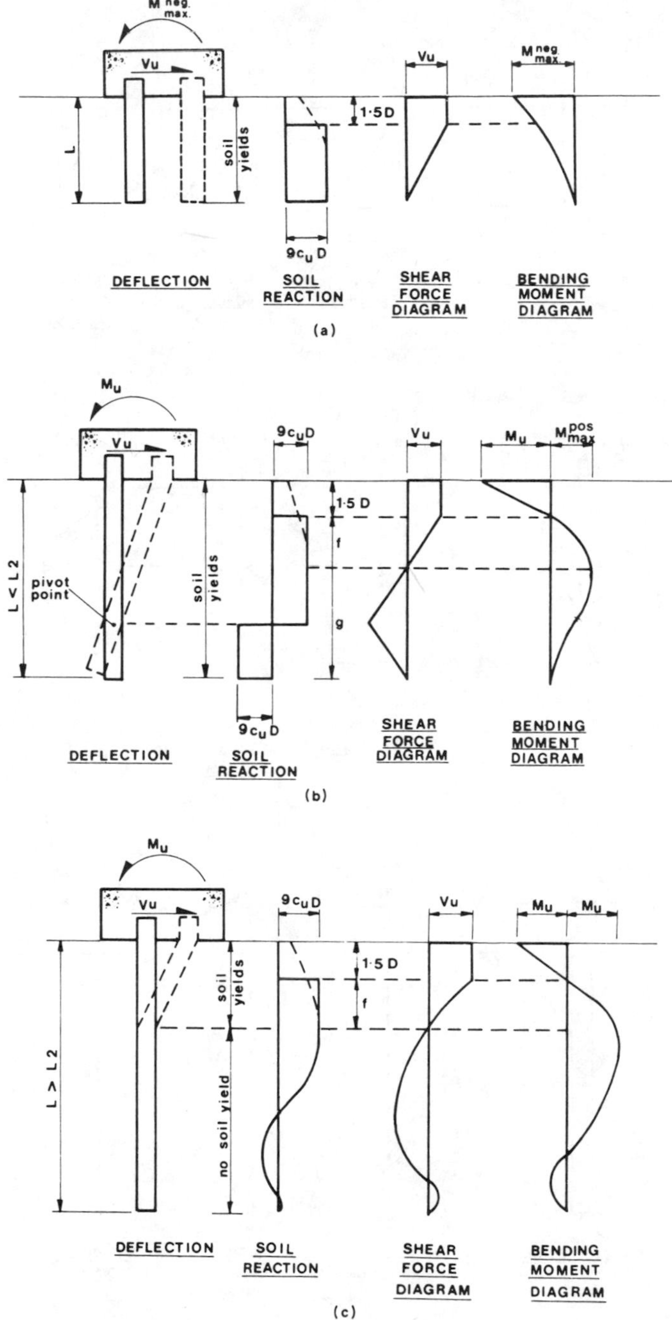

Figure 6.22 Lateral resistance of fixed-head piles in cohesive soil. (a) Short pile; (b) intermediate pile; (c) long pile (developed from Broms[98])

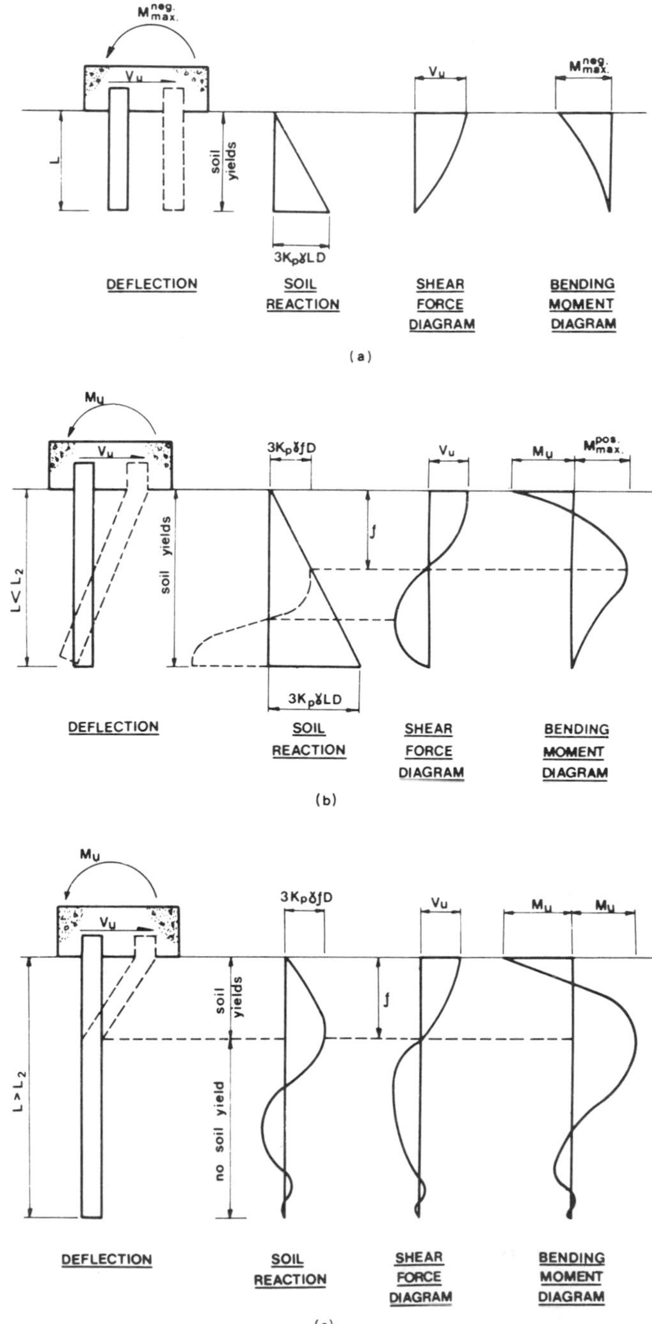

Figure 6.23 Lateral resistance of fixed-head piles in cohesionless soil. (a) Short pile; (b) intermediate pile; (c) long pile (developed from Broms[98])

and bending strength. Where plastic hinges are likely to form in concrete piles, suitable confinement reinforcement must be provided (Sections 7.2.9 and 7.5.4.1)

6.4.5.2 Dynamic response of piles

In response to horizontal ground motions it appears that piles generally follow the deformations of the ground[95] and do not cut through the soil. It also seems that piles are subject to two distinct failure mode zones:

(1) In the upper part of the pile, say the top $10D$ (D = diameter), the response is affected by the presence of the free soil surface which permits the soil adjacent to the pile to yield and move upwards in a wedge (Figure 6.24). Also the upper part of the pile has inertia loads induced in it by the surrounding soil and the structure above.
(2) In the lower part of the pile the surrounding soil dominates the response, and flexibility or ductility is required to permit the pile to safely conform to the curvatures imposed by the soil deformations.

In the dynamic response analysis of piled foundations for design purposes, because the soil–pile interaction is so complex it is usual to simplify the structural modelling problem, often as much as in the following opposing options:

(1) Ignore the soil entirely, using only the stiffness of the pile, after having first defined some depth to pile fixity based on soil stiffness (see Equivalent Cantilever Method, Section 6.4.5.3(iii); or
(2) Ignore the horizontal and rotational stiffnesses of the pile, using only the stiffness of the soil.

The development of more sophisticated research-orientated analytical techniques, similar to the shear beam model described in Section 6.2.2.2, has been along two main lines:

(1) A continuous elastic model (for example, Novak,[96] and Gazetas and Dobry[59]);
(2) A discrete model with lumped masses, springs, and dashpots (for example, Penzien[55] and Blaney *et al.*[97]).

Such techniques permit pile stresses as well as stiffnesses to be estimated. The linear analyses are generally conducted in the frequency domain, pile stiffnesses and damping being expressed in frequency dependent terms, as discussed in Section 6.3.3.3. Significant differences between the results of such studies need to be resolved, particularly for the low-frequency properties.

Method (2) above is used for non-linear as well as linear analyses. Some of the non-linear analyses studied that have been done have not been true dynamic earthquake analyses of the shear beam type noted above, but have been of either repeated or quasi-static cyclic loading nature, in which the soil–pile system has been loaded by a horizontal load and perhaps a moment applied at the top. Clearly, such a loading model closely represents wind or wave loads, but for

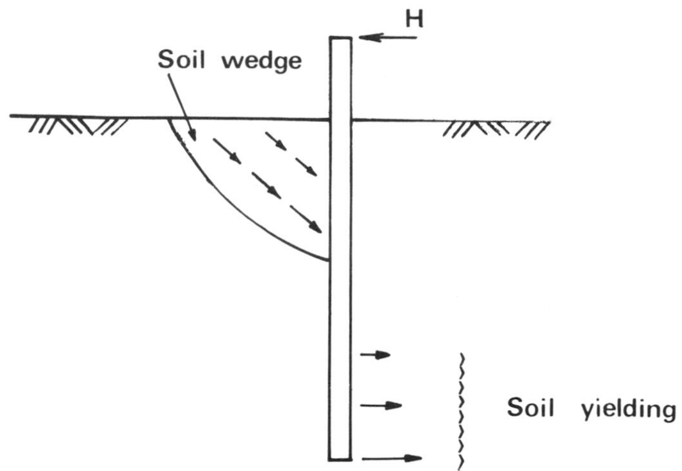

Figure 6.24 Limiting lateral load on a pile defined by two separate failure modes

seismic loading it would give better estimates of stiffness than of worst deformations in the pile.

In an example of a full dynamic earthquake response analysis, some light has been thrown on the likely behaviour of long piles in deep sensitive clay in a sophisticated non-linear analysis of a bridge described by Penzien.[55] In this case it was found that if subjected to an earthquake like that of El Centro (1940), the piles would have been deformed to their yield curvatures.

6.4.5.3 Equivalent static lateral loads on piles

In the majority of design projects, pile design and foundation modelling for superstructure analysis will be carried out with reference to (separate) equivalent static load analyses of the piles. The latter comprise many methods which may be divided into three categories:

(1) Limiting (or ultimate) loads;
(2) Elastic continuum;
(3) Non-linear discontinuum.

Each of these analytical concepts is discussed below.

6.4.5.3(i) *Limiting lateral loads on single piles*

In limiting load analyses the soil is considered to have two distinct failure mode zones as described in Section 6.4.5.2. A leading example of the application of these assumptions is that due to Broms,[98] who assumed that the ultimate lateral resistance of a pile is developed when the soil yields and plastic hinges develop in the pile. The design method for friction piles set out below uses Broms recommendations for the ultimate bearing capacity of the soil, i.e. $9c_u$ for cohesive soil and $3K_p$ for cohesionless soil. It should be noted that other research has suggested different values and soil stress profiles, e.g. Matlock[99] and Brinch Hansen.[100]

The method, however, is thought to give a reasonable estimate of the ultimate moments in piles, and may be adapted to suit *end bearing piles* with moment fixity at the bottom. Its main shortcoming from a design point of view is that the method does not provide an estimate of deflections, unlike the other methods outlined below.

Lateral resistance of friction piles

It is assumed that there are three different modes of pile failure, depending on the pile length L.

Short pile (Figure 6.22a). When the pile is short, failure occurs as the soil yields along the full length of the pile. The limiting case is reached when a hinge develops in the top of the pile at the same time as the soil yields. This occurs when $L=L_1$, and for $L<L_1$ the ple is classed as short. L_1 can be evaluated in terms of the soil properties and the moment capacity of the pile.

Long pile (Figure 6.22c). If the pile is sufficiently long, failure occurs when two hinges develop in the pile. This occurs if the pile is firmly enough embedded in the bearing stratum, otherwise the soil will fail and the pile will rotate (Figure 6.22b). A limiting length L_2 can be determined such that when $L>L_2$ the pile is classed as long.

Intermediate length pile. If the pile length is between L_1 and L_2 the pile is classed as intermediate.

The design procedure is as follows.

(1) Decide pile lengths on the basis of vertical load bearing considerations.
(2) For the given soil conditions and pile dimensions establish L_1 and L_2.
(3) Establish the type of lateral failure by comparing L to L_1 and L_2.
(4) Having classed the pile, calculate the ultimate lateral resistance V_u.
(5) Check that the lateral deflection of the piles is tolerable.

When the pile is of intermediate length it may be easiest to calculate V_u and L_1 and L_2 and interpolate for V_u at L.

Evaluation of limit lengths L_1 and L_2

Let M_u be the ultimate moment capacity of the pile for the given axial load and reinforcement layout, and let D be the pile diameter.

(1) *Friction piles in cohesive soils* (*Figure 6.22*). Let the ultimate bearing capacity of the soil be $9c_u$, where c_u is the undrained shear strength, and assume $c_u=0$ to a depth of $1.5D$. Referring to Figure 6.22(a) when $L=L_1$:

$$M_u=9c_uD(L_1-1.5D)\left\{\frac{(L_1-1.5D)}{2}+1.5D\right\}$$

i.e.

$$M_u = 4.5c_u D(L_1^2 - 2.25D^2)$$

Therefore

$$L_1 = \left\{ 2.25D^2 + \frac{M_u}{4.5c_u D} \right\}^{1/2} \qquad (6.48)$$

Similarly, referring to Figures 6.22(b) and (c), when $L = L_2$:

$$L_2 = f + g + 1.5D \qquad (6.49)$$

where f can be found from

$$2.25c_u Df^2 + 6.75c_u D^2 f - M_u = 0$$

and

$$g = \left[\frac{M_u}{2.25c_u D} \right]^{1/2}$$

(2) *Friction piles in cohesionless soils* (*Figure 6.23*). Assume the lateral earth pressure coefficient at failure is $3K_p$, where K_p is the Rankine passive pressure coefficient. Referring to Figure 6.23(a), when $L = L_1$:

$$M_u = 2K_p\gamma \frac{DL_1^2}{2} \cdot \frac{2L_1}{3}$$

$$L_l = \left[\frac{M_u}{K_p\gamma D} \right]^{1/3} \qquad (6.50)$$

where γ is the weight per unit volume of the soil.

Similarly, referring to Figures 6.23(b) and (c), when $L = L_2$, L_2 may be found from

$$M_u = V_u L_2 - \frac{K_p\gamma L_2^3 D}{2} \qquad (6.51)$$

using

$$V_u = 1.5K_p\gamma Df^2 \qquad (6.52)$$

and

$$f = \left[\frac{2M_u}{\gamma K_p D} \right]^{1/3}$$

6.4.5.3(ii) *Elastic continuum analysis of single piles*

In this method the pile is assumed to be a thin rectangular vertical strip of width D, and length L, with constant stiffness $E_p I_p$. The pile is divided vertically into elements each of which is acted on by a uniform horizontal stress p, while the soil is assumed to be a homogeneous, isotropic, elastic half-space with properties E_s and υ_s which are unaffected by the presence of the pile. Poulos[101–3] and Mathewson[104] have both used the above assumptions and their results compare well with displacement profiles measured in tests,[104] and also with results of the subgrade reaction method discussed below.

In order to extend the method from uniform soil properties to the usual situation where soil stiffness increases with depth, Poulos[105] developed an approximate modification of the above method. In the same paper, he describes a further use of the elastic theory to allow for elastoplastic behaviour and shows the importance of allowing for soil yield, especially for flexible piles.

As the term 'continuum' implies, this method takes account of the influence of the relative displacement of adjacent points in the soil, which is a feature not incorporated into the discontinuum methods of the following section.

6.4.5.3(iii) *Non-linear discontinuum analyses of single piles*

This method represents the soil as a series of independent horizontal springs interacting with the pile which is considered as a beam. Hence the method is often referred to as the *Winkler spring* method. Also as the spring stiffness may be related to the coefficient of horizontal subgrade reaction, the method is sometimes referred to as the *subgrade reaction* method.

The force p_i in each spring is

$$p_i = k_{si} \Delta_{si} \tag{6.53}$$

where Δ_s is the horizontal deflection and k_s is the spring stiffness, which is

$$k_s = k_h \, D \, \delta L = K_h \, \delta L \qquad \text{cohesive soil} \tag{6.54}$$

or $$K_s = k_h \, D \, \delta L = n_h \, z \, \delta L \qquad \text{cohesionless soil} \tag{6.55}$$

where k_h = coefficient of horizontal subgrade reaction;
k_h = modulus of horizontal subgrade reaction;
n_h = constant of horizontal subgrade reaction;
D = diameter of pile;
δL = height of pile related to the soil spring;
z = depth of spring below ground surface.

Typical values of subgrade reaction parameters for sands derived by Terzaghi[106] are given in Table 6.14, while values specific to a given site may be determined by various field or laboratory methods.

Because the stress–strain behaviour of soil is highly non-linear, the modulus of subgrade reaction is non-linear, and is generally derived from p–y curves (i.e. force-deflection curves) in a method introduced by Matlock.[99] Obviously,

Table 6.14. Constant of horizontal subgrade reaction, n_h, for sands (kN/m^3) after Terzaghi[106]

Loose		Medium		Dense	
Dry	Submerged	Dry	Submerged	Dry	Submerged
2500	1400	7500	5100	20 000	12 000

an iterative procedure is required to find the equivalent linear system when solving the structural model of equation (6.53) for all values of p_i when using a linear analytical technique.

The *equivalent cantilever method* is another procedure for analysing pile behaviour which is based on the above method. Earlier work on this method by Kocsis[107] has been adapted by the New Zealand Ministry of Works and Development[108] to give the equivalent depths to fixity L_m, L_d, that give the design moment or deflection, respectively, in the pile, assuming the pile to be free of the soil.

In this method the pile is modelled as a free-standing cantilever or a built-in beam. Different lengths of equivalent cantilever are selected depending on whether moment developed within the pile or deflection of the pile is being investigated. Rules to estimate equivalent cantilever lengths are given in Table 6.15 for free- and fixed-headed piles. Bending moments and displacement are quite sensitive to the degree of fixity at the pile head. Piles of intermediate pile head flexibility may develop maximum bending moments as little as 60 percent of that of the free head case, while for displacement the free and fixed-headed cases provide upper and lower bounds, respectively. Bending moments applied at the head of a free-headed pile in conjunction with a lateral load may be modelled by increasing the effective length of the pile and calculating moment and deflection accordingly.

The method can be relied upon only where there is little variation in soil properties to a depth of $4R$ below the surface, where R is a relative stiffness factor. Also, in deriving the expression in Table 6.15, the piles have been assumed to be flexible such that

$$L/R > 4 \tag{6.56}$$

where L = length of pile below ground surface;

$$R = \left[\frac{EI}{K_h}\right]^{1/4} \quad \text{for cohesive soils} \tag{6.57}$$

$$R = \left[\frac{EI}{n_h}\right]^{1/5} \quad \text{for cohesionless soils} \tag{6.58}$$

where E and I are the usual stiffness properties of the pile, and K_h and n_h are soil properties as defined above. For the purposes of evaluating R, the soil parameters are averaged to a depth of 6 to 8 pile diameters.

6.4.5.4 Pile groups

The resistance of pile groups to horizontal loads is more complex than for single piles, due to interaction between piles and also due to geometry in the case of raking piles. As pile spacing decreases, so does the effective stiffness of the trailing piles in a group. From the limited research that has been done on this subject[109-11] it appears that this effect occurs when pile spacing in the direction of loading is less than about $8D$, where D is the pile diameter, and that the parameters of horizontal subgrade reaction (k_h or n_h) used above should be reduced for cohesionless soils as shown in Figure 6.25.

For piles in cohesionless soils it seems that no reduction in K_h is required for trailing piles where the pile spacing exceeds about $4D$.

6.4.6 Foundations in liquefiable ground

When a structure is to be built on liquefiable ground there are a number of procedures which may be used to ensure adequate foundation safety, which fall into two categories;

(1) Pile foundations; and
(2) Ground improvement.

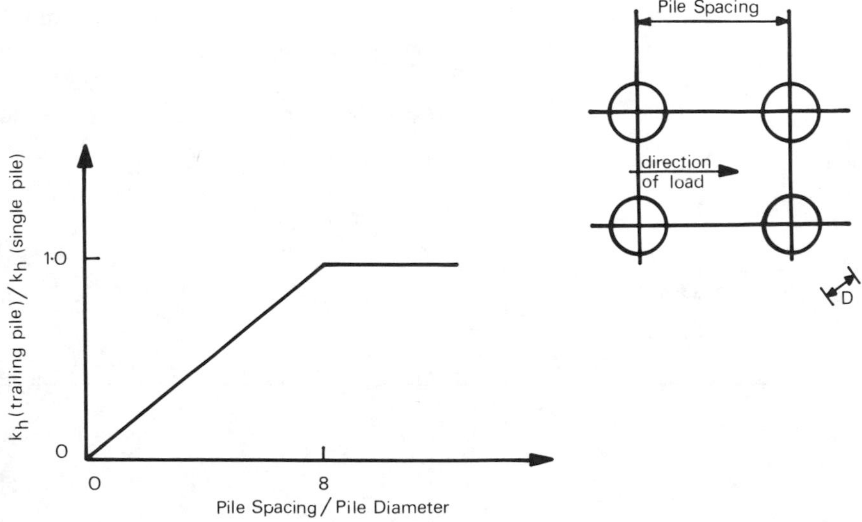

Figure 6.25 Effect on horizontal subgrade reaction of interaction between piles subject to lateral load (after Hughes *et al.*[157])

Table 6.15. Equivalent cantilever expressions for design of piles for lateral loads (from reference[108])

Type	Equivalent structure	Moment and deflection
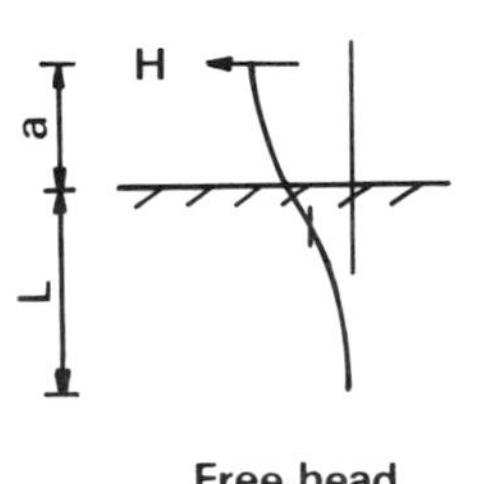 Free head	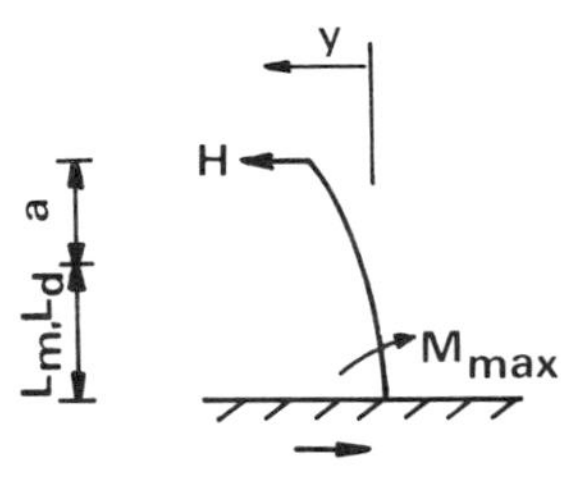	$M_{max} = H(L_m + a)$ $y = \dfrac{H(L_d + a)^3}{3EI}$
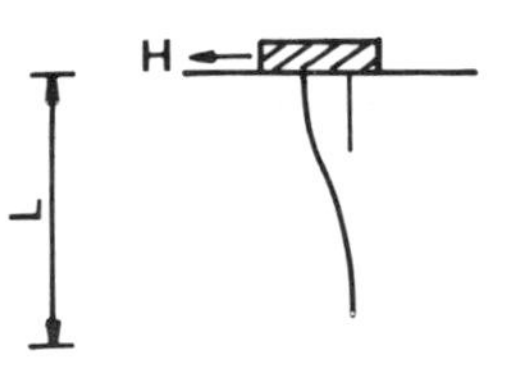 Fixed head		$M_{max} = \dfrac{H\ L_m}{2}$ $y = \dfrac{H\ D_d^3}{12\ EI}$

DEPTHS TO FIXITY, L_d or L_m		
Soil type	(For finding L_d displacement y)	(For finding L_m moment M_{max})
Cohesive where $R=4\sqrt{\left(\frac{EI}{K_h}\right)}$	(i) Free-head $L_d=1.4R$ for $\frac{a}{R}>2$ $L_d=1.6R$ for $\frac{a}{R}<2$ (ii) Fixed-head $L_d=2.2R$	(i) Free-head $L_m=0.5R$ (ii) Fixed head $L_m=1.5R$
Cohesionless where $R=5\sqrt{\left(\frac{EI}{n_h}\right)}$	(i) Free-head $L_d=1.8R$ for $\frac{a}{R}>1$ $L_d=2.2R$ for $\frac{a}{R}<1$ (ii) Fixed-head $L_d=2.5R$	(i) Free-head $L_m=0.8R$ (ii) Fixed head $L_m=2.0R$

When using piles in liquefiable ground, the piles should be designed for the conditions induced by liquefaction, as the loss of soil support in the liquefied layer may cause large forces in the piles.[112] The piles should be detailed for strength and ductility as described for concrete in Section 7.2. However, sole reliance on piles should be practised with caution, because of the difficulty of determining the location and thickness of potential liquefaction layers. In some cases it may be wise to combine piling with a degree of ground improvement to reduce the probability of liquefaction occurring.

Ground improvement techniques to overcome the liquefaction hazard include (1) compaction to increase the density out of the dangerous range or (2) use of drainage to reduce the build-up of pore-water pressure during ground shaking. Ohsaki[113] reported on the effectiveness of compaction using the vibroflotation technique in preventing liquefaction in the Tokachioki earthquake of 1968, which caused liquefaction at similar uncompacted nearby sites. Seed and Booker[114] describe a design procedure for the use of gravel drains in liquefiable sand. The drains consist of vertical columns of gravel or stones which are installed by driving a steel casing into the sand and filling it with the drainage material. The spacing of the drains is governed by the length of drainage path and the corresponding time required to permit safe dissipation of pore-pressure build-up.

6.5 ASEISMIC DESIGN OF EARTH-RETAINING STRUCTURES

6.5.1 Introduction

As the non-seismic design of earth-retaining structures is well discussed elsewhere little other than seismic considerations are dealt with here. The principal types of structure covered in this section are retaining walls and basement walls.

The magnitude of the seismic earth pressures acting on an earth-retaining structure in part depends on the relative stiffness of the structure and the associated soil mass. Two main categories of soil–structure interaction are usually defined, namely;

(1) Flexible structures which move away from the soil sufficiently to minimize the soil pressures, such as slender free-standing retaining walls;
(2) Rigid structures such as basement walls or tied-back retaining walls.

In case (1) above, active pressures will occur, and the amount of movement required to produce the active state is of the order indicated in Table 6.16. The amount of wall movement which will occur during earthquakes depends mainly on the foundation fixity and the wall flexibility. Unless a more exact analysis is made, the following earth pressure states may be used:

(1) *Flexible*. Walls founded on non-rock materials or cantilever walls higher than 5 m, assume active soil state.

Table 6.16. Movement of wall required to produce active state

Soil	Wall movement/height
Cohesionless, dense	0.001
Cohesionless, loose	0.001–0.002
Firm clay	0.01–0.02
Loose clay	0.02–0.05

(2) *Intermediate*. Cantilever walls less than 5 m high founded on rock.
(3) *Rigid*. Counterfort or gravity wall founded on rock or piles; at-rest soil state.

In general wall stiffnesses will lie somewhere between the extremes of *flexible* and *rigid*, and interpolation between the forces generated in these states may be appropriate.

6.5.2 Seismic earth pressures

In the present state of knowledge, the recommended method of obtaining seismic earth forces is that using equivalent-static coefficients.[115,116] Only for exceptional structures would dynamic analyses using finite elements seem warranted for design.

In the equivalent-static method a horizontal earthquake force equal to the weight of the soil wedge multiplied by a seismic coefficient is assumed to act at the centre of gravity of the soil mass. This earthquake force is additional to the static forces on the wall.

In general, the total earth pressure on a wall during an earthquake equals the sum of three possible components:

(1) Static pressure due to gravity loads;
(2) Dynamic pressure due to the earthquake;
(3) Pressure due to the wall being displaced into the backfill by an external force, e.g. by the horizontal sway of a bridge deck at a monolithic abutment. Design recommendations for this condition are given by Matthewson *et al.*[117]

The soil pressures may be estimated by the following methods:

(1) Elastic theory;
(2) Approximate plasticity theory, e.g. Coulomb and Mononobe–Okabe;
(3) Numerical methods, modelling the soil as Winkler springs (Section 6.4.5.3(iii)) or as finite elements.

It should be noted that it is not appropriate to design all earth-retaining structures for earthquake soil pressures. For example, external retaining walls of modest height with no significant consequences of failure are generally not designed for earthquakes in many countries. Using the Mononobe–Okabe methods, it is readily seen[116] that it requires an effective ground acceleration of about 0.3g to produce an earthquake force increment equal to the static earth pressure for cohesionless soils. So, clearly a safety factor of 2.0 on a non-seismic design should permit walls to survive moderate earthquakes, with acceptable displacements. This rationale is applied to bridge design in New Zealand, where it is recommended[117] that higher risks should be accepted for lesser structures, by designing the walls (as well as the decks) for earthquakes of lower return period.

6.5.2.1 Active seismic pressures in unsaturated cohesionless soils

The most commonly used solution is that derived by Mononobe and Okabe[118,119] based on Coulomb's theory. The effect of an earthquake is represented by a static horizontal force equal to the weight of the wedge of soil multiplied by the seismic coefficient. Referring to Figure 6.26, the Mononobe–Okabe equations are as follows:

The total force on a wall due to the static and earthquake active earth pressures due to *unsaturated* cohesionless soils is

$$P_{AE} = \tfrac{1}{2} K_{AE} \gamma_d H^2 (1 - \alpha_v) \tag{6.59}$$

where

$$K_{AE} = \frac{\cos^2(\phi o' - \beta - \theta)}{\cos\theta \cos^2\beta \cos(\delta + \beta + \theta) \left\{ 1 + \sqrt{\left[\dfrac{\sin(\phi o' + \delta)\sin(\phi - i - \theta)}{\cos(\delta + \beta + \theta)\cos(\beta - i)} \right]} \right\}^2} \tag{6.60}$$

and

$$\cot(\alpha_{AE} - i) = -\tan(\phi' + \delta + \beta - i) + \sec(\phi' + \delta + \beta - i) \times \sqrt{\left\{ \frac{\cos(\beta + \delta + \theta)\sin(\phi' + \delta)}{\cos(\beta - i)\sin(\phi' - \theta - i)} \right\}} \tag{6.61}$$

where

α_{AE} = slope angle of failure plane in an earthquake (Figure 6.26)
β = the angle of the back face of the wall to the vertical;
γ_d = the unit weight of the soil;
δ = the angle of wall friction;
ϕ' = the effective angle of shearing resistance;
i = the slope angle of the backfill;
$\theta = \tan^{-1}[\alpha_h/(1 - \alpha_v)]$
α_h = seismic coefficient = $\frac{1}{g}$ (horizontal ground acceleration);
α_v = seismic coefficient = $\frac{1}{g}$ (vertical ground acceleration).

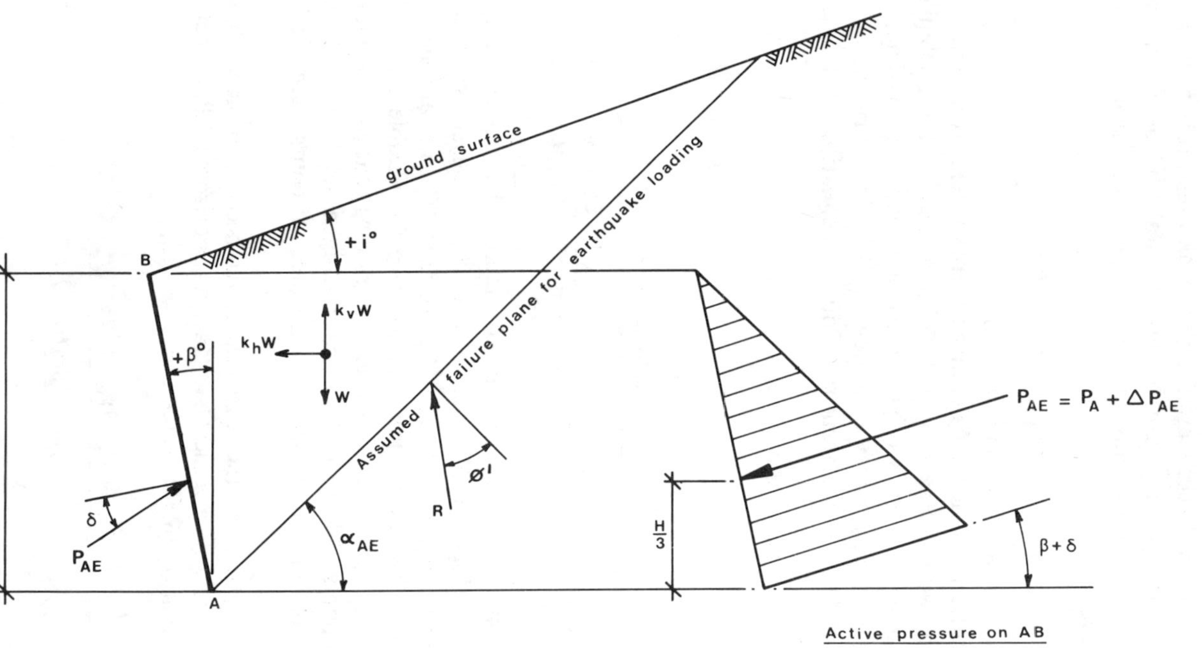

Figure 6.26. Active pressure due to unsaturated cohesionless soil on a flexible retaining wall during an earthquake, for use with Mononobe–Okabe equations (Coulomb conditions)

The ground accelerations $\alpha_h g$ and $\alpha_v g$ would normally correspond to those for the *design earthquake* ground motions (Chapter 4), except as modified to allow for wall inertia effects, as discussed below.

The effect of vertical acceleration on the wall pressures has been shown to be small,[116] except in the case of gravity walls,[120] as discussed below. Therefore, for non-gravity walls, the term α_v disappears from equation (6.59) which thus reduces to

$$P_{AE} = \tfrac{1}{2} K_{AE} \gamma_d H^2 \tag{6.62}$$

and the expression for θ becomes $\theta = \tan^{-1}\alpha_h$.

In the conditions assumed in Coulomb's theory where the shearing resistance is mobilized between the back of the wall and the soil, the earthquake earth pressure is calculated directly (Figure 6.26). For concrete walls against formwork, the wall friction δ may be taken as $\tfrac{1}{2}\phi'$. The static active force P_A (Figure 6.26) may be found from the Coulomb equation

$$P_A = \tfrac{1}{2} K_A \gamma_d H^2 \tag{6.63}$$

where

$$K_{AE} = \frac{\cos^2(\phi' - \beta)}{\cos^2\beta \cos(\delta + \beta)\left[1 + \sqrt{\left\{\dfrac{\sin(\phi' + \delta)\sin(\phi - i)}{\cos(\phi' + \beta)\cos(i - \beta)}\right\}}\right]^2} \tag{6.64}$$

Mononobe and Okabe apparently considered that the earthquake force $\Delta P_{AE} = P_{AE} - P_A$ (Figure 6.26), calculated by their analysis, would act on the wall at the same position as the initial static force P_A, i.e. at a height of $H/3$ above the base. This assumption is reasonable for flexible walls[117,121] which rotate as required for the active state. Suggestions that the earthquake force from the Mononobe–Okabe analysis acts at a higher level, appear to have been based on tests of more rigid construction which, of course, are not applicable to the active state.[116,120]

Equation (6.60) describes the general case, and become considerably simplified in the case of a wall with a vertical back face and horizontal fill, so that $\beta = \omega = 0$. In this case equation (6.60) reduces to

$$K_{AE} = \frac{\cos^2(\phi' - \theta)}{\cos^2\theta\left[1 + \sqrt{\left\{\dfrac{\sin\phi'\sin(\phi - \theta)}{\cos\theta}\right\}}\right]^2} \tag{6.65}$$

A simple way of obtaining K_{AE} from K_A (for which design charts are available) has been derived by Arango and is described by Seed and Whitman.[116]

While the Mononobe–Okabe analysis is widely accepted as the basis for seismic design of retaining walls in Coulomb conditions, Richards and Elms[120] have shown that for *gravity walls* it needs modification to allow for the effect of wall inertia, which causes pressures of the same size as the dynamic pressure derived from the Mononobe–Okabe analysis given above. Their design method, based on a deflection criterion rather than stresses or stability, involves the following steps:

(1) Select values of peak ground acceleration Ag and the ground velocity V (e.g. for the USA see reference 120, or for New Zealand see reference 122).
(2) Select the maximum allowable permanent horizontal displacement, d_L.
(3) Find the resistance factor N (where Ng is the acceleration at which the wall begins to slide) such that the actual permanent displacement will just equal d_L. For finding N, Richards and Elms recommend using an expression for the dimensionless parameter N/A, as follows:

$$\frac{N}{A} = \left[\frac{0.087V^2}{d_L Ag} \right]^{1/4} \tag{6.66}$$

(4) However, N is equivalent to the limiting value of the seismic coefficient acting on the wall, α_h. Hence we may find the active pressure coefficient K_{AE} from equation (6.60). The horizontal force due to the wall (weight W_w) is $\alpha_h W_w$, and the effective design weight of the wall for sliding is $(1-\alpha_v)W_w$. The resistance to sliding is

$$F = (1-\alpha_v)W_w \tan\theta_b \tag{6.67}$$

where θ_b is the friction angle for the base of the wall.

Equating F to $\alpha_h W_w$ plus the horizontal components of P_{AE} it is found that

$$W_w = \frac{\frac{1}{2}\lambda_d H^2[\cos(\delta+\beta) - \sin(\delta+\beta)\tan\theta_b]}{\tan\theta_b - \tan\theta} K_{AE} \tag{6.68}$$

In order to allow for uncertainties in their design method, Richards and Elms[120] suggest using a safety factor of 1.5 on the above wall weight so that the weight of wall as built should be $1.5W_w$.

As an improvement on the above design procedure for gravity walls, in an enlightening study of the uncertainties involved Whitman and Liao[123] propose replacing equation (6.66) with

$$\frac{N}{A} = -\frac{1}{9.4} \ln \left[\left(\frac{d_L}{F_c} \right) \frac{Ag}{130V^2} \right] \tag{6.69}$$

where F_c is a safety factor on the allowable displacement d_L. An appropriate value of F_c may be found from the probability distribution of d_R $(=d_L/F_c)$, which appears to be log-normal.[123] Thus, if a 95 percentile value is required, a value of $F_c = 3.8$ should be used. Then, using the value of N obtained from equation (6.69) would lead to the value of W_w from equation (6.68), which would be used directly in the design without applying the factor 1.5 as in step (4) above.

6.5.2.2 *Active seismic pressures in cohesionless soils containing water*

For cohesionless soils containing water the above solution using the Mononobe–Okabe equations is not realistic, and attempts to use them by applying factors to the densities and using the apparent angle of internal friction ϕ_u may be grossly unconservative. In the case of loose saturated sands liquefaction leads to virtually zero values of ϕ_u and the Mononobe–Okabe equations would be used to solve the wrong problem.

The undrained situation is not only undesirable physically but also difficult to analyse, hence it is recommended that good drainage should be provided to obviate the problem. Such drainage should be effective to well below the potential failure zone behind the wall, and also in front of the wall if cohesionless soils exist there in order that the required passive resistance is available.

6.5.2.3 *Active seismic pressures in cohesive soils or with irregular ground surface*

The trial wedge method (Figure 6.27) offers the easiest derivation of seismic earth pressure when the material is cohesive or the surface of the ground is irregular. This figure is drawn for Rankine conditions, and where the ground surface is very irregular the direction of P_{AE} may be taken as approximately parallel to a line drawn between points A and C. For Coulomb conditions the principles of the trial wedge method are similar and the direction of P_{AE} will be at an angle δ to the surface on which the pressure is calculated, similar to Figure 6.26.

Note that in seismic conditions tension cracks may be ignored on the assumption that this introduces relatively small errors compared with others involved in the analysis. For saturated soils the appropriate density will have to be taken in determining W on Figure 6.27.

6.5.2.4 *Completely Rigid Walls*

Where soil is retained by a rigid wall, pressures greater than active develop. In this situation the static and earthquake earth pressures may be taken as

$$\begin{aligned} P_E &= P_o + P_{oE} \\ &= \tfrac{1}{2}K_o \gamma H^2 + \alpha_h \gamma H^2 \qquad (6.70) \end{aligned}$$

where γ is the total unit weight of the soil and K_o is the coefficient of at-rest earth pressure.

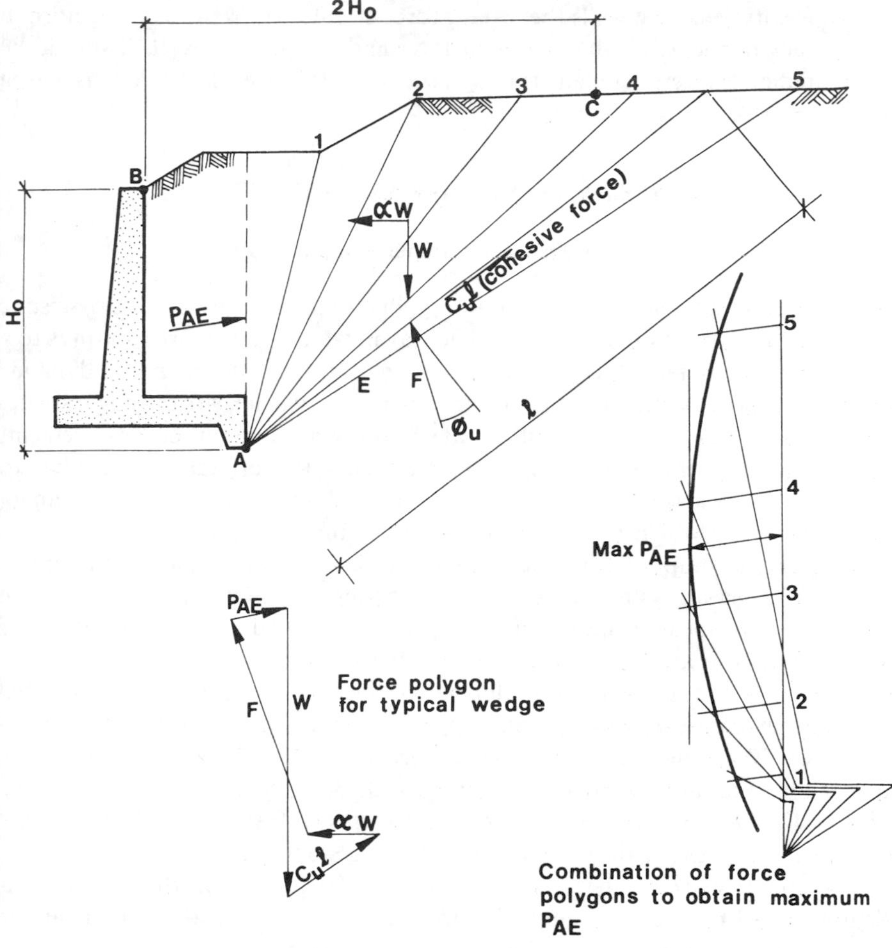

Figure 6.27 Trial wedge method for earthquake loading in Rankine conditions for cohesive soil or irregular ground surface

As with the active pressure case discussed above, this equation should not be applied to saturated sands. For a vertical wall and horizontal ground surface, and for all normally consolidated materials, K_o may be taken as

$$K_o = 1 - \sin\phi' \tag{6.71}$$

where ϕ' is the effective angle of shearing resistance. For other wall angles and ground slopes, K_o may be assumed to vary proportionally to K_A. The at-rest earth pressure force $P_o = \frac{1}{2}K_o\gamma H^2$ may be assumed to act at a height $H/3$ above the base of the wall, while the dynamic pressure $\Delta P_{OE} = \alpha_h\gamma H^2$ may be assumed[117] to act at a height $0.58H$.

For gravity retaining walls the at-rest force should be taken as acting normal to the back of the wall, while for cantilever and counterfort walls it should be calculated on the vertical plane through the rear of the heel and taken as acting parallel to the ground surface.

6.6 SEISMIC RESPONSE OF STRUCTURES

6.6.1 Elastic seismic response of structures

Dynamic loading comprises any loading which varies with time, and seismic loading is a complex variant of this. The way in which a structure responds to a given dynamic excitation depends on the nature of the excitation and the dynamic characteristics of the structure, i.e. on the manner in which it stores and dissipates vibrational energy. Seismic excitation may be described in terms of displacement, velocity, or acceleration varying with time. When this excitation is applied to the base of a structure it produces a time-dependent response in each element of the structure which may be described in terms of motions or forces.

Perhaps the simplest dynamical system which we can consider is a single-degree-of-freedom system (Figure 1.5) consisting of a mass on a spring which remains in the linear elastic range when vibrated. The dynamic characteristics of such a system are simply described by its natural period of vibration T (or frequency ω) and its damping ξ (defined in Section 1.4.2.5). When subjected to a harmonic base motion described by $u_g = a \sin \omega t$, the response of the mass at top of the spring is fully described in Figure 6.28. The ratios of response amplitude to input amplitude are shown for displacement R_d, velocity R_v, and acceleration R_a, in terms of the ratio between the frequency of the forcing function ω and the natural frequency of the system ω_n.

The significance of the natural period or frequency of the structure is demonstrated by the large amplifications of the input motion at or near the resonance conditions, i.e. when $\omega/\omega_n = 1$. Figure 6.28 also shows the importance of damping particularly near resonance. When the damping $\xi = 0.01$, the resonant amplification of the input motion is fiftyfold for this system, but if the damping is increased to $\xi = 0.05$ the resonant amplification is reduced to five times the input motions.

The response of a structure to the irregular and transient excitation of an earthquake will obviously be much more complex than in the simple harmonic steady-state motion discussed above. Consider the ground motion of the 1940 El Centro earthquake, the accelerogram for which is shown in Figure 1.4. If we apply this motion to a series of single-degree-of-freedom structures with different natural periods for damping, we can plot the maximum acceleration response of each of these structures as in Figure 6.29.

As with simple harmonic ground motion, the natural period and degree of damping is again evident in Figure 6.29. While no simple periodicity occurs in the ground motion of Figure 1.4, the dominance of the shorter periods is seen from the region of magnified acceleration responses on the left of Figure 6.29.

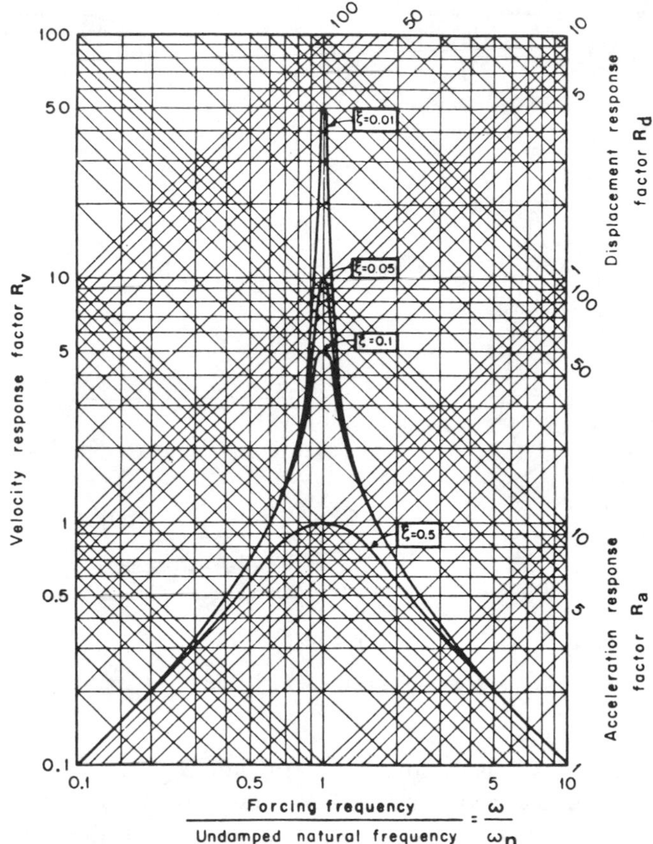

Figure 6.28 Response of linear elastic single-degree-of-freedom system to a harmonic forcing function (after Blake[124])

For example, a single-degree-of-freedom structure with a period of 0.8 s and damping $\xi = 0.02$ has a maximum acceleration of approximately $0.9g$ compared with a peak input ground motion of about $0.33g$. This represents an amplification of 2.7 at $\xi = 0.02$, whereas if the damping is $\xi = 0.05$ the amplification can be seen to reduce to 1.8.

Most structures are more complex dynamically than the single-degree-of-freedom system discussed above. Multi-storey buildings, for example, are better represented as multi-degree-of-freedom structures, with one degree of freedom for each storey, and one natural mode and period of vibration for each storey (Figure 1.12). The response history of any element of such a structure is a function of all the modes of vibration, as well as of its position within the overall structural configuration.

For many multi-degree-of-freedom structures the linear elastic responses can be computed with a high degree of mathematical accuracy. For example,

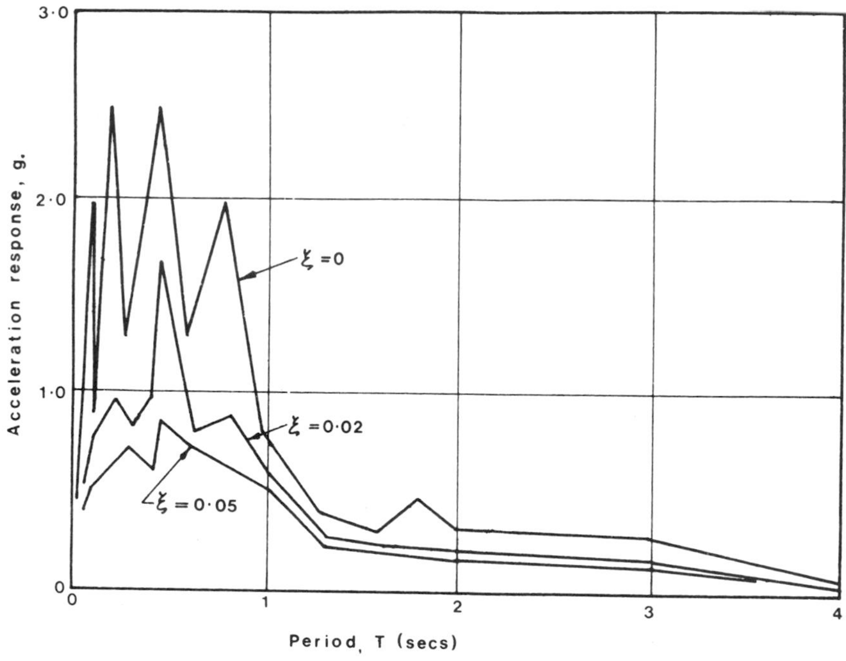

Figure 6.29 Elastic acceleration response spectra of north–south component of the 1940 El Centro earthquake

assuming linear elastic behaviour, in the dynamic analysis of a thirty storey building subjected to a ground motion 1.5 times that of Figure 1.4, the maximum horizontal shears at each floor level were computed to be as shown in Figure 6.30. Notice the considerable difference in response between the elastic case assuming 2 percent damping (curve 1) and that for 5 percent damping (curve 3). Further discussion of damping follows in Sections 6.6.2 and 6.6.4.

6.6.2 Non-linear seismic response of structures

For economical resistance against strong earthquakes most structures must behave inelastically. In contrast to the simple linear elastic response model examined in the previous section, the pattern of inelastic stress–strain behaviour is not constant, varying with the member size and shape, the materials used, and the nature of the loading.

The typical stress–strain curves for various materials under repeated and reversed direct loading shown in Figure 6.31 illustrate the chief characteristics of inelastic dynamic behaviour, namely:

Plasticity;
Strain hardening and strain softening;
Stiffness degradation;
Ductility;
Energy absorption.

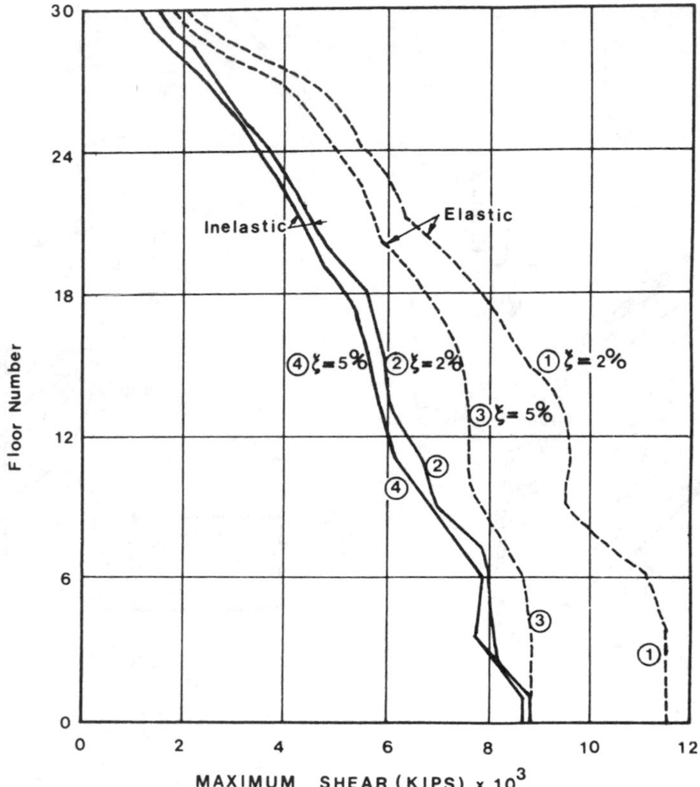

Figure 6.30 Maximum horizontal shear response for Bank of New Zealand Building, Wellington, subjected to 1.5 times El Centro (1940) north–south component (after Albert C. Martin and Associates,[125] reproduced by permission of the New Zealand National Society for Earthquake Engineering)

Plasticity, as exhibited by mild steel (Figure 6.31a), is a desirable property in that it is easy to simulate mathematically and provides a convenient control on the load developed by a member. Unfortunately the higher the grade of steel, the shorter the plastic plateau, and the sooner the *strain hardening* effect shown in Figure 6.31(a) sets in. *Strain softening* is the opposite of strain hardening, involving a loss of stress or strength with increasing strain as seen in Figure 6.31(a), or in the stress–strain envelope for concrete (Figure 6.31).

In the reversed loading of steel, the *Bauschinger effect* occurs, i.e. after loading past the yield point in one direction the yield stress in the opposite direction is reduced. Another characteristic of the cyclic loading of steel is the increased non-linearity in the elastic range which occurs with load reversal (Figure 6.31b). *Stiffness degradation* is an important feature of inelastic cyclic loading of concrete and masonry materials. The stiffness as measured by the overall stress/strain ratio of each hysteresis loop of Figure 6.31(c) to (f) is clearly reducing with each successive loading cycle.

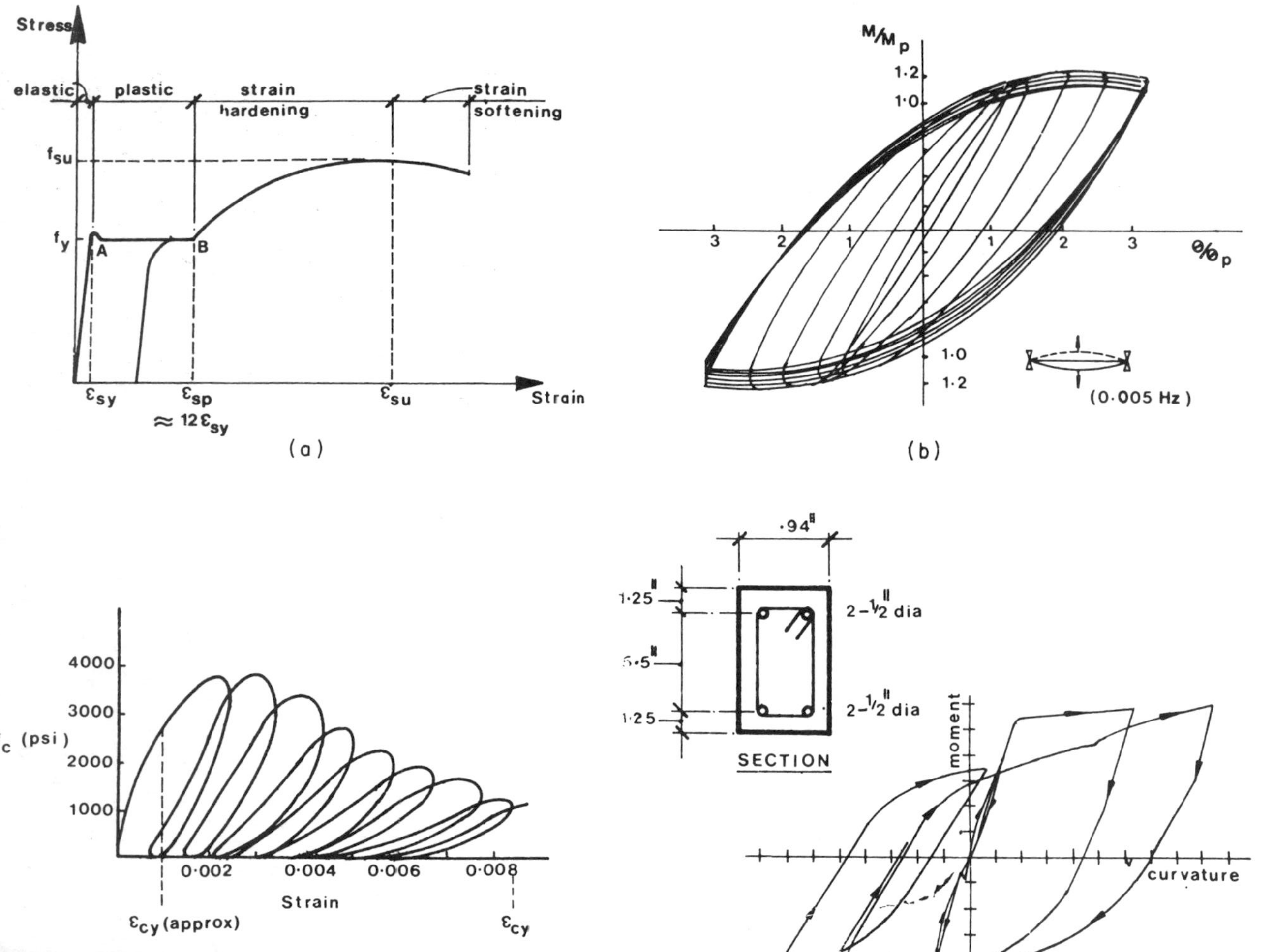
Stress
elastic
plastic
strain
hardening
strain
softening
f_{su}
f_y
A
B
ε_{sy}
ε_{sp}
$\approx 12\varepsilon_{sy}$
ε_{su}
Strain
(a)
M/M_p
1·2
1·0
3
2
1
1
2
3
θ/θ_p
1·0
1·2
(0·005 Hz)
(b)
4000
3000
f_c (psi)
2000
1000
0·002
0·004
0·006
0·008
Strain
ε_{cy} (approx)
ε_{cy}
·94"
1·25"
5·5"
1·25
2–½" dia
2–½" dia
SECTION
moment
curvature

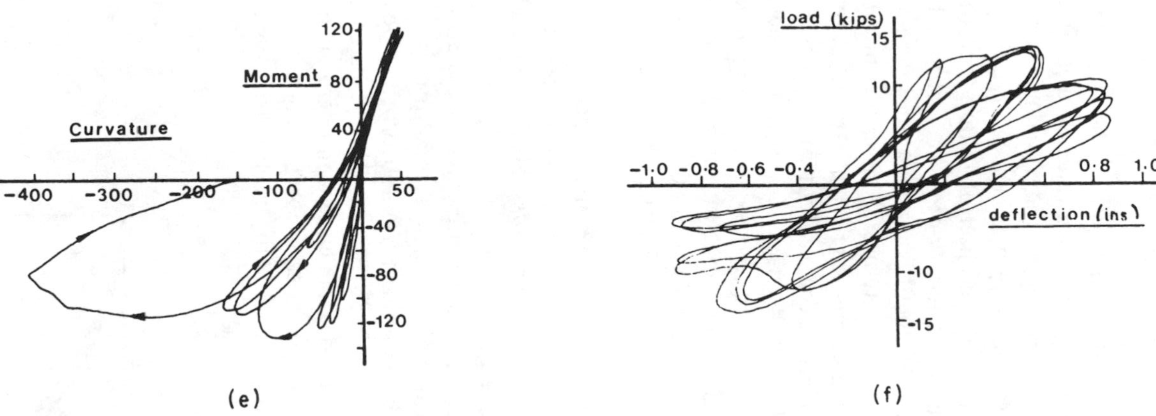

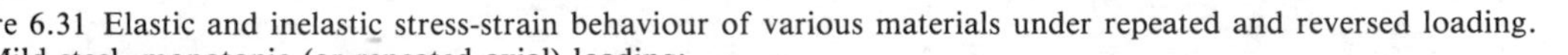

Figure 6.31 Elastic and inelastic stress-strain behaviour of various materials under repeated and reversed loading.
(a) Mild steel, monotonic (or repeated axial) loading;
(b) Structural steel under cyclic bending (after reference 130);
(c) Unconfined concrete, repeated loading (after reference 126);
(d) Doubly reinforced concrete beam, cyclic loading (after reference 127);
(e) Prestressed concrete column, cyclic bending (after reference 128);
(f) Masonry wall, cyclic lateral loading (after reference 129).
(Part (c) reproduced by permission of the American Concrete Institute)

The *ductility* of a member or structure may be defined in general terms by the ratio

$$\text{ductility} = \frac{\text{deformation at failure}}{\text{deformation at yield}}$$

In various uses of this definition, 'deformation' may be measured in terms of deflection, rotation or curvature. The numerical value of ductility will also vary depending on the exact combination of applied forces and moments under which the deformations are measured. Ductility is generally desirable in structures because of the gentler and less explosive onset of failure than that occurring in brittle materials. The favourable ductility of mild steel may be seen from Figure 6.31(a) by the large value of ductility in direct tension measured by the ratio $\epsilon_{su}/\epsilon_{sy}$. This ductility is particularly useful in seismic problems because it is accompanied by an increase in strength in the inelastic range. By comparison the high value of compressive ductility for plain concrete expressed by the ratio $\epsilon_{cu}/\epsilon_{cy}$ in Figure 6.31(c) is far less useful because of the inelastic loss of strength. Steel has the best ductility properties of normal building materials, while concrete can be made moderately ductile with appropriate reinforcement. The ductility of masonry, even when reinforced, is much more dubious. Further discussion of the ductility of the various materials is found elsewhere in this book, particularly in Chapter 7.

A high *energy absorption* capacity is often mentioned rather loosely as a desirable property of earthquake-resistant construction. Strictly speaking, a distinction should be made between *temporary* absorption and *permanent* absorption or dissipation of energy.

Compare the simple elastoplastic system represented by OABD in Figure 6.32 with the hypothetical non-linear mainly elastic system of curves $\widehat{OB}$ and $\widehat{BE}$; after loading each system to B the total energy 'absorbed' by each system is nearly equal, as represented by the area $OABC$ and $\widehat{OB}C$, respectively. However, the ratio between temporarily stored strain energy and permanently dissipated energy for the two systems are far from equal. After unloading to zero stress it can be seen that the energy dissipated by the elastoplastic system is equal to

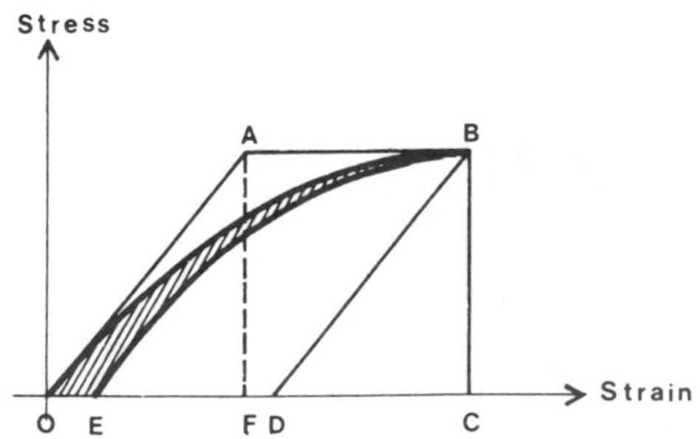

Figure 6.32 Energy stored and dissipated in idealized systems

the hysteretic area *OABD*, while the energy dissipated by the non-linear system is equal to the much smaller hysteretic area $\widehat{OBE}$ (shaded in Figure 6.32). The energy absorbed in *hysteresis* may be expressed as an *equivalent viscous damping* as described for soils in equation (6.5).

As further elucidation of the seismic energy absorption of structures, consider Figure 6.33 derived from the inelastic seismic analysis of the Bank of New Zealand building, Wellington. A substantial part of the energy is temporarily stored by the structure in elastic strain energy and kinetic energy. After three seconds the earthquake motion is so strong that the yield point is exceeded in parts of the structure and permanent energy dissipation in the form of inelastic strain (or hysteretic) energy begins. Throughout the whole of the earthquake, energy is also dissipated by viscous damping (from effects other than hysteresis), which is of course the means by which the elastic energy is dissipated once the forcing ground motion ceases. The value of damping reducing the energy available for damage is suggested by Figure 6.33. It has been shown by Housner and Jennings[131] that 2 percent of critical damping reduces the energy available for damage to about 50 percent of the total energy input, while 5 percent damping reduces it to about 30 percent.

It is evident from the large proportion of hysteretic energy dissipated by this building that considerable ductility without excessive strength loss is required. A brittle building with the same yield strength, but with no inelastic behaviour ($\epsilon_u/\epsilon_y = 1$), would have begun to fail after three seconds of the earthquake. In other words stronger members would have been necessary in a purely elastic design. This can be seen in another way in Figure 6.30, showing the reduction in storey shears achieved when assuming inelastic behaviour (curve 2) as compared with the elastic case (curve 1).

We note that as energy absorption is achieved from post-elastic deformations it means that the structure is being damaged, and a good earthquake resistant structure designed to yield exhibits high *damageability* without collapsing. Alternatively, as discussed in Section 5.5, energy-dissipating devices may be used to protect the structure from damage (e.g. yielding), by absorbing energy in elements which are replaceable. Thus structures incorporating energy-absorbing devices, by definition have high damageability.

It will be apparent that *ductility* and *energy absorption* are related, both being functions of post-yield behaviour, and are strength criteria for good seismic response. It is difficult, however, to use these terms clearly and quantitatively to measure seismic resistance. For example, the 1984 New Zealand earthquake loadings code[132] describes good seismic resistance by stating that 'structural systems intended to dissipate seismic energy by ductile flexural yielding shall have *adequate ductility*'. The code commentary then offers an approximate criterion for adequate ductility such that the building as a whole shall be able to sustain four cycles of loading with a ductility $\mu = 4$ (i.e. deflecting horizontally to four times the yield deflection) without the horizontal strength being reduced by more than 20 percent. While this definition is apparently clear in itself, in practice it has so far not proved possible to rate all structures

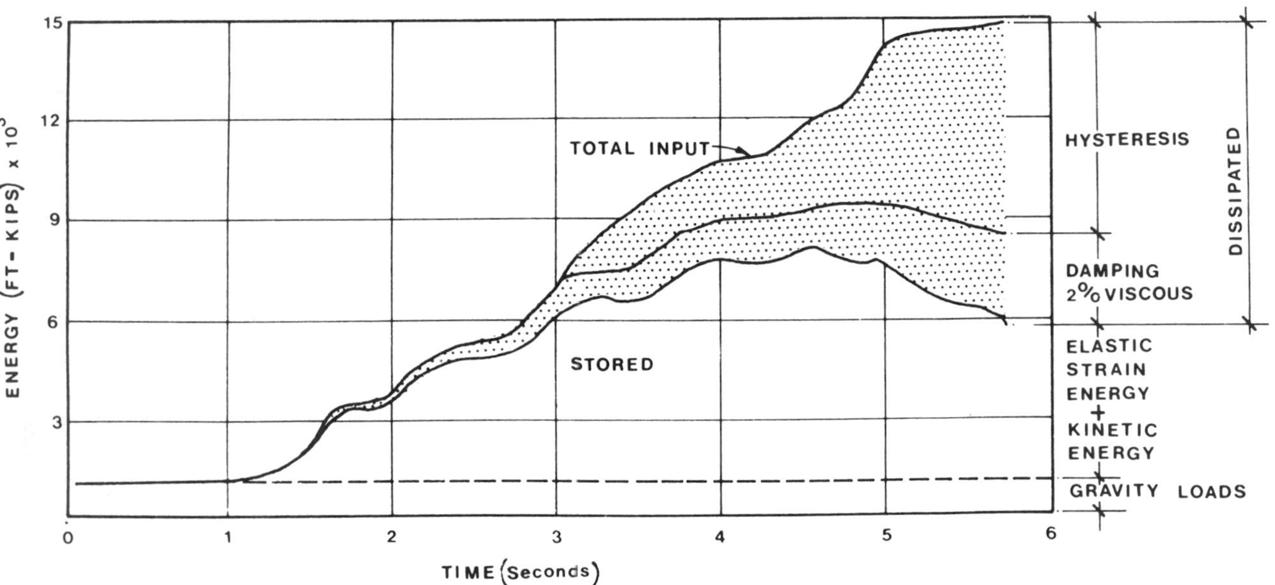

Figure 6.33 Energy expenditure in Bank of New Zealand Building, Wellington, computed for the first part of an earthquake equal to 1.5 times the El Centro (1940) north–south component (after Albert C. Martin and Associates,[125] reproduced by permission of the New Zealand National Society for Earthquake Engineering)

objectively in terms of this yardstick. This difficulty can easily be seen by trying to compare the results of different cyclic load tests, such as those shown in Figure 6.31. It is also noted that the above definition of adequate ductility is appropriate for elastoplastic behaviour (or similar), but not for hysteretic behaviour with heavily pinched loops or where slackness develops. The definition also relates to highly ductile structures, and not those of *limited ductility* discussed in Section 7.2.3.2.

Recognizing the above problem of quantification, a less bold approach to describing the above good earthquake resistant qualities is that of the ACI concrete code,[133] which simply states that 'the lateral-force resisting system' should 'retain a substantial portion of its strength when subjected to displacement reversals into the inelastic range'. It labels this property *toughness*.

The above criterion (called *adequate ductility* or *toughness*) represents in essence an extension to the definition of the *strength* criterion for *reliability* in earthquakes, as discussed in Section 5.2.2. However, as shown in Table 5.1, the prescription for a 'good' or reliable structure is further complicated by the need for other design criteria such as deformation control and repairability. It will be observed that strength, toughness, deflection control, and repairability are interrelated and hard to define, and we await more powerful definitions of them to arise from research and experience.

6.6.3 Mathematical models of non-linear seismic behaviour

When examining the range and complexity of the hysteretic behaviour shown in Figure 6.31 the problems involved in establishing usable mathematical stress–strain models are obvious. It follows that many hysteresis models have been developed, such as:

(1) Elastoplastic;
(2) Bilinear;
(3) Trilinear;
(4) Multilinear;
(5) Ramberg–Osgood;
(6) Degrading stiffness;
(7) Pinched loops;
(8) Slackness developing.

Three of the simplest of these hysteresis models are illustrated in Figure 6.34, namely;

(1) Elastoplastic;
(2) Bilinear (non-degrading);
(3) Ramberg–Osgood (modified).

The elastoplastic and bilinear models are the most commonly used because of their simplicity and relative computational efficiency, and the bilinear form in Figure 6.34(b) is found to be suitable for most response analyses of reinforced

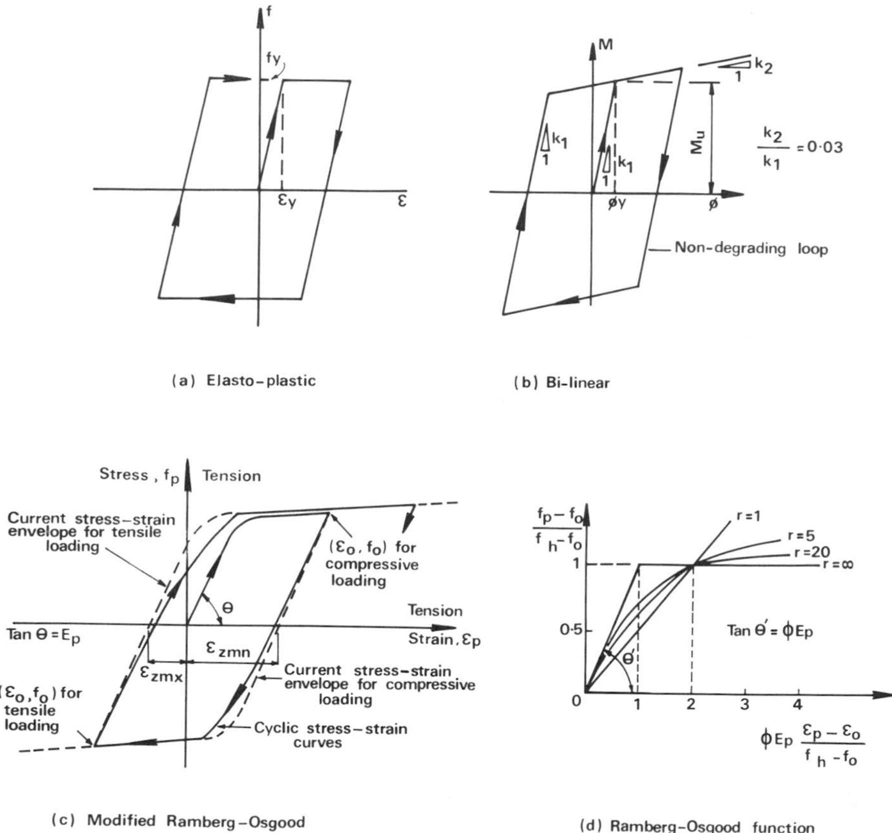

Figure 6.34 Three of the simpler idealized models of hysteretic behaviour under reversed cyclic loading. (The Ramberg–Osgood model is from Thompson and Park[134])

concrete and steel structures. The Ramberg–Osgood model shown in Figure 6.34(c) was developed by Thompson and Park[134] for closely modelling the hysteretic behaviour of prestressing steel, using the Ramberg–Osgood function for obtaining any desired shape of curve by varying the parameter r, as shown in Figure 6.34(d).

The best type of model depends heavily on the structural materials, for example as described for concrete by Park[135] and Otani.[136] However, compromises have to be made in selecting a suitable model, partly because the choice may be governed by the models incorporated into the available computer programs. For further discussions on methods of analysis see Section 6.6.7.

6.6.4 Level of damping in different structures

The general influence of damping upon seismic response is discussed in Sections 6.6.1 and 6.6.2 above, but when choosing the level of damping for use in the dynamic analyses of a structure, the following factors should be considered.

Damping varies with the materials used, the form of the structure, the nature of the subsoil and the nature of the vibration. Large-amplitude post-elastic vibration is more heavily damped than small-amplitude vibration, while buildings with heavy shear walls and heavy cladding and partitions have greater damping than lightly clad skeletal structures. The overall damping of a structure is clearly also related to the damping characteristics of the subsoil as discussed in Sections 6.2 and 6.3, and if the soil damping is to be incorporated into the effective damping of the structure the method given in Section 6.3.4.3 may be appropriate.

The many experimentally determined values of damping reported in the literature are generally derived either for individual structural components or for low-amplitude vibration of buildings. Hence for whole structures subject to strong ground motion some extrapolation of such damping data is necessary.

Table 6.17 indicates representative values of damping for a range of construction. These values are suitable for normal response spectrum or modal analysis in which viscous damping, equal in all modes, is assumed. These damping values also assume that the structure is a normal risk structure expected to yield in the design earthquake (i.e. it conforms to Box 2(a) of Table 5.1), and hence the damping due to hysteresis is included, but no allowance for radiation damping has been made. Except for a few forms of construction which have been specifically tested, insufficient evidence exists to warrant any more detailed allowance for differences in structural and non-structural form, and designers will need to use their own judgement to interpret the table.

For steel and concrete buildings, Hart and Vasudevan[137] offer a systematic method for estimating damping as a function of spectral velocity and modal frequency, based on an analysis of the response of buildings in the San Fernando earthquake. While the influence of structural form is not indicated in the above paper, the further development of this approach is desirable. For use with dynamic analysis programs which do not permit differences in modal damping, a weighted average based on modal contributions to base shear would be appropriate. Some further data on damping of different types of structure is given in later chapters of this book and in references 38 and 138–141.

6.6.5 Periods of vibration of structures

As indicated in various parts of this book, the periods of vibration of structures are primary tools in determining seismic response. While the periods of vibration are found during dynamic analyses by the solution of the eigenvalue equation (as described in Section 1.4, equation (1.38)), it is often desirable to make a quick estimate of the fundamental period T of a structure using approximate formulae. Various such formulae exist, the simplest being one for moment-resisting framed buildings built of steel or concrete:

$$T = \frac{H}{30}\ \mathrm{s} \tag{6.72}$$

where H is the height of the building (in metres).

Table 6.17. Typical damping ratios for structures

Type of construction	Damping ξ, percentage of critical
Steel frame, welded, with all walls of flexible construction	2
Steel frame, welded, with normal floors and cladding	5
Steel frame, bolted, with normal floors and cladding	10
Concrete frame, with all walls of flexible construction	5
Concrete frame, with stiff cladding and all internal walls flexible	7
Concrete frame, with concrete or masonry shear walls	10
Concrete and/or masonry shear wall buildings	10
Timber shear wall construction	15

Notes
(1) The term 'frame' indicates beam and column bending structures as distinct from shear structures.
(2) The term 'concrete' includes both reinforced and prestressed concrete in buildings. For isolated prestressed concrete members such as in bridge decks damping values less than 5 percent may be appropriate, e.g. 1–2 percent if the structure remains substantially uncracked.

As alternatives, the ATC[13] gives for multi-storey moment-resisting frames

$$T = 0.085\,H^{3/4} \qquad \text{(steel)} \tag{6.73}$$
$$T = 0.061\,H^{3/4} \qquad \text{(concrete)} \tag{6.74}$$

Braced frames or shear wall structures are stiffer and have smaller values of T for the same height than moment-resisting frames. A formula for such buildings (built of steel, masonry, or concrete) has been given[14] as

$$T = \frac{0.09H}{\sqrt{L}} \tag{6.75}$$

where L is the overall length (in metres) of the base of the building in the direction under consideration.

The above empirical formulae have the advantage that no structural analyses are required for their use, and they are thus suitable for preliminary design purposes. However, they are obviously insensitive to the actual mass and stiffness distributions of a given building and thus are subject to significant error. Also the above formulae are not reliable for flexible construction such as portal frames, and are not applicable to timber structures which are very varied in form, e.g. timber portal frame structure may have very long periods of 1.0 s and more. Hence, in cases where these formulae are likely to be insufficiently accurate, once a static linear elastic load analysis which calculates deflections has been done, T may be conveniently estimated by a more reliable formula based on Rayleigh's method for a structure with masses of N levels, as follows:

$$T = 2\pi \left[\frac{\sum_{i=1}^{N} m_i u_i^2}{\sum_{i=1}^{N} F_i u_i} \right] \quad (6.76)$$

where F_i is the seismic lateral force at level i;
m_i is the mass assigned to level i;
u_i is the static lateral deflection at level i due to the forces F_i.

Finally it is noted that all the above formulae are for the initial elastic state. With the stiffness degradation that occurs in the post-elastic range, the period progressively increases and may become twice the elastic value (or more) prior to failure. This may increase or decrease the response, depending on where the elastic period is on the response spectrum, and such effects may deserve inclusion in the analysis.

6.6.6 Interaction of frames and infill panels

6.6.6.1 Introduction

Walls are often created in buildings by infilling parts of the frame with stiff construction such as bricks or concrete blocks. Unless adequately separated from the frame (Section 12.2), the structural interaction of the frame and infill panels must be allowed for in the design. This interaction has a considerable effect on the overall seismic response of the structure and on the response of the individual members. Many instances of earthquake damage to both the frame members and infill panels have been recorded.[142,143] The currently available analytical techniques used in studying frame/panel interaction are briefly discussed below.

6.6.6.2 The effect of infill panels on overall seismic response

The principal effects of infill panels on the overall seismic response of structural frames are;

(1) To increase the stiffness and hence increase the base shear response in most earthquakes;
(2) To increase the overall energy absorption capacity of the building;
(3) To alter the shear distribution throughout the structure.

The more flexible the basic structural frame, the greater will be the above-mentioned effects. As infill is often made of brittle and relatively weak materials, in strong earthquakes the response of such a structure will be strongly influenced by the damage sustained by the infill and its stiffness-degradation characteristics.

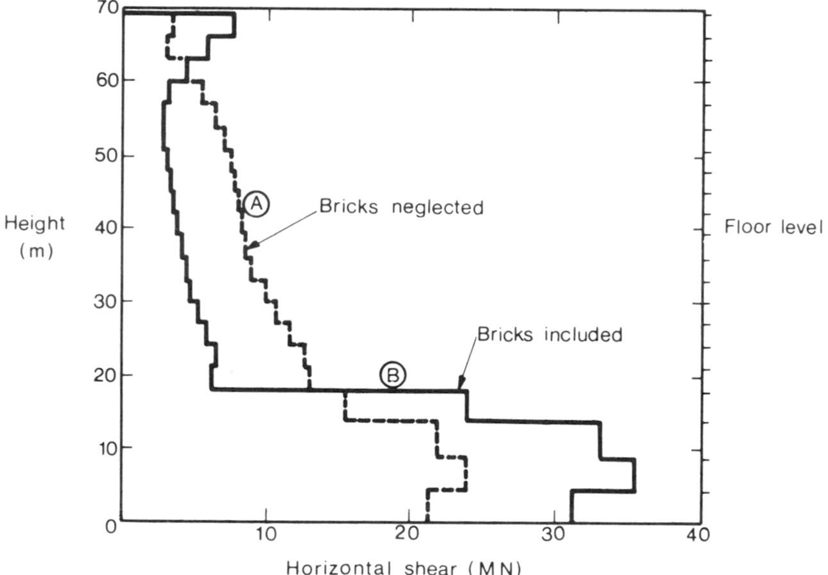

Figure 6.35 Horizontal seismic shear diagram for lift core of 20-storey building showing effect of brick partitions above the fourth floor

In order to fully simulate the earthquake response of an infilled frame, a complex non-linear time-dependent finite-element dynamic analysis would be necessary. At present no *practical* computer program capable of such an analysis has been published, and such are the problems involved in modelling the structural behaviour of normal masonry infill[144–6] that only rudimentary dynamic analyses are likely to be warranted for some time to come.

For many structures a response spectrum analysis in which the infill panels are simulated by simple finite elements, will be very revealing. Figure 6.35 shows the results of such an analysis of a multi-storey hotel building, in which all of the bedroom floors (fourth to twentieth) have alternate partitions in brickwork. Curve A shows the horizontal earthquake shear distribution up the shear core ignoring the brickwork, while curve B shows the shears when an approximate allowance for the brick walls is made.

Allowing for the brick reduces the fundamental period of the structure from 1.96 s to 1.2 s, and correspondingly increases the base shear on the shear core from 21.0 MN to 31.0 MN. The effect on the distribution of shear is particularly dramatic; it can be seen how the brick walls carry a large portion of the shear until they terminate at the fourth-floor level; below this level the shear walls of the core must, of course, take the total load (see also Section 5.3.5).

In carrying out this simple type of dynamic analysis difficulty may be experienced in selecting a suitable value of shear modulus G for the infill material. Not only is the G value notoriously variable for bricks, but the infill material may not even have been chosen at the time of the analysis. Either a single representative value may have to be assumed or it may be desirable to take a lower and a higher likely value of G in two separate analyses for purposes of comparison.

Further examples of the effect of infill on mode shapes and periods of vibration of structural frames are briefly reported by Lamar and Fortoul.[144] In their examples the period of the first mode is generally reduced by a factor of three or four when comparing the 'infill-included' with the 'infill-excluded' cases. The comparable mode shapes also vary considerably.

Interesting use of the dynamic properties of infill panels has been made in Japan by Muto.[147] Tall steel-framed buildings have been fitted with precast concrete panels, which not only stiffen the otherwise highly flexible frames against excessive sway under lateral loading but also considerably improve the energy-absorption and ductility characteristics of the structures in strong earthquakes. The precast panels are specially designed and detailed with slits and reinforcement to have prescribed elastic and post-elastic deformation behaviour, and are attached to the beams of the steel frame at discrete points. These provisions, coupled with considerable experimental performance data, make the Muto slitted shear panels readily amenable to rational analysis in distinct contrast to ordinary masonry infill.

6.6.6.3 The effect of infill panels on member forces

As mentioned above, there is no practical means of predicting accurately the seismic interaction between infill panels and structural frames. Only fairly crude assessments can be made of the stresses in the panels and in the adjacent members. However, from a straightforward finite-element response spectrum analysis some basic design information may be derived. While to take such data as definitive seismic design criteria would be misleading, sensible use of the computed forces in design would nevertheless be much better than ignoring the presence of the infill, as has often been the case in the past. By carrying out comparative analyses with and without the infill panels, at least a qualitative idea of the effect of the infill can be obtained.

(1) *Infill panels*. The shear stresses computed in the infill panels should give a reasonable indication whether or not the infill will survive the design earthquake. Despite being very approximately determined, the shear stress level will also help in determining what reinforcement to use in the panel and whether to tie the panel to the frame (Section 8.5).

(2) *Frame members*. The design of the beams and columns abutting the infill is generally the least satisfactory aspect of this form of aseismic construction. Because of the approximations in the analytical model, the stresses in the frame members are ill-defined. Failures tend to occur at the tops and bottoms of columns, due to shears arising from interaction with the compression diagonal which exists in the infill panel during the earthquake.[143] Unfortunately, no comprehensive design criteria for this problem have yet been established, and further research examining the frame rather than the panel stresses is required.

If the analysis indicates the failure of the infill panels, the frame should be analysed with any failed panels deleted, so that appropriate frame stresses may be taken into consideration. The fail-safe structure will not necessarily be less highly stressed in individual frame members than the original undamaged structure, although the resulting more flexible structure will generally have a reduced overall response.

6.6.7 Methods of seismic analysis for structures

The many methods for determining seismic forces in structures fall into two distinct categories;

(1) Equivalent static force analysis;
(2) Dynamic analysis.

6.6.7.1 Equivalent static force analysis

These are approximate methods which have been evolved because of the difficulties involved in carrying out realistic dynamic analysis. Codes of practice inevitably rely mainly on the simpler static force approach, and incorporate varying degrees of refinement in an attempt to simulate the real behaviour of the structure. Basically they give a crude means of determining the 'total' horizontal force (base shear) V, on a structure:

$$V = ma$$

where m is the mass of the structure and a is the seismic horizontal acceleration. a is generally in the range $0.05g$ to $0.20g$. V is applied to the structure by a simple rule describing its vertical distribution. In a building this generally consists of horizontal point loads at each concentration of mass, most typically at floor levels (Figure 6.36). The seismic forces and moments in the structures are then determined by any suitable statical analysis and the results added to those for the normal gravity load cases.

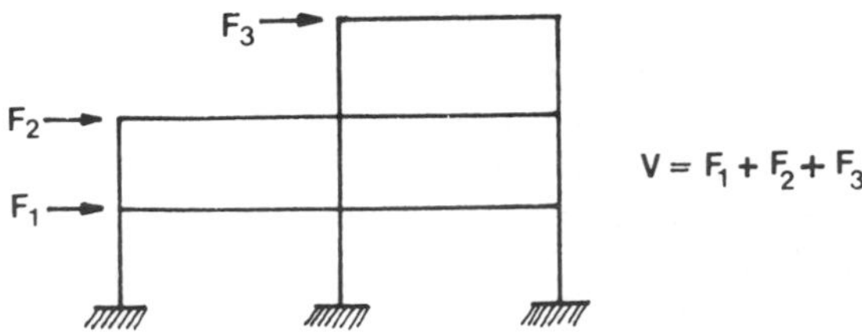

Figure 6.36 Example of frame with equivalent static forces applied at floor levels

In the subsequent design of structural sections an increase in permissible elastic stresses of 33–50 percent is usually permitted, or a smaller load factor than normal is required for ultimate load design. In regions of high winds and moderate earthquake requirements, the worst design loads of taller structures may well arise from wind rather than earthquake forces. Even so, the form and detail of the structure should still be governed by seismic considerations.

An important feature of equivalent static load requirements in most codes of practice, up till 1986, is the fact that the calculated seismic forces are considerably less than those which would actually occur in the larger earthquakes likely in the area concerned. The forces calculated in more rigorous dynamic analyses based on a realistic earthquake excitation can be as much as ten times greater than those arising from the static load provisions of some codes. This state of affairs has been 'justified' by arguing that the forced discrepancy will be taken up by inelastic behaviour of the structure, which should therefore be detailed to be specially ductile, and some codes do have specific ductility requirements. It is widely recognized that more stringent static load requirements should be laid down, and recent developments in some codes have begun to set equivalent-static loadings at a realistic level. For example, as discussed in Section 4.3.3.2, New Zealand bridge code loadings propose that the design earthquake in the form of a uniform risk elastic response spectrum is the basis of the loadings, factored according to the design ductility level, and that both dynamic and static analysis use the same family of spectra as loadings (Figure 4.5).

6.6.7.2 Dynamic analysis

For large or complex structures static methods of seismic analysis are often deemed to be not accurate enough and many authorities demand dynamic analyses for certain types and size of structure. Various methods of differing complexity have been developed for the dynamic seismic analysis of structures. They all have in common the solution of the equations of motion as well as the usual statical relationships of equilibrium and stiffness. For structures with more than three degrees of freedom such analyses are, of course, generally carried out using computers.

The three main techniques currently used for dynamic analysis are;

(1) Direct integration of the equations of motion by step-by-step procedures;
(2) Normal mode analysis;
(3) Response spectrum techniques.

Direct integration provides the most powerful and informative analysis for any given earthquake motion. A time-dependent forcing function (earthquake accelerogram) is applied and the corresponding response-history of the structure during the earthquake is computed. That is, the moment and force diagrams at each of a series of prescribed intervals throughout the applied motion can be found. Computer programs have been written for both linear elastic and nonlinear inelastic material behaviour, using step-by-step integration procedures.

Linear behaviour is seldom analysed by direct integration, unless mode coupling is involved, as normal mode techniques are easier, cheaper, and nearly as accurate. Three-dimensional non-linear analyses have been devised which can take the three orthogonal accelerogram components from a given earthquake, and apply them simultaneously to the structure.[148] In principle, this is the most complete dynamic analysis technique so far devised, and is unfortunately correspondingly expensive to carry out.

Normal mode analysis is a more limited technique than direct integration, as it depends on artificially separating the normal modes of vibration and combining the forces and displacements associated with a chosen number of them by superposition. As with direct integration techniques, actual earthquake accelerograms can be applied to the structure and a stress-history determined, but because of the use of superposition the technique is limited to linear material behaviour. Although modal analysis can provide any desired order of accuracy for linear behaviour by incorporating all the modal responses, some approximation is usually made by using only the first few modes in order to save computation time. Problems are encountered in dealing with systems where the modes cannot be validly separated, i.e. where mode coupling occurs.

The most serious shortcoming of linear analyses is that they do not accurately indicate all the members requiring maximum ductility. In other words the pattern of highest elastic stresses is not necessarily the same as the pattern of plastic deformation in an earthquake structure.[149] For important structures in zones of high seismic risk, non-linear dynamic analysis is sometimes called for.

The response spectrum technique is really a simplified special case of modal analysis. The modes of vibration are determined in period and shape in the usual way and the maximum response magnitudes corresponding to each mode are found by reference to a response spectrum. An arbitrary rule is then used for superposition of the responses in the various modes. The resultant moments and forces in the structure correspond to the envelopes of maximum values, rather than a set of simultaneously existing values. The response spectrum method has the great virtues of speed and cheapness.

Although this technique is strictly limited to linear analysis because of the use of superposition, simulations of non-linear behaviour have been made using pairs of response spectra, one for deflections and one for accelerations.[150,151] The expected ductility factor is chosen in advance and the appropriate spectra are used. This is clearly a fairly arbitrary procedure, and appears unlikely to be more realistic than the linear response spectrum method. Lai and Biggs[151] have shown that Newmark and Hall's method[150] can be unconservative and have developed an improved procedure.

6.6.7.3 Selection of method of analysis for structures

In the past there has generally been little choice in the method of analysis, mainly because suitable and economical computer programs have not been readily

Table 6.18

Type of structure	Method of analysis (two dimensional or three-dimensional)
Small simple structures	(1) Equivalent static forces
↓	(2) Response spectra
Progressively more demanding structures	(3) Modal analysis
↓	(4) Direct integration
Large complex structures	(5) Non-linear soil–structure

available. To date most earthquake-resistant structures, even in California, have been analysed with an equivalent static load derived from a code of practice. However, this situation is changing. An increasing number of efficient and economical dynamic analysis programs are being written for faster computers, and many design offices have access to such programs, especially since the advent of microcomputers. Dynamic analyses are demanded now by some owners, and by the regulations of more countries.

It is difficult to give clear general advice on selecting the means of analysis, as each structure will have its own requirements, technical, statutory, economic, and political. Broadly speaking, however, the larger and/or more complex the structure, the more sophisticated the dynamic analysis used. Table 6.18 gives a very simple indication of the applicability of the main methods of analysis.

At present, except in special projects, designers are unlikely to do a three-dimensional dynamic analysis, whether elastic or inelastic, which allows for two orthogonal horizontal components of ground motion simultaneously. In order to make some allowance for the resultant diagonal response some codes are now stipulating arbitrary means of adding the separately computed orthogonal components. Until further research has been done, there are great problems in doing this without the risk of being too conservative.

It is important to note that the methods of analysis in Table 6.18 become successively more realistic *only* if the appropriate seismic loadings and a suitable model of material behaviour are used. Even with the best input, research shows[152] that non-linear analysis is likely to be accurate in global terms but not at local member behaviour level. Response spectrum analysis is not only far simpler and cheaper to use than non-linear analysis but the use of smoothed spectra by definition gives more reliable allowance for applied loads than do response history methods.

6.6.7.3(i) Method of analysis and material behaviour

Dynamic analysis techniques relating to the behaviour of soils in the soil–structure interaction problem have been discussed in Section 6.3.3. The following discussion is complementary to that for soils, having structures more specifically in mind. The problem of selecting an analysis method, of course, depends largely on whether the materials are intended to be elastic or inelastic during the design earthquake. The usual methods of analysis for these two states are set out in Table 6.19.

Elastic behaviour during the design earthquake obviously has the advantage of making linear analysis entirely appropriate. It may arise because:

(1) The material is brittle, i.e. has zero ductility (e.g. porcelain, mass concrete, some masonry);
(2) The material has limited ductility (e.g. wood, ill-confined concrete, most reinforced masonry);
(3) The designer chooses to keep a ductile material within the elastic range (e.g. steel, reinforced concrete) where greater stiffness is required for functional reasons or greater safety is desired (Boxes (2b) or (4) of Table 5.1).

Materials in the brittle category, such as masonry or porcelain, may be realistically analysed by method (1a) in Table 6.19 because their linear elastic behaviour matches the analytical assumptions. The chief problems lie in choosing an adequate safety margin (load factor) within the elastic range, to cover the normal errors involved in assessing the loading, the geometric modelling, and the ultimate strength. For cases (2) and (3) above, designed to be elastic in the design earthquake (methods (1b) and 3(b) of Table 6.19), it is common practice to enhance the safety factor by developing as much inelastic deformation prior to collapse as practicable with nominal ductility reinforcement. This increases the safety against strong shaking, particularly of longer duration. In porcelain structures it is good practice to increase the earthquake protection by deliberately enhancing the damping. For further discussion of masonry and porcelain see Chapter 8 and Section 11.2, respectively.

Materials which become inelastic in the design earthquake are more satisfactory for earthquake resistance than brittle ones because of their inelastic deformability, but are less convenient and more expensive to analyse for the same reason. Of the methods used to analyse inelastic behaviour, only method (5) attempts to model the hysteretic stress–strain behaviour directly. This cannot always be done reliably because suitable hysteresis models are available for only limited types of element and stress condition in each construction material.

For steel structures which can be realistically reduced to regular plane frames, such as that referred to in Figures 6.30 and 6.33, method (5) may be economic from an analytical as well as from a construction point of view. It follows that most structures which will behave inelastically in the design earthquake are designed by methods (2) and (4) of Table 6.19 assuming linear elastic material behaviour in the analysis. Both methods imply the application of an artificially reduced earthquake loading within the elastic capacity of the structure, and an approximate or arbitrary allowance for the inelastic deformations by ensuring certain ductility levels in highly stressed zones.

Table 6.19. Seismic analysis and design procedures

Method of analysis	Material behaviour	Design provisions	Seismic loading
Static linear	Elastic	(1a) Permissible stress or factored ultimate design (1b) Sometimes detailed for nominal ductility	Full
	Inelastic	(2) Permissible stress or factored ultimate design. Detailed for ductility	Reduced by factor R (Figure 6.37)
Dynamic linear	Elastic	(3a) Permissible stress or factored ultimate design (3b) Sometimes detailed for nominal ductility	Full
	Inelastic	(4) Permissible stress or factored ultimate design. Detailed for ductility	Reduced by factor R (Figure 6.37)
Dynamic non-linear	Inelastic	(5) Hysteresis loops required. Ultimate strength design. Ductility demands found from plastic deformations.	Full

This design process, which attempts to do inelastic design by an 'equivalent-elastic' method, has great difficulties in:

(1) Allowing for inelastic deflection;
(2) Allowing for stiffness degradation;
(3) Determining the distribution of ductility demands; and
(4) Allowing for the duration of strong shaking.

These four factors are not independent and vary with the nature of the material, the structural form, and the loading, as discussed below.

In endeavouring to define an equivalent-elastic loading two alternative methods of equating elastic and inelastic response are commonly considered (Figure 6.37).

For simple elastoplastic systems two possible reduction factors for the loading, $R=c/d$, are obtained in terms of the deflection ductility factor $\mu=b/a$, as shown. Here $\mu=b/a$ is the ratio of the total deflection to the elastic deflection of the system. In the case of buildings of other real structures μ is generally referred to as the structure deflection ductility factor, where μ is the ratio of the deflection at ultimate to the deflection at first yield, measured at the top of the structure, as shown in Figure 5.5. Despite being known to be unreliable at lower periods vibration T, the equal deflection reduction factor $R=1/\mu$ is widely used because

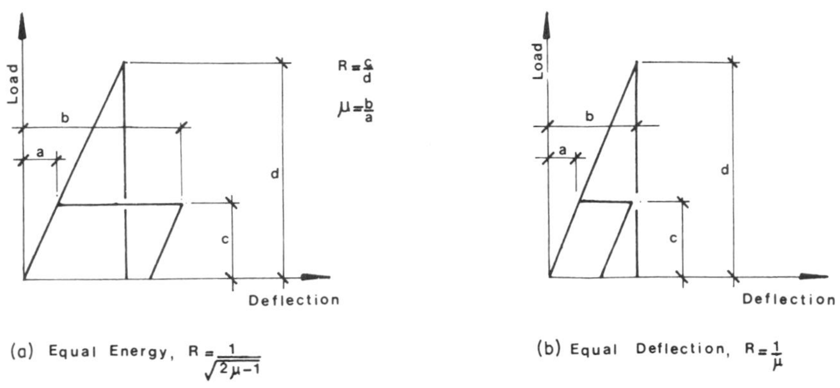

Figure 6.37 Reduction factors R for seismic loading equating elastic and inelastic response in terms of (a) energy and (b) deflection

of its simplicity. Although R is relatively insensitive to the shape of the hysteresis loop, it is strongly dependent on the nature of the accelerogram used as input,[153] and it appears from the work of Moss *et al.*[153] that the use of $R = 1/\mu$ may not be appropriate for T less than about 1.5 s.

The factor R is reflected in the variation in value of the structural factors in formulae for obtaining base shear forces in earthquake codes, e.g. the factor K in the Californian code[14] and S and M in the New Zealand code.[132] As indicated by the absence of recommended values of such factors for various types of structure, more research is still needed to evaluate reliable μ and R values for different structural configurations and materials. Generally codes do not recommend the use of values of μ greater than about 6 because this would imply excessive damage occurring in events smaller than the design earthquake, although values of μ as high as 10 may be achievable in ductile steelwork, partly because of the difficulty of detailing for high ductilities and partly because of the higher deformations that are implied.

The above reduction factors and ductility factors, appropriate to averaged earthquake response spectra, are fairly well established. However, problems arise when considering individual earthquake motions for equivalent-elastic design criteria. This is illustrated by comparing the El Centro earthquake with that of Parkfield (Figure 6.38).

It can be seen that the elastic response of a single degree-of-freedom system is much greater for the Parkfield than for the El Centro earthquake. Yet El Centro at magnitude 6.4 did much more damage than Parkfield at magnitude 5.6. The elastic response analysis has taken account of the higher peak acceleration of the Parkfield accelerogram (about 0.5g) compared with El Centro's peak acceleration of 0.33g, without being able to allow for the effect of the much greater duration of strong shaking which occurred at El Centro.

Clearly great care is necessary in selecting design earthquakes of different magnitude to ensure an appropriate relationship between elastic and inelastic

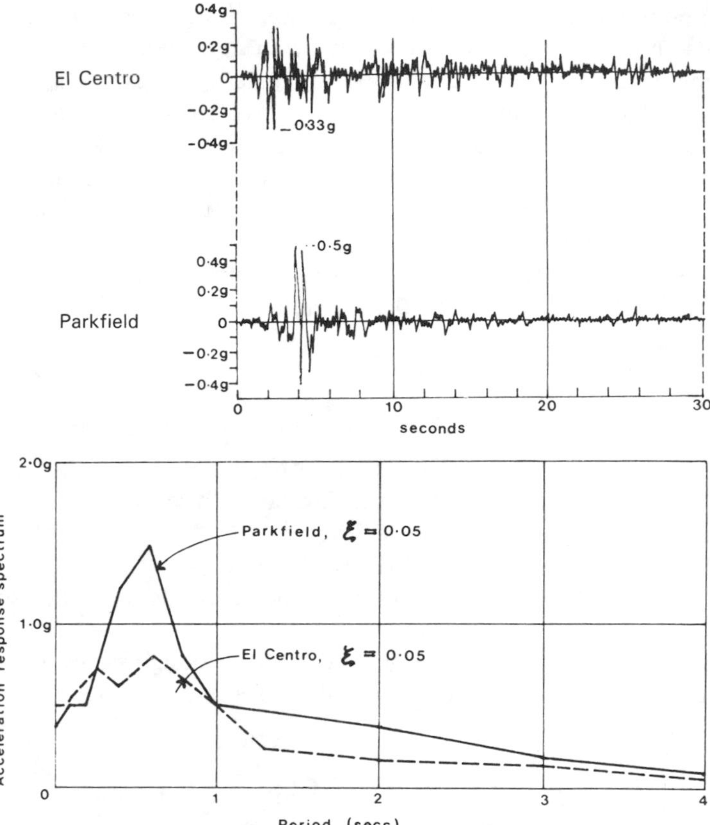

Figure 6.38 Accelerograms and elastic response spectra for the El Centro (1940) ($M=6.4$) and Parkfield (1966) ($M=5.6$) earthquakes

response. A much greater reduction factor R for Parkfield would be required than for the El Centro earthquake.

The determination of the distribution of ductility demand throughout a multi-redundant structure using an equivalent-elastic analysis is also unreliable. The positions of maximum moment in a frame, determined elastically, will not necessarily indicate the order of plastic hinge formation.[149] However, in ideally regular plane frames this approximation may be reasonable, and is often taken in practice.

6.6.7.3(ii) P-delta effect

When structures sway horizontally in earthquakes an overturning moment $M=P\Delta$ exists equal to the weight P multiplied by the horizontal displacement Δ. While usually negligible globally, the P-delta moment at individual vertical members is sometimes significant, particularly in moment-resisting frames supporting reasonably substantial gravity loads, and where fully ductile design permits inelastic displacements, or in tower structures (Section A.2.6).

In normal buildings the P-delta moments are not significant if the drift does not exceed about 1.5 percent of the storey height. Hence where the amount of sway is controlled by the drift limitations of building codes, the P-delta effect may safely be ignored except where the weights involved are exceptionally heavy.

In cases where the P-delta effect is likely to be significant its effect needs to be derived from the analysis. As most structural analysis computer programs do not find the P-delta moments they may have to be specially calculated, such as by the method noted below.

The ATC[13] states that P-delta effects in buildings may be ignored if the stability factor θ does not exceed 0.10, where

$$\theta = \frac{P_x \Delta}{V_x h_{sx} C_d} \tag{6.77}$$

where Δ = the calculated design storey drift;
V_x = the seismic shear force acting between levels x and x-1;
h_{sx} = the storey height below level x;
P_x = the total gravity load above level x;
C_d = a deflection amplification factor dependent on the structural form.[13]

If θ exceeds 0.10, the ATC[13] recommends that the design storey drifts should be multiplied by the factor $1/(1-\theta)$ to allow for P-delta effects.

REFERENCES

1. Shannon and Wilson, Agbabian–Jacobson Associates, Seattle and Los Angeles, *Soil behaviour under earthquake conditions*, State of the art evaluation for US Atomic Energy Commission (1972).
2. Seed, H. B., and Idriss, I. M., 'Influence of soil conditions on ground motions during earthquakes', *J. Soil Mechanics and Foundations Division, ASCE*, **95**, No. SM1, 99–137 (1969).
3. Sugimura, Y., and Ohkawa, I., 'Seismic microzonation of Tokyo area', *Proc. 8th. World Conf. on Earthq. Eng., San Francisco*, **II**, 721–8 (1984).
4. Seed, H. B., and Idriss, I. M., 'Soil moduli and damping factors for dynamic response analysis', *Report No. EERC 70–10*, Earthquake Engineering Research Centre, University of California, Berkeley (1970).
5. Seed, H. B., Wong, R. T., Idriss, I. M., and Tokimatsu, K., 'Moduli and damping factors for dynamic analyses of cohesionless soils', *Report No. UCB/EERC-84/14*, Earthquake Engineering Research Center, University of California, Berkeley (1984).
6. Zen, K., and Higuchi, Y., 'Prediction of vibratory shear modulus and damping ratio for cohesive soils', *Proc. 8th World Conf. on Earthq. Eng., San Francisco*, **III**, 23–30 (1984).
7. Whitman, R. V., 'Soil structure interaction', and 'Evaluation of soil properties for site investigation and dynamic analysis of nuclear plants', *Seismic Design for Nuclear Power Plants* (Ed. R. J. Hansen), MIT Press (1970), pp. 245–305.
8. Whitman, R. V., and Richart, F. E., 'Design procedures for dynamically loaded foundations', *J. Soil Mechanics and Foundations Division, ASCE*, **93**, No. SM6, 169–91 (1967).

9. D'Appolonia, D. J., Poulos, H. G., and Ladd, C. C., 'Initial settlement of structures on clay', *J. Soil Mechanics and Foundations Division, ASCE*, **97**, No. SM10, 1359–77 (1971).
10. Lambe, T. W., and Whitman, R. V., *Soil Mechanics*, John Wiley and Sons, New York (1969), p. 155.
11. Deere, D. U., 'Geological considerations', in *Rock Mechanics in Engineering Practice* (Eds K. G. Stagg and O. C. Zienkiewicz), Wiley, London, (1968), pp. 1-20.
12. Ambraseys, N. N., 'Dynamics and response of foundation materials in epicentral regions of strong earthquakes', *Proc. 5th. World. Conf. on Earthq. Eng., Rome*, **1**, CXXVI-CXLVIII (1973).
13. Applied Technology Council, 'Tentative provisions for the development of seismic regulations for buildings', *ATC 3-06, NBS SP-510, NSF 78-8*, National Bureau of Standards, USA, (1982).
14. Seismology Committee, SEAOC, *Recommended lateral force requirements and commentary*, Structural Engineers Association of California, (1980).
15. Newmark, N. M., Robinson, A. R., Ang, A. H. S., Lopez, L. A., and Hall, W. J., 'Methods for determining site characteristics', *Proc. Int. Conf. on Microzonation for Safer Construction, Seattle*, **1**, 113–29 (1972).
16. Tsai, N. C., 'Influence of local geology on earthquake ground motion', *PhD Thesis*, California Institute of Technology, 1969.
17. Whitman, R. V., Roesset, J. M., Dobry, R., and Ayestaran, L., 'Accuracy of modal superposition for one-dimensional soil amplification analysis', *Proc. Int. Conf. on Microzonation, Seattle*, **2**, 483–98 (1972).
18. Schnabel, P. B., Seed, H. B., and Lysmer, J., 'Shake—A computer program for earthquake response analysis of horizontally layered sites', *Report No. EERC 72-12*, Earthquake Engineering Research Center, University of California, Berkeley (1972).
19. Kausel, E., and Roesset, J. M., 'Soil amplification: some refinements', *Soil Dynamics and Earthquake Engineering*, **3**, No. 3, 116–23 (1984).
20. Tazoh, T., Nakahi, S., Shimizu, K., and Yokata, H., 'Vibration characteristics of soil and their applicability to the field', *Proc. 8th World Conf. on Eathq. Eng., San Francisco*, **II**, 679–86 (1984).
21. Roesset, J. M., 'Fundamentals of soil amplification', in *Seismic Design of Nuclear Reactors*, (Ed. R. J. Hansen), MIT Press (1972), pp. 183–244.
22. Schnabel, P. B., Seed, H. B., and Lysmer, J., 'Modification of seismograph records for effects of local soil conditions', *Report No. EERC 71-8*, Earthquake Engineering Research Center, University of California, Berkeley (1971).
23. Seed, H. B., 'Earthquake effects on soil-foundation systems', in *Foundation Engineering Handbook* (Eds H. F. Winterkorn and H.-Y. Fang), Van Nostrand Reinhold Company, New York (1975), pp. 700–32.
24. Valera, J. E., and Donovan, N. C., 'Incorporation of uncertainties in the seismic response of soils', *Proc. 5th World Conf. on Earthq. Eng., Rome*, **1**, 370–79 (1973).
25. Sugimura, Y., and Ohkawa, I., 'Seismic microzonation of Tokyo area', *Proc. 8th World Conf. on Earthq. Eng., San Francisco*, **II**, 721–8 (1984).
26. Reimer, R. B., Clough, R. W., and Raphael, J. M., 'Evaluation of the Pacoima dam accelerogram', *Proc. 5th World Conf. on Earthq. Eng., Rome*, **2**, 2328–37 (1973).
27. Boore, D. M., 'The effect of simple topography on seismic waves: Implications for the recorded accelerations at Pacoima Dam', *Bull. Seism. Soc. Amer.*, **63**, No. 5, 1603–09 (1973).
28. Brune, J. N., 'Preliminary results on topographic seismic amplification effect on a foam rubber model of the topography near Pacoima dam', *Proc. 8th World Conf. on Earthq. Eng., San Francisco*, **II**, 663–70 (1984).
29. Dowrick, D. J., 'Preliminary field observations of the Chilean earthquake of 3 March, 1985', *Bull. NZ Nat. Soc. for Earthq. Eng.*, **18**, No. 2, 119–27 (1985).

30. McVerry, G. H., Hodder, S. B., Hefford, R. T., and Heine, A. J., 'Records of engineering significance from the New Zealand strong-motion network', *Proc. 8th World Conf. on Earthq. Eng., San Francisco*, **II**, 199–206 (1984).
31. Bouchon, M., 'Effect of topography on surface motion', *Bull. Seism. Soc. Amer.*, **63**, No. 2, 615–32 (1973).
32. Smith, W. D., 'The application of finite element analysis to body wave propagation problems', *Geophys. J. R. Astron. Soc.*, **42**, 747–68 (1974).
33. Wong, H. L., 'Effect of surface topography on diffraction of P, SV, and Rayleigh waves', *Bull. Seism. Soc. Amer.*, **72**, No. 4, 1167–83 (1982).
34. Sánchez-Sesma, F. J., Chávez-Pérez, S., and Avilés, J., 'Scattering of elastic waves by three-dimensional topographies', *Proc. 8th World Conf. on Earthq. Eng., San Francisco*, **II**, 639–46 (1984).
35. Lee, V. W., 'Three-dimensional diffraction of plane P, SV and SH waves by hemispherical alluvial valley', *Soil Dynamics and Earthquake Engineering*, **3**, No. 3, 133–44 (1984).
36. Ayala, G., Munoz, C., and Esteva, L., 'Determination of earthquake design parameters for different local soil and topographical conditions', *Proc. 8th World Conf. on Earthq. Eng., San Francisco*, **II**, 621–268 (1984).
37. Silver, M. L., and Seed, H. B., 'The behaviour of sands under seismic loading conditions', *Report No. EERC 69-16*, Earthquake Engineering Research Center, University of California, Berkeley (1969).
38. Newmark, N. M., and Rosenblueth, E. *Fundamentals of Eathquake Engineering*, Prentice-Hall, Englewood Cliffs, NJ (1971).
39. Seed, H. B., and Idriss, I. M., 'Analysis of soil liquefaction; Niigata earthquake', *J. Soil Mechanics and Foundations Division, ASCE*, **93**, No. SM3, 83–108 (1967).
40. Seed, H. B., and Idriss, I. M., 'Simplified procedure for evaluating soil liquefaction potential', *J. Soil Mechanics and Foundations Division, ASCE*, **97**, No. SM9, 1249–73 (1971).
41. Kishida, H., 'A note on liquefaction of hydraulic fill during the Tokachi–Oki earthquake', *Second Seminar on Soil Behaviour and Ground Response during Earthquakes*, University of California, Berkeley (1969).
42. Lee, K. L., and Fitton, J. A., 'Factors affecting the cyclic loading strength of soil', *Special Technical Publication No. 450, Symposium on Vibration Effects of Earthquakes on Soils and Foundations, ASTM*, 71–95 (1968).
43. Ambraseys, N. N., *ESEE Research Report No. 85.5*, Engineering Seismology and Earthquake Engineering Section, Imperial College, London (1985).
44. Tokimatsu, K., and Yoshimi, Y., 'Criteria of soil liquefaction with SPT and fines content', *Proc. 8th World Conf. on Earthq. Eng., San Francisco*, **III**, 255–62 (1984).
45. Seed, H. B., and Idriss, I. M., *Ground motions and soil liquefaction during earthquakes*, Earthquake Engineering Research Institute, California (1982).
46. Seed, H. B., Tokimatsu, K., Harder, L., and Chung, R., 'The influence of SPT procedures in soil liquefaction resistance evaluations', *Report No. UCB/EERC-84/15*, University of California, Earthquake Engineering Research Center, Berkeley (1984).
47. Penzien, J., and Tseng, W. S., 'Seismic analysis of gravity platforms including soil–structure interaction effects', *Proc. Offshore Technology Conference, Houston, Texas, Paper No. 2674* (1976).
48. Vaish, A. K., and Chopra, A. K., 'Earthquake finite element analysis of structure-foundations systems', *J. Engineering Mechanics Division, ASCE,* **100**, No. EM6, 1101–16 (1974).
49. Wolf, J. P., *Dynamic Soil–structure Interaction*, Prentice-Hall, Englewood Cliffs, NJ (1985).
50. Seed, H. B., Lysmer, J., and Hwang, R., 'Soil–structure interaction analysis for seismic response', *J. Geotechnical Engineering Division, ASCE*, **101**, No. GT5, 439–57 (1975).

51. Poulos, H. G., and Davis, E. H., *Elastic Solutions for Soil and Rock Mechanics*, John Wiley and Sons, New York (1974).
52. Bycroft, G. N., 'Forced vibrations of a rigid circular plate on a semi-infinite elastic space and on an elastic stratum', *Phil. Trans. Roy. Soc., London, Series A*, **248**, 327–68 (1956).
53. Clough, R. W., and Penzien, J., *Dynamics of Structures*, McGraw-Hill, New York (1975).
54. Roesset, J. M., Whitman, R. V., and Dobry, R., 'Modal analysis for structure with foundation interaction', *J. Structural Division, ASCE*, **99**, No. ST3, 399–416 (1973).
55. Penzien, J., 'Soil–pile foundation interaction', in *Earthquake Engineering* (Ed. R. L. Weigel), Prentice-Hall, Englewood Cliffs, NJ (1970), pp. 349–81.
56. Luco, J. E., and Westmann, R. A., 'Dynamic response of circular footings, *J. Engineering Mechanics Division, ASCE*, **97**, 1381–95 (1971).
57. Veletsos, A. S., and Wei, Y. T., 'Lateral and rocking vibrations of footings', *J. Soil Mechanics and Foundations Division, ASCE*, **97**, No. SM9, 1227–48 (1971).
58. Veletsos, A. S., and Verbic, B., 'Vibration of viscoelastic foundations', *Earthquake Engineering and Structural Dynamics*, **2**, No. 1, 87–102 (1973).
59. Gazetas, G., and Dobry, R., 'Simple radiation damping model for piles and footings', *J. of Engineering Mechanics, ASCE*, **110**, No. 6, 937–56 (1984).
60. Danay, A., 'Vibrations of rigid foundations', *The Arup Journal, London*, **12**, No. 1, 19–27 (1977).
61. Veletsos, A. S., and Nair, V. V. D., 'Seismic interaction of structures on hysteretic foundations', *J. Structural Division, ASCE*, **101**, No. ST1, 109–29 (1975).
62. Watt, B. J., Boaz, I. B., Ruhl, J. A., Shipley, S. A., Dowrick, D. J., and Ghose, A., 'Earthquake survivability of concrete platforms', *Proc. Offshore Technology Conference, Houston, Texas, Paper No. 3159*, 957–73 (1978).
63. Luco, J. E., 'Impedance functions for a rigid foundation on a layered medium', *Nuclear Engineering and Design*, **31**, No. 2, 204–217 (1974).
64. Kausel, E., 'Forced vibrations of circular foundations on layered media', *Research Report R74-11*, Dept of Civil Engineering, Massachusetts Institute of Technology (1974).
65. Lysmer, J., and Kuhlemeyer, R. L., 'Finite dynamic model for infinite media', *J. Engineering Mechanics Division, ASCE*, **95**, No. EM4, 859–77 (1969).
66. Ang. A. H., and Newmark, N. M., 'Development of a transmitting boundary for numerical wave motion calculations', *Report to Defence Atomic Support Agency, Contract DASA-01-0040*, Washington, DC (1971).
67. Lysmer, J., and Waas, G., 'Shear waves in plane infinite structures', *J. Engineering Mechanics Division, ASCE*, **98**, No. EM1, 85–105 (1972).
68. Kausel, E., 'Nonlinear behaviour of soil–structure interaction', *J. Geotechnical Engineering Division*, **102**, No. GT11, 1159–70 (1976).
69. Kausel, E., and Roesset, J. M., 'Dynamic stiffness of circular foundations', *J. Engineering Mechanics Division, ASCE*, **101**, No. EM6, 771–85 (1975).
70. Bayo, E., and Wilson, E. L., 'Solution of the three dimensional soil–structure interaction problem in the time domain', *Proc. 8th World Conf. on Earthq. Eng., San Francisco*, **III**, 961–8 (1984).
71. Bayo, E., and Wilson, E. L., 'Numerical techniques for the evaluation of soil–structure interaction effects in the time domain', *Report No. UCB/EERC-83/04*, Earthquake Engineering Research Center, University of California, Berkeley (1983).
72. Gupta, S., Penzien, J., Lin, T. W., and Yeh, C. S., 'Three dimensional hybrid modelling of soil–structure interaction', *Earthquake Engineering and Structural Dynamics*, **10**, 1, 69–87 (1982).
73. Veletsos, A. S., and Meek, J. W., 'Dynamic behaviour of building-foundation systems', *Earthquake Engineering and Structural Dynamics*, **3**, No. 2, 121–38 (1974).

74. Watt, B. J., Boaz, I. B., and Dowrick, D. J., 'Response of concrete gravity platforms to earthquake excitations', *Proc. Offshore Technology Conference, Houston, Texas, Paper No. 2673* (1976).
75. Jennings, P. C., and Bielak, J., 'Dynamics of building-soil interaction', *Bull. Seism. Soc. Amer.*, **63**, 1, 9–48 (1973).
76. Veletsos, A. S. 'Dynamics of structure-foundation systems', in *Structural and Geotechnical Mechanics, A volume honoring N. M. Newmark* (Ed. W. J. Hall), Prentice-Hall, Englewood Cliffs, NJ (1977), pp. 333–61.
77. Bielak, J. 'Dynamic behaviour of structures with embedded foundations', *Earthquake Engineering and Structural Dynamics*, **3**, No. 3, 259–74 (1975).
78. Luco, J. E., Wong, H. L., and Trifunac, M. D., 'A note on the dynamic response of rigid embedded foundations', *Earthquake Engineering and Structural Dynamics*, **4**, No. 2, 119–27 (1975).
79. Novak, M., and Beredugo, Y. O., 'Vertical vibration of embedded footings', *J. Soil Mechanics and Foundations Division, ASCE*, **98**, No. SM12, 1291–1310 (1972).
80. Novak, M., 'Effect of soil on structural response to wind and earthquake', *Earthquake Engineering and Structural Dynamics*, **3**, No. 1, 79–96 (1974).
81. Novak, M., and Sachs, K., 'Torsional and coupled vibrations of embedded footings', *Earthquake Engineering and Structural Dynamics*, **2**, No. 1, 11–33 (1973).
82. Elsabee, F., Kausel, E., and Roesset, J. M., 'Dynamic stiffness of embedded foundations', *Proc. ASCE 2nd Annual Eng. Mech. Div. Specialty Conf., Nth Carolina*, 40–3 (1977).
83. Erden, S. M., 'Influence of shape and embedment on dynamic foundations response', *PhD Thesis*, University of Massachusetts, Amherst (1974).
84. Chen, J. C., Chun, R. C., Goudreau, G. L., Maslenikov, O. R., and Johnson, J. J., 'Uncertainty in soil–structure interaction analysis of a nuclear power plant due to different analytical techniques', *Proc. 8th World Conf. on Earthq. Eng., San Francisco*, **III**, 905–12 (1984).
85. Seed, H. B., 'Soil strength during earthquakes', *Proc. 2nd World Conf. on Earthq. Eng., Tokyo*, **1**, 183–94 (1960).
86. New Zealand Ministry of Works, 'Design of public buildings', *Code of Practice PW 81/10/1* (1970).
87. Ishihara, K., Koyamachi, N., and Kasuda, K., 'Strength of a cohesive soil in irregular loading', *Proc. 8th World Conf. on Earthq. Eng., San Francisco*, **III**, 7–14 (1974).
88. Zeevaert, L., *Foundation Engineering for Difficult Sub-soil Conditions*, Van Nostrand Reinhold, New York (2nd edn) (1983).
89. Allardice, N. W., Fenwick, R. C., Taylor, P. W., and Williams, R. L., 'Foundations for ductile frames', *Bull. NZ Nat. Soc. for Earthq. Eng.*, **11**, No. 2, 122–8 (1978).
90. Taylor, P. W., and Williams, R. L., 'Foundations for capacity designed structures', *Bull. NZ Nat. Soc. for Earthq. Eng.*, **12**, No. 2, 101–13 (1979).
91. US Department of Navy, *Design Manual—Soil Mechanics, Foundations and Earth Structures*, Navfac DM-7 (1971).
92. Barnes, S. B., 'Some special problems in the design of deep foundations', *Proc. 4th World Conf. on Earthq. Eng., Chile*, **III**, A-6, 29–36 (1969).
93. Wyllie, L. A., McClure, F. E., and Degenkolb, H. J., 'Performance of underground structures at the Joseph Jensen filtration plant', *Proc. 5th World Conf. on Earthq. Eng., Rome*, **1**, 66–75 (1973).
94. Iwasaki, T., and Hagiwara, R., 'Seismic response of highway bridges with deep caisson foundations embedded into soft soils', *Proc. 8th World Conf. on Earthq. Eng., San Francisco*, **III**, 569–76 (1984).
95. Margason, E., 'Pile bending during earthquakes', Lecture series on Design, Construction and Performance of Deep Foundations, held at University of California, Berkeley (1975).

96. Novak, M., 'Dynamic stiffness and damping of piles', *Canadian Geotechnical Journal*, **11**, 574 (1974).
97. Blaney, G. W., Kausel, E., and Roesset, J. M., 'Dynamic stiffness of piles', *2nd Int. Conf. on Numerical Methods in Geomechanics, Blacksburg, USA, ASCE* (1976).
98. Broms, B. B., 'Design of laterally loaded piles', *J. Soil Mechanics and Foundations Division, ASCE*, **91**, No. SM3, 79–98 (1965).
99. Matlock, H., 'Correlations for design of laterally loaded piles in soft clay', *Proc. Offshore Technology Conference, Houston, Texas, Paper No. OTC* 1204 (1970).
100. Brinch Hansen, J., 'The ultimate resistance of rigid piles against transversal forces', The Danish Geotechnical Institute, Copenhagen, *Bulletin No. 12* (1961).
101. Poulos, H. G., 'Analysis of the displacement of laterally loaded piles: I—Single piles', *Research Report No. R105*, Dept of Civil Engineering, University of Sydney, 1968.
102. Poulos, H. G., 'Behaviour of laterally loaded piles: I—Single piles', *J. Soil Mechanics and Foundations Division, ASCE*, **97**, No. SM5, 711–32 (1971).
103. Poulos, H. G., 'Behaviour of laterally loaded piles: II—Pile groups', *J. Soil Mechanic and Foundations Division, ASCE*, **97**, No. SM5, 733–52 (1971).
104. Matthewson, C. D., 'The elastic behaviour of a laterally loaded pile', *PhD Thesis*, University of Canterbury, New Zealand (1969).
105. Poulos, H. G., 'Load-deflection prediction for laterally loaded piles', *Australian Geomechanics Journal*, **G3**, No. 1, 1–8 (1973).
106. Terzaghi, K., 'Evaluation of coefficient of subgrade reaction', *Geotechnique*, **5**, No. 4 (1955).
107. Kocsis, P., *Lateral Loads on Piles*, Bureau of Engineering, Chicago (1968).
108. Ministry of Works and Development, 'Pile foundations design notes', Civil Engineering Division, Wellington, New Zealand, *CDP 812/B:1981*.
109. Prakash, S., 'Behaviour of pile groups subjected to lateral loads', *PhD Thesis*, University of Illinois, USA (1962).
110. Davisson, M. T., and Salley, J. R., 'Model study of laterally loaded piles'. *J. Soil Mechanics and Foundations Division, ASCE*, **96**, No. SM5, 1605–28 (1970).
111. Poulos, H. G., 'Group factors for pile-deflection estimation', *J. Geotechnical Division, ASCE*, **105**, No. GT12, 1489–1510 (1979).
112. Nishizawa, T., Tajiri, S., and Kawamura, S., 'Excavation and response analysis of damaged rc piles by liquefaction', *Proc. 8th World Conf. on Earthq. Eng., San Francisco*, **III**, 593–600 (1984).
113. Ohsaki, Y., 'Effects of sand compaction on liquefaction during the Tokachioki earthquake', *Soils and Foundations*, **X**, No. 2, 112–28 (1970).
114. Seed, H. B., and Booker, J. R., 'Stabilization of potentially liquefiable sand deposits using gravel drains', *J. Geotechnical Engineering Division, ASCE*, **103**, No. GT7, 757–768 (1977).
115. Japanese Society of Civil Engineers, *Earthquake resistant design for civil engineering structures, earth structures and foundations in Japan* (1973).
116. Seed, H. B., and Whitman, R. V., 'Design of earth retaining structures for dynamic loads', *ASCE Specialty Conference on Lateral Stresses and the Design of Earth Retaining Structures*, New York, 103–48 (1970).
117. Matthewson, M. B., Wood, J. H., and Berrill, J. B., 'Earth retaining structures', *Bull. NZ Nat. Soc. for Earthq. Eng.*, **13**, No. 3, 280–93 (1980).
118. Mononobe, N., 'Earthquake-proof construction of masonry dams', *Proc. World Engineering Conference*, **9**, 275 (1929).
119. Okabe, S., 'General theory of earth pressures', *J. Japanese Society of Civil Engineers*, **12**, No. 1 (1926).
120. Richards, R., and Elms, D. G., 'Seismic behaviour of gravity retaining walls', *J. Geotechnical Engineering Division, ASCE*, **105**, No. GT4, 449–64 (1979).

121. Wood, J. H., 'Earthquake-induced soil pressures on structures', *Report No. EERL 73-05*, Earthquake Engineering Research Laboratory, California Institute of Technology, Pasadena (1973).
122. Elms, D. G., and Richards, R., 'Seismic design of gravity retaining walls', *Bull. NZ Nat. Soc. for Earthq. Eng.*, **12**, No. 2, 114–21 (1979).
123. Whitman, R. V., and Liao, S., 'Seismic design of gravity retaining walls', *Proc. 8th World Conf. on Earthq. Eng., San Francisco*, **III**, 533–40 (1984).
124. Blake, R. E., 'Basic vibration theory, in *Shock and Vibration Handbook*, (Eds C. M. Harris and C. E. Crede), Vol. I, McGraw-Hill, New York (1961).
125. Albert C. Martin and Associates, *Inelastic dynamic analysis, Bank of New Zealand Headquarters Building, Wellington*, Report prepared for Brickell, Moss, Rankine and Hill, Wellington, A. C. Martin and Associates, Los Angeles (1973).
126. Sinha, B. P., Gerstle, K. H., and Tulin, L. G., 'Stress–strain relationships for concrete under cyclic loading', *J. American Concrete Institute*, **61**, No. 2, 195–211 (1964).
127. Park, R., Kent, D. C., and Sampson, R. A., 'Reinforced concrete members with cyclic loading', *J. Structural Division, ASCE*, **98**, No. ST7, 1341–60 (1972).
128. Blakeley, R. W. G., 'Prestressed concrete seismic design', *Bull. NZ Soc. for Earthq. Eng.*, **6**, No. 1, 2–21 (1973).
129. Williams, D., and Scrivener, J. E., 'Response of reinforced masonry shear walls to static and dynamic cyclic loading', *Proc. 5th World Conf. on Earthq. Eng., Rome*, **2**, 1491–4 (1973).
130. Takanashi, K., 'Inelastic lateral buckling of steel beams subjected to repeated and reversed loadings', *Proc. 5th World Conf. on Earthq. Eng., Rome*, **1**, 795–8 (1973).
131. Housner, G. W., and Jennings, P. C., 'The capacity of extreme earthquake motions to damage structures', in *Structural and Geotechnical Mechanics—A volume honoring N. M. Newmark* (Ed. W. J. Hall), Prentice-Hall, Englewood Cliffs, NJ (1977), pp. 102–16.
132. Standards Association of New Zealand, 'Code of practice for general structural design and design loadings for buildings', *NZS 4203: 1984*.
133. American Concrete Institute, 'Building code requirements for reinforced concrete', *ACI 318-83* (1983).
134. Thompson, K. J., and Park, R., 'Stress–strain model for prestressing steel with cyclic loading', *Bull. NZ Nat. Soc. for Earthq. Eng.*, **11**, No. 4, 209–18 (1978).
135. Park, R., 'Theorisation of structural behaviour with a view to defining resistance to ultimate deformability', *Bull. NZ Soc. for Earthq. Eng.*, **6**, No. 2, 52–70 (1973).
136. Otani, S., 'Hysteresis models of reinforced concrete for earthquake response analysis, *J. Faculty of Engineering, University of Tokyo*, **XXXVI**, No. 2, 125–59 (1981).
137. Hart, G. C., and Vasudevan, R., 'Earthquake design of buildings: damping', *J. Structural Division, ASCE*, **101**, No. ST1, 11–30 (1975).
138. Jennings, P. C., Matthiesen, R. B., and Hoerner, J. B., 'Forced vibrations of a tall steel-frame building', *Int. J. Earthq. Eng. and Struct. Dynamics*, **1**, No. 2, 107–32 (1972).
139. Blume, J. A., 'The motion and damping of buildings relative to seismic response spectra', *Bull. Seism. Soc. Amer.*, **60**, No. 1, 231–259 (1970).
140. Despeyroux, J., 'Applications of precast prestressed concrete in seismic-zone structures', Report presented at FIP Symposium on Seismic Structures, Tbilisi, 1972, London (1972).
141. Reay, A. M., and Shepherd, R., 'Steady state vibration tests of a six storey reinforced concrete building', *Bull. NZ Soc. for Earthq. Eng.*, **4**, No. 1, 94–107 (1971).
142. Esteva, L., Rascón, O. A., and Gutiérrez, A., 'Lessons from some recent earthquakes in Latin America', *Proc. 4th World Conf. on Earthq. Eng., Chile*, **III**, J2, 65,66, and 73 (1969).

143. Stratta, J. L., and Feldman, J., 'Interaction of infill walls and concrete frames during earthquakes', *Bull. Seism. Soc. Amer.*, **61**, No. 3, 609–12 (1971).
144. Lamar, S., and Fortoul, C., 'Brick masonry effect in vibration of frames', *Proc. 4th World Conf. on Earthq. Eng., Chile*, **II**, A3, 91–8 (1969).
145. Moss, P. J., and Carr, A. J., 'Aspects of the analysis of frame-panel interaction', *Bull. NZ Soc. for Earthq. Eng.*, **4**, No. 1, 126–44 (1971).
146. Wen, R. K., and Natarajan, P. S., 'Inelastic seismic behaviour of frame-wall systems', *Proc. 5th World Conf. on Earthq. Eng., Rome*, **1**, 1343–52 (1973).
147. Muto, K., 'Earthquake resistant design of 36-storied Kasumigaseki building', *Proc. 4th World Conf. on Earthq. Eng., Chile*, **III**, J4, 15–33 (1969).
148. Nigam, N. G., and Housner, G. W., 'Elastic and inelastic response of framed structures during earthquakes', *Proc. 4th World Conf. on Earthq. Eng., Chile*, **II**, A4, 89–104 (1969).
149. Clough, R. W., Benuska, K. L., and T. Y. Lin and Associates, 'FHA study of seismic design criteria for high-rise buildings', US Dept of Housing and Urban Development, *FHA, HUD TS-3*, 168–170 (1966).
150. Newmark, N. M., and Hall, W. J., 'Rational approach to seismic design standards for structures', *Proc. 5th World Conf. on Earthq. Eng., Rome*, **2**, 2266–75 (1973).
151. Lai, S. S. P., and Biggs, J. M., 'Inelastic response spectra for aseismic building design, *J. Structural Division, ASCE*, **106**, No. ST6, 1295–1310 (1980).
152. Tang, D. T., and Clough, R. W., 'Shaking table earthquake response of steel frame', *J. Structural Division, ASCE*, 105, No. ST1, 221–43 (1979).
153. Moss, P. W., Carr, A. J., and Buchanan, A. H., 'Seismic design loads for low-rise steel buildings', *Proc. Pacific Structural Steel Conference, Auckland*, **1**, 149–65 (1986).
154. Seed, H. B., Whitman, R. V. Dezfulian, H., Dobry, R., and Idriss, I. M., 'Soil conditions and building damage in 1967 Caracas earthquake', *J. Soil Mechanics and Foundations Division, ASCE*, **98**, SM8, 787–806 (1972).
155. Werkle, H., and Waas, G., 'Dynamic stiffness of foundations in inhomogeneous soils', *Proc. 8th European Conference on Earthquake Engineering*, Lisbon (1986).
156. Seed, H. B., and Romo, M. P., 'Analytical modelling of dynamic soil response', *Proc. Int. Conf. on the 1985 Mexico Earthquakes*, Mexico City, American Society of Civil Engineers, September (1986).
157. Hughes, J. M. O., Goldsmith, P. R., and Fendall, H. D. W., 'The behaviour of piles subjected to lateral loads', *Report No. 178*, Dept of Civil Engineering, University of Auckland, New Zealand (1978).

Chapter 7

Concrete Structures

7.1 INTRODUCTION

As can be seen from the simplified flow chart, Figure (i), at the beginning of the book, this chapter concerns parts of the final step in the design process, i.e. the detailed design of the structure. In choosing concrete as the structural material, considerations such as those set out in Chapter 5 should have been taken into account. In this chapter reinforced concrete (*in situ* and precast) and prestressed concrete are considered.

In the mid-1980s an internationally valid state-of-the-art review for resistance of concrete was exceptionally difficult to write, partly because of the enormous amount of literature on the subject and partly because of the large differences in attitude in different countries towards certain aspects of design. It is to be hoped that the outcome of the US–Japan Co-operative Research Program and the US–Japan–New Zealand one on beam–column joints will greatly reduce the latter source of divergence. In the meantime it is hoped that the following discussion is reasonably balanced.

7.2 *IN SITU* REINFORCED CONCRETE DESIGN AND DETAIL

7.2.1 Introduction

There is more information available about the seismic performance of reinforced concrete than any other material. No doubt this is because of its widespread use and because of the difficulties involved in ensuring its adequate ductility (toughness). Well-designed and well-constructed reinforced concrete is suitable for most structures in earthquake areas, but achieving both these prerequisites is problematical even in areas of advanced technology.

Reinforced concrete is generally desirable because of its availability and economy, and its stiffness can be used to advantage to minimize seismic deformations and hence reduce the damage to non-structure. Difficulties arise due to reinforcement congestion when trying to achieve high ductilities in framed structures, and the problem of detailing beam–column joints to withstand strong cyclic loading remains a difficult and contentious problem. It should be recalled that no amount of good detailing will enable an ill-conceived structural form to survive a strong earthquake. The principles for determining good seismic-resistant structural form are discussed in Chapter 5.

The following notes provide an introduction to earthquake resistance of reinforced concrete, which has been further discussed by Park and Paulay.[1]

7.2.2 Seismic response of reinforced concrete

The seismic response of structural materials has been discussed generally in Section 6.6, where some stress–strain diagrams were presented. The hysteresis loops of Figure 6.31(d) indicate that considerable ductility without strength loss can be achieved in doubly reinforced beams having adequate confinement reinforcing. This is in distinct contrast to the loss of strength and stiffness degradation exhibited by plain unconfined concrete under repeated loading as shown in Figure 6.31(c). Because the hysteretic behaviour of reinforced concrete is so dependent on the amount and distribution of the longitudinal and transverse steel, mathematical models of hysteresis curves need to be chosen with care to reflect the details of the actual construction, using methods such as those outlined by Park[2] and Otani.[3,4] Unfortunately, most testing on beams and columns has not included floors or lateral beams, so that the response characteristics of complete buildings have not been properly described, and the strength of complete buildings may significantly exceed that predicted by codes, as reported by Bertero.[5] The US–Japan Co-operative Research Program has addressed these problems and interim reports at the 8th World Conference on Earthquake Engineering confirm that floors and lateral beams have significant effects.

Reinforcement controls and delays failure in concrete members, the degradation process generally being initiated by cracking of the concrete. Inelastic elongation of reinforcement within a crack prevents the latter from closing when the load direction is reversed and cyclic loading leads to progressive crack widening and steel yielding (Figure 7.1). Fenwick[6] argued that shear in plastic hinge regions of beams is resisted by truss action until the phase of rapid strength degradation in which large shear displacements occur.

7.2.3 Reliable seismic behaviour of concrete structures

7.2.3.1 Introduction

For obtaining reliable seismic response behaviour the principles concerning choice of form, materials, and failure mode control discussed in Section 5.3 should be applied to concrete structures. Designing for failure mode control requires consideration of the structural form used, with most of the forms discussed in Section 5.4 being appropriate for concrete, i.e.:

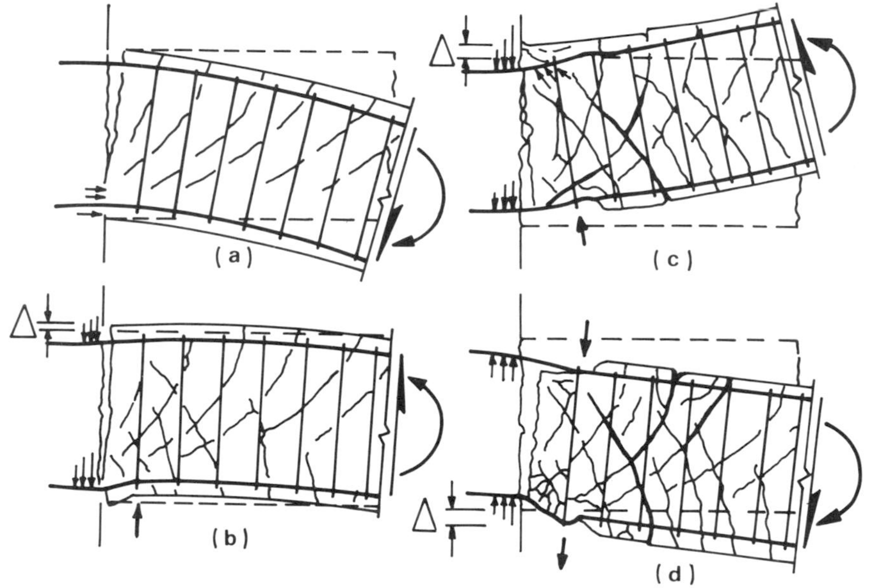

Figure 7.1 Significant stages of development of a plastic hinge in reinforced concrete during cyclic flexural and shear loading (after Paulay[39], Reproduced by permission of the Canadian Journal of Civil Engineering)

(1) Moment-resisting frames;
(2) Structural walls (i.e. shear walls);
(3) Concentrically braced frames;
(4) Hybrid structures.

For concrete structures, in addition to the discussion in Section 5.3.8, it should be noted that the essential objectives of failure mode control are:

(a) Beams should fail before columns (unless extra column strength is provided);
(b) Brittle failure modes should be suppressed;
(c) An appropriate degree of ductility (toughness) should be provided.

In order to help fulfil these three objectives, some concrete codes have[7,8] specific strength factors for enhancing column strength in relation to beams and for enhancing shear strength in relation to flexural strength. Also for highly ductile (tough) structures, the 1981 New Zealand concrete code[7] seeks to attain objectives (a) to (c) above by requiring a *capacity design* procedure to be followed, wherein greater strength capacity has to be supplied in the brittle modes than in the ductile ones. Unfortunately, the full rigour of this capacity design approach outlined in the code commentary is impractical to apply in all but the simplest of structures. This is because there are too many variable factors, in a design procedure involving as many as 15 steps,[9] and too much hand calculation work that cannot be practically computerized. Because the capacity design procedure appears to lead to very reliable failure mode control[10] it is to be hoped that a simpler and more usable version of it becomes available, such as by increasing the column design forces by a single easily determined factor.

Returning specifically to objective (b) above, the best-known brittle failure mode in concrete which should be suppressed is shear failure. In order to prevent shear failure occurring before bending failure it is good practice to design so that the flexural steel in a member yields while the shear reinforcement is working at a stress less than yield (say 90 percent). In beams a conservative approach to safety in shear is to make the shear strength equal to the maximum shear demands which can be made on the beam in terms of its bending capacity.

Referring to Figure 7.2, the shear strength of the beam should correspond to

$$V_{\max} = \frac{M_{u1} - M_{u2}}{l} + V_g \tag{7.1}$$

where V_g is the dead load shear force and

$$M_u = A_s f_{su} z$$

where A_s is all the steel in the tension zone (Figure 7.2(b)) f_{su} is the maximum steel strength after strain hardening, say the 95 percentile for the steel samples (Figure 7.2(c)), and z is the lever arm.

7.2.3.2 Required ductility (toughness) of concrete structures

Referring again to objective (c) above, the degree to which ductility should be enhanced is debatable. Until the 1980s research and codes had rightly been preoccupied with overcoming the excessive brittleness and unreliability of ill-reinforced concrete. However, there may have been too much emphasis on creating ductility for ductility's sake. The high cost of design and the complexity of some of the reinforcement of highly ductile concrete has raised the valid question, 'How do we design less ductile structures which are sufficiently reliable in earthquakes?' This question has long been raised regarding structures in regions of lesser seismicity, but there has only been arbitrary answers such as the recommendations in the Uniform Building Code.[8] In any seismic region the question applies not only to whole structures but also to parts of structures, as recognized by Selna *et al.*,[12] e.g. beams and columns in buildings where the primary earthquake resisting elements are structural walls.

Methods of adjusting the design loading for different degrees of ductility have been discussed in Chapters 4 and 6, such that the value of the ductility factor may be chosen in the range from $\mu = 1$ (non-ductile) to about $\mu = 6$ (highly ductile). However, the best design office method of achieving levels of ductility intermediate between non-ductile and highly ductile is unclear. The New Zealand concrete code[7] gives recommendations for the design of structures of *limited ductility* implying a value of $\mu = 2$ or thereabouts, but these rules represent early thoughts on the subject, and further development is required.

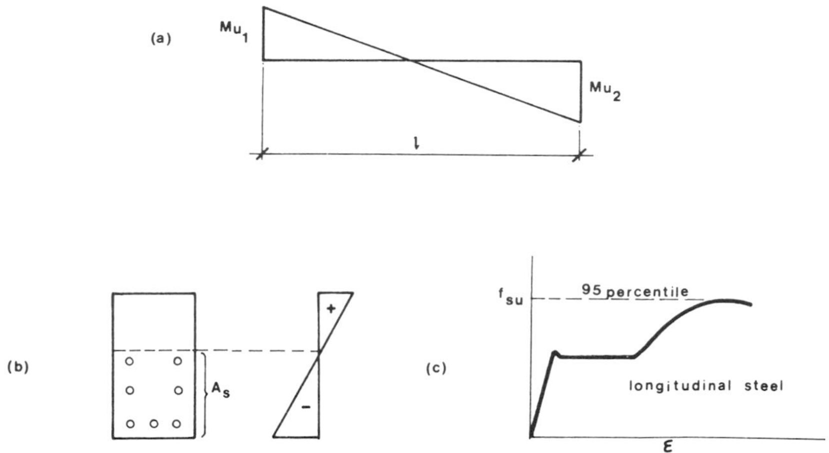

Figure 7.2 Shear strength considerations for reinforced concrete beams

Ductility and toughness have been discussed in the general terms of inelastic behaviour in Section 6.6.2, and the problems of analysing inelastic behaviour and hence assessing the required (*ductility demand*) in a structure have been considered in Section 6.6.7.3(i). While most concrete structures are designed by equivalent static analysis and codified reinforcing rules aimed at providing ductility, it is important for designers to understand how the ductility demand arises. This is now discussed using a simplified method of determining hinge rotations in reinforced concrete frames, which involves the assumption of a hinge mechanism (Figure 5.5) and the imposition of an arbitrary lateral deflection ductility factor μ on the frame.

As mentioned above, it is preferable that beams should fail before columns (for safety reasons). Considering ten storeys above the column hinges of a column sidesway mechanism, Park[11] found that for an overall frame deflection ductility factor $\mu = 4$, the required section ductility ratio was $\phi_u/\phi_y = 122$, which is impossibly high as shown by Figure 7.8. ϕ_u and ϕ_y are the hinge curvatures at ultimate and first yield, respectively. On the other hand, for a beam sidesway mechanism the required section ductility was found to be less than 20.

Having made an estimate of the ductility demands in the structure, the members should be detailed to have the appropriate section ductility, the theory for which is discussed below.

7.2.3.3 Available ductility for reinforced concrete members

The available section ductility of a concrete member is most conveniently expressed as the ratio of its curvature at ultimate moment ϕ_u to its curvature at first yield ϕ_y. The expression ϕ_u/ϕ_y may be evaluated from first principles, the answers varying with the geometry of the section, the reinforcement arrangement, the loading, and the stress–strain relationships of the steel and the concrete. Various idealizations of the stress–strain relationships give similar values for ductility, and the following methods of determining the available ductility should be satisfactory for most design purposes. It should be noted that the ductility of walls is discussed elsewhere (Section 7.2.4).

7.2.3.3(i) Singly reinforced sections

Consider conditions at first yield and ultimate moment as shown in Figure 7.3. Assuming an under-reinforced section, first yield will occur in the steel, and the curvature

$$\phi_y = \frac{\epsilon_{sy}}{(1-k)d} = \frac{f_y}{E_s(1-k)d} \tag{7.2}$$

where

$$k = \surd\{(\rho n)^2 + 2\rho n\} - \rho n \tag{7.3}$$

where $\rho = A_s/bd$ = tensile reinforcement ratio;

n = modular ratio $= E_s/E_c$, where E_s and E_c are the modulus of elasticity of the steel and the concrete, respectively.

Strictly, this formula for k is true for linear elastic concrete behaviour only, i.e. for

$$f_{cu} = \frac{2\rho f_y}{k} \leqslant 0.7 f_c'$$

where f_y is the steel yield stress and f_c' is the concrete cylinder compressive strength. For higher concrete stresses the true non-linear concrete stress block should be used. However, according to Blume *et al.*[13] equation (7.3) provides a reasonable estimate of k even if the computed concrete stress is as high as f_c'. Referring again to 7.3 it can be shown that the ultimate curvature is

$$\phi_u = \frac{\epsilon_{cu}}{c} = \frac{\beta_1 \epsilon_{cu}}{a} \tag{7.4}$$

where

$$a = \frac{A_s f_y}{0.85 f_c' b} \tag{7.5}$$

and β_1, which describes the depth of the equivalent rectangular stress block, may be taken as in UBC[8]

$$\beta_1 = 0.85$$

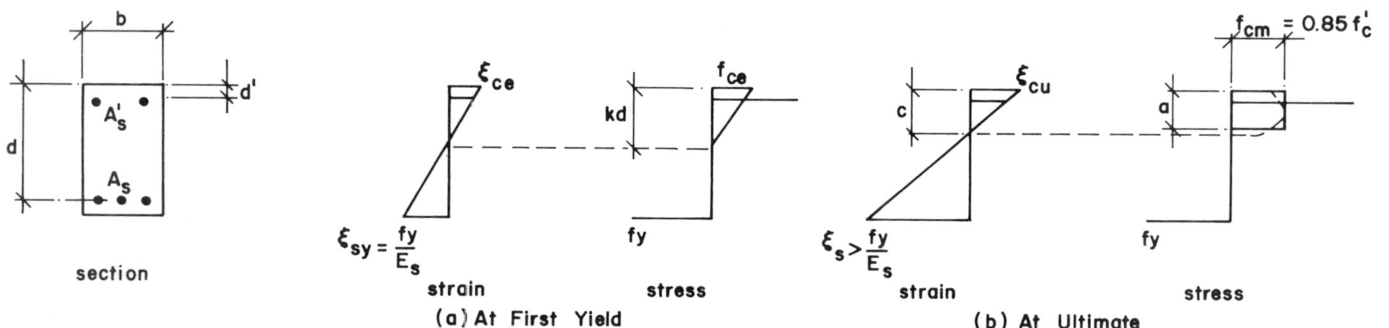

Figure 7.3 Reinforced concrete section in flexure

for $f_c' = 27.6$ N/mm^2 (4000 psi), otherwise

$$\beta_1 = 0.0308 f_c' - 0.0072(f_c' - 27.6) \tag{7.6}$$

From the above derivation the available section ductility may be written as

$$\frac{\phi_u}{\phi_y} = \frac{\epsilon_{cu} d(1-k)E_s}{c f_y} \tag{7.7}$$

The ultimate concrete strain ϵ_{cu} is given various values in different codes for different purposes. For estimating the ductility available from reinforced concrete in a strong earthquake a value of 0.004 was taken as representing the limit of useful concrete strain by Blume *et al.*,[13] although some codes[7,8] conservatively recommend a value of 0.003.

7.2.3.3(ii) *Doubly reinforced sections*

The ductility of doubly reinforced sections (Figure 7.4) may be determined from the curvature in the same way as for singly reinforced sections above.

Once again the expression for available section ductility is

$$\frac{\phi_u}{\phi_y} = \frac{\epsilon_{cu} d(1-k)E_s}{c f_y} \tag{7.7}$$

but to allow for the effect of compression steel ratio ρ', the expressions for c and k become

$$c = \frac{a}{\beta_1}$$

i.e.

$$c = \frac{(\rho - \rho')f_y d}{0.85 f_c' \beta_1} \tag{7.8}$$

and

$$k = \sqrt{\{(\rho + \rho')^2 n^2 + 2\,[\rho + (\rho' d/d)]\, n\}} - (\rho + \rho')n \tag{7.9}$$

The above equations assume that the compression steel is yielding, but if this is not so, the *actual* value of the steel stress should be substituted for f_y. As k has been found assuming linear elastic concrete behaviour, the qualifications mentioned for singly reinforced members also apply.

Computations of ductility are clearly best done with the aid of a small computer, and may be presented graphically as in Figure 7.5.

In Figure 7.5 the ultimate compressive force in the concrete was taken as $0.7 f_c'\, bc$ acting at a distance of $0.4c$ from the extreme compressive fibre. It can be seen from Figure 7.5 that ductility:

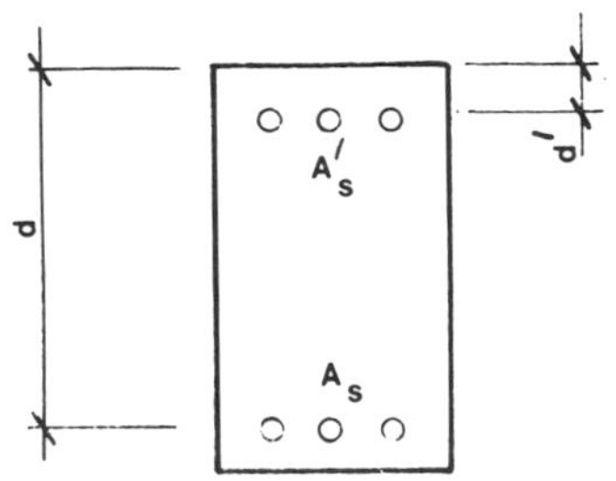

Figure 7.4 Doubly reinforced section

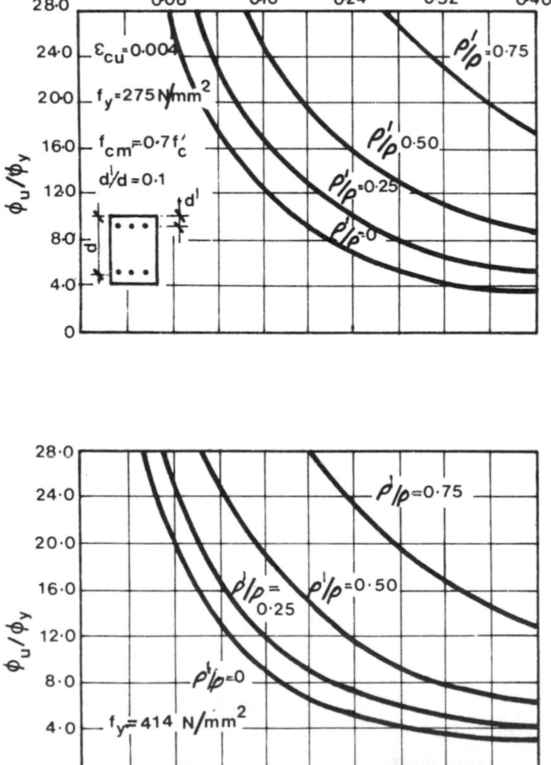

Figure 7.5 Variation of ϕ_u/ϕ_y for singly and doubly reinforced unconfined concrete (after Blume *et al.*[13])

(1) Reduces with increasing tension steel ρ;
(2) Increases with increasing compression steel ρ';
(3) Reduces with increasing yield stress f_y.

7.2.3.3(iii) The effect of confinement on ductility

That the ductility and strength of concrete is greatly enhanced by confining the compression zone with closely spaced lateral steel has been demonstrated by various workers.[14–16] In order to quantify the ductility of confined concrete, a number of stress–strain curves for monotonic loading of confined concrete have been derived from research.[13,17–20]

Probably the best of these for our purposes is the modified Kent and Park model shown in Figure 7.6, the relationships for which are given in references 20 and 21. Figure 7.6 illustrates the beneficial effect on ductility of confinement, with curve (d) being for unconfined concrete and curves (a) to (c) being for $\rho_s = 2.55$ percent, 1.7 percent, and 0.85 percent of confining reinforcement content, respectively. Using the modified Kent and Park stress–strain model for concrete and an appropriate stress–strain model for the longitudinal reinforcement, flexural strengths and moment-curvature diagrams of the type shown in Figure 7.22 can be reliably predicted for a wide variety of member properties.[20,21]

In addition, much research has been done[21] on the response to cyclic loading of various shapes of beams and columns with different arrangements and details of confining steel in an endeavour to find construction methods that ensure strength retention under inelastic cycling, as discussed in Section 7.2.9.2.

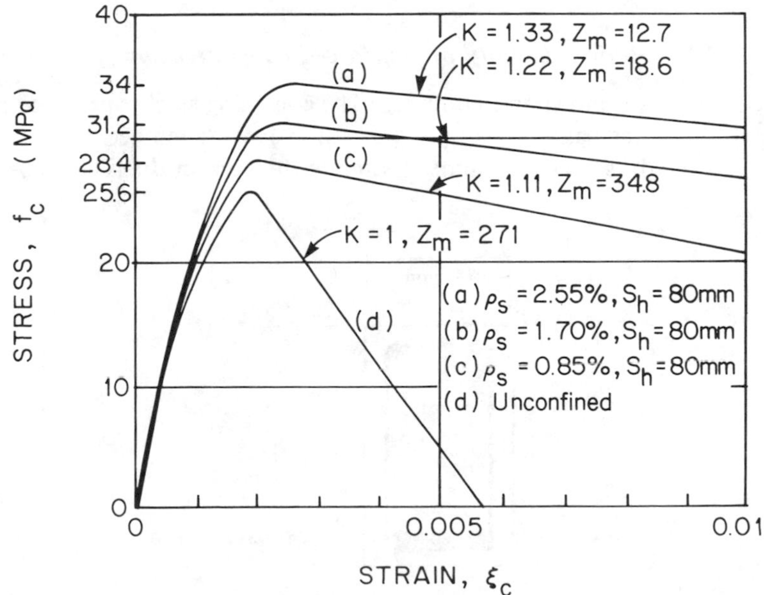

Figure 7.6 Modified Kent and Park stress–strain model for concrete in compression (confined: curves a, b, c; unconfined: curve d)[20,21]

The procedure for calculating the section ductility ϕ_u/ϕ_y is the same as that for unconfined concrete described above, the only difference being in determining an appropriate value of ultimate concrete strain ϵ_{cu} for use in equation (7.7). Corley[19] has recommended that a lower bound for the maximum concrete strain for concrete confined with rectangular links is

$$\epsilon_{cu} = 0.003 + 0.02\,(b/l_c) + \left(\frac{\rho_v f_{yv}}{138}\right)^2 \tag{7.10}$$

where b/l_c is the ratio of the beam width to the distance from the critical section to the point of contraflexure, ρ_v is the ratio of volume of confining steel (including the compression steel) to volume of concrete confined, and f_{yv} is the yield stress of the confining steel (N/mm^2).

Because of the high strains at ultimate curvature, the increased tensile force due to strain hardening should be taken into account, or the calculated ultimate curvature may be too large and the estimated ductility will be unconservative. Spalling of the concrete in compression is ignored in Corley's method.

As will be apparent from the following example, it is easy to increase the ultimate concrete strain to 0.01 or higher. As confinement and shear reinforcement are generally provided by the same bars, and as it is necessary for controlling the width of diagonal shear cracks to limit the strength of shear reinforcement to $f_y \approx 415$ N/mm^2, only modest advantage may be taken by increasing f_{yv} in equation (7.10).

Example 7.1—*Section ductility of reinforced concrete beam*

Consider the beam shown in Figure 7.7. The confining steel consists of 12 mm diameter mild steel bars ($f_y = f_{yv} = 275$ N/mm^2) at 75 mm centres, and the concrete strength is $f_c' = 21$ N/mm^2. Estimate the section ductility ϕ_u/ϕ_y.

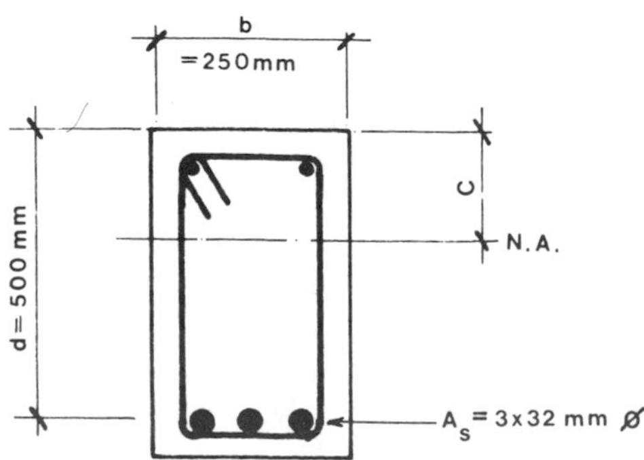

Figure 7.7 Reinforced concrete beam for ductility calculation example

To find the curvature at first yield, first estimate the depth of the neutral axis using equation (7.3), the section being effectively singly reinforced. As the modular ratio $n=9$, and $\rho=0.0193$

$$\rho n = 0.174$$

and

$$k = \sqrt{\{(\rho n)^2 + 2\rho n\}} - \rho n$$

$$= 0.441$$

Although this implies a computed maximum concrete stress greater than f'_c, the triangular stress block gives a reasonable approximation. Using equation (7.2) the yield curvature is found to be

$$\phi_y = \frac{f_y}{E_s(1-k)d} = \frac{275}{2\times 10^5 \times (1-0.441)500}$$

i.e.

$$\phi_y = 4.92 \times 10^{-6} \text{ radian/mm}$$

To find the ultimate curvature for the confined section first determine the ultimate concrete strain from equation (7.10). Assume that for this beam

$$\frac{b}{l_c} = \frac{1}{8}$$

and

$$\rho_v = \frac{113 \times 2(490+190)}{490 \times 190 \times 75}$$

and

$$\epsilon_{cu} = 0.003 + 0.02(b/l_c) + \left(\frac{\rho_v f_{yv}}{138}\right)^2 \qquad (7.10)$$

$$= 0.003 + \frac{0.02}{8} + \left(\frac{0.022 + 275}{138}\right)^2$$

i.e.

$$\epsilon_{cu} = 0.00742$$

Next find the depth of the neutral axis at ultimate from

$$c = \frac{a}{\beta_1}$$

$$= \frac{A_s f_y}{\beta_1 \times 0.85 f'_c b}$$

$$= \frac{2412 \times 275}{0.85 \times 0.85 \times 21 \times 250}$$

i.e.

$$c = 175\text{mm}$$

Hence the ultimate curvature is

$$\phi_u = \frac{\epsilon_{cu}}{c}$$

$$= \frac{0.00742}{175}$$

i.e.

$$\phi_u = 4.24 \times 10^{-5} \text{ radian/mm}$$

The available curvature ductility for the confined section can now be found:

$$\frac{\phi_u}{\phi_y} = \frac{4.24 \times 10^{-5}}{4.92 \times 10^{-6}} = 8.6$$

It is of interest to observe that the ultimate concrete strain $\epsilon_{cu} = 0.00742$, computed in the above example, is about twice the figure of 0.004 noted earlier for unconfined concrete. Hence the available section ductility has been roughly doubled by the use of confinement steel. This can be checked by reference to the curves of Figure 7.5, which gives value of ϕ_u/ϕ_y for unconfined flexural members. Now for the example beam

$$\frac{\rho f_y}{0.7 f'_c} = \frac{0.193 \times 275}{0.7 \times 21} = 0.36$$

and as the beam is singly reinforced, $\rho'/\rho = 0$. Hence from Figure 7.5 it can be seen that for the unconfined section

$$\phi_u/\phi_y \approx 4.25$$

which is about half the figure of $\phi_u/\phi_y = 8.6$ determined above.

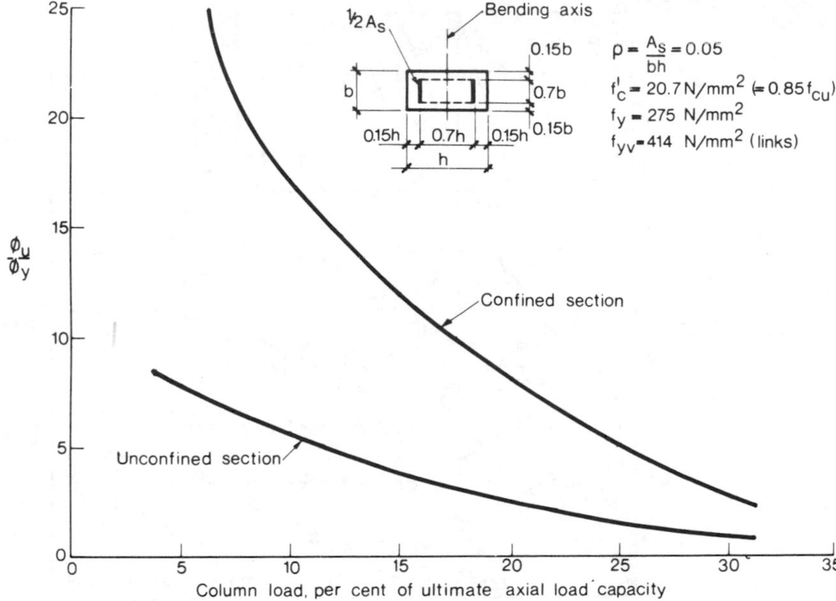

Figure 7.8 ϕ_u/ϕ_y for columns with confined or unconfined concrete (after Blume *et al.*[13])

7.2.3.3(iv) Ductility of reinforced concrete members with flexure and axial load

Axial load unfavourably affects the ductility of flexural members, as can be seen from Figure 7.8. Indeed it has been shown[22] that only with axial compression less than the balanced load does ductile failure occur.

It is evident from Figure 7.8 that for practical levels of axial load, columns must be provided with confining reinforcement. For rectangular columns with closely spaced hoops, and in which the longitudinal steel is mainly concentrated in two opposite faces, the ratio ϕ_u/ϕ_y may be estimated from Figure 7.9.

In Figure 7.9, A_s is the area of tension reinforcement and

$$\beta_h = \frac{A_h f_{yh}}{s h_h f_c'} \tag{7.11}$$

where A_h is the cross-sectional area of the links, f_{yh} is the yield stress of the hoop reinforcement, s is the spacing of the hoop reinforcement, and h_h is the longer dimension of the rectangle of concrete enclosed by the hoops.

The value ϕ_u/ϕ_y for a particular section is obtained by following a path parallel to the arrowed zigzag on the diagram.

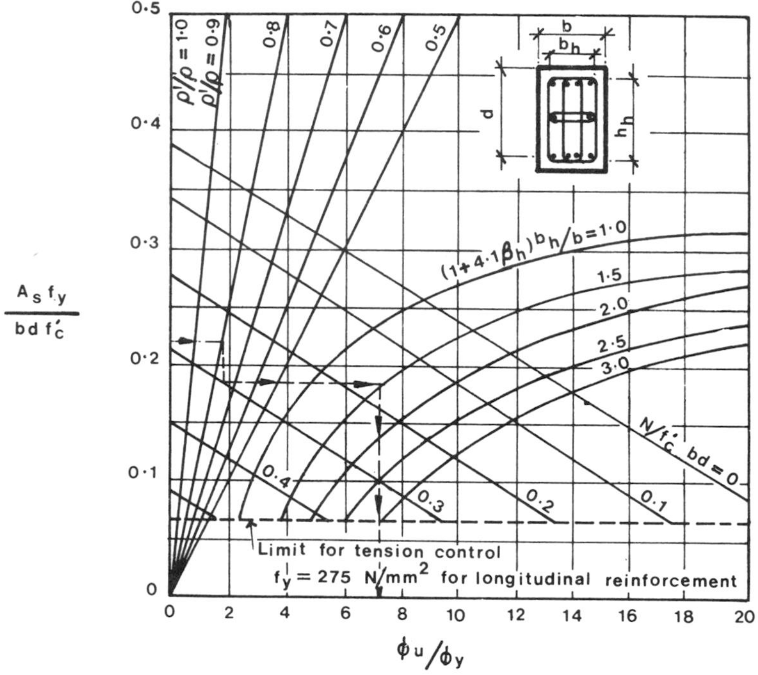

Figure 7.9 ϕ_u/ϕ_y for columns of confined concrete (after Blume *et al.*[13])

7.2.4 Reinforced concrete structural walls

7.2.4.1 Introduction

Great structural advantage may be taken from reinforced concrete structural walls (shear walls) in aseismic construction, provided they are properly designed and detailed for strength and ductility. Favourably positioned structural walls can be very efficient in resisting horizontal wind and earthquake loads. The considerable stiffness of walls not only reduces the deflection demands on other parts of the structure, such as beam–column joints, but may also help to ensure development of all available plastic hinge positions throughout the structure prior to failure. A valuable bonus of structural wall stiffness is the protection afforded to non-structural components in earthquakes due to the small storey drift compared with beam and column frames. Further discussion of stiff and flexible construction can be found in Section 5.3.6.

A notable example of the confidence being accorded concrete structural wall construction is the forty-four-storey Parque Central apartment buildings in Caracas,[23] built after the 1967 Caracas earthquake.

It should be noted that simpler methods of analysis, particularly equivalent-static seismic analysis, may give markedly inaccurate force distributions, especially in upper storeys due to the interaction of structural walls with rigid-jointed frames. This interaction may have undesirable effects, resulting in greater than expected ductility demands such as in captive spandrel columns.[12]

Most walls are fairly lightly loaded vertically and behave essentially as cantilevers, i.e. as vertical beams fixed at the base. A discussion of the basic design criteria for structural walls follows under the headings of tall and squat cantilevers. Coupled walls are then discussed as a special case of cantilever walls. Irregular arrangements of openings in structural walls require individual consideration and may require analysis by finite element techniques which, however, may not lead to adequate prediction of ductility requirements. Such structures may invite disaster by concentrating energy absorption in a few zones which are unable to develop the strength or ductility necessary for survival.

7.2.4.2 Cantilever walls

A single cantilever wall can be expected to behave as an ordinary flexural member if its height to depth ratio H/h is greater than about 2.0. Some distinctions between the two types of wall are made in the following sections.

Having obtained the design ultimate axial force N_u, moment M_u, and shear force V_u for a given wall it will usually be appropriate first to check the wall size and reinforcement for bending strength. This should be followed by a check that its ductility is adequate, and then that the wall's shear strength is somewhat greater than its bending strength (these two procedures may be implicit in detailing rules of advanced concrete codes, e.g. references 7 and 8). While considering the shears it should be ensured that the safe maximum applied shear stress is never exceeded and that the construction joints are adequately reinforced. These considerations are discussed more fully below.

7.2.4.2(i) Bending strength of cantilever walls

Rectangular walls. When walls of rectangular section are designed for small bending moments, the designer may be tempted to use a uniform distribution of vertical steel as for walls in non-seismic areas, but it may be shown from first principles that with this steel arrangement the ductility reduces as the total steel content increases.

When the flexural steeel demand is larger, it will be better to place much of the flexural steel near the extreme fibres, while retaining a minimum of 0.25 percent vertical steel in the remainder of the wall (see also Section 7.2.4.4). Apart from efficient bending resistance this steel arrangement will considerably enhance the rotational ductility.

In rectangular wall sections in which the reinforcement is concentrated at the extremities, the bending strength may be calculated from first principles following accepted codes of practice, or use may be made of column design charts which are frequently available. As design charts for *uniformly* reinforced members are not so readily available, their bending strength is discussed below.

If the contribution of the reinforcement in the elastic core of a uniformly reinforced wall with $H/h > 1.0$ is neglected, the following simple, conservative expression found by Cardenas *et al.*[24] for ultimate bending strength arises:

$$M_u = 0.5 A_s f_y h \left(1 + \frac{N_u}{A_s f_y}\right)\left(1 - \frac{c}{h}\right) \tag{7.12}$$

where

$$\frac{c}{h} = \frac{\alpha + \beta}{2\beta + 0.85\beta_1};$$

$$\alpha = \frac{A_s f_y}{b h f'_c};$$

$$\beta = \frac{N_u}{b h f'_c};$$

M_u = design resisting moment (ultimate) (N mm);
A_s = total area of vertical reinforcement (mm^2);
f_y = yield strength of vertical reinforcement (N/mm^2);
h = horizontal length of shear wall (mm);
c = distance from extreme compression fibre to neutral axis (mm);
b = thickness of shear wall (mm);
N_u = design axial load (ultimate), positive if compressive (N);
f'_c = characteristic cylinder compressive strength of concrete (N/mm^2);
β_1 = 0.85 for strength up to 27.6 N/mm^2 and reduced continuously at a rate of 0.05 for each 6.9 N/mm^2 of strength in excess of 27.6 N/mm^2.

Alternatively the bending strength of uniformly reinforced rectangular walls can be predicted from non-linear beam theory as discussed by Salse and Fintel,[25] who derived the axial load–moment interaction curves shown in Figure 7.10.

Flanged structural walls. Flanged walls are desirable for their high bending resistance and ductility, and arise in the form of I-sections or as channel sections which may be coupled together as lift shafts. As for rectangular walls, the derivation of interaction curves for axial load and bending of flanged walls is relatively easy working from first principles and with the aid of a small computer. Figure 7.16 shows the interaction curves for a channel section with bending about the minor axis.

Behaviour effects on different reinforcement arrangements can be seen in Figure 7.11, which shows axial load–moment interaction curves for I-sections or channel sections derived from non-linear beam theory. The curves are general for all values of b and h, and the web reinforcement is 0.25 percent in all cases except curve (1). It should be noted that curves (1) and (3) both represent sections containing 3 percent of steel in the flanges, the considerably enhanced strength

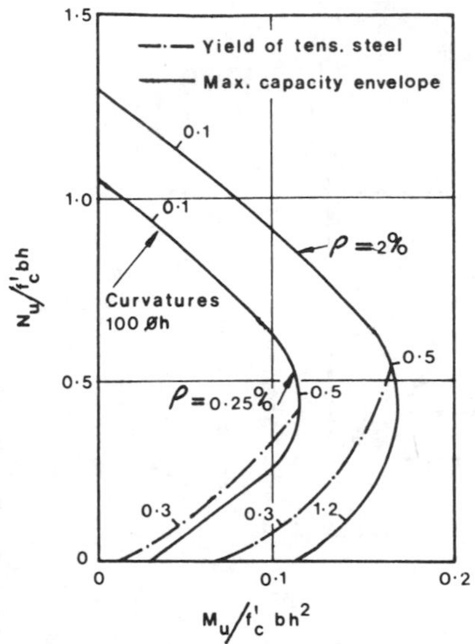

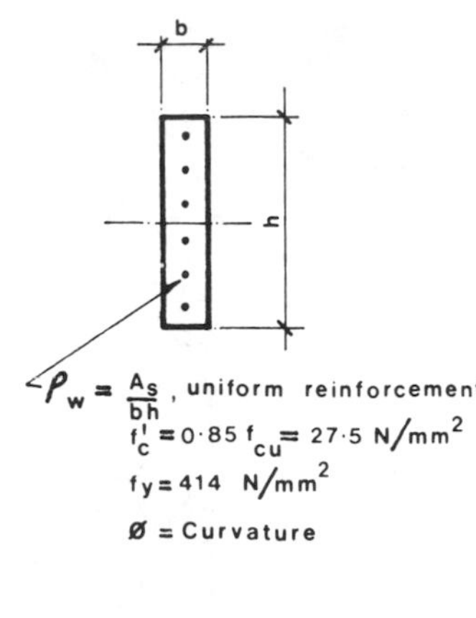

Figure 7.10 Axial load-moment interaction curves for rectangular uniformly reinforced walls (after Salse and Fintel[25])

of curve (1) being largely due to the assumptions of *high concrete confinement* in the flanges in this case.

Squat walls. Park and Paulay[26] have summarized the design procedure for squat walls as follows:

> In low-rise buildings the height of a structural wall may be less than its length. Such walls cannot be designed with the customary techniques of reinforced concrete theory. However, because the earthquake load for squat walls is seldom critical, an approximate design that ensures at least limited ductility will often suffice. The strength of many low-rise shear walls will be limited by the capacity of the foundations to resist the overturning moments. In such cases a rocking structure results and thus ductility becomes irrelevant.
>
> As Figure 7.12 indicates, after diagonal cracking the horizontal shear introduced at the top of a low-rise cantilever will need to be resolved into diagonal compression and vertical tensile forces. Thus, distributed vertical flexural reinforcement will also enable the shear to be transmitted to the foundations. The equilibrium condition of the free body marked ② shows this in Figure 7.12. Where the diagonal compression field does not find a support at foundation level, as is the case with the triangular free body marked ①, an equal amount of horizontal shear reinforcement will be required. Figure 7.12 thus shows that for a squat shear wall a steel mesh with equal area in both directions will be required if a compression field acting at 45° is conservatively assumed. The flexural strength at the base must be carefully evaluated, taking the contribution of all vertical bars into account, to ensure

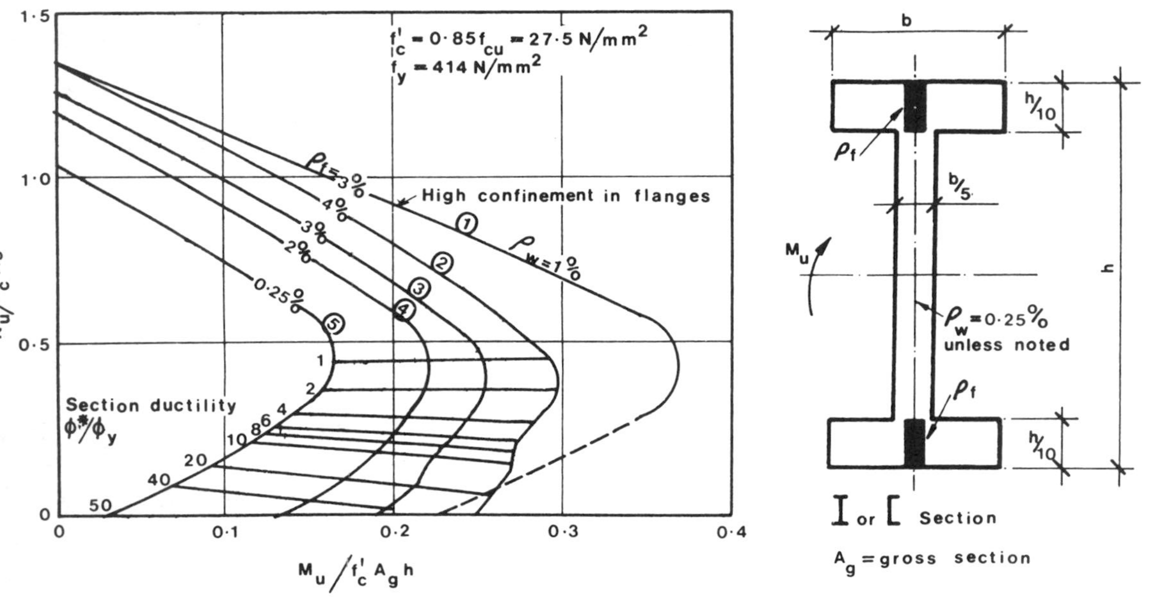

Figure 7.11 Axial load–moment interaction curves for I-section or [-section reinforced concrete walls (after Salser and Fintel[25])

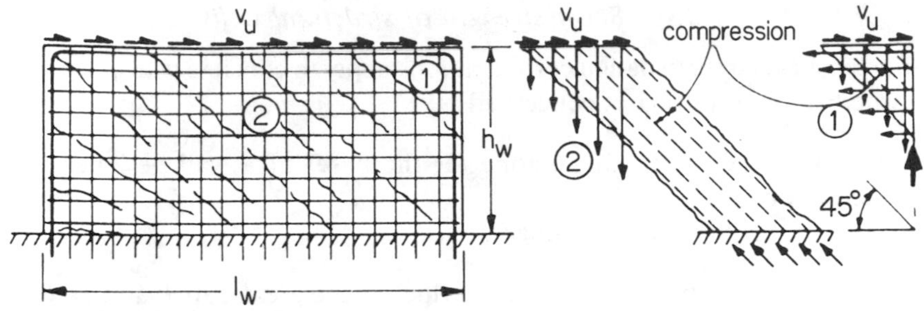

Figure 7.12 The shear resistance of squat shear walls (after Paulay[27])

that the required shear strength can be provided. This way most squat shear walls can be made ductile and a brittle failure will be avoided.

7.2.4.2(ii) Ductility of cantilever walls

The general problem of ductility in concrete structures is discussed elsewhere (Section 7.2.3.2), but suffice it here to say that adequate ductility under seismic loadings implies inelastic cyclic deformations without appreciable loss of strength.

As mentioned above, walls will exhibit greater ductility in bending if much of the reinforcement is concentrated near the extreme fibres, and consequently flanged sections are more ductile than rectangular walls. A comparison of the ductility of rectangular and I (or [ ) sections is given in Figure 7.13 where it was taken that

$$\text{available section ductility} = \frac{\phi^*}{\phi_y} \tag{7.13}$$

where ϕ^* = curvature at maximum moment;
ϕ_y = curvature at initiation of tension steel yield.

The ductility calculation was based on monotonic loading only, and hence Figure 7.13 serves better for comparative purposes than quantitative; the true ductility under reversible loading may be less than that shown, depending on the reinforcement quantities and disposition.

From Figure 7.13 it can be seen that both increasing steel percentages and increasing axial loads will decrease ductility. By comparing curve A with B, and curve C with D, it can be seen that the section ductility for I shapes is three to four times greater than that for uniformly reinforced rectangular sections. By comparing curve E with the remainder in Figure 7.13 the great effect on ductility of concrete confinement in the flanges can be seen.

In design situations it may be convenient to refer to an interaction diagram as shown on Figure 7.11, which incorporates ductility factors, thus allowing suitable strength and ductility to be chosen simultaneously.

Squat walls, i.e. those with height to depth ratio $H/h \lesssim 1.0$, are not amenable to the above ductility calculations as discussed in Section 7.2.4.2(i).

7.2.4.3 *Shear strength of structural walls*

Different shear strength design considerations apply to concrete walls, depending on whether the wall region concerned is:

(1) In a potential plastic hinge zone (at the base) of a wall designed to be ductile; or
(2) Not in a potential plastic hinge zone.

In plastic hinge zones, the height of which is the greater of h and $H/6$ (but not greater than $2h$), the shear strength is affected by yielding of the wall vertical reinforcement during reversed cyclic loading, and the New Zealand concrete code[7] requires that the ideal (nominal) shear strength provided by the concrete be not greater than

$$V_c = 0.6bd\sqrt{\left(\frac{N_e}{A_g}\right)} \tag{7.14}$$

where N_e is the design vertical load in compression acting simultaneously with the maximum design shear force and A_g is the gross sectional area of the wall. Also the ideal shear strength V_i should not exceed $0.9bd\sqrt{(f'_c)}$ for walls of limited ductility demand, the permissible V_i reducing to as low as $0.5bd\sqrt{(f'_c)}$ for walls required to be highly ductile.[7]

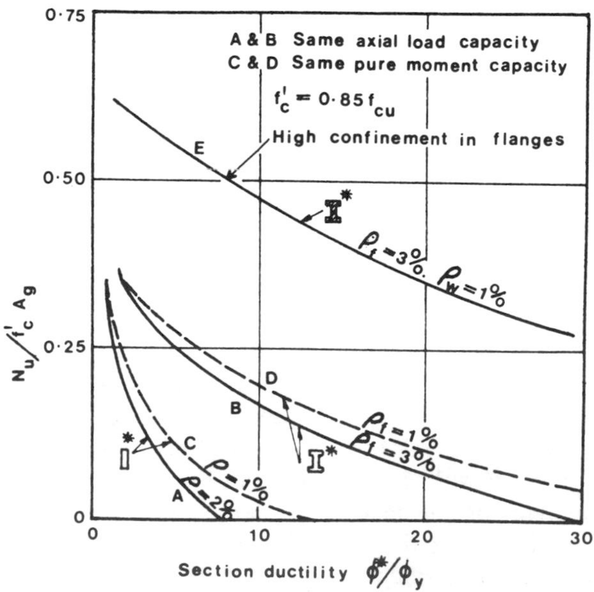

Figure 7.13 Ductility of walls as affected by cross-sectional shape, steel distribution, and concrete confinement (after Salse and Fintel[25])

In regions away from plastic hinge zones the restrictions on V_c are more relaxed than equation (7.14), and may be taken as the lesser of

$$V_c = \frac{bd}{4}\left(\sqrt{(f'_c)} + \frac{N_u}{bh}\right) \tag{7.15}$$

or

$$V_c = \left\{0.05 f'_c + \frac{h(\sqrt{(f'_c)} + 2N_u/bh)}{10(M_u/V_u - h/2)}\right\} bd \tag{7.16}$$

where N_u is negative for tension. When $(M_u/V_u - h/2)$ is negative, equation (7.16) does not apply.

For both cases (1) and (2) above, in addition to the shear strength provided by the concrete as in equations (7.14) and (7.15), the contribution made to the ideal shear strength by the horizontal reinforcement is

$$V_s = \frac{f_{yh} A_h d}{s} \tag{7.17}$$

In plastic hinge zones, in order to ensure that flexural yielding of the vertical reinforcement occurs before diagonal tension occurs, the minimum amount of horizontal reinforcement[7] should be

$$\rho_h = \frac{4}{3}\left[\frac{d_s V_u}{M_u}\frac{f_{yn}}{f_{yh}}\rho_n - \frac{V_c}{bdf_{yh}}\right] \tag{7.18}$$

where ρ_n is the total vertical steel content and f_{yn} and f_{yh} are the yield strengths of the vertical and horizontal steel, respectively.

Equation (7.18) may control the quantity of horizontal reinforcement for squat walls with aspect ratios, H/h, less than unity. Squat walls are further discussed in Section 7.2.4.2(i).

7.2.4.4 Horizontal construction joints in structural walls

Earthquake damage in walls has often occurred at horizontal construction joints in the form of sliding movements. In order to reliably prevent sliding due to reversible seismic shear forces a rough concrete interface combined with sufficient vertical reinforcement must be provided; allowance may be made for the beneficial effect of gravity loading.

The vertical reinforcement required across a joint should be estimated using the shear-friction method, for which various expressions have been derived, of the form

$$A_{vf} = \frac{1.25 V_u - N_u}{\mu f_y} \tag{7.19}$$

where μ is the coefficient of friction along the joint. For normal density concrete μ varies from about 0.7 for untreated surfaces to 1.4 for conditions simulating monolithic concrete.[7,8] In conjunction with equation (7.19), the ideal shear strength of the joint[7] should not be greater than $0.2bd\,f_c'$ or $6bd$ Newtons.

The above equation shows that with the minimum vertical steel content of 0.25 percent, moderate shear stress can be resisted without any assistance from vertical loads.

In computing the value of N_u for use in the above equation, care should be taken to deduct a suitable upwards seismic axial force from the gravity loading. In strong ground motion areas an upwards acceleration of 20–30 percent of gravity may be appropriate. N_u is positive for compression.

7.2.4.5 Coupled walls

7.2.4.5(i) Design approach

It is common practice nowadays to utilize the inherent lateral resistance of adjacent walls by coupling them together with beams at successive floor levels. Vertical access shafts punctured by door openings, as shown in Figure 7.14, form the classical example of this type of member. The analysis of coupled shear walls requires consideration of axial deformations of the walls and shear distortions of the coupling beams.

Ideally the designer would like the coupled walls to act as a box or I-unit as if the openings did not exist; such a structure would be much stronger than the two constituent channel units acting independently. In an efficiently coupled pair of walls the beam stiffnesses will be such that less than about one third of the total overturning moment

$$M_o = M_1 + M_2 + Nl \tag{7.20}$$

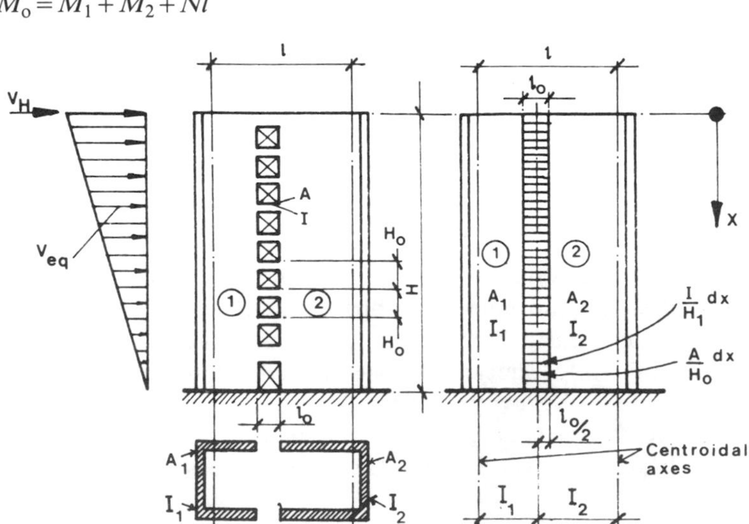

Figure 7.14 A typical coupled wall structure and its mathematical laminar model (after Paulay[27])

is resisted by the fixity moments of the walls ($M_1 + M_2$), the majority of the overturning moment being taken by the push–pull couple Nl due to the vertical reactions N at the base of the walls and their lever arm l. This implies the development of high shears in the coupling beams acting as a web, and the existence of large longitudinal forces in each wall unit. The failure of the coupling beams in coupled walls exposed to strong earthquakes indicates insufficient ductility of the beams. This has been due partly to inadequate detailing of the beams and partly to the use of elastic analysis which has not been adjusted to model the behaviour adequately. Standard frame analysis may suffice as long as the extra stiffness of the beam ends (within the walls) is taken into account, and redistribution of beam moments due to inelastic effects is properly done.[28]

Alternatively, Paulay has suggested a deterministic elastoplastic analysis based on a laminar model,[29] in which a desirable sequence of plastic hinge formations in the coupling beams is dictated. The plastic hinges in the walls, which are also major gravity load carrying units, should be the last ones to form. As the two walls are subjected to axial forces generated by the lateral load, their ductility will be restricted, so that any delay in the formation of their hinges should be beneficial in terms of reduced ductility demand.

Using the above-mentioned deterministic design approach, Paulay[27] analysed the core of a twenty-storey building with the dimensions and loads shown in Figure 7.15(a).

The ductilities required in the coupling beams are shown in Figure 7.15(b) in terms of maximum and yield rotations θ_p and θ_y. It can be seen that the maximum ductility was about $\theta_p/\theta_y = 11$, if all the beams were of equal strength (curve A); while when the beams were given strengths proportional to the elastic strength demand (curve B) the maximum ductility required was virtually the same at 12.6. Note that these ductility demands are a revision[1] to those originally published by Paulay.[27]

7.2.4.5(ii) The strength of coupled walls

Having derived the bending moments and forces acting on the wall elements of the coupled system, it will be necessary to design the walls to withstand those forces. The bending moment pattern will be similar to that of simple cantilever walls. In addition, because of the coupling system there will be considerable axial forces which may produce net tensions in the walls.

It is evident that the design considerations are as for cantilever walls discussed above. In the design of a high-rise structure with many similar horizontal sections to consider it may be worth producing a family of axial load-moment interaction diagrams (Figure 7.16) with the aid of a small computer program. It is to be noted that similar diagrams for different ratios of biaxial bending may be necessary for the same section.

7.2.5.5(iii) The strength and ductility of coupling beams

The classical failure mode of coupling beams in earthquakes is that of diagonal tension. To avoid this brittle type of failure two alternative methods of

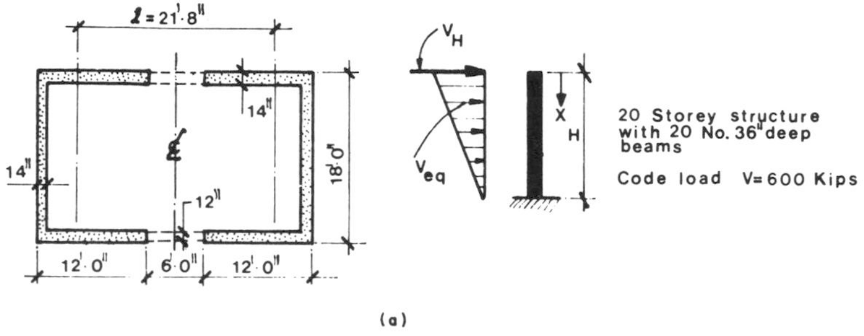

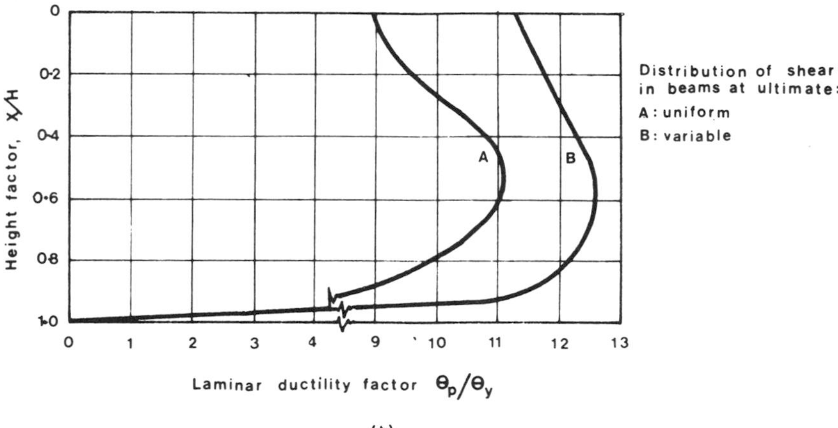

Figure 7.15 Example core wall structure, dimensions, loads, and calculated ductility demand in coupling beams (after Park and Paulay[1])

reinforcing the coupling beams are available. It has been recommended[7,28] that when the earthquake shear stress in the beam is less than

$$v = 0.1\,\frac{l_n}{h}\sqrt{(f_c')} \tag{7.21}$$

then the beam may be detailed in the normal manner, otherwise all of the shear force should be resisted by diagonal reinforcement. The danger of shear failure and the inhibition of ductility increases with increasing depth to span ratio h/l_n (where l_n is clear span of the beams), as reflected in equation (7.21). This severe limitation is recommended because coupling beams of shear walls can be subjected to very large rotational ductility demands, as noted below.

Where coupling beams may experience high seismic stresses, diagonal reinforcement of the type shown in the reinforced concrete Detail 7.6 provides far greater seismic resistance than conventional steel arrangements, as the comparison of ductilities in Figure 7.17 shows.

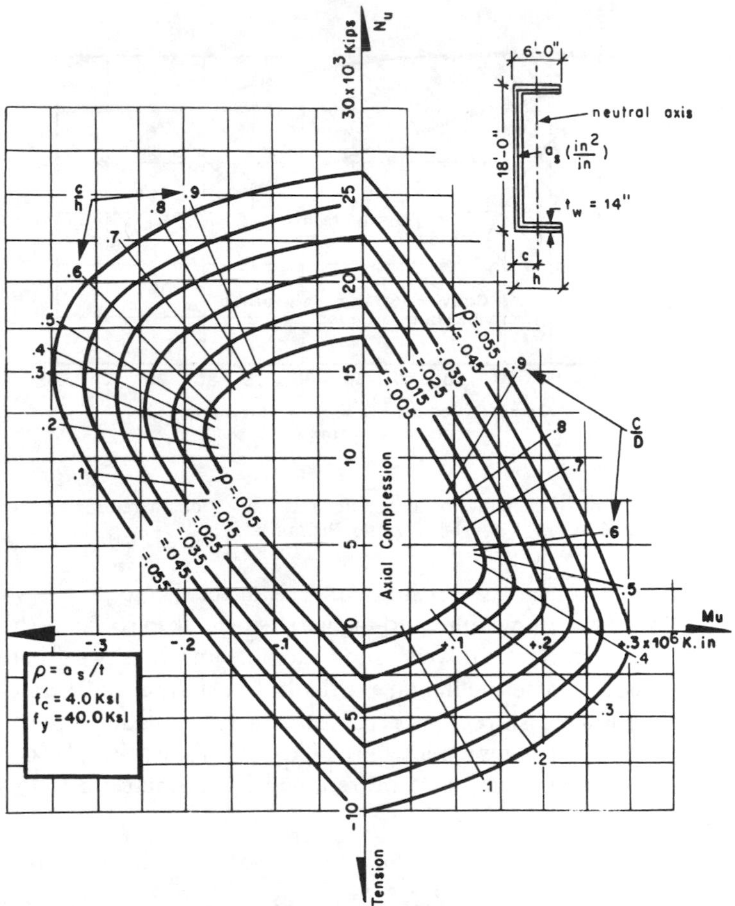

Figure 7.16 Axial force–moment interaction diagram for channel-shaped wall section (after Paulay[27])

Conventionally reinforced deep coupling beams having a ductility ratio of 4–5 (Figure 7.17) clearly would be unsatisfactory for the structure examined in Figures 7.14 and 7.15, whereas the diagonal reinforcement arrrangement easily provides the required ductility ratio of about 12. Thus in moderate or strong ground motion areas, diagonal reinforcement of deep coupling beams is seen to be required. The importance of restraining the diagonals against buckling in compression must, however, be realized, and careful detailing to suit this and still allow the proper placement will be necessary.

7.2.5 *In situ* concrete design and detailing—general requirements

The following notes and the associated detail drawings have been compiled to enable the elements of reinforced concrete structures to be detailed in a consistent

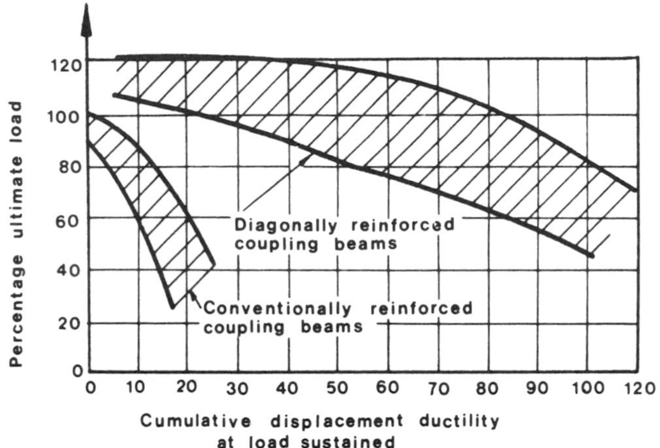

Figure 7.17 Comparison of ductilities of diagonally and conventionally reinforced deep coupling beam (after Paulay[27])

and satisfactory manner for earthquake resistance. These details should be satisfactory in regions of medium and higher seismic risk in so far as they reflect the present state-of-the-art. However, considerable uncertainty exists regarding effective details for some members, particularly columns and beam–column connections. In lower risk regions relaxations may be made to the following requirements, such as recommended in the UBC,[8] but the principles of splicing, containment and continuity must be retained if adequate ductility is to be obtained.

7.2.5.1 Splices

Splices in earthquake resisting frames must continue to function while the members or joints undergo large deformations. As the stress transfer is accomplished through the concrete surrounding the bars, it is essential that there be adequate space in a member to place and compact good quality concrete.

Splice laps should never be made in joints or in plastic hinge zones.[7,8] Tensile reinforcement in beams or columns should not be spliced in regions of tension or reversing stress unless the spliced region is confined by hoops or stirrups so that the area of the confinement steel is not less than[8]

$$A_t = 8\,d_b\,s/f_y$$

where d_b is the diameter of the bars being spliced and s is the spacing on the confining steel.

Tests have shown that contact laps perform just as well as spaced laps, because the stress transfer is primarily through the surrounding concrete. Contact laps (as with welded splices) reduce the congestion and give better opportunity to obtain well compacted concrete over and around the bars.

Laps should preferably be staggered but where this is impracticable and large numbers of bars are spliced at one location (e.g. in columns) adequate links or ties should be provided to minimize the possibility of splitting the concrete. In columns and beams even when laps are made in regions of low stress at least two links should be provided as shown in the details.

7.2.5.2 Development (Anchorage)

Satisfactory development may be achieved by extending bars as straight lengths, or by using 90 degree and 180 degree bends, but development efficiency will be governed largely by the state of stress of the concrete in the anchorage length. Tensile reinforcement should not be anchored in zones of high tension. If this cannot be achieved, additional reinforcement in the form of links should be added, especially where high shears exist, to help to confine the concrete in the development length.[7] It is especially desirable to avoid anchorage bars in the 'panel' zone of beam-column connections. Large amounts of the reinforcement should not be curtailed at any one section.

7.2.5.3 Bar bending

The minimum bend radius depends on the ductility of the steel being used (Appendix B.2) and upon the stress in the bar, so that earthquake-related codes[7,8] have a range of values on this subject. Alternatively, the use of BS 4466[30] will lead to standardization of bar shapes but due attention must be made to the bearing stresses in bends as follows. The bearing stress inside a bend in a bar which does not extend or is not assumed to be stressed beyond a point four times the bar size past the end of the bend need not be checked, as the longitudinal stresses developed in the bar at the bend will be small.

The bearing stress inside a bend in any other bar should be calculated from the equation

$$\text{bearing stress} = \frac{F_t}{r\phi}$$

where F_t is the tensile force due to ultimate loads in a bar or group of bars, r is the internal radius of the bend, and ϕ is the diameter of the bar or, in a bundle, the diameter of a bar of equivalent area.[31]

This stress should not exceed $1.75f_c'/(1 + 2\phi/a_b)$ where a_b for a particular bar or group of bars in contact should be taken as the centre to centre distance perpendicular to the plane of the bend between bars or groups of bars. For a bar or group of bars adjacent to the face of the member, a_b should be taken as the cover plus ϕ as defined above; f_c' is the characteristic cylinder crushing strength of the concrete.

7.2.5.4 *Cover*

Minimum cover to reinforcement should comply with local codes of practice.

7.2.5.5 *Concrete quality*

For earthquake resistance[6,7] the minimum recommended characteristic cylinder crushing strength for structural concrete is 20.0 N/mm^2.

The use of lightweight aggregates for structural purposes in seismic zones should e very cautiously proceeded with, as many lightweight concretes prove very brittle in earthquakes. Appropriate advice should be sought in selecting the type of aggregate and mix proportions and strengths in order to obtain a suitably ductile concrete. It cannot be overemphasized that quality control, workmanship, and supervision are of the utmost importance in obtaining earthquake-resistant concrete.

7.2.5.6 *Reinforcement quality*

For adequate earthquake resistance, suitable quality of reinforcement must be ensured by both specification and testing. As the properties of reinforcement vary greatly between countries and manufacturers, much depends on knowing the source of the bars, and on applying the appropriate tests. Particularly in developing countries the role of the resident engineer may be decisive and indeed onerous.

The following points should be observed (as amplified in Appendix B.2):

(1) An adequate minimum yield stress may be ensured by specifying steel to an appropriate standard, such as BS 4449,[32] or ASTM A706.[33]

(b) The variability of the strength of reinforcing steels as currently manufactured is so great as to inhibit design control of failure modes. For example, the best that the UBC[8] can realistically require for steel to ASTM A615[34] is that the actual yield stress should not exceed the minimum specified yield stress (characteristic strength) by more than 124 N/mm^2 (18 000 psi). Retests should not exceed this value by more than an additional 20 N/mm^2 (3000 psi).

(c) Grades of steel with characteristic strength in excess of 415 N/mm^2 (60 000 psi) are not permitted in some earthquake areas, e.g. California and New Zealand, but slightly greater strengths may be used if adequate ductility is proven by tests.

(d) Cold worked steels are not recommended in California or New Zealand, but cold worked steel to BS 4461[35] is sufficiently ductile (Appendix B.2).

(e) Steel of higher characteristic strength than that specified should *not* be substituted on site.

(f) The elongation test is particularly important for ensuring adequate steel ductility. In BS 4449, BS 4461, ASTM A615, and ASTM A706 appropriate requirements are set out for steels conforming to those standards. Steels to other standards require specific consideration.

(g) Bending tests are most important for ensuring sufficient ductility of reinforcement in the bent condition. In BS 4449, BS 4461, ASTM A706, and ASTM A615 appropriate requirements are set out for steels conforming to those standards. Steels to other standards require specific consideration.

(h) The minimum bend radius for bars to ASTM A615, and ASTM A706 is sometimes greater than for British steels.

(i) Resistance to brittle fracture should be checked by a notch toughness test conducted at the minimum service temperature, where this is less than about 3–5°C.

(j) Strain–age embrittlement should be checked by rebend tests, similar to those for British steels.

(k) Welding of reinforcing bars may cause embrittlement and hence should only be allowed for steel of suitable chemical analysis and when using an approved welding process.

(l) Galvanizing may cause embrittlement and needs special consideration.

(m) Welded steel fabric (mesh) is unsuitable for ductile earthquake resistance because of its potential brittleness. However, mesh to BS 4483[36] or similar may be used in slabs or walls where little ductility is required.

7.2.5.7 Codes and standards

The reinforcing details recommended in this book are derived from a wide range of experience. Greatest reliance has been placed on American and New Zealand opinion, and their codes and leading research results have been applied.

In some earthquake countries, local codes may overrule some of the recommendations given in this book, but generally the requirements herein reflect the mainstream of current good aseismic detailing. As such they are imperfect and generalized and will need updating from time to time and at the discretion of earthquake-experienced engineers.

7.2.6 Foundations

(See Details 7.1 and 7.2)

7.2.6.1 Column bases and pile caps

The following rules apply;

(1) Minimum percentage of steel = 0.15 percent each way;

(2) Bars should be anchored at the free end as shown on the detail sheet;

(3) Piles and caps should be carefully tied together to ensure integral action in earthquakes and sufficient reinforcement should be provided in non-tension piles to prevent separation of pile and cap due to ground movements.

7.2.6.2 Foundation tie-beams

In the absence of a thoroughgoing dynamic analysis of the substructure, tie-beams may be designed for arbitrary longitudinal forces of up to 10 percent of the maximum vertical column load into which the particular beam connects (Section 6.4.2). As the axial loads may be either tension or compression, the following rules are appropriate;

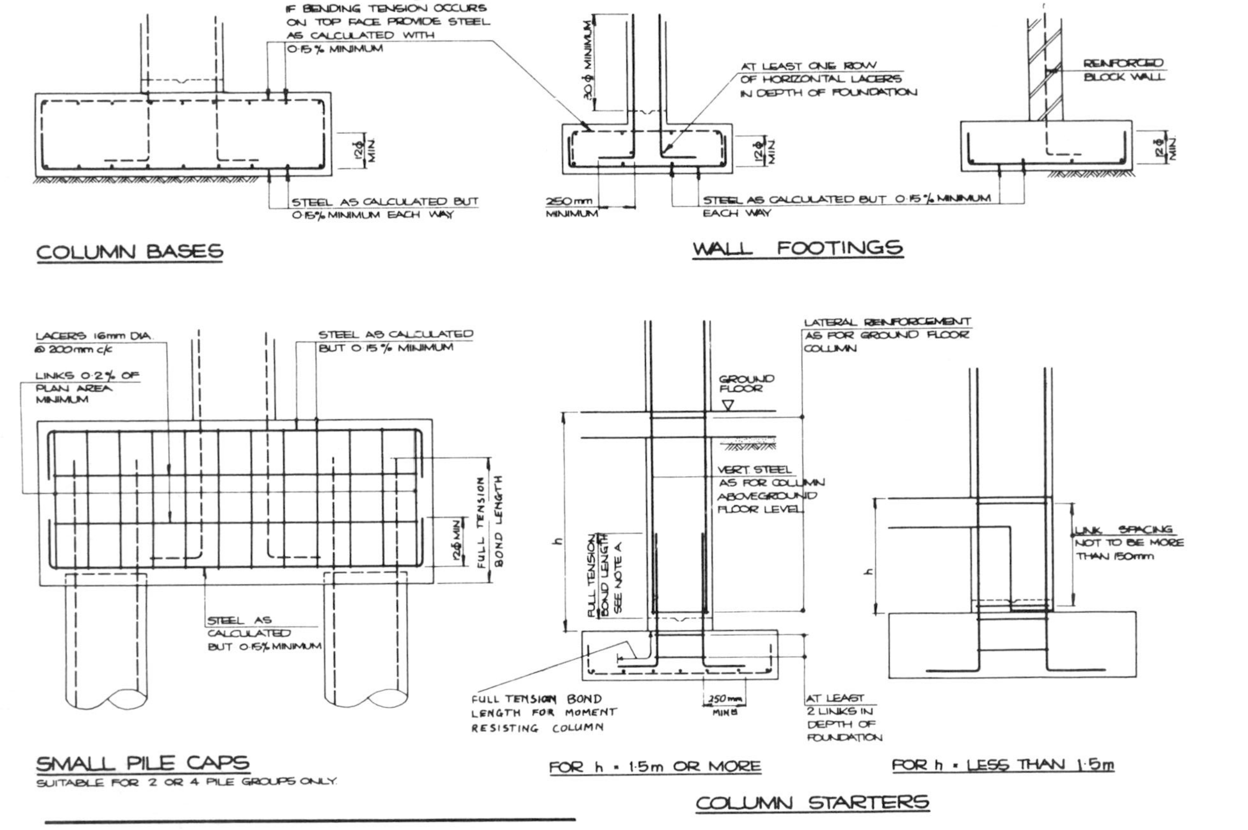

Detail 7.1 Foundations sheet 1

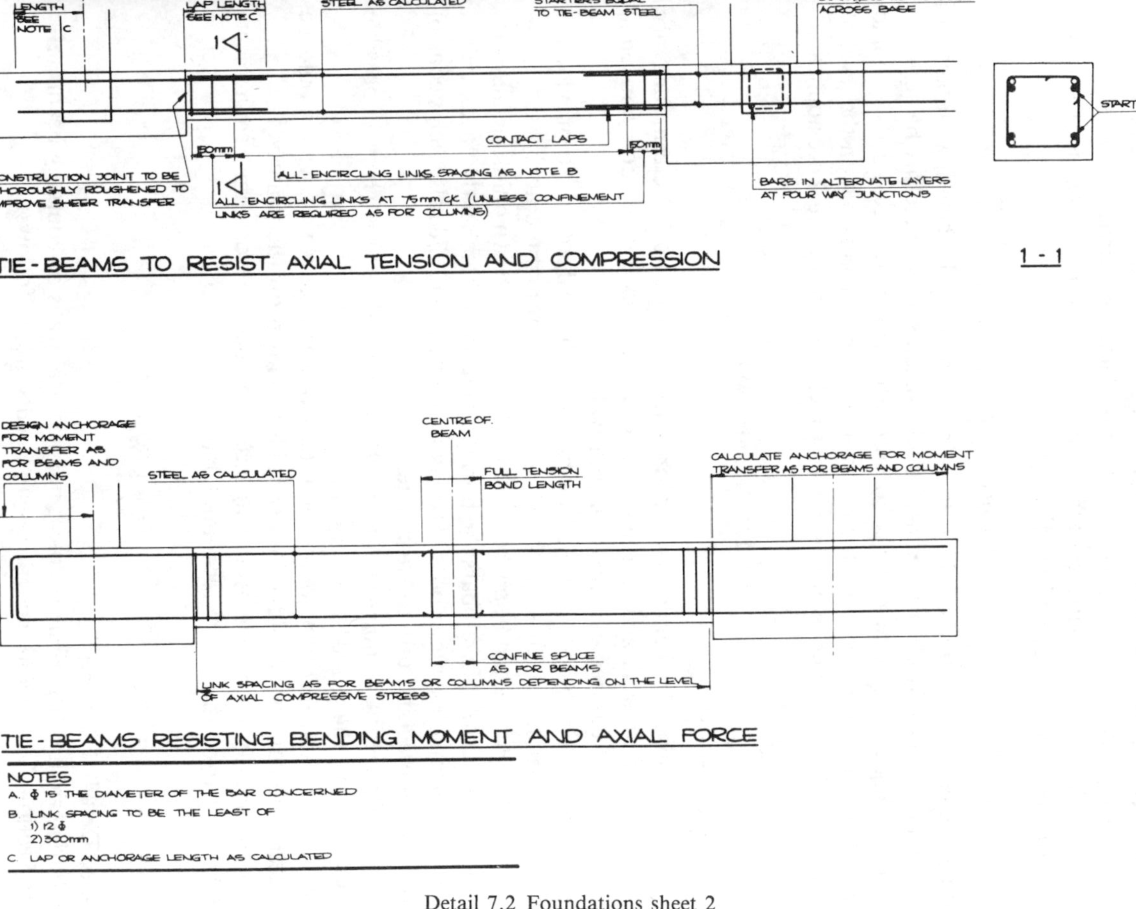

Detail 7.2 Foundations sheet 2

(1) Minimum percentage of longitudinal steel = 0.8 percent;
(2) Maximum percentage of longitudinal steel see Section 7.2.9.1;
(3) Maximum and minimum spacing as for columns;
(4) To enable foundation bases or footings to be cast before tie beams, beam starter bars from the footings should be detailed as shown on the detail sheet;
(5) The design check for the compressive case should be carried out as for design of columns with regard to such items as permissible compressive stresses, slenderness effects, and confining links.

7.2.6.3 Tie-beams taking bending

In some cases it may be required to transmit part of the bending moment at the column base into the tie-beams. Such tie-beams must therefore be designed for bending combined with axial compression or tension. The design should be carried out using the rules for beams or columns depending on the level of compressive stress. The requirements (a) to (c) in Section 7.2.6.2 are applicable.

7.2.7 Retaining walls

(See Detail 7.3.)

(a) The minimum percentage of reinforcement should be 0.15 percent each face each way in both walls and footings. For adequate crack control more horizontal steel may be required especially in thin walls.

(b) Top and bottom steel should be provided in the footings to provide for bending tensions which may not be apparent from a static analysis.

(c) Footing bars should be anchored at the free end as shown on the detail sheet.

(d) To ease fixing of wall bars the horizontal layer of bars on the face exposed to the air should wherever possible be in the outer layer. This also helps control of vertical cracking due to shrinkage.

(e) The staggering of horizontal and vertical laps is desirable wherever possible as shown on the detail sheet.

(f) Vertical construction joint positions along the length of wall should be selected to suit the detailing and should be shown on the drawings.

(g) Basement walls or water-retaining structures may have detailing requirements which override the above recommendations.

7.2.8 Walls

(See Details 7.4–7.6.)

For determining the reinforcement in structural walls or coupling beams refer to Section 7.2.4. More general requirements are as follows:

(1) The minimum content of vertical and horizontal steel should be 0.25 percent.
(2) The detailing around openings is important, and the details applicable to holes through suspended slabs may also be appropriate for smaller holes in walls.
(3) Horizontal construction joints should be cleaned and roughened to match the design assumptions.

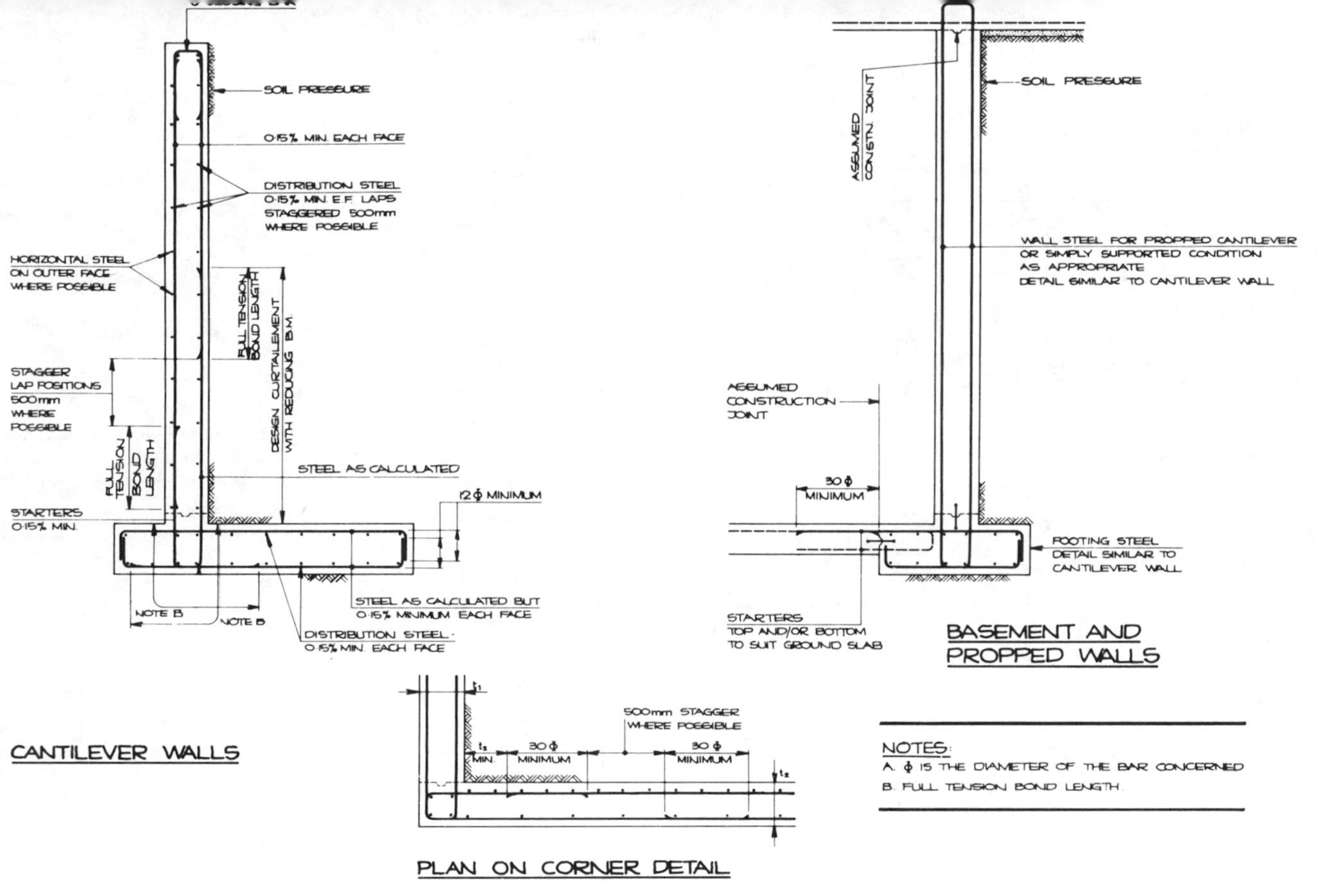

Detail 7.3 Retaining walls

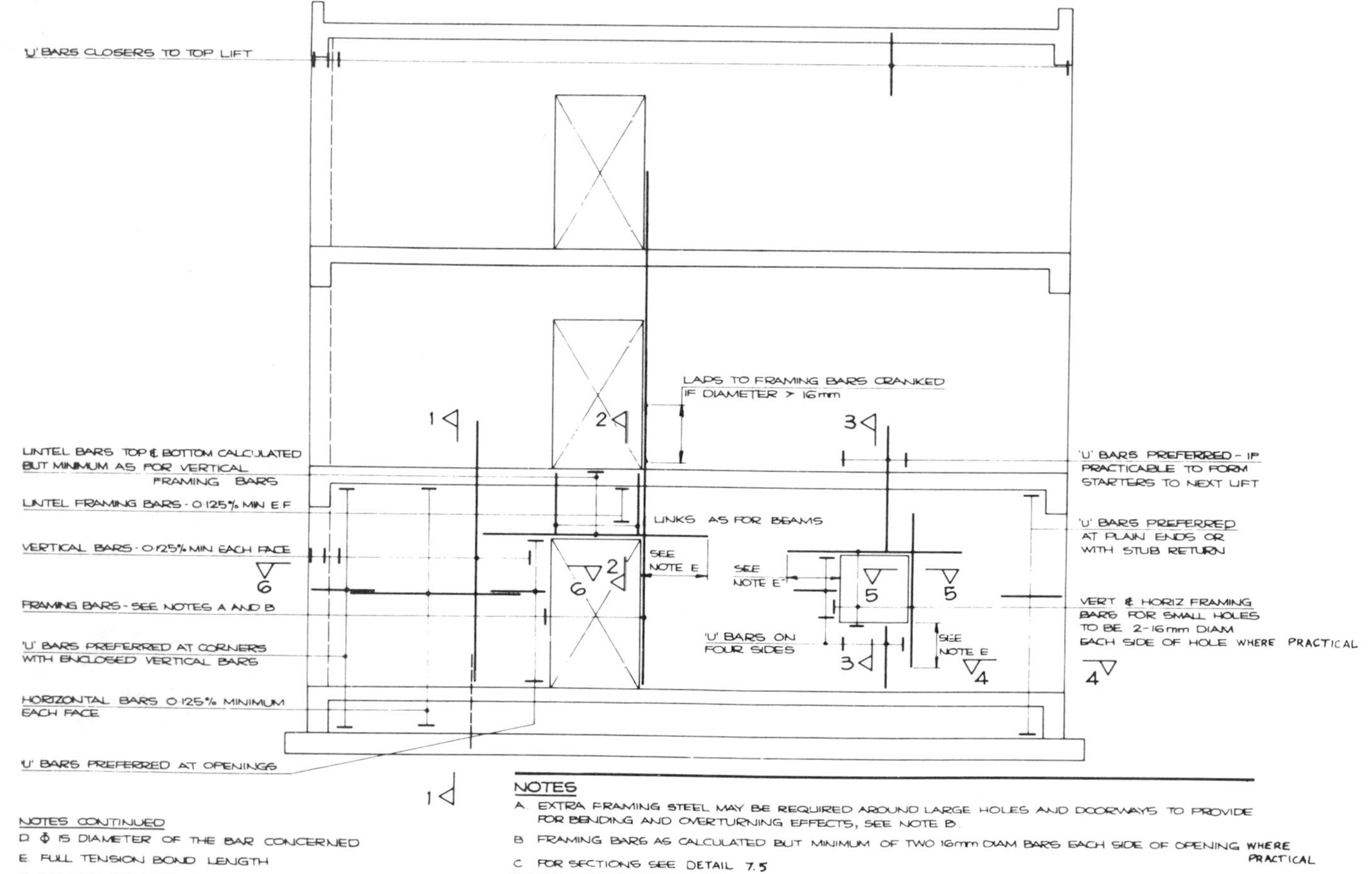

Detail 7.4 Walls sheet 1

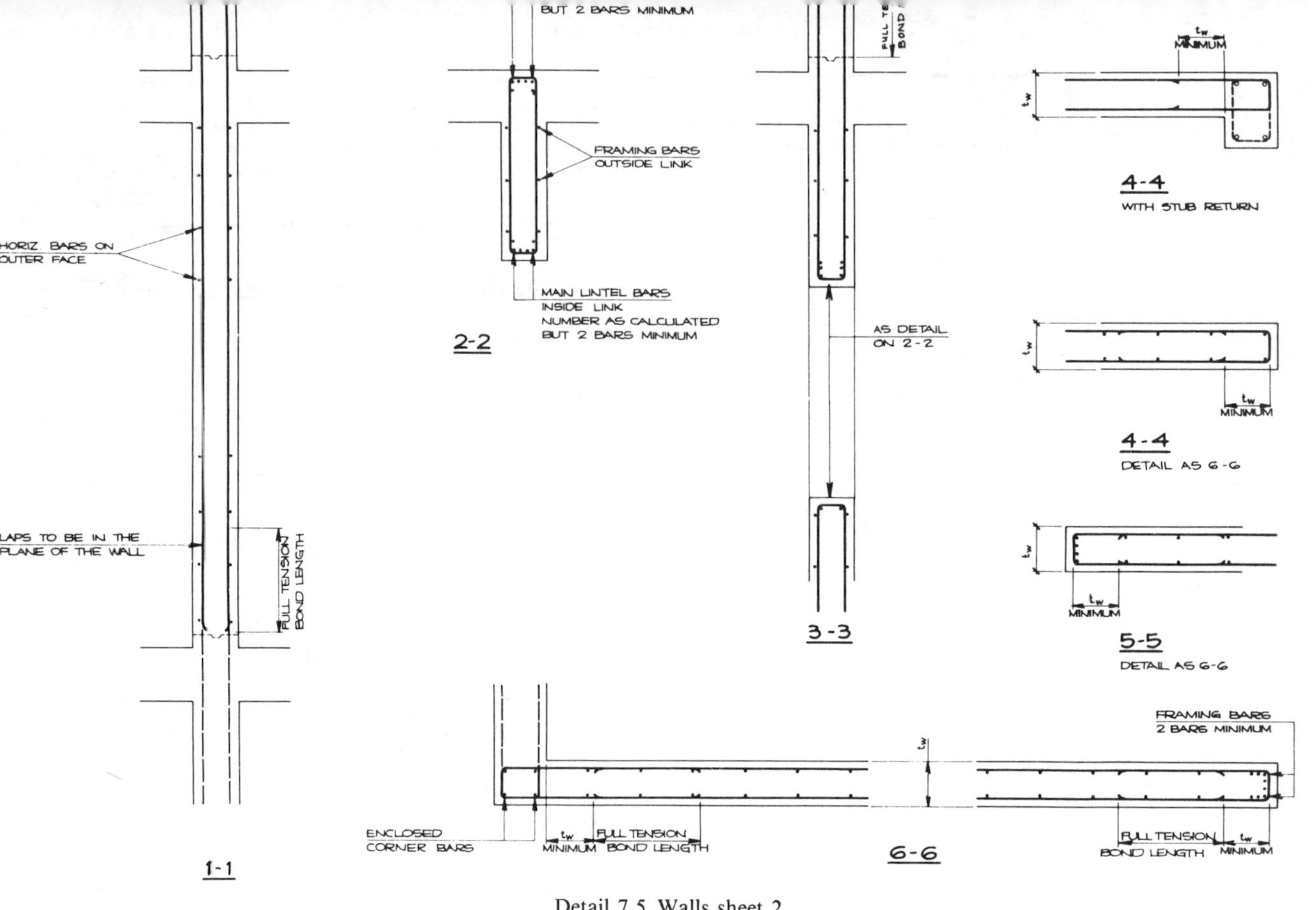

Detail 7.5 Walls sheet 2

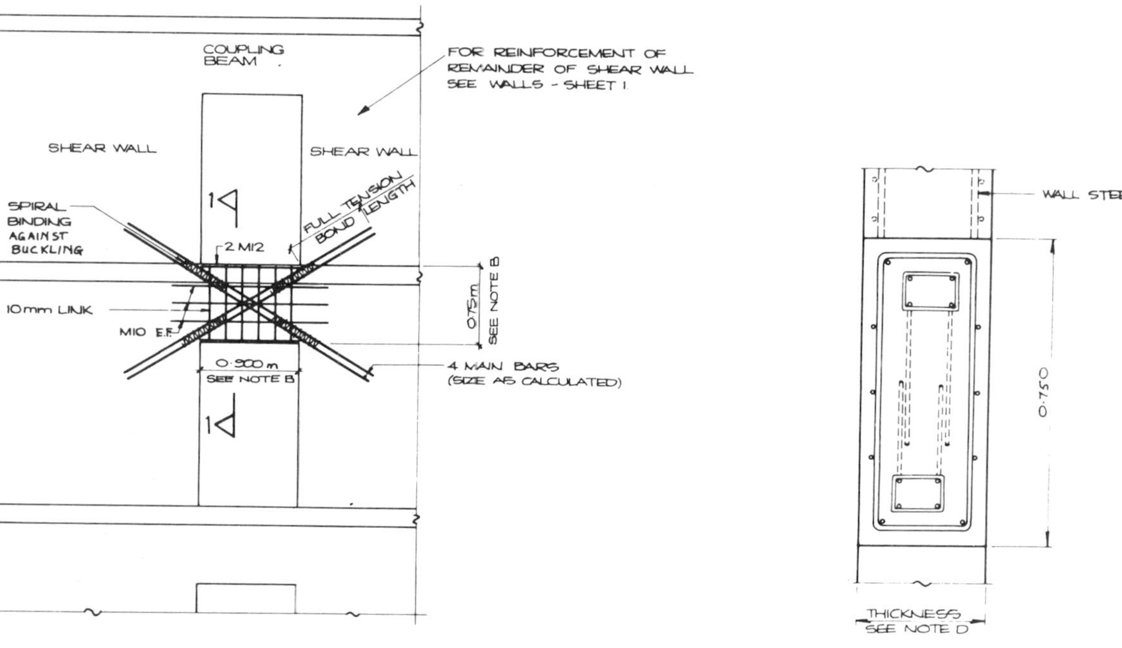

HIGH-STRENGTH COUPLING BEAM

1 - 1

NOTES

A. TO BE USED WHERE COUPLED SHEAR WALL ACTION IS ESSENTIAL.

B. THIS DETAIL IS APPROPRIATE FOR DEEP COUPLING BEAMS OF SIMILAR SPAN/DEPTH RATIO TO THAT SHOWN

C. CARE IS NECESSARY IN DETAILING TO EASE THE STEEL FIXING AND CONCRETING PROBLEMS.

D. USE WITH WALLS THINNER THAN 0.300m IS NOT RECOMMENDED.

Figure 7.6 Walls sheet 3

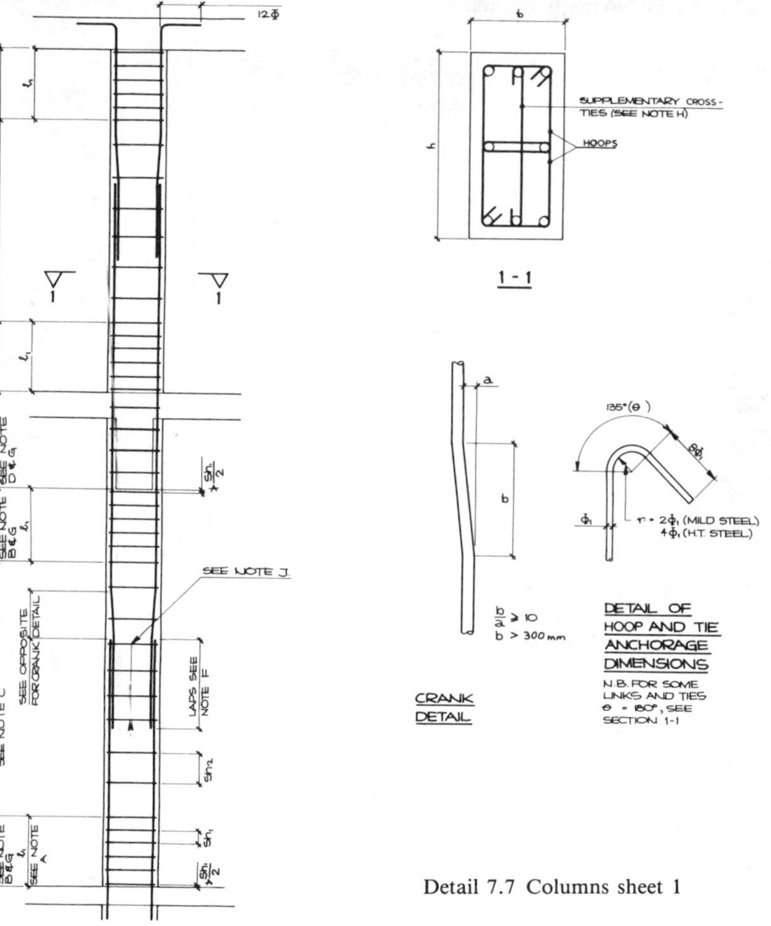

NOTES

A. l_1 = LARGEST OF 1) h (h > b)
2) $l_o/6$
3) 450mm

B. FOR COLUMNS REQUIRING SPECIAL CONFINEMENT STEEL
Sh_1 = SPACING OF ALL CONFINEMENT HOOPS AND CROSS TIES IN CONFINEMENT ZONE. USING A MAX. SPACING OF 0.2b, 6Φ OR 200mm. THE AREA OF CONFINEMENT LINKS SHOULD BE DETERMINED ACCORDING TO SECTION

C. Sh_2 = SPACING OF ALL LINKS & TIES IN INTERMEDIATE ZONE:- SPACING TO BE LEAST OF:- 1) 12 Φ
2) 400mm
3) 0.4b

Φ = DIAMETER OF SMALLEST LONGITUDINAL BAR IN COLUMN.

D. SPACING AND TOTAL CROSS-SECTIONAL AREA OF HOOPS & TIES THROUGH THE BEAM COLUMN CONNECTION (OR ANY OTHER CONNECTING MEMBER) REFER TO SECTION 7.2.11

E. ALL HOOP ARRANGEMENTS MUST ALSO BE CAPABLE OF RESISTING THE APPLIED SHEARS THROUGHOUT THE WHOLE COLUMN LENGTH INCLUDING THE BEAM COLUMN CONNECTION ZONE.

F. IN MEDIUM AND HIGH RISK EARTHQUAKE ZONES SPLICES SHOULD BE MADE OUTSIDE OF PLASTIC HINGE ZONES AND ARE TO BE LARGER OF:-
1) CALCULATED BOND LENGTH
2) 30 Φ, FOR fy=415MPa
3) 20 Φ, FOR fy=275MPa

G. FOR COLUMNS NOT REQUIRING SPECIAL CONFINEMENT OR SHEAR HOOPS PROVIDE MINIMUM HOOPS AS NOTE C THROUGHOUT

H. DIAM. OF SUPPLEMENTARY TIES TO BE SAME DIAM. AS HOOPS

J. SPLICES TO BE CONFINED BY MIN. OF 3 LINKS. TOPMOST LINK TO BE AT TOP OF LOWER SPLICE BAR.

K. DIAMETER OF HOOPS AND CROSS-TIES (min^m):
6mm, IF ϕ_m < 20mm
10mm IF 20mm ≤ ϕ_m ≤ 32mm
12mm IF ϕ_m > 32mm
ϕ_m = DIAM LONG L BARS

Detail 7.7 Columns sheet 1

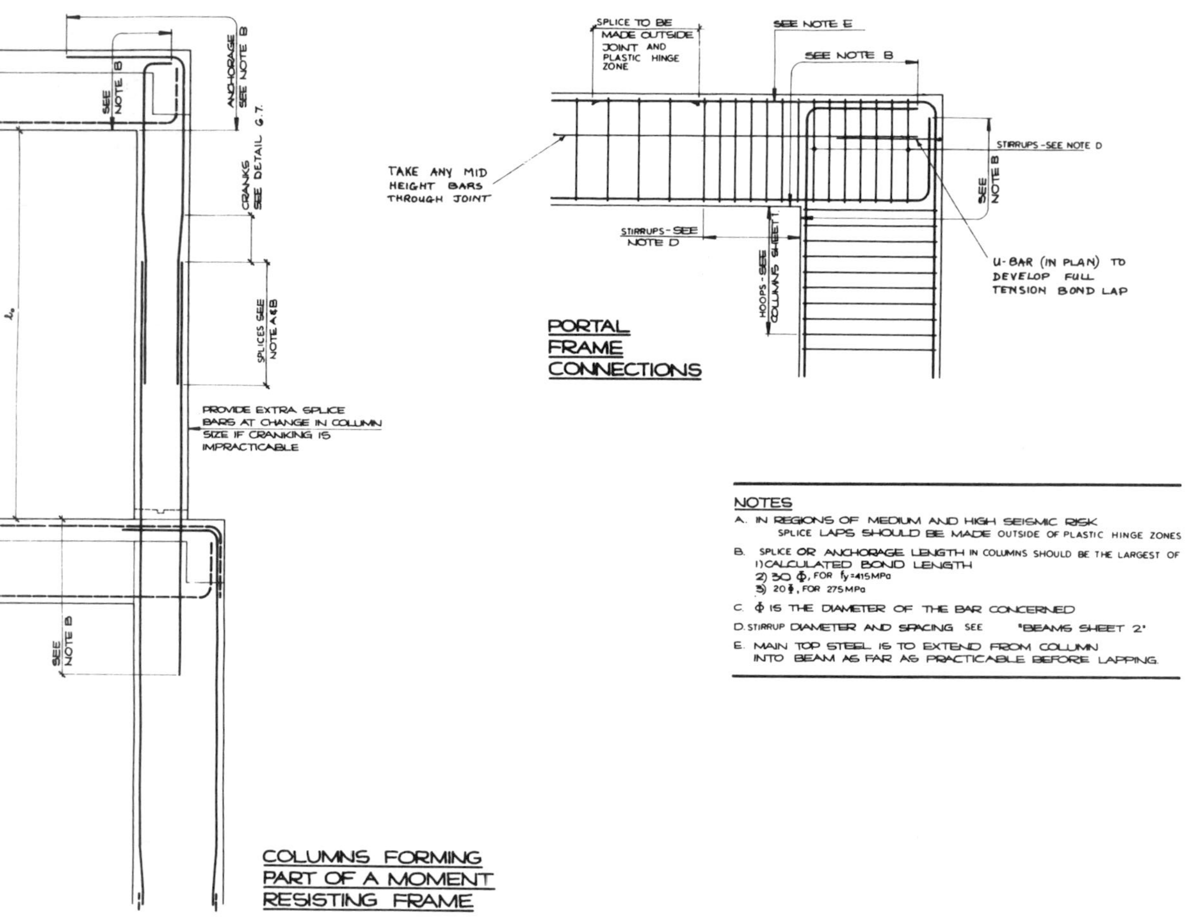

Detail 7.8 Columns sheet 2

7.2.9 Columns

(See Details 7.7 and 7.8.)

7.2.9.1 *General*

The design notes given in this section are aimed primarily at columns which form part of ductile moment-resisting frames. Columns in other situations, such as

(1) Trapped spandrel columns in wall/frame systems;
(2) Columns in flat slab structures; and
(3) Pilaster columns

require specific consideration, as outlined by Selna *et al.*[12] Other general design requirements for columns are as follows:

(1) The minimum width of the compression face of a member should be 200 mm.[7]
(2) The minimum content of longitudinal steel should be 0.8 percent of the gross sectional area. The maximum content should be 6 percent for grade 300 steel (8 percent at lap splices) and 4.5 percent for Grade 400 steel (6 percent at lap splices).

7.2.9.2 *Column confinement reinforcing*

Transverse reinforcement required for confinement should be provided unless a larger amount is required for shear. In potential plastic hinge zones of columns, when spirals or circular hoops are used the volumetric ratio of the transverse reinforcement should not be less than either[8]

$$\rho_s = 0.45 \left(\frac{A_g}{A_c} - 1 \right) \frac{f'_c}{f_{yh}} \tag{7.22}$$

or $$\rho_s = 0.12 \frac{f'_c}{f_{yh}} \tag{7.23}$$

where A_g = gross cross-sectional area of the concrete;
A_c = area of core of column measured to outside of the hoops;
f_{yh} = yield strength of confining steel; and
ρ_s = (volume of confining steel)/(volume of core).

Alternatively, if rectangular hoops are used, the total area of hoop bars (and supplementary cross-ties if used) in each principal direction within spacing s should not be less than either[8]

$$A_{sh} = 0.35\, s\, h_c \left(\frac{A_s}{A_c} - 1 \right) \frac{f'_c}{f_{yh}} \tag{7.24}$$

or

$$A_{sh} = 0.12\, s\, h_c \frac{f_c'}{f_{yh}} \tag{7.25}$$

where h_c = dimension of concrete core measured perpendicular to the direction of the hoop bars and centre to centre of the peripheral hoop.

The confinement steel zone shown in Detail 7.7 has the length specified by the UBC.[8] In New Zealand[7] the length defined is similar except that an increase in length of 50 percent or more is required when axial compression is high, i.e. when $N_e > 0.33 f_c' A_g$. Outside of potential plastic hinge zones, of course, less confinement reinforcement is usually needed, the requirements being according to equations (7.22) and (7.24).

Care is necessary in selecting the details for anchorage of the transverse reinforcement, as some commonly used details have ineffective anchorage in reversed cyclic loading, i.e. bars with only 90 percent hooks are unsatisfactory in most situations. This has been demonstrated by Tanaka *et al.*,[21] who studied four rectangular sections incorporating a range of confinement details used in California and New Zealand, as shown in Figure 7.18. It was concluded:

(1) The hoop and cross-tie arrangements in Units 1 and 2 were satisfactory;
(2) Perimeter hoop U bars (Unit 3) were unsatisfactory;
(3) J bar interior cross ties with a $24d_b$ tension splice were satisfactory in the columns tested, with a measured stress in the ties of $0.6f_y$. Their effectiveness at higher stresses was not known.

It is noted that the satisfactory extension lengths in Units 1 and 2 are slightly less than currently recommended in some codes[7,8] (e.g. Detail 7.7).

7.2.9.3 Shear strength of columns

For members subject to axial compression, the nominal (ideal) shear strength provided by the concrete should not be greater than either[8]

$$V_c = \frac{bd}{7}\left(\sqrt{(f_c')} + \frac{120\rho V_u d}{M_m}\right) \tag{7.26}$$

or

$$V_c = 0.3bd \sqrt{(f_c')} \sqrt{\left(1 + \frac{0.3N_u}{bd}\right)} \tag{7.27}$$

where b is the width or diameter of column section, and $\rho = A_s/bd$, where A_s is the area of tensile steel.

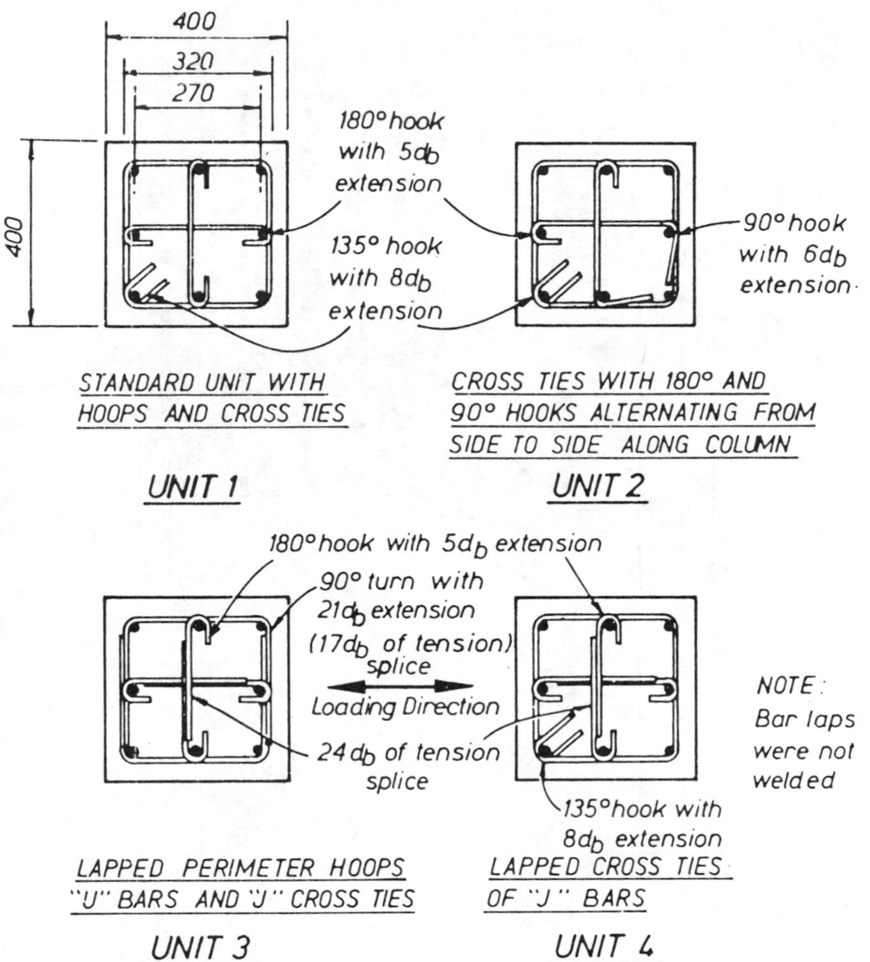

Figure 7.18 Alternative column confinement details of different effectiveness (see text) (after Tanaka *et al.*[21])

In equation (7.26),

$$M_m = N_u \frac{(4h-d)}{8}$$

where if M_m is negative, equation (7.26) should not be used.

For columns subjected to significant tension[8]

$$V_c = \frac{bd}{6}\sqrt{(f_c')}\left(1 + \frac{0.3N_u}{bd}\right) \tag{7.28}$$

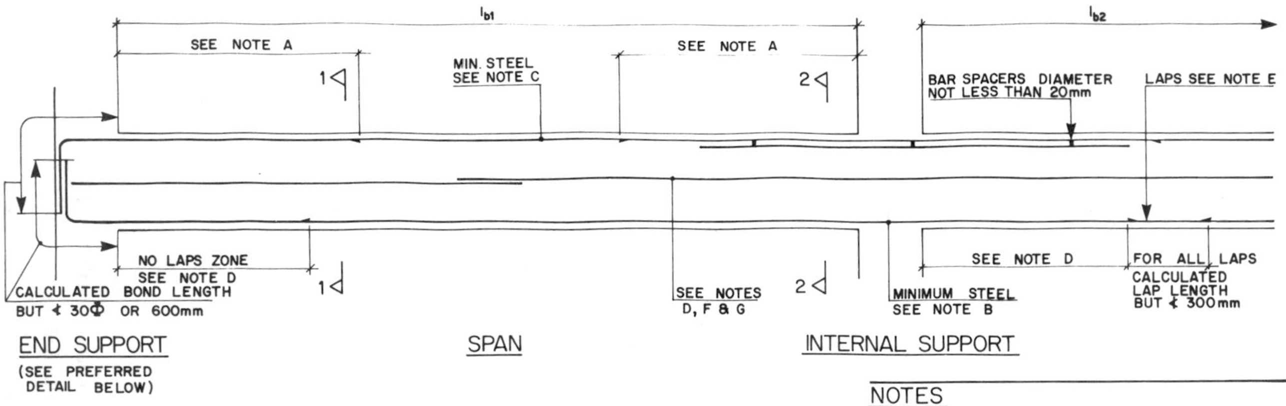

l_{b1} l_{b2}

SEE NOTE A

MIN. STEEL
SEE NOTE C

SEE NOTE A

BAR SPACERS DIAMETER
NOT LESS THAN 20mm

LAPS SEE NOTE E

NO LAPS ZONE
SEE NOTE D

CALCULATED BOND LENGTH
BUT ≮ 30Φ OR 600mm

SEE NOTES
D, F & G

MINIMUM STEEL
SEE NOTE B

SEE NOTE D

FOR ALL LAPS
CALCULATED
LAP LENGTH
BUT ≮ 300mm

END SUPPORT
(SEE PREFERRED
DETAIL BELOW)

SPAN

INTERNAL SUPPORT

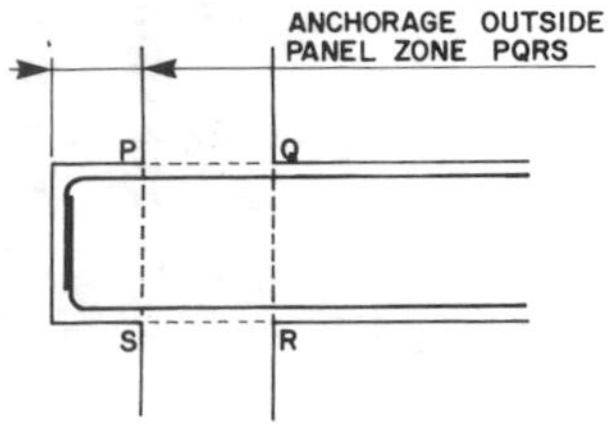

ANCHORAGE OUTSIDE
PANEL ZONE PQRS

TRANSVERSE SLAB STEEL

ALTERNATIVE PREFERRED
END SUPPORT DETAIL

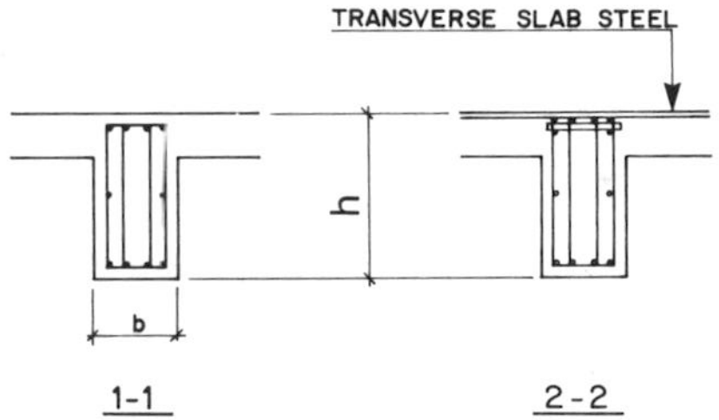

1-1

2-2

BEAMS FORMING PART OF A
DUCTILE MOMENT RESISTING FRAMEWORK

NOTES

A CURTAILMENT OF TOP BARS TO BE TENSION BOND LENGTH BEYOND THE POINT OF CONTRAFLEXURE BUT NOT LESS THAN $l_b/4$ FROM SUPPORT. IF SOME BARS ARE CURTAILED AT SHORTER DISTANCES AT LEAST A THIRD OF TOTAL STEEL MUST EXTEND THIS DISTANCE.

B THE AREA OF BOTTOM STEEL PROVIDED AT THE SUPPORTS MUST EQUAL AT LEAST HALF THE AREA OF THE TOP STEEL AT THAT SECTION.

C A MIN OF 1/4 OF THE LARGER AMOUNT OF TOP STEEL REQUIRED AT EITHER END MUST CONTINUE FOR THE WHOLE LENGTH OF THE SPAN.

D NO LAPS IN TOP & BOTTOM TO OCCUR WITHIN A DISTANCE OF 2h FROM FACE OF SUPPORT.

E LAPS ARE TO BE KEPT TO A MINIMUM AND CRANKS IF REQUIRED TO BE AS FOR COLUMNS.

F IF h > 750 mm PROVIDE EXTRA LONGITUDINAL BARS FOR 2/3 OF THE BEAM DEPTH FROM THE TENSION FACE WITH SPACING ≯ 250mm.

G LAPS IN FRAMING BARS TO BE STAGGERED RELATIVE TO LAPS IN MAIN TOP AND BOTTOM BARS.

Detail 7.9 Beams sheet 1

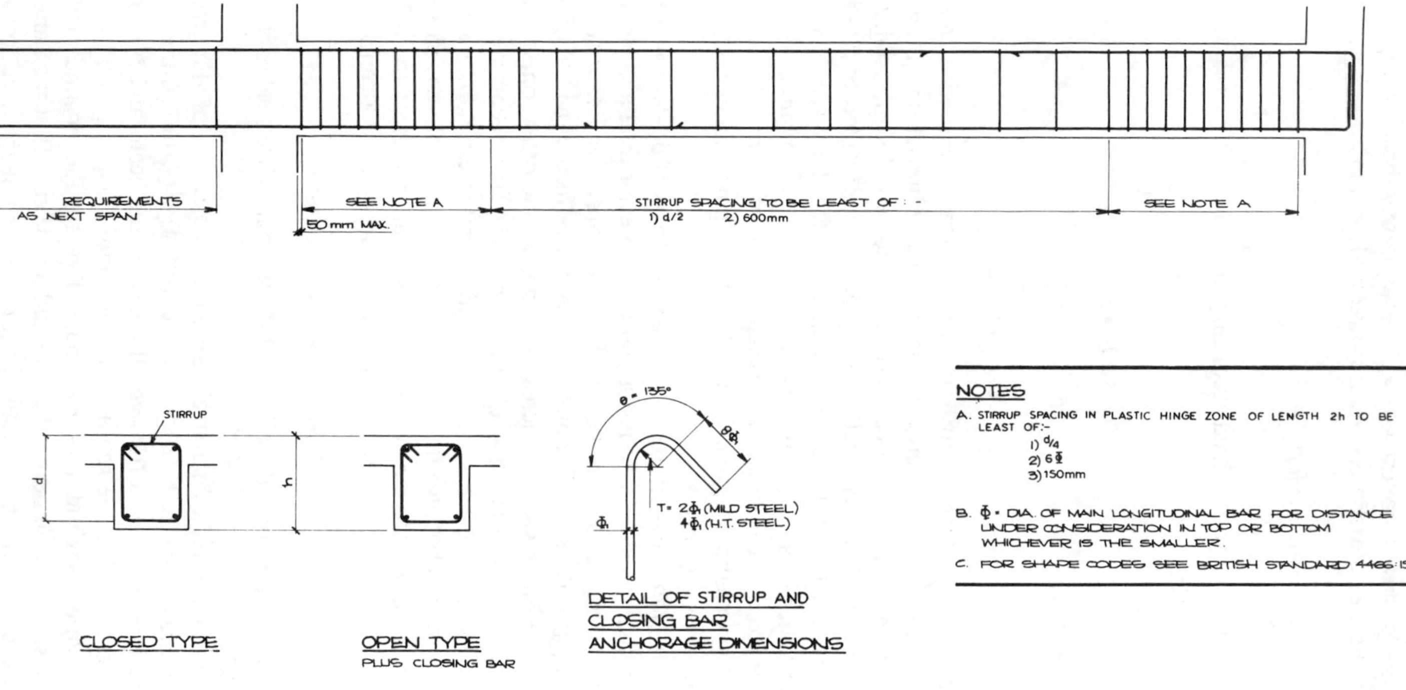

Detail 7.10 Beams sheet 2

Where the design shear force exceeds V_c, the additional shear strength required is provided by the area of transverse reinforcement A_v within a spacing s, such that[8]

$$V_s = \frac{A_v f_y d}{S} \leqslant \frac{2 b_w d}{3} \sqrt{(f_c')} \tag{7.29}$$

Confinement reinforcing provided according to Section 7.2.9.2 may be assumed also to act in shear.

7.2.10 Beams

(See Details 7.9 and 7.10.)

7.2.10.1 General

In moment-resisting frames forming the generally desired beam-hinging mechanism of Figure 5.5(b) potential plastic hinge zones of length $2h$ occur at the ends of beams. In order to ease the reinforcement design problems of the beam–column joint (Section 7.2.11) it may be desirable to locate plastic hinges a short distance away from the column face. This is feasible when the earthquake moments are much greater than the gravity moments, e.g. in buildings of perhaps five or more storeys, otherwise insufficient length of beam will have negative moment.[37] Two methods of relocating the plastic hinge zones away from the columns are shown in Figure 7.19.

The *effect of floor slabs* on the strength and stiffness of beams has been found to be greater than previously believed and much greater than allowed in codes, according to several reports on the US–Japan Co-operative Research Program. For instance, Suzuki *et al.*[38] observed that the effective width of the slab contributing to beam resistance increases with beam deformation, and could be as wide as the entire slab width at ultimate load. The final recommendations of the above research program should be significant for estimation not only of the strength of beams but also for the design of beam and column failure modes.

The *width of beams* should preferably be similar to that of the columns to ease reinforcement detailing, and should not be less than 200 mm.[7]

7.2.10.2 Beam longitudinal steel

(1) To help ensure adequate ductility the longitudinal tensile steel content should not exceed $f_y/7$.
(2) The minimum longitudinal steel content as a fraction of the gross cross-sectional area of the web ($h \times b$) should be $1.4/f_y$ (f_y in N/mm^2) or $200/f_y$ (f_y in lb/in^2), where h is the overall depth of the beam and b is the width of the web for both the top and bottom reinforcement.
(3) Curtailment of longitudinal steel should allow for the most adverse loading conditions. Large numbers of bars should not be cut off at the same section.
(4) The distance between bars should be according to the code adopted but not less than 25 mm.

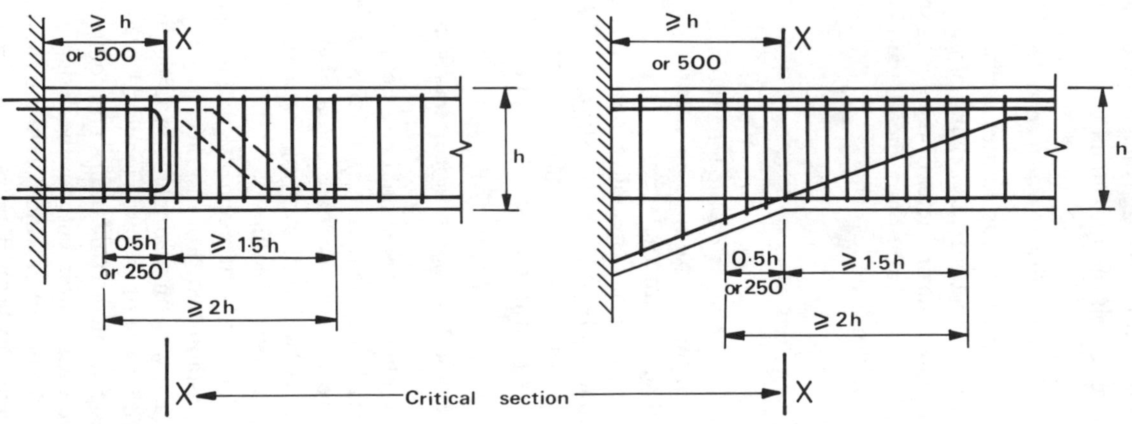

Figure 7.19 Relocation of plastic hinges in reinforced concrete beams (after Part 2 of reference 7. Reproduced with permission of Standards Assn. of New Zealand)

(5) In beams forming part of a moment-resisting framework, the positive moment capacity at columns should not be less than half the negative moment capacity provided. At least two bars of 16 mm diameter should be provided both top and bottom throughout the length of the member, and the bending strength (of either sign) at any section along the beam should not be less than one fourth the maximum bending strength at either end.

7.2.10.3 Beam transverse steel

In potential plastic hinge zones of beams, located as discussed in Section 7.2.10.1, stirrups (hoops) are required to control buckling of the longitudinal steel and to confine the concrete core. The diameter of the hoops should not be less than 6 mm and the area of one leg of a hoop in the direction of potential buckling of longitudinal bar should not be less than[7]

$$A_{te} = \frac{\Sigma A_b f_y}{16 f_{yt}} \left(\frac{s}{100} \right) \tag{7.30}$$

where ΣA_b is the sum of the areas of the longitudinal bars being restrained, f_y is the yield strength of the longitudinal bars, and f_{yt} and s are, respectively, the yield strength and spacing of the stirrups.

In designing against shear forces the above transverse confinement steel may be assumed to contribute fully to the shear strength of the beam. The use of diagonal reinforcement for shear resistance may be advantageous when the reversible shear stress is high in plastic hinge zones.[7,39]

7.2.11 Beam–column joints in moment-resisting frames

The strength of beam–column cores should be at least as great, and preferably greater than, the strengths of the members it joins. This is because the joint area is subject to failure under cyclic loads and is obviously difficult to repair. Also as it is part of the vertical load-carrying system it should comply with the principle that beams fail before columns.

Unfortunately, the proper reinforcement of this zone is constructionally difficult and not fully established. For instance, the provisions of ACI 318–77 appear to be inadequate, with an example joint failing in shear and slip of the beam bars after the first cycle of inelastic loading.[1] In an attempt to solve this problem, based on tests of intersecting beams and columns[1,40] the New Zealand concrete code[7] has specific requirements for the design of beam–column joints, particularly for resistance of vertical and horizontal shear and for confinement. Also, for the development of beam flexural steel which passes through a joint there are limits on bar diameter as a fraction of the column depth. These code requirements are onerous and means for easing them are desirable, such as the use of bond plates on the beam flexural steel investigated by Fenwick.[41] A missing factor in most of the experimental work so far carried out has been the effect of the floor slab on joint behaviour. In the US–Japan

Co-operative Research Program structures tested incorporated slabs, and showed that their effects were significant.[42] It is hoped that the planned US–Japan–New Zealand–China co-operative study of beam–column joints will favourably resolve the present marked differences in national code provisions.

7.2.12 Slabs

(See Detail 7.11.)
(a) Reinforcement designed for gravity loads in slabs forming part of a normal beam and slab system will generally be adequate to ensure that the slabs behave satisfactorily both as flexural members and as horizontal diaphragms transmitting earthquake forces. Certain elements such as flat slabs and waffle slabs, which may form part of the earthquake resisting framework, must, of course, be designed and detailed accordingly. Their capacity to transfer earthquake moments between columns and slabs must be assured.
(b) A minimum steel content of 0.15 percent is required in most slabs or diaphragms, for the control of temperature effects and shrinkage.
(c) For cantilevering slabs, bottom steel should be provided to counteract bending tensions which may occur during earthquakes.
(d) Holes through slabs should be framed with extra steel as shown on the detail sheet because of the diaphragm action of slabs during earthquakes.
(e) For ground floor or basement slabs which are designated as 'ground bearing', and are not part of the foundation structure, special seismic considerations may not exist and it is usual merely to place one layer of nominal steel in each direction to prevent cracking and shrinkage. This is usually placed in the top of the slab. Tie steel between column bases may also be placed in the ground slab in some instances, instead of in foundation tie beams (Section 7.2.6.2).

7.2.13 Staircases

(See Detail 7.12.)
(a) Generally, the rules for slabs apply. As shown on the detail sheet, top steel should be provided at each landing to provide for bending tensions which may not be apparent from a simple analysis.
(b) If stairs are part of a horizontal diaphragm or moment resisting framework they should be reinforced accordingly. Due care must then be taken at the changes in slope to restrain the longitudinal bars.

7.2.14 Upstands and parapets

(See Detail 7.13.)
(a) Upstands and parapets should be carefully designed against seismic accelerations, which may considerably exceed those occurring elsewhere in the structure due to dynamic effects.

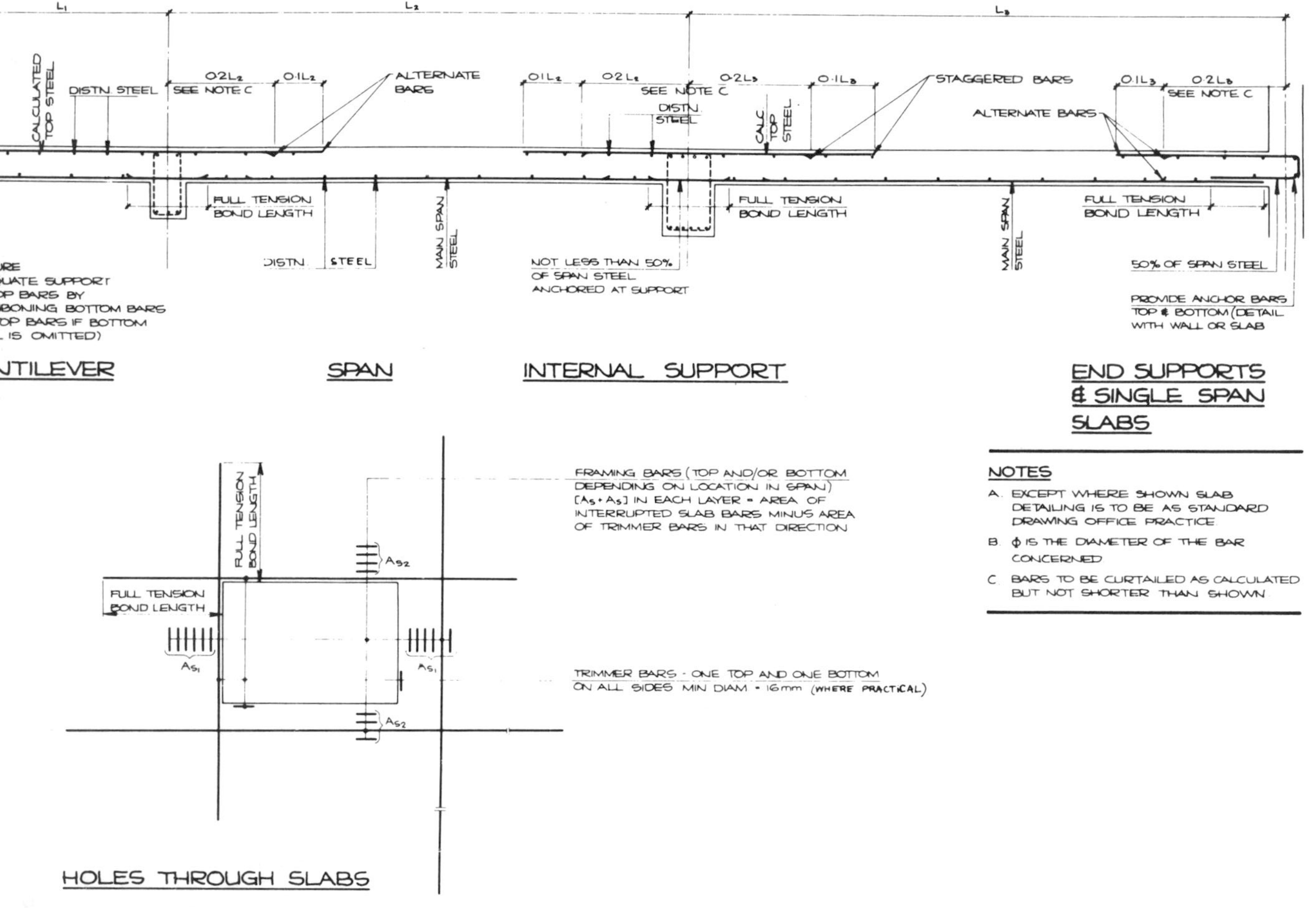

Detail 7.11 Slabs

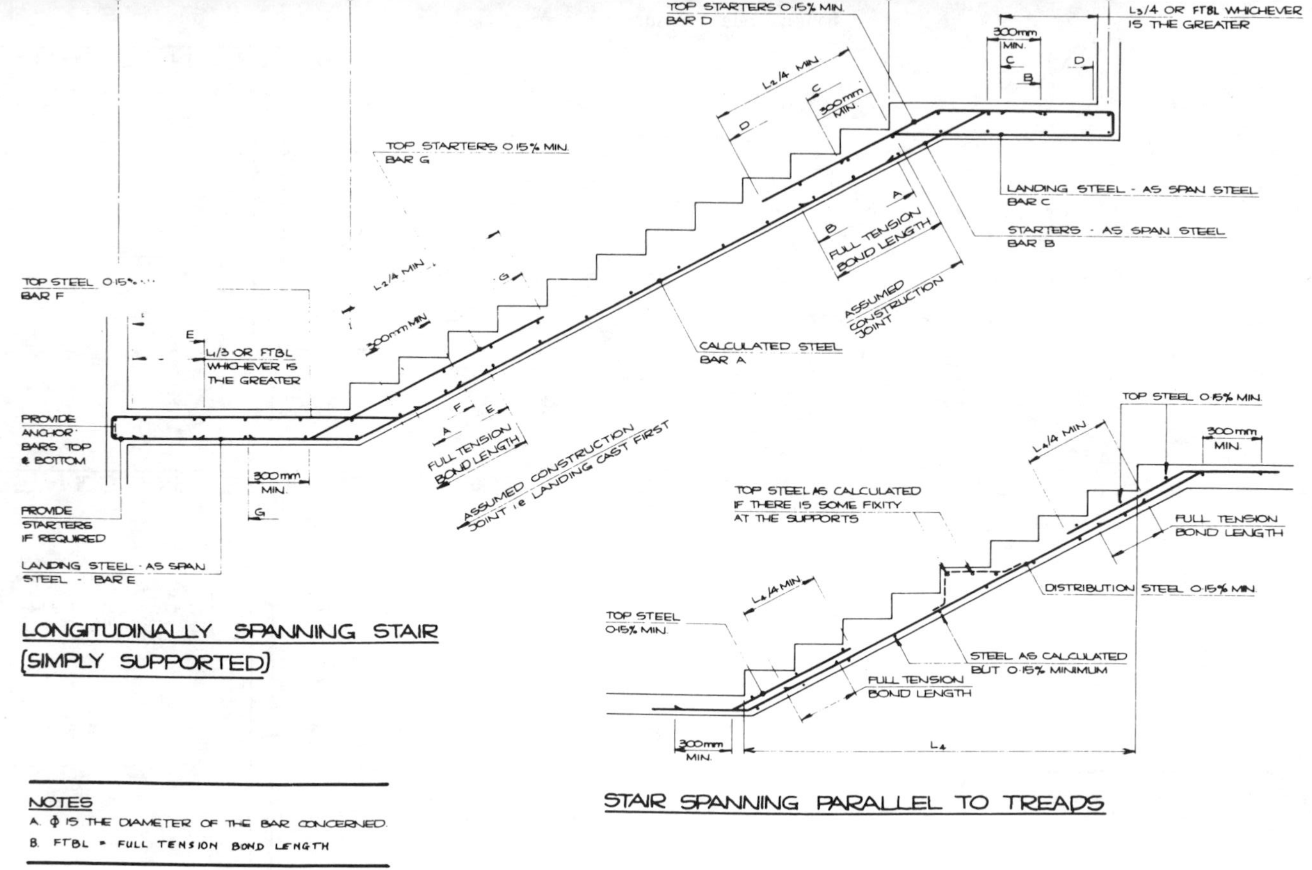

Detail 7.12 Staircases

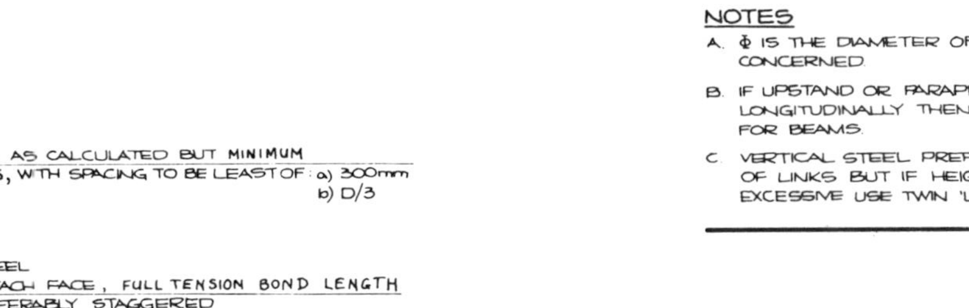

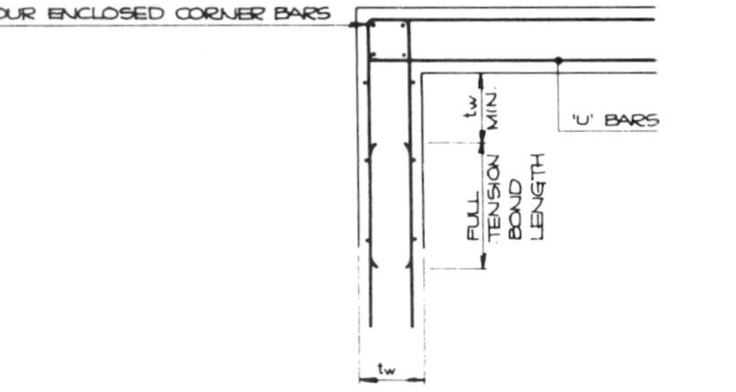

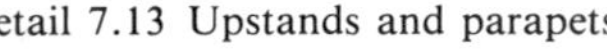
Detail 7.13 Upstands and parapets

(b) The arrangement of reinforcement at corners and junctions should be as for walls.

7.3 STRUCTURAL PRECAST CONCRETE DETAIL

7.3.1 Introduction

Precast concrete structures have given mixed performance in earthquakes, difficulties mainly being experienced at connections between members. Prior to the mid-1980s a negligible number of properly documented test results have been published on the behaviour of connections under cyclic loading. Much of the testing which has been done has been related to specific proprietary precast systems, and such results as have been published are usually either lacking in essential detail or are not readily applicable to other precasting assemblies. As a result the development of precasting has yet to reveal its full potential in aseismic construction. However, this situation seems to be improving, because several papers on experimental studies of large panel construction were delivered at the 1984 World Conference on Earthquake Engineering,[43] together with some papers on design of precast structures.[44–6]

Nevertheless precast concrete, either reinforced or prestressed, has been used to some extent in most forms of structure in earthquake areas, often in conjunction with a cautiously large amount of unifying *in situ* concrete. The nature of the seismic response of a precast structure must be inferred from the response of the reinforced or prestressed members involved (Sections 7.2 and 7.5). Allowance for the effect of the connection on the stress flow must also be made. This is particularly important when adapting proprietary precast products made for general purposes especially when intended originally for non-seismic areas. Without appropriate dynamic test results the effect of the connections may be difficult or impossible to assess, especially if they depart substantially from providing full continuity and homogeneity between adjacent members.

Dealing with building tolerances is a major problem in the design of connections. Constructional eccentricities may result in large secondary stresses in earthquakes, and should be either designed for or minimized by the manner of connection. It may be advantageous to design structural joints which permit generous constructional tolerances and restrict the expensive fine tolerance work to the cladding for visual or drainage purposes.

In order to overcome the connection problem, partial precasting is often done. For example, precast beams may be used with *in situ* columns, or precast walls may be used with *in situ* floors, or vice versa.

As the basic detailing of reinforced and prestressed concrete has been discussed elsewhere in this book, only the essential problem of precast construction, that of *connection*, is considered in this section. For further reading in addition to those quoted above, see references 47–53.

7.3.2 Connections between bases and precast columns

The following typical details must be individually designed for the forces acting on the joint. The base considered may be at foundation level or on suspended members higher up the structure. Member reinforcement is not shown.

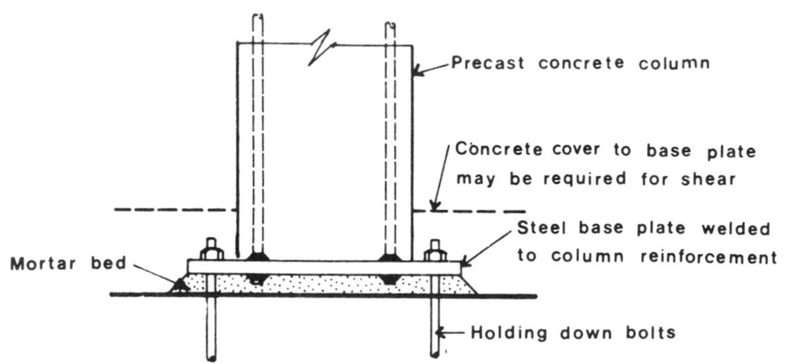

Detail 7.14 Site bolted. Moment transfer controlled by base plate

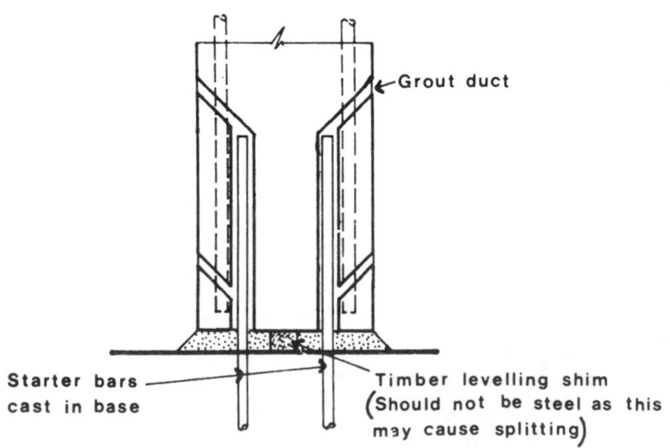

Detail 7.15 Site grouted. Effectiveness depends on grouting

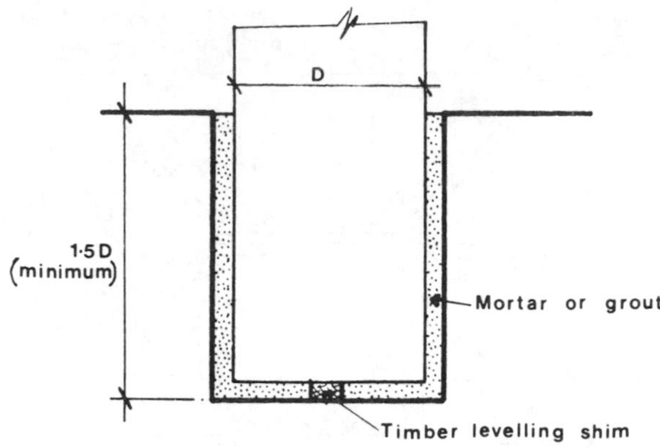

Detail 7.16 Site grouted. Best all-round joint of this type. Method of transfer of vertical load to base must be checked

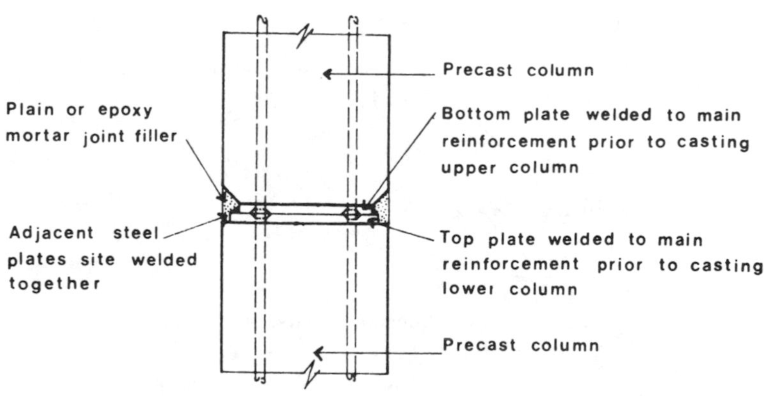

Detail 7.17 Site welded

7.3.3 Connections between precast columns and beams

The following typical details must be individually designed for the forces acting on the joint under construction. Variations on these connections may be made to suit the circumstances. Member reinforcement is not shown. Note that welding of bent bars should only be done with suitable steels. High carbon steels, such as those to ASTM A615,[20] are prone to brittle fracture in this situation.

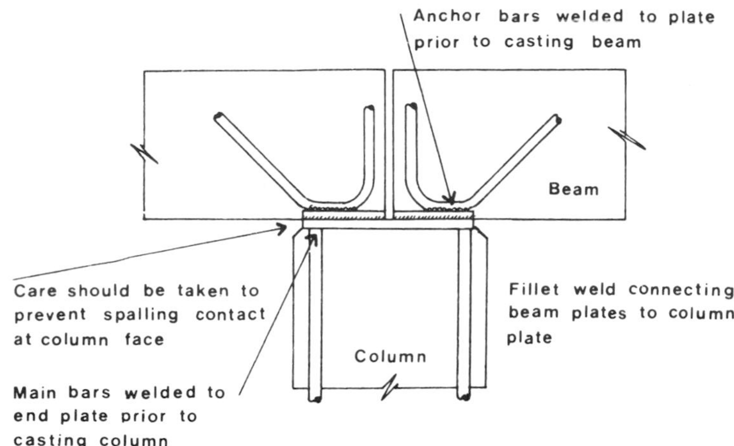

Detail 7.18 Site welded. Low moment capacity

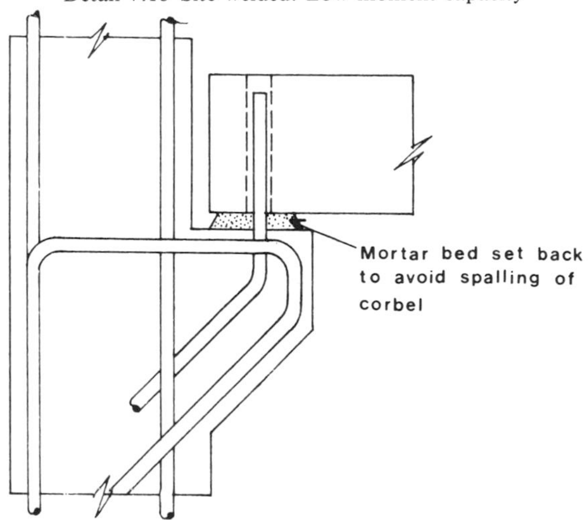

Detail 7.19 Site grouted. Low moment capacity, poor in horizontal shear

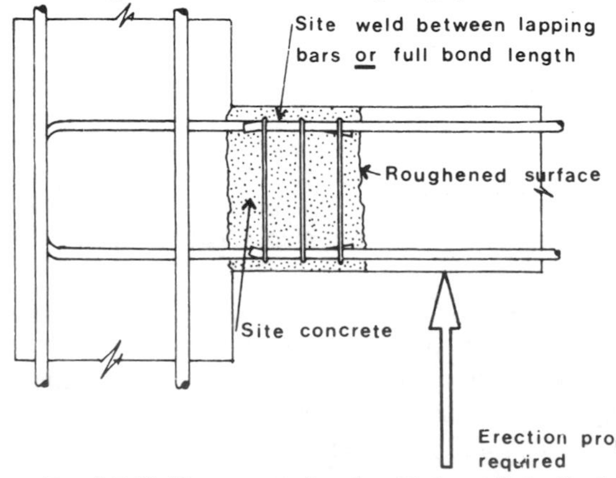

Detail 7.20 Site concreted and welded and links fixed

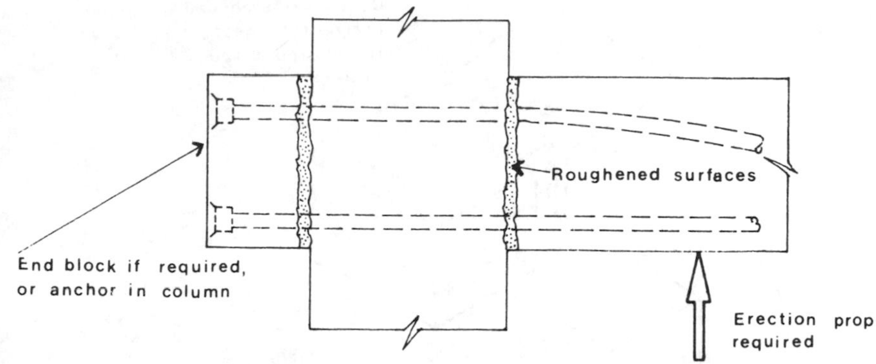

Detail 7.21 Site mortared and post-tensioned

7.3.4 Connections between precast floors and walls

The following typical details must be designed for the forces acting on the joint under consideration. Member reinforcement is not shown.

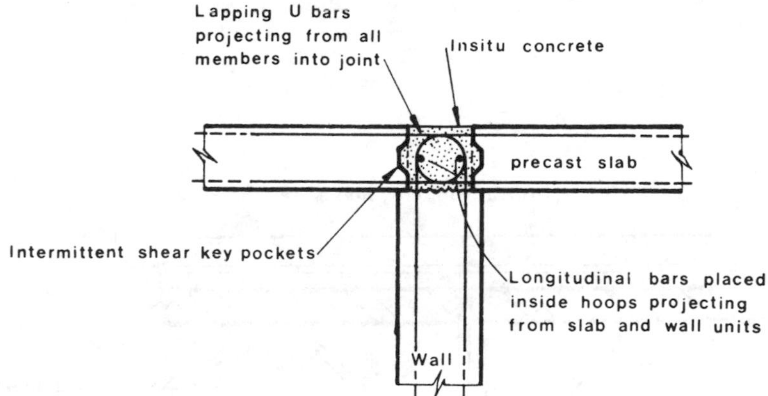

Detail 7.22 Site concrete and reinforcement

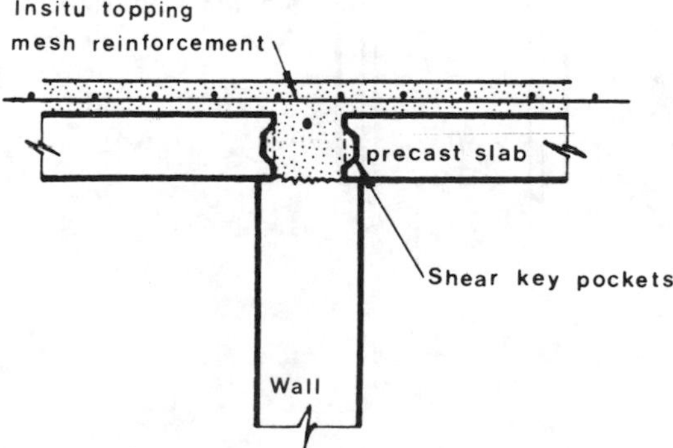

Detail 7.23 Site concrete and reinforcement

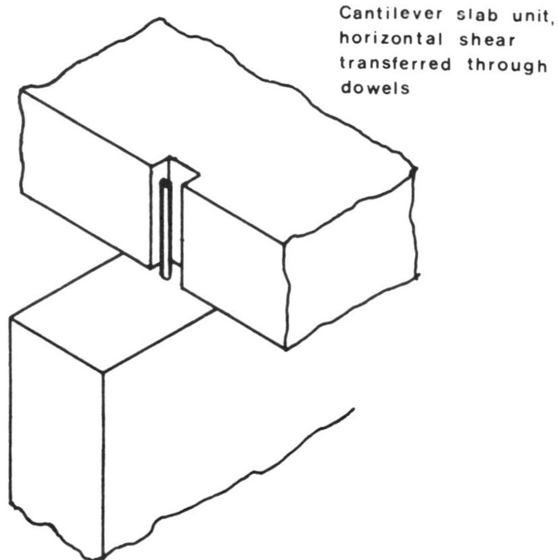

Detail 7.24 Site grouting

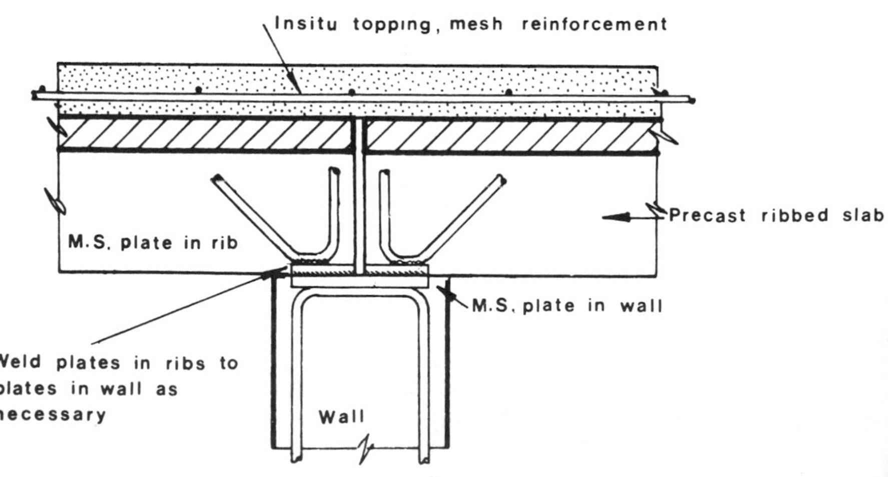

Detail 7.25 Site concrete, reinforcement, and welding

7.3.5 Connections between adjacent precast floor and roof units

The following typical details must be designed for the forces acting on the joint under consideration. Floor slabs should be designed as a whole, to act as diaphragms distributing the shear between the vertical members of the structure. The unifying effect of perimeter beams should be taken into account. Member reinforcement is not shown.

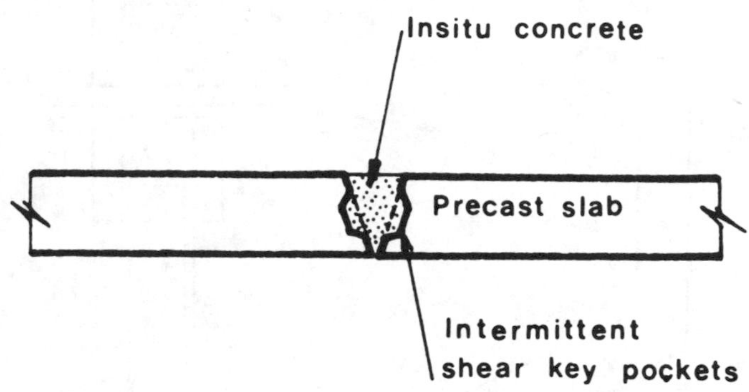

Detail 7.26 Site concreting. This joint depends on perimeter reinforcement to complete shear transfer system

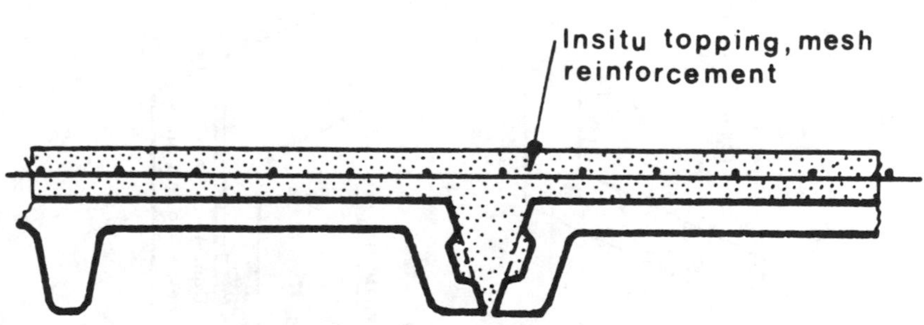

Detail 7.27 Site concreting and reinforcing

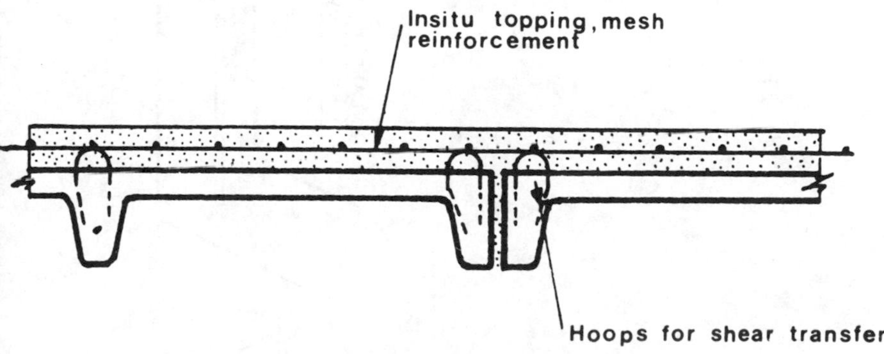

Detail 7.28 Site concreting and reinforcing

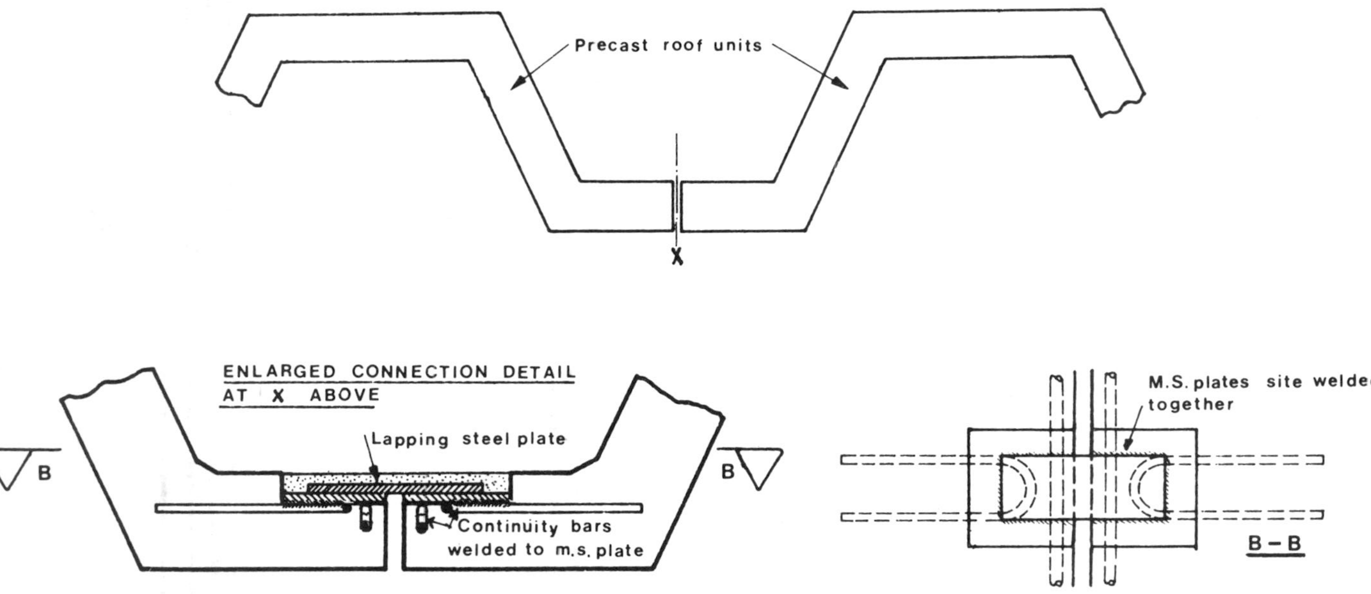

Detail 7.29 Site welding and mortaring. Lapping steel plate bent on site to suit differential camber of adjacent precast units

7.3.6 Connections between adjacent precast wall units

The following typical details should be designed for the forces acting on the joint under consideration. Great problems occur in producing a ductile and easily-erected precast shear wall, and no universal solution has as yet been evolved. The details below may be adapted for use with internal or external walls, i.e. cladding. Member reinforcement is not shown.

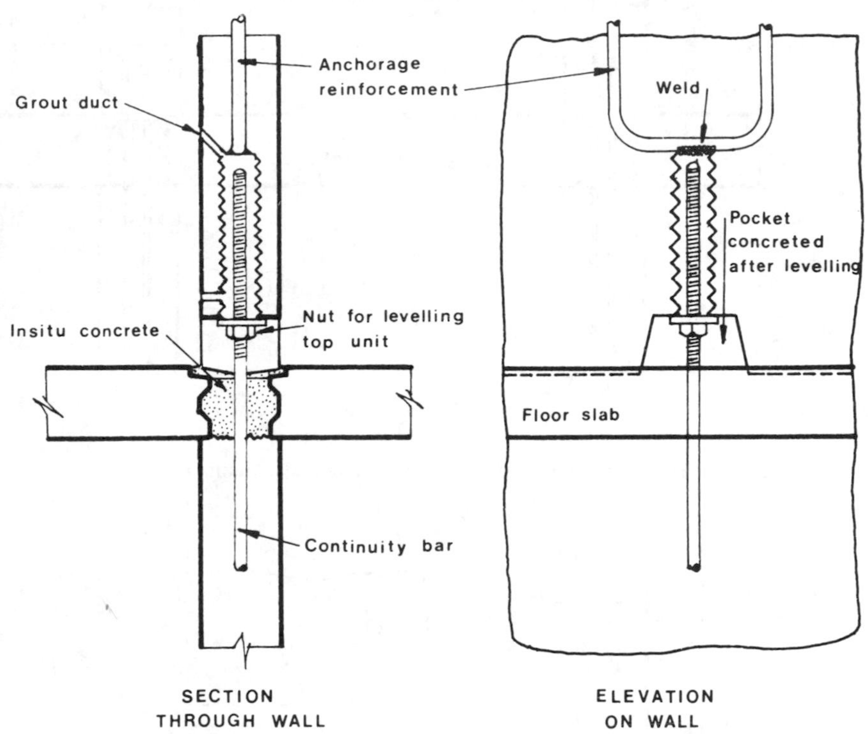

Detail 7.30 Site concreting and grouting

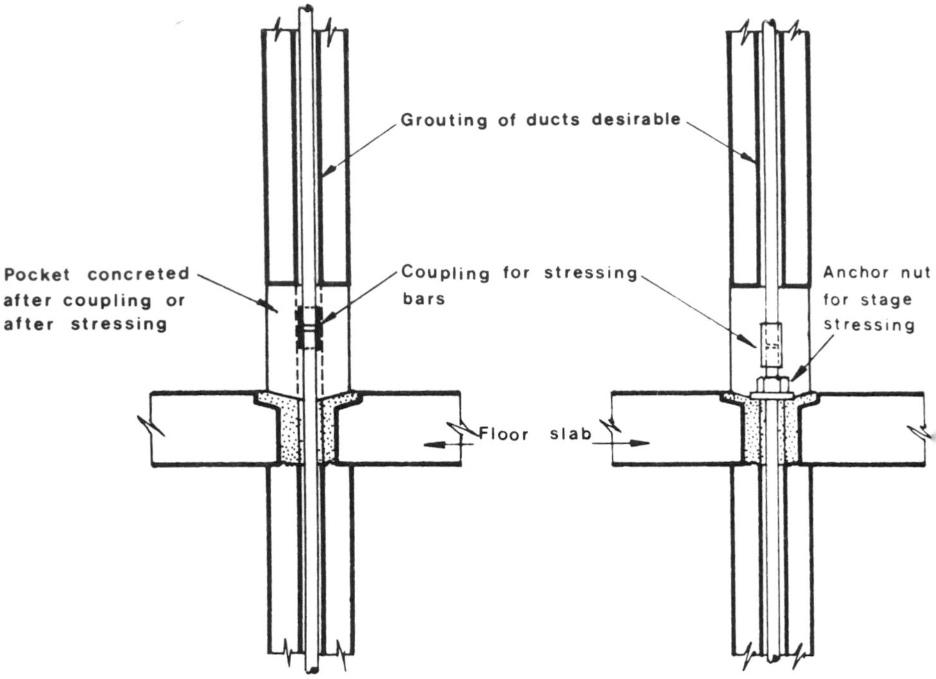

Grouting of ducts desirable
Pocket concreted after coupling or after stressing
Coupling for stressing bars
Anchor nut for stage stressing
Floor slab

Detail 7.31 Site concreting and post-tensioning and grouting of ducts

Continuity reinforcement welded to M.S. plate
Floor
Overlapping M.S. plates welded together on site
Wall
SECTION
ELEVATION

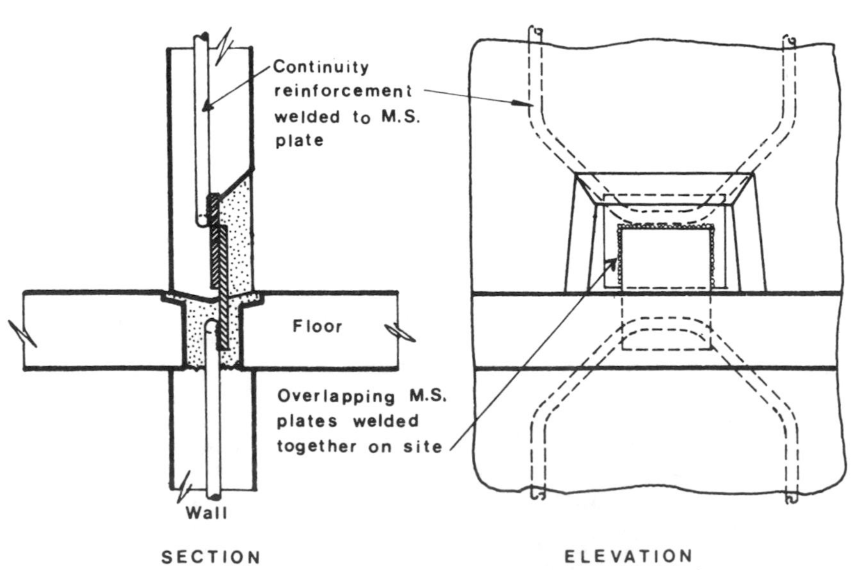

Detail 7.32 Site welded and concreted

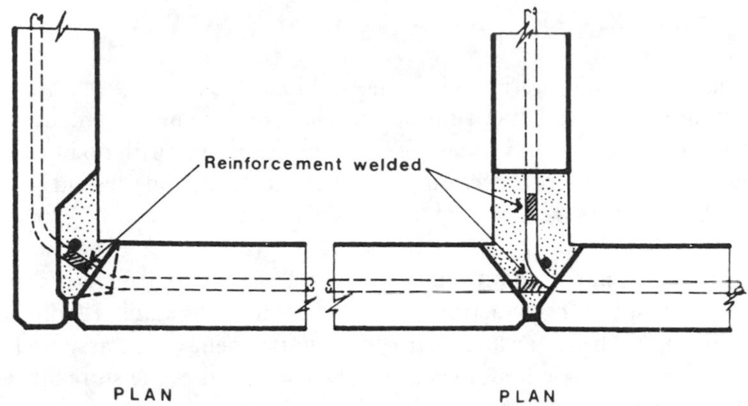

Detail 7.33 Site welded and concreted

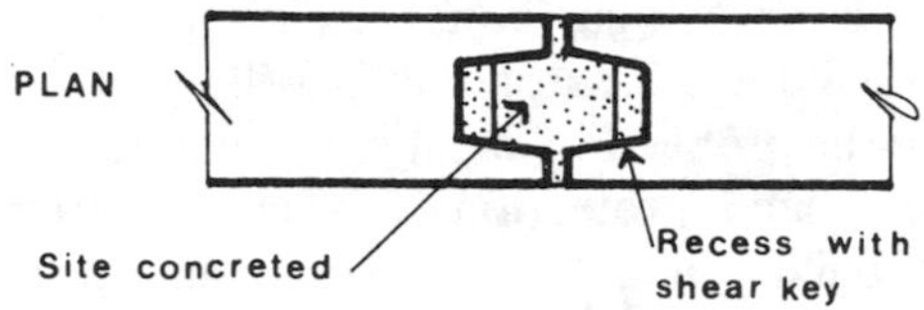

Detail 7.34 Site concreted

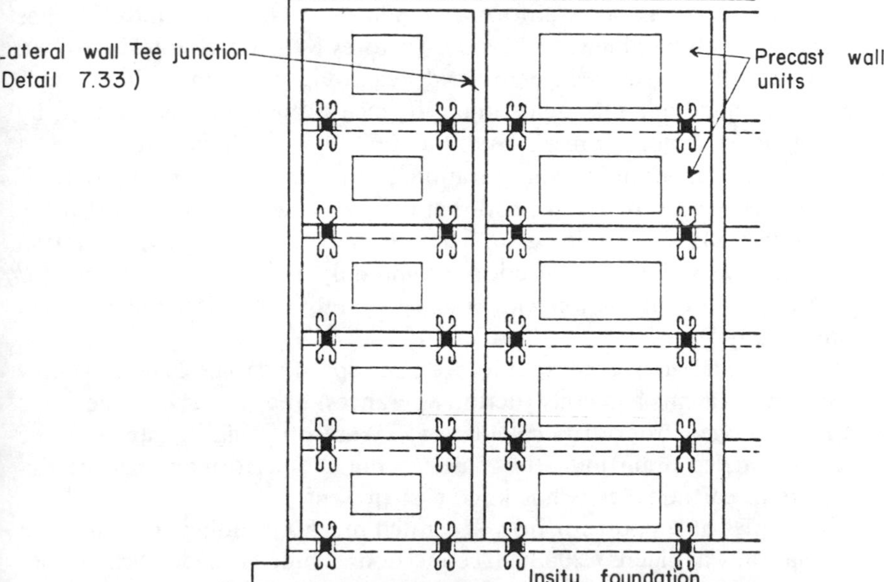

Detail 7.35 Layout of joints in wall elevation

7.4 PRECAST CONCRETE CLADDING DETAIL

Precast concrete cladding varies in its relationship to the building structure, from being fully integrated to being fully separated from frame action. Ideally the cladding should be either fully integrated or fully separated, with no intermediate conditions. Fully integrated *structural* precast concrete cladding should be treated like any other precast structural element, as discussed in Section 7.3. Cladding which is not considered as part of the structure is considered below.

In flexible beam and column buildings it is desirable to effectively separate the cladding from the frame action, both to protect the cladding from seismic deformations and also to ensure that the structure behaves as assumed in the analysis. For very flexible buildings in strong earthquakes the storey drift may be so large as to make full separation difficult to achieve, and some interaction of frame and cladding through bending of the connections may have to be accepted. Ductile behaviour of the cladding and of its connections to the structure is most important in such cases to ensure that the cladding does not fall from the building during an earthquake.

In stiff (shear wall) buildings the storey drift will generally be small enough to significantly reduce the problem of detailing of connections which give full separation. On the other hand, protection of the cladding from seismic motion is less necessary in stiff buildings, and connections permitting movement through bending may be satisfactory as long as the interaction between cladding and frame can be allowed for in the frame analysis.

It is common for precast cladding to be fully separated from the frame in strong motion areas like California, Japan, and New Zealand. This has been done on major buildings, such as the 47-storey Keio Plaza Hotel in Tokyo. Unfortunately little has been published regarding the connection details for separated cladding, although some reference is made to this problem by Uchida *et al.,*[54] Brooke-White[55] and the PCI.[56] In Uchida's structure, a 25-storey steel framed building, separation of the cladding was only partial, and the connections were designed so that the panels would not fall off if the storey drift was 50 mm. Goodno *et al.*[57] have found that some of the separation details, particularly slotted connections, commonly used as shown in the *PCI Manual*[56] have more interaction with frame action than is desired and assumed in seismic analysis.

Gaps between adjacent precast units are often specified to be 20 mm to allow for seismic movements and construction tolerances, but smaller or larger gaps may be determined from drift calculations. Waterproofing of gaps may be effected by baffled drain joints or mastic,[58] but the performance of mastic-filled joints in earthquakes is not known at present.

The principles of support for fully separated precast cladding are illustrated diagrammatically in Figure 7.20. Such connections should be made of corrosion-resistant materials, and must be designed to carry the gravity and wind loads of the cladding back into the structure as well as to allow the free movement of the frame to take place.

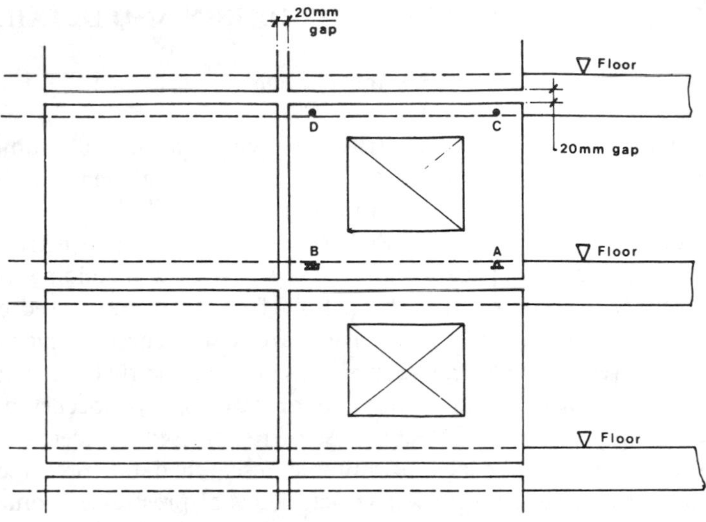

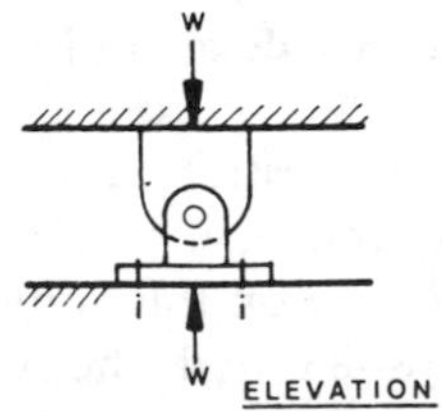

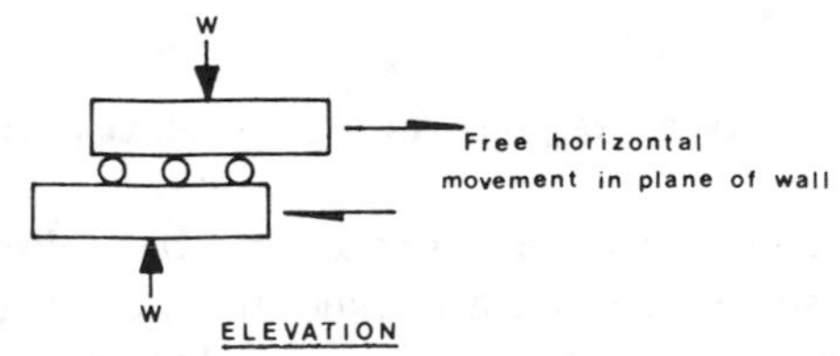

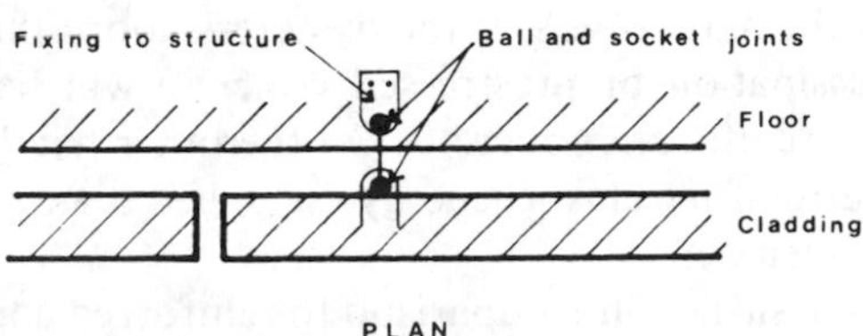

Figure 7.20 Schematic illustration of supports for precast concrete cladding fully separated from frame action

7.5 PRESTRESSED CONCRETE DESIGN AND DETAIL

7.5.1 Introduction

Prestressed concrete elements in structures which have been subjected to earthquakes have mostly performed well. Failures have been mainly due to inadequate connection details or supporting structure.[59,60]

Although prestressed concrete is well established in bridge construction and various civil engineering applications, it is less widely used in building structures, and relatively few structures have been fully framed in prestressed concrete. This is true in both seismic and non-seismic areas. The comparative neglect of prestressed concrete for building structures has occurred partly for constructional and economic reasons, and in earthquake areas it has also occurred because of divergent opinions on the effectiveness of prestressed concrete in resisting earthquakes. Also initially lack of suitable research data prevented proper assessment of the seismic response characteristics of prestressed concrete, but more recent research work has clarified the situation.

7.5.2 Official recommendations for seismic design of prestressed concrete

Some organisations interested in the use of prestressed concrete have published seismic design recommendations. For example, the FIP,[61] in addition to the New Zealand concrete code,[7] gives guidance on this subject. In contrast, the major USA concrete codes (ACI 318–83 and the UBC[8]) only discuss prestressed concrete in non-seismic terms, which unfortunately implies a reluctance to assess the aseismic design requirements of this valuable material.

7.5.3 Seismic response of prestressed concrete

The seismic response of structural materials has been discussed generally in Section 6.6, where some stress–strain diagrams were presented. The main characteristics of prestressed concrete under cyclic loading may be inferred from Figure 6.31(e), from which Blakeley[59] has proposed the idealized hysteresis loop shown in Figure 7.21.

It is evident from the narrowness of the hysteresis loops that the amount of hysteretic energy dissipation of prestressed concrete will be relatively small compared to steel or reinforced concrete. On the other hand, the capacity of prestressed concrete to store elastic energy is higher than for a comparable reinforced concrete member.

Prestressed concrete suffers in comparison to reinforced concrete because of its lack of compression steel, so that its performance is poorer once concrete crushing begins. When compared to reinforced concrete, prestressed concrete undergoes relatively more uncracked deformation and relatively less deformation

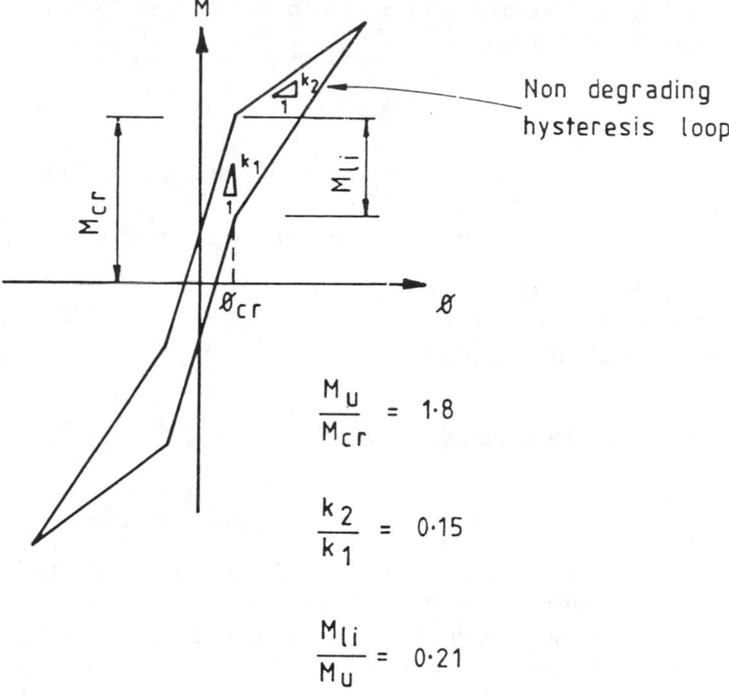

Figure 7.21 Moment curvature idealization for plastic hinge regions in prestressed concrete (after Blakeley[59])

in the cracked state. This means that prestressed concrete structures should exhibit less structural damage in moderate earthquakes. In the event of structural repairs being necessary after an earthquake, there are obvious difficulties in restoring the prestress to sections of replaced concrete, and conversion of the failure zones to reinforced concrete may be necessary.

It has been suggested that prestressed concrete buildings may be more flexible than comparable reinforced concrete structures, and that more non-structural damage may occur. However, differences in flexibility will be small in practical design terms, and structures in either material will generally be less flexible than steelwork. In any case proper detailing of the non-structure will be necessary regardless of the materials used in the structure.

For notes on the damping of prestressed concrete structures see Table 6.17.

7.5.4 Factors affecting ductility of prestressed concrete members

For the satisfactory seismic resistance of prestressed concrete members brittle failure must be avoided by the creation of sufficient useful ductility, as discussed in Section 6.6.2. In the case of prestressed concrete the useful available section ductility may be defined as

$$\frac{\phi_{0.004}}{\phi_{cr}}$$

where $\phi_{0.004}$ is the curvature at a nominal maximum concrete strain of 0.004, and ϕ_{cr} is the curvature at first cracking.

The ductility or rotation capacity of prestressed concrete is affected by:

(1) The longitudinal steel content;
(2) The transverse steel content;
(3) The distribution of longitudinal steel;
(4) The axial load.

Each of these variables is discussed below.

7.5.4.1 Longitudinal and transverse steel content

From Figure 7.22 it may be seen that ductility decreases markedly with increasing prestressing steel content. As seen in Section 7.2.3.3, *unstressed* longitudinal reinforcement also reduces the ductility. Thus, to ensure that reasonable ductility is obtained in potential plastic hinge zones the content of prestressed plus non-prestressed flexural steel should be such that at the flexural capacity of the section, the depth of the equivalent rectangular stress block[7]

$$a \leqslant 0.2h \tag{7.31}$$

where h is the overall depth of the section, unless confinement reinforcing is provided, in which case

$$a \leqslant 0.3h \tag{7.32}$$

The confinement should be provided as for reinforced concrete columns, Park *et al.*[63] reporting good cyclic behaviour using half the amount of confinement steel required for columns by the New Zealand code.[7] As with reinforced concrete, the use of confinement steel greatly increases the ductility of prestressed members compared with unconfined ones.

7.2.4.2 *Distribution of longitudinal steel*

At positions of moment reversal where the greatest ductility requirements exist, the required distribution of prestress will usually be nearly axial. Blakeley[59] demonstrated that a single axial tendon produced a less ductile member than that achieved by multiple tendons placed nearer the extreme fibres. At points in structures where stress reversals do not occur, eccentric prestress may be used. Where no unstressed reinforcement exists, an eccentrically prestressed beam is notably less ductile than a concentrically stressed one with equal prestressing steel content (Figure 7.23).

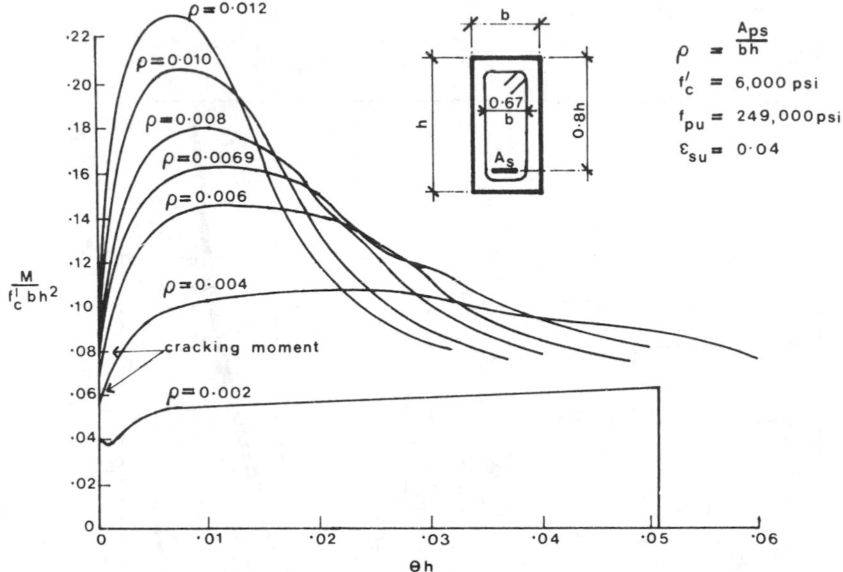

Figure 7.22 Moment–curvature relationship for rectangular prestressed concrete beams showing the effect of prestressing steel content on ductility (after Blakeley and Park[62])

The tendon distribution three in Figure 7.23 is not only as ductile as two but has the advantage that the axial tendon will be practically unharmed by large rotations, and should hold the structure together after the tendons near the extreme fibres have failed.

7.5.4.3 *Effect of axial load on ductility of prestressed concrete columns*

The section ductility $\phi_{0.004}/\phi_{cr}$ decreases rapidly with increasing column axial load N. This effect is seen in Figure 7.24, where ductility is plotted against the level of prestress for columns carrying varying axial loads.

7.5.5 Detailing summary for prestressed concrete

For the adequate performance of prestressed concrete in earthquakes, its ductility and continuity should be maximized by careful consideration of the following items;

(1) Transverse and longitudinal steel content (Section 7.5.4.1);
(2) Longitudinal steel distribution (Section 7.5.4.2);
(3) Continuity, ensured by adequate lapping of prestressing tendons or reinforcing bars;
(4) Anchorages in post-tensioned construction, carefully positioned to avoid congestion and stress-raising in highly stressed zones. They should be situated as far from potential plastic hinge positions as possible;

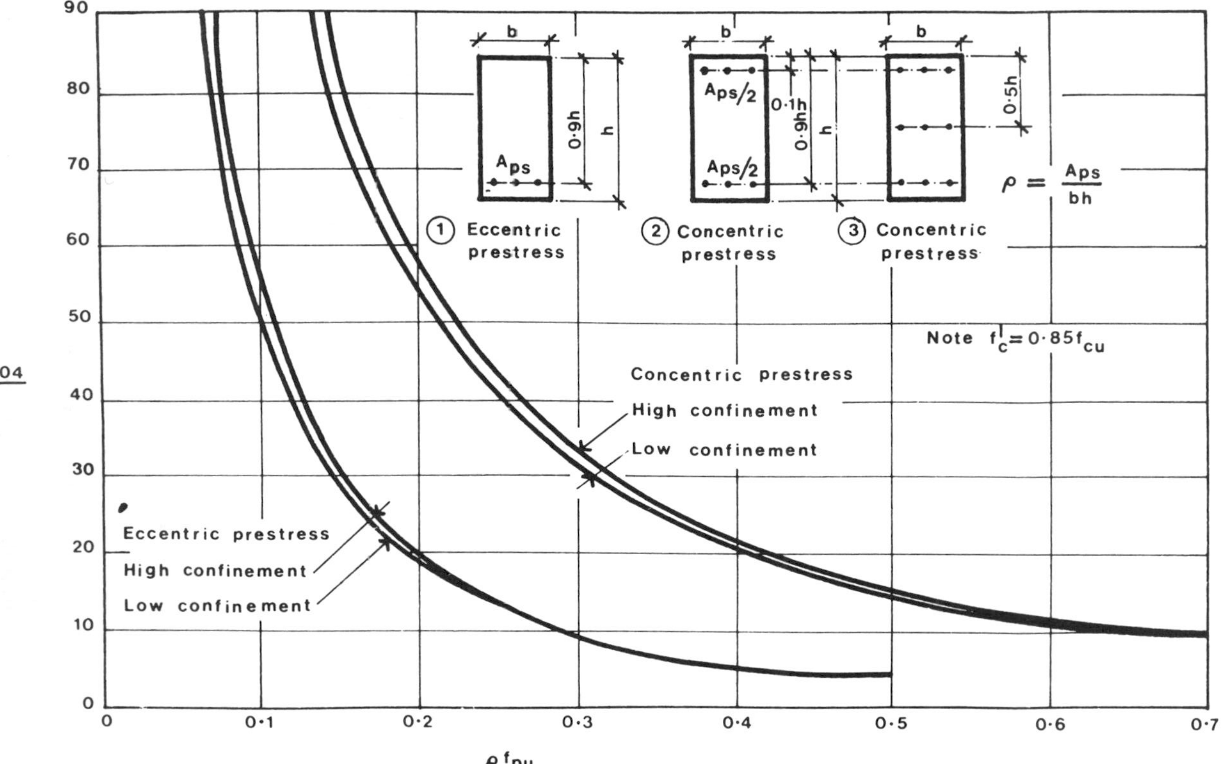

Figure 7.23 Variation of curvature ratio at crushing (section ductility) for prestressed concrete beams (after Blakeley[59])

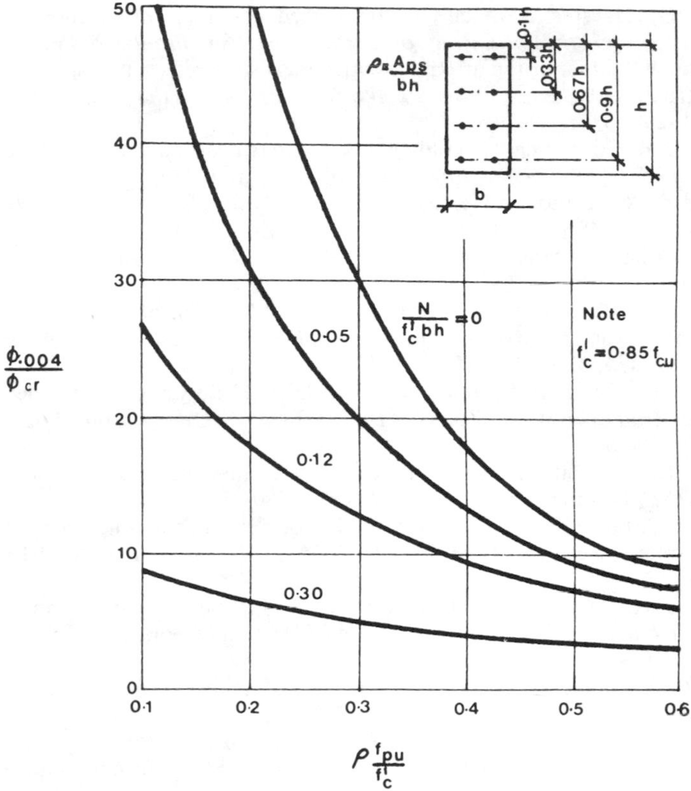

Figure 7.24 Variation of curvature ratio at crushing (section ductility) for columns with varying prestress and varying axial load (after Blakeley[59])

(5) Joints between prestressed members involving ordinary reinforced concrete, properly designed as outlined in Sections 7.2 and 7.3;
(6) Joints using mechanical details, as suggested in Section 7.3;
(7) Unbonded tendons are acceptable for use in primary earthquake resistant members if a substantial quantity of bonded steel is also present.[8] In general they should be used with anchorages of proven realiability under cyclic loading, and proper crack control measures should be taken.

REFERENCES

1. Park, R., and Paulay, T. *Reinforced Concrete Structures*, John Wiley and Sons, New York (1975).
2. Park, R., 'Theorisation of structural behaviour with a view to defining resistance and ultimate deformability', *Bull. NZ Soc. for Earthq. Eng.*, **6**, No. 2, 52–70 (1973).
3. Otani, S., 'Hysteresis models of reinforced concrete for earthquake response analysis', *J. of the Faculty of Engineering, University of Tokyo*, **XXXVI**, No. 2 (1981).

4. Otani, S., 'Hysteresis models of reinforced concrete for earthquake response analysis', *Proc. 8th World Conf. on Earthq. Eng., San Francisco*, **IV**, 551–8 (1984).
5. Bertero, V., 'State of the art and practice in seismic resistant design of r/c frame-wall structural systems', *Proc. 8th World Conf. on Earthq. Eng., San Francisco*, **V**, 613–20 (1984).
6. Fenwick, R. C., 'Strength degradation of concrete beams under cyclic loading', *Bull. NZ Nat. Soc. for Earthq. Eng.*, **16**, No. 1, 25–38 (1983).
7. Standards Association of New Zealand, 'Code of practice for the design of concrete structures', *NZS 3101*, Parts 1 and 2, 1982.
8. International Conference of Building Officials, *Uniform Building Code*, ICBO, Whittier, California (1985).
9. Goodsir, W. J., Paulay, T., and Carr, A. J., 'A design procedure for interacting wall-frame structures under seismic actions', *Proc. 8th World Conf. on Earthq. Eng., San Francisco*, **V**, 621–8 (1984).
10. Paulay, T., Carr, A. J., and Tompkins, D. N., 'Response of ductile reinforced concrete frames located in Zone C', *Bull. NZ Nat. Soc. for Earthq. Eng.*, **13**, No. 3, 209–25 (1980).
11. Park, R., 'Ductility of reinforced concrete frames under seismic loading, *New Zealand Engineering*, **23**, No. 11, 427–35 (1980).
12. Selna, L., Martin, I., Park, R., and Wyllie, L., 'Strong and tough concrete columns for seismic forces', *J. of the Structural Division, ASCE*, **106**, No. ST8, 1717–34 (1980).
13. Blume, J. A., Newmark, N. M. and Corning, L. H., *Design of multi-storey reinforced concrete buildings for earthquake motions*, Portland Cement Association, Skokie, Illinois (1961).
14. Base, G. D., and Read, J. B., 'Effectiveness of helical binding in the compression zone of concrete beams', *ACI J.*, **62**, 763–781 (1985).
15. Bertero, V. B., and Felippa, C., Discussion to paper by Roy, H., and Sozen, M., 'Ductility of concrete', *Proc. of the Int. Symp. on Flex. Mechanics for Reinforced Concrete*, ASCE–ACI, Miami, 213–35 (1964).
16. Nawy, E. G., Danesi, R. F., and Grosko, J. J., 'Rectangular spiral binders, effect on plastic hinge rotation capacity in reinforced concrete beams', *ACI J.*, **65**, 1001–16 (1968).
17. Baker, A. L. L., and Amarakone, A. M. N., 'Inelastic hyperstatic frame analysis', *Proc. Int. Symp. on Flex. Mechanisms of Reinforced Concrete*, ASCE–ACI, Miami, 85–142 (1984).
18. Soliman, M. J. M., and Yu, C. W., 'The flexural stress–strain relationship of concrete confined by rectangular transverse reinforcement', *Magazine of Concrete Research*, **61**, 223–38 (1967).
19. Corley, W. G., 'Rotational capacity of reinforced concrete beams', *J. of the Structural Division, ASCE*, **92**, No. ST5, 121–146 (1966).
20. Park, R., Priestley, N. J. M., and Gill, W. D., 'Ductility of square confined concrete columns', *J. of the Structural Division, ASCE*, **108**, No. ST4, 929–50 (1982).
21. Tanaka, H., Park, R., and McNamee, B., 'Anchorage of transverse reinforcement in rectangular reinforced concrete columns in seismic design', *Bull. NZ Nat. Soc. for Earthq. Eng.*, **18**, No. 2, 165–90 (1985).
22. Pfrang, E. O., Seiss, C. P., and Sozen, M. A., 'Load-moment-curvature characteristics of r.c. cross sections', *ACI J.*, **61**, 763–78 (1964).
23. Paparoni, M., Ferry Borges, J., and Whitman, R. V., 'Seismic studies of Parque Central buildings', *Proc 5th World Conf. on Earthq. Eng., Rome*, **2**, 1991–2000 (1973).
24. Cardenas, A. E., Hanson, J. M., Corley, W. G., and Hognestad, E., 'Design provisions for shear walls', *ACI J.* **70**, No. 3, 221–30 (1973).
25. Salse, E. A. B., and Fintel, M., 'Strength, stiffness and ductility properties of slender shear walls', *5th World Conf. on Earthq. Eng., Rome*, **1**, 919–28 (1973).

26. Park, R., and Paulay, T., 'Concrete structures', in *Design of Earthquake Resistant Structures*, (Ed. E. Rosenblueth), Pentech Press, London (1980), Chapter 5.
27. Paulay, T., 'Some aspects of shear wall design', *Bull. NZ Soc. for Earthq. Eng.*, **5**, No. 3, 89–105 (1972).
28. Paulay, T., and Williams, R. L., 'The analysis and design of and the evaluation of design actions for reinforced concrete ductile shear wall structures', *Bull. NZ Nat. Soc. for Earthq. Eng.*, **13**, No. 2, 108–43 (1980).
29. Paulay, T., 'An elasto-plastic analysis of coupled shear walls', *ACI J.*, **67**, No. 11, 915–22 (1970).
30. British Standards Institution, 'Bending dimensions and scheduling of bars for reinforcement of concrete', BS 4466: 1981.
31. British Standards Institution, 'The structural use of concrete', British Standard Code of Practice, CP110: 1972.
32. British Standards Institution, 'Hot rolled steel bars for the reinforcement of concrete', BS 4449: 1978.
33. American Society for Testing and Materials, 'Low-alloy deformed bars for concrete reinforcement', ASTM A706, 1984.
34. American Society for Testing and Materials, 'Deformed and plain billet-steel bars for concrete reinforcement', ASTM A15, 1984.
35. British Standards Institution, 'Cold worked steel for the reinforcement of concrete', BS 4461: 1969.
36. British Standards Institution, 'Steel fabric for reinforcement of concrete', BS 4483: 1969.
37. Park, R., and Milburn, J. R., 'Comparison of recent New Zealand and United States seismic design provisions for reinforced concrete beam–column joints and test results from units designed according to the New Zealand code', *Bull. NZ Nat. Soc. for Earthq. Eng.*, **16**, No. 1, 3–24 (1983).
38. Suzuki, N., Otani, S., and Kabayashi, Y., 'Three dimensional beam column subassemblages under bidirectional earthquake loadings', *Proc. 8th World Conf. on Earthq. Eng., San Francisco*, **VI**, 453–60 (1984).
39. Paulay, T., 'Developments in the seismic design of reinforced concrete frames in New Zealand, *Canadian J. of Civil Engineering*, **8**, No. 2, 91–113 (1981).
40. Blakeley, R. W. G., 'Design of beam–column joints', *Bull. NZ Nat. Soc. for Earthq. Eng*, **10**, No. 4, 226–37 (1977).
41. Fenwick, R. C., 'Seismic resistant joints for reinforced concrete structures', *Bull. NZ Nat. Soc. for Earthq. Eng.*, **14**, No. 3, 145–59 (1981).
42. Joglekar, M. R., Murray, P. A. Jirsa, J. O., and Klingner, R. E., 'Full scale tests of beam–column joints', *Proc. 8th World Conf. on Earthq. Eng., San Francisco*, **VI**, 691–8 (1984).
43. Various papers, *Proc. 8th World Conf. on Earthq. Eng., San Francisco*, **VI**, 717–88 (1984).
44. Sauter, F., 'Earthquake resistant criteria for precast concrete structures', *Proc. 8th World Conf. on Earthq. Eng., San Francisco*, **V**, 629–36 (1984).
45. Clough, D. P., 'Development of seismic design criteria for connections in jointed precast concrete structures', *Proc. 8th World Conf. on Earthq. Eng., San Francisco*, **V**, 645–52 (1984).
46. Spencer, R. A., and Tong, W. K. T., 'Design of a one storey precast concrete building for earthquake loading', *Proc. 8th World Conf. on Earthq. Eng., San Francisco*, **V**, 653–60 (1984).
47. Forrest, E. J., 'Seismic resistance of industrial building', in *Concrete for the 70's, Industrialisation, Proc. Nat. Conf., Wairakei, New Zealand*, Portland Cement Association, New Zealand, 70–5 (1972).
48. Wood, B. J., 'Structural jointing', in *Concrete for the 70's, Industrialisation, Proc. Nat. Conf. Wairakei, New Zealand*, Portland Cement Association, New Zealand, 76–9 (1972).

49. Mast, R. F., 'Seismic design of 24-storey building with precast elements', *J. Prestressed Concrete Institute*, **17**, 45–59 (1972).
50. Seminar under the Japan–US Co-operative Seismic Program—*Construction and Behaviour of Precast Concrete Structures*, Reports by Japanese Participants, Parts I and II, 23–7 August, Seattle, USA (in English) (1971).
51. Petrovic, B., Muravljov, M., and Dimitrievic, R., 'The IMS assembly framework system and its resistance to seismic influences, Earthquake Engineering, Proc. 3rd European Symposium on Earthquake Engineering, Sofia, 513–20 (1970).
52. PCI Committee on Connection Details, *Connection details for precast-prestressed concrete buildings*, Prestressed Concrete Institute, USA, 1963.
53. Tankersley, R. N., and Sewell, J. M. F., 'Design and construction of a precast multistorey parking station', *New Zealand Engineering*, **25**, No. 11, 291–9 (1970).
54. Uchida, N., Aoyagi, T., Kawamura, M., and Nakagawa, K., 'Vibration test of steel frame having precast concrete panels', Proc. 5th World Conf. on Earthq. Eng., Rome, **1**, 1167–76 (1973).
55. Brooke-White, C. J., 'An architectural application of precast concrete', in *Concrete for the 70's—Industrialisation, Proc. Nat. Conf., Wairakei, New Zealand*, Portland Cement Association, New Zealand (1972).
56. *PCI manual for structural design of architectural precast concrete*, PCI Publication No. MNL-121-77, Prestressed Concrete Institute, Chicago (1977).
57. Goodno, B. J., Palsson, H., and Pless, D. G., 'Localized cladding response and implications for seismic design', *Proc. 8th World Conf. on Earthq. Eng., San Francisco*, **V**, 1143–50 (1984).
58. Miller, M. S., 'Joints in architecture', *in Concrete for the 70's—Industrialisation, Proc. Nat. Conf., Wairakei, New Zealand*, Portland Cement Association, New Zealand (1970).
59. Blakeley, R. W. G., 'Prestressed concrete seismic design', *Bull. NZ Nat. Soc. for Earthq. Eng.*, **6**, No. 1, 2–21 (1973).
60. Pond, W. F., 'Performance of bridges during the San Fernando earthquake', *J. Prestressed Concrete Institute*, **17**, No. 4, 65–75 (1972).
61. Federation Internationale de la Precontrainte, *Recommendations for the design of aseismic prestressed concrete structures*, FIP/7/1, November (1977).
62. Blakeley, R. W. G., and Park, R., 'Ductility of prestressed concrete members', *Bull. NZ Soc. for Earthq. Eng.*, **4**, No. 1, 145–70 (1971).
63. Park, R., Priestley, N. J. M., Falconer, T. J., and Joen, P. H., 'Detailing of prestressed concrete piles for ductility', *Bull. NZ Nat. Soc for Earthq. Eng.*, **17**, No. 4, 251–71 (1984).

Chapter 8

Masonry Structures

8.1 INTRODUCTION

Masonry is a term covering a very wide range of materials such as adobe, brick, stone, and concrete blocks; and each of these materials in turn varies widely in Form and mechanical properties. Also masonry may be used with or without reinforcement or in conjunction with other materials. As well as its use for primary structure, masonry is used for infill panels creating partitions or cladding walls.

The variety available in form, colour and texture makes masonry a popular construction material, as does its widespread geographic availability and, in some cases, its comparative cheapness. Properly used, it also has reasonable resistance to horizontal forces. However, masonry has a number of serious drawbacks for earthquake resistance. It is naturally brittle; it has high mass and hence has high inertial response to earthquakes; its construction quality is difficult to control; and relatively little research has been done into its seismic response characteristics compared with steel and concrete.

Because of the poor performance of some forms of masonry in earthquakes official attitudes towards masonry are generally cautious in most moderate or strong motion seismic areas. For example, in Japan masonry has not been permitted for buildings of more than three storeys in height but the outcome of the US–Japan Co-operative Research Program on masonry which started *c.* 1985 may change attitudes in Japan (and elsewhere).

By way of contrast, carefully designed apartment buildings of 15 storeys or more have been built in California. No doubt there is considerable inherent strength in the 'egg-crate' form of apartment buildings, but there has been little dynamic testing or seismic field experience of tall masonry structures to date.

The other types of construction in which masonry is most popular are low-rise housing and industrial buildings (Appendices A.3 and A.4).

8.2 SEISMIC RESPONSE OF MASONRY

This discussion is supplementary to the general introduction to seismic response given in Section 6.6.

The tendency to fail in a brittle fashion is the central problem with masonry. While unreinforced masonry may be categorically labelled as brittle, uncertainty exists as to the degree of ductility which should be sought in reinforced masonry.

Based on static load-reversal tests, Meli[1] contended that 'for walls with interior reinforcement, where failure is governed by bending, behaviour is nearly elasto-plastic with remarkable ductility and small deterioration under alternating load except for high deformation. . . . If failure is governed by diagonal cracking, ductility is smaller and, when high vertical loads are applied, behaviour is frankly brittle. Furthermore . . . important deterioration (occurs) after diagonal cracking'. Meli concluded that bending failure was the most favourable design condition for walls. In dynamic tests, Williams and Scrivener[2] confirmed this conclusion. Their tests on brick walls showed fairly stable hysteretic behaviour at drifts up to about 1 percent, despite the absence of horizontal reinforcement (Figure 8.1).

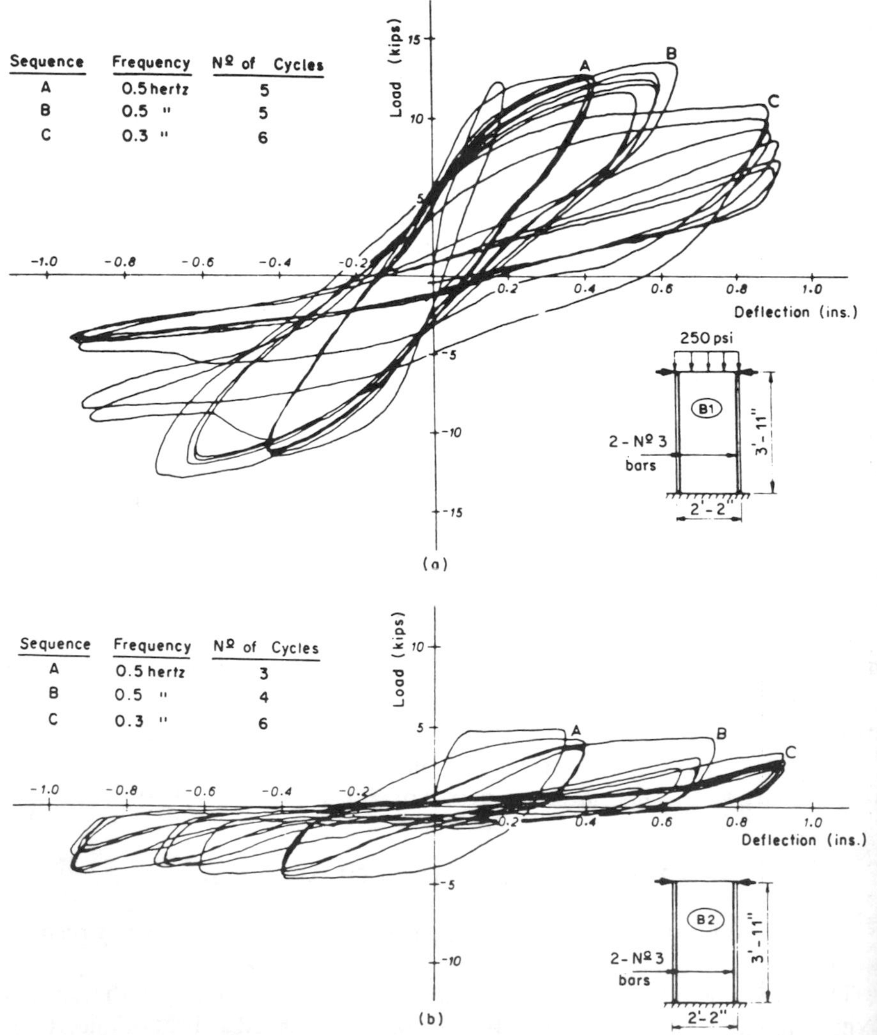

Figure 8.1 Response of reinforced brick walls to dynamic loading (after Williams and Scrivener[2])

Subsequent research in various countries (for example, references 3–8) has examined various masonry products and wall-reinforcing layouts, sometimes under slow cyclic reversed loading and shake-table dynamic tests. The value of having vertical and horizontal reinforcement distributed throughout walls is apparent, but the use of reinforcement only at the perimeter of wall panels is surprisingly effective for both in-plane and out-of-plane loading (Reference 7, UCB/EERC-79/23). The latter is more true for masonry of higher tensile strength (i.e. concrete blocks) and also is probably more true for stiff structures with low lateral displacements. Perimeter-only reinforcement is very cost-effective as a minimum provision for low-cost construction and for strengthening of existing buildings.

The adequacy of very light reinforcement in low-rise construction has been remarkably demonstrated in shake-table tests at the EERC[7,8] of hollow concrete block houses, where three simultaneous components of shaking with peak accelerations up to about 0.9g were applied. The walls had no intermediate height horizontal reinforcement, but were well connected at top and bottom to continuous horizontal perimeter members of the foundation and roof construction. The vertical reinforcement content varied from zero to a maximum of four bars (of 10 and 12 mm diameter) in walls about 4 m long.

8.3 RELIABLE SEISMIC BEHAVIOUR OF MASONRY STRUCTURES

8.3.1 Introduction

For obtaining reliable seismic response behaviour the principles concerning choice of form, materials, and failure mode control discussed in Section 5.3 apply to masonry structures, while further factors specific to masonry are discussed below.

The wide range of masonry products, of clay and concrete types, means a wide range of material behaviour and hence of seismic reliability. Probably the most reliable type is reinforced hollow concrete blocks, which have been more studied than other masonry materials. However, with the growing research interest in reinforced clay bricks[6] and other masonry products the full reliability potential and relative merits of the various masonry materials are becoming better understood. Where a choice between relatively unresearched masonry materials has to be made, those which are weaker in compression and tension will obviously tend to be less reliable in earthquakes.

In considering reliability of seismic behaviour of masonry structures through structural forms and failure mode control fewer alternatives need be considered than for other structural materials. Masonry is best suited to forming walls and less suited to columns and lintel beams, and is constructionally and aseismically ill-suited for forming other structural members. Thus this discussion mainly relates to the reliable seismic behaviour of walls.

While quite high repeatable ductilities can be achieved in masonry walls and columns by using thin steel plates between block courses[4] the constructional complications and costs of such measures suggest that seeking high ductilities for masonry structures is another example of seeking high ductility for its own sake, as discussed for concrete (Section 7.2.3.2). The more pragmatic traditional approach of seeking limited ductility, so well demonstrated as successful (at least for single-storey buildings) by the EERC tests (Section 8.2), seems likely to remain appropriate for most masonry structures, namely:

(1) Suppress the more brittle failure modes (e.g. shear);
(2) Design for limited ductility and adequate strength (e.g. the UBC[9] approach);
(3) Use sound structural forms (as discussed below).

8.3.2 Structural form for masonry buildings

The general principles of earthquake-resistant structural form have been given in Section 5.3, but additional guidance peculiar to masonry is given here. Five interrelated criteria for consideration in masonry construction are that:

(1) The aspect ratio H/B for the structure should be a minimum;
(2) The aspect ratio H'/B' for vertical members should be a minimum;
(3) The ratio of aperture area to wall area, $\Sigma A_a/HB$, should be a minimum;
(4) The distribution of apertures should be as uniform as possible;
(5) Stress-raising apertures should be located away from highly stressed zones.

Considering these criteria in relation to a single storey structure with zero or minimum reinforcement, Figure 8.2(a) represents a good structure and Figure 8.2(b) represents a bad one. Neither case has a bad overall aspect ratio, as is typical for single-storey buildings. For unreinforced buildings with a maximum aperture area, a value of $H/B \ngtr 2/3$ might be taken.

The aspect ratio of vertical members, particularly those at the ends of walls (H_1'/B_1' and H_2'/B_4' in Figure 8.2(a)), should be little greater than unity in buildings of minimum reinforcement. This is clearly not so in Figure 8.2(b).

It is commonly recommended that the total area of holes should not exceed one third of the wall area, i.e. $\Sigma A_a/HB \ngtr 1/3$. If criteria (2) and (3) above have been satisfied it is likely that the distribution of apertures will be reasonably uniform. Small holes, such as those used for the passage of services pipes and ducts, should be kept away from corners of load-bearing members; A_{a3} in Figure 8.2(b) is badly placed compared with that in Figure 8.2(a).

The main objective of the above criteria is to distribute the strength as uniformly as possible; in brittle structures the early failure of one member causes the remaining members to share the total load, and often leads to incremental collapse.

8.3.3 Structural form for reinforced masonry

The criteria set out above are also applicable to reinforced masonry, although some relaxation of the suggested limits may be made. The degree of relaxation

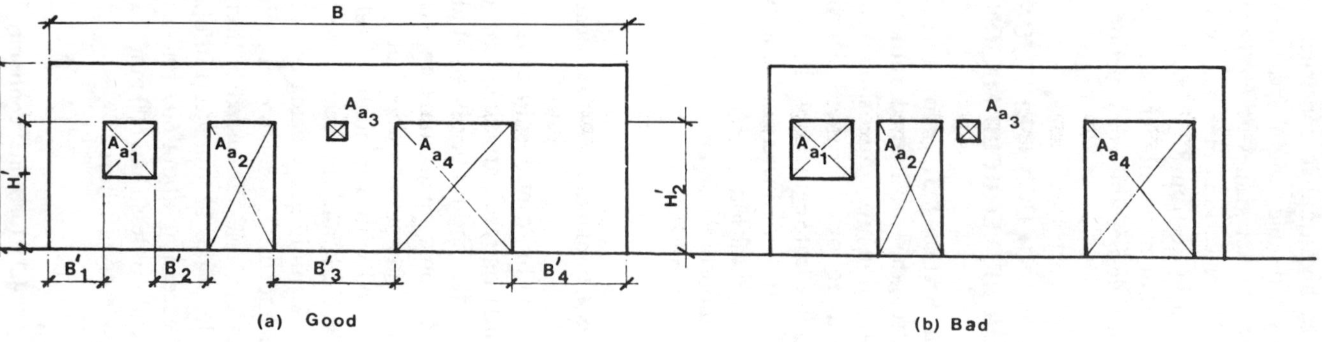

Figure 8.2 Structural form of low-rise masonry buildings

will depend on the degree of protection against early brittle failure afforded by the reinforcement. In high-quality construction, building aspect ratios $H/B=2$, or $H/B=3$ are reasonable, and in apartment block construction with small aperture ratios even higher values of H/B have been used.

The aspect ratio H'/B' for vertical members when reinforced may also be increased over those given in Section 8.3.2, perhaps to values of 2 or 3 for members of low strength, but much higher values may be taken for column members of higher quality design and construction.

8.4 DESIGN AND CONSTRUCTION DETAILS FOR REINFORCED MASONRY

For the reasons given in Section 8.3.1, most masonry structures should be designed primarily for strength, with limited, rather than high, ductility being sought. Design and construction procedures should conform to well-established codes of practice[9] such as followed in the design handbook by Amrhein.[10] Seismic loadings set out in local regulations should be used with larger load factors than those used for steel or concrete unless a specific factor for materials is used in finding the horizontal loads.

Some of the more important details of aseismic reinforced masonry construction are commented on below.

8.4.1 Minimum reinforcement

Although code recommendations on this subject vary somewhat, the following American practice[9] is fairly representative. At least one vertical bar, not less than 12 mm ($\frac{1}{2}$ in) in diameter, should be placed at all corners, wall ends, and wall junctions. Such bars should be anchored into the upper and lower walls beams or foundations, and adequately lapped at splices.

Walls should be reinforced with a minimum steel content[9] of 0.07 percent both horizontally and vertically, and the total of the two directions should not be less than 0.2 percent. A minimum spacing of 1.2 m between bars in both directions is also recommended.[9] These nominal requirements seem to be well supported by the EERC test series discussed in Section 8.2.

At least one bar, not less than 12 mm ($\frac{1}{2}$ in) in diameter, should be placed on all sides of apertures exceeding about 600 mm (2 ft) in any direction. Such framing bars should extend not less than 600 mm (2 ft) beyond each corner of the aperture or be equivalently anchored.

8.4.2 Horizontal continuity

Horizontal continuity around the perimeter of the building should be ensured at least at the levels of the base, the floors, and the roof. The walls should be

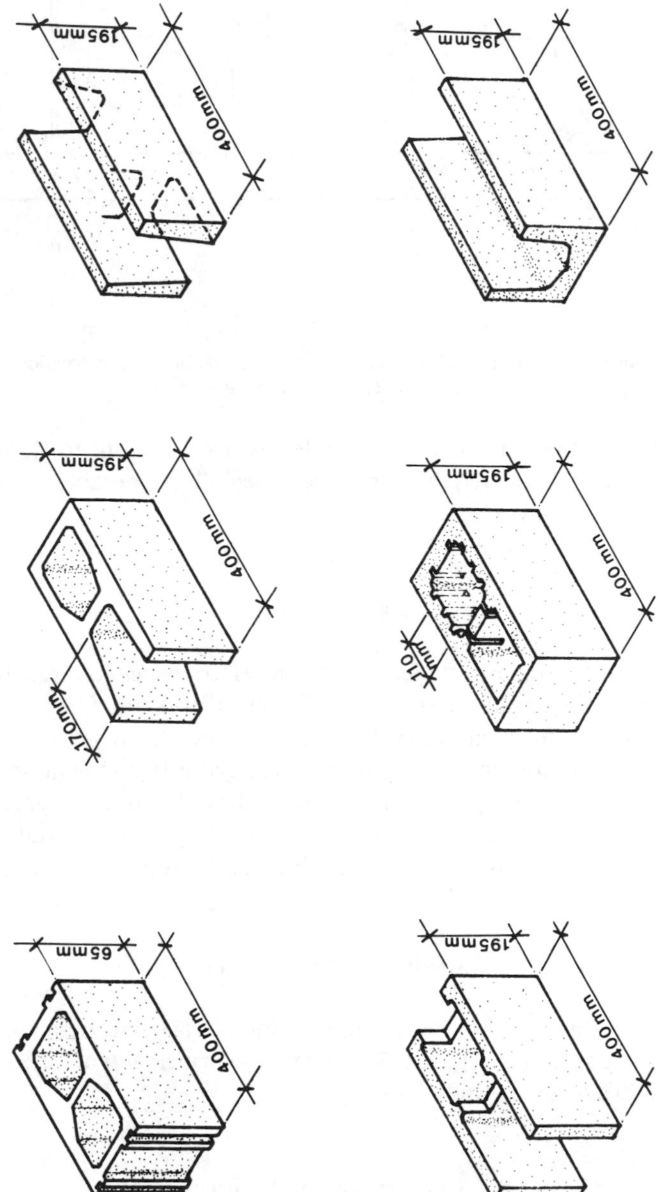

Figure 8.3 Principal structural concrete blocks used in New Zealand for reinforced masonry construction, 150 mm to 200 mm wall thicknesses

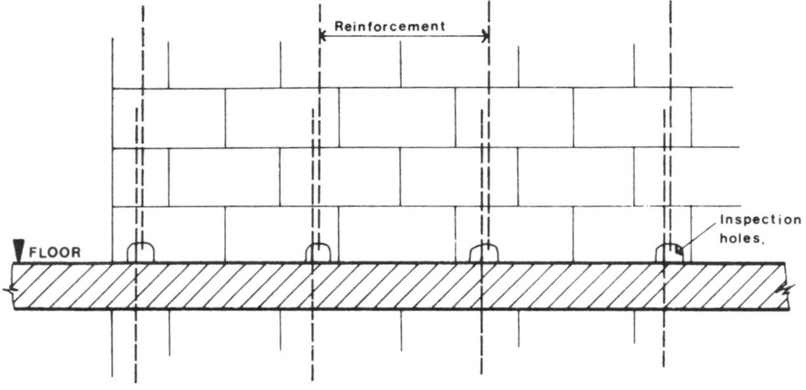

Figure 8.4 Inspection holes in hollow reinforced concrete block construction for checking reinforcement and grouting (see text)

tied into an effective ring beam at these levels. Connections to floors or roofs other than of *in situ* concrete have proved especially vulnerable in earthquakes (Appendices A.3, A.4).

8.4.3 Grouting

The reinforcement of masonry depends for its effectiveness on transfer of stress through the grout from steel to masonry. Every effort should be made to ensure that compacted grout completely fills the cavities. A low-shrinkage grout is essential in order to minimize separation of the grout from the masonry. Grout for cavities of up to 60 mm ($2\frac{1}{2}$ in) in width should contain aggregates up to 5 mm in size, while for larger cavities coarser aggregate may be suitable. Grout should have a characteristic cube strength of 20 N/mm^2 (3000 psi) at 28 days.

8.4.4 Hollow concrete blocks

Although the shape of hollow concrete blocks varies in detail in different countries, those shown in Figure 8.3 illustrate the principal types used in 150 mm (6 in) and 200 mm (8 in) thick walls.

8.4.5 Supervision of construction

In order to ensure adequate standards of construction more supervision is required for reinforced masonry than on equivalent projects in other materials. The following points in particular need watching;

(1) Cavities should be clean and free from mortar droppings;
(2) Reinforcement should be placed centrally or properly spaced from the masonry;
(3) Reinforcement should be properly lapped;
(4) The grouting procedure should be properly carried out;
(5) The grout mix should conform to the specification.

In multi-storey hollow concrete block construction, inspection holes at the bottom of walls on the line of vertical reinforcement are advisable to facilitate the checking of items (1) to (4) above (Figure 8.4).

8.5 CONSTRUCTION DETAILS FOR STRUCTURAL INFILL WALLS

Masonry is often used as structural infill, either as cladding or as interior partitions. It should be either effectively separated from frame action or fully integrated with it as discussed in Section 12.2.2. The analysis of the interaction between frames and integrated infill panels has been discussed in Section 6.6.6. Very few data on the detailing of masonry in this situation exist, but the following points may be made.

(1) No gap should be left between the infill and the frame, so as to prevent accidental pounding damage during earthquakes.
(2) The top of the panel should be structurally connected to the structure above to ensure lateral stability of the infill in earthquakes.
(3) Ideally the form of the structure and the strength of the infill panel would be such that shear failure of the masonry infill would not occur. When this is not the case, the reinforcement required in each infill panel for resisting shear is difficult to determine. It may be calculated in the same way as normal masonry shear walls, although it has been suggested by some workers such as Meli[1] that peripheral reinforcement (provided in this case by the frame) is more effective in creating ductile masonry than internal reinforcement.

Placing of full-height vertical reinforcement is obviously difficult in hollow block or brick construction of the form shown in Figure 8.3, when the infill is erected after the upper frame member has been constructed, as is usually the case in this form of construction. This difficulty is obviated by the use of external reinforcement in the form of expanded metal sheets bonded to the sides of the block wall by means of a layer of mortar. Tso *et al.*[11] have reported favourable behaviour of this type of construction in cyclic loading tests, using washers and bolts through the wall to improve the bonding of the expanded metal to the wall.

REFERENCES

1. Meli, R., 'Behaviour of masonry walls under lateral loads', *Proc. 5th World Conf. on Earthq. Eng., Rome*, **1**, 853–62 (1973).

2. Williams, D. and Scrivener, J. C., 'Behaviour of reinforced masonry shear walls under cyclic loading', *Bull. NZ Nat. Soc. for Earthq. Eng.*, **4**, No. 2, 316–32 (1971).
3. Priestley, N. J. M., 'Seismic resistance of reinforced concrete masonry shear walls with high steel percentages', *Bull. NZ Nat. Soc. for Earthq. Eng.*, **10**, No. 1, 1–16 (1977).
4. Priestley, N. J. M., 'Ductility of confined concrete masonry shear walls', *Bull. NZ Nat. Soc. for Earthq. Eng.*, **15**, No. 1, 22–6 (1982).
5. Thurston, S. J., and Hutchison, D. L., 'Reinforced masonry shear walls: cyclic load tests in contraflexure', *Bull. NZ Nat. Soc. for Earthq. Eng.*, **15**, No. 1, 27–45 (1982).
6. Wakabayashi, M., and Nakamura, T., 'Reinforcing principle and seismic resistance of brick masonry walls', *Proc. 8th World Conf. on Earthq. Eng., San Francisco*, **V**, 661–8 (1984).
7. Various authors, 'Shaking table studies of single story masonry houses; (etc)', Earthquake Engineering Research Center Reports Nos UCB/EERC-79/23 (1979); UCB/EERC-79/24 (1979); UCB/EERC-79/25 (1979); UCB/EERC-83/11 (1983).
8. Manos, G. C., Clough, R. W., and Mayes, R. L., 'A three component shaking table study of the dynamic response of a single story masonry house', *Proc. 8th World Conf. on Earthq. Eng., San Francisco*, **VI**, 855–62 (1984).
9. International Conference of Building Officials, *Uniform Building Code*, ICBO, Whittier, California (1985).
10. Amrhein, J. E., *Reinforced Masonry Engineering Handbook*, The Masonry Institute of America, Los Angeles (3rd edn) (1978).
11. Tso, W. K., Pollner, E., and Heidebrecht, A. C., 'Cyclic loading on externally reinforced masonry walls', *Proc. 5th World Conf. on Earthq. Eng.*, Rome, **1**, 1177–86 (1973).

Chapter 9

Steel structures

9.1 INTRODUCTION

Because of the inherent ductility available in appropriately manufactured steel, structures made from this material have been less liable to collapse in earthquakes than traditionally designed concrete or masonry ones. Presumably because of the relative ease with which a reasonably good aseismic performance may be obtained with steel, surprisingly little research has been done on the response of steel structures to cyclic load reversals of either slow or true dynamic nature. This has meant that the full earthquake resisting potential of steel structures was much further from realization in the mid-1980s than need have been the case, and there was less insight available into the earthquake resistance capabilities of steelwork than there was for concrete at that time.

However, the advent of eccentrically braced frames in the late 1970s gave an impetus to the interest in research into earthquake resistance of steelwork, together with a renewal of interest in concentrically braced frames arising partly from the construction of large steel offshore platforms in seismic areas. It is hoped that a proliferation of such research will continue, particularly in the widespread subject of instability of members during earthquake motions. As well as identifying problem areas, a comprehensive set of design procedures has been outlined in papers from a New Zealand study[1] of the seismic design of structures, which will be referred to below.

9.2 SEISMIC RESPONSE OF STEEL STRUCTURES

The seismic response of steel structures depends mainly on:

(1) The onset of instability (local or global);
(2) The nature of the steel members;
(3) The nature of the connections;
(4) The nature of other components interacting with the frame.

Under ideal conditions of lateral restraint, repeatable high ductility of a very stable nature can be obtained, as shown by the hysteresis loops for bending illustrated in Figure 6.31(b), but with less restraint to webs or flanges marked loss of strength and stiffness may occur as shown in Figures 9.4 and 9.6. Similar deterioration in strength occurs as a result of damage from *low cycle fatigue*, a phenomenon which increases with increasing strain. For example, Bertero and Popov[2] found that a 100 mm × 100 mm WF beam failed after 607 cycles when strained to ± 1.0 percent, while strain amplitudes of ± 2.5 percent produced failure after only 16 cycles.

Commonly members fail at plastic hinges after local buckling has increased the strains at the surface sufficiently to initiate cracking which rapidly reaches the critical size for fast fracture. In members which are well detailed for local buckling (Section 9.4.2), with workmanship that keeps notches or notch-like defects to a minimum, good welding details,[3] and the use of normally available weldable steels (Section 9.3.2.6), low-cycle fatigue should be controlled. This problem, however, is not fully understood, although Kaneta *et al.*[4] and Krawinkler[5] have proposed methods of assessing the degree of low-cycle fatigue damage to steel structures. Neither method is comprehensive and both are too sophisticated for use on normal design projects, and, in common with most fatigue-life prediction problems, no simple alternative appears to exist.

Different types of connection affect response, depending on the damping that they produce. For example, elastically responding steel structures typically have up to 2 percent of critical damping if fully welded compared with up to 7 percent if fully bolted.

The response of steel structures, particularly in terms of damping, is greatly influenced by *components interacting* with the steel structure itself, e.g. cladding, partitions, floors, and chimney linings. The combined effects of stress level, connections, and interacting components (clad or unclad) are indicated by the typical design values[6] of equivalent viscous damping given in Table 9.1. (Damping is further discussed in Section 6.6.4.)

A further feature of the seismic response of steel is the increase in yield strength exhibited with increase in *rate of loading*. Normal quasi-static tests of yield stress f_y are conducted at low strain rates of about 10^{-3}/s. Under seismic loading conditions in short period structures local strain rates may be in excess of 1.0/s, causing increases in f_y of 30 percent or so over the quasi-static value. While no increase in strength exists at fracture, the increase in yield strength is reflected in the stress–strain relationship throughout the usable inelastic strain range. Unfortunately the scatter in the results reported by different researchers is large.

Table 9.1. Typical damping ratios for steel structures, ξ, percentage of critical[6]

Type of structure	Stress state	Welded connections	Bolted connections
Clad*	Elastic	2.5	5
Clad	Inelastic	5	10
Unclad*	Elastic	2	5
Unclad	Inelastic	5	7.5

**Unclad* refers to open industrial frameworks (perhaps with web grating steel flooring or platforms). *Clad* refers to most other structures such as offices, car parking buildings.

For example, Symonds[7] has shown that the dynamic yield stress f_{yd} is given by

$$f_{yd}/f_y = 1 + \left(\frac{\dot{\epsilon}}{D}\right)^{1/p} \tag{9.1}$$

where $\dot{\epsilon}$ is the strain rate and D and p are constants, depending on the type of steel. For mild steel $D = 40.4$ and $p = 5.0$.

Wakabayashi *et al.*[9] found smaller dynamic enhancements than those given by Symonds, from tests which resulted in the expression

$$f_{yd}/f_y = 1 + 0.0473 \log\left(\frac{\dot{\epsilon}}{\dot{\epsilon}_o}\right) \tag{9.2}$$

where $\dot{\epsilon}_o = 50 \times 10^{-6}/\mathrm{s}$

Because equations (9.1) and (9.2) are derived from tests at constant strain rates it is not clear how to factor their results for application under the cyclically varying strain rate conditions of earthquake response. However, it is reasonable to assume that the response is approximately sinusoidal, and let us conjecture that the *average* rate of change of strain applies. Using these assumptions, equations (9.1) and (9.2) may be used to find the dynamic yield stress f_{yd} in a plastic hinge as a function of the fundamental period of the structure and the maximum strain ϵ_{max} occurring in a response cycle (Figure 9.1). The curve for $k = 1$ represents very limited ductility demand, and $k = 10$ may be typical of fully ductile behaviour. It thus appears that substantial enhancement of strength due to dynamic effects will always occur, even if (as seems likely) the

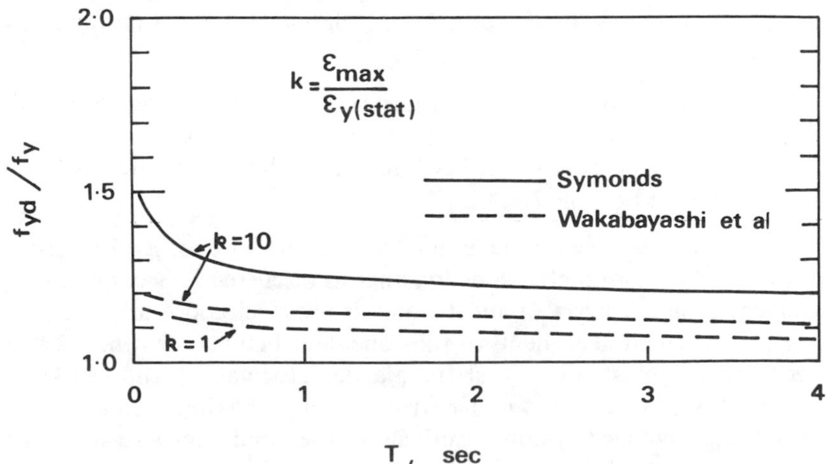

Figure 9.1 Dynamic enhancement of yield stress of steel as a function of period of vibration and maximum strain reached

results of Wakabayashi *et al.*[9] prove to be more reliable than those of Symonds.[7]

Separate studies by Udagawa *et al.*[8] and Wakabayashi *et al.*[9] found that the monotonic and cyclic loading strength of steel members was up to 20 percent greater at their rates of loading than the strength under quasi-static conditions. While these results tend to confirm the enhancement of the static f_y predicted by Figure 9.1, the mechanism and magnitude of the strength enhancement under earthquake conditions needs to be clarified.

9.3 RELIABLE SEISMIC BEHAVIOUR OF STEEL STRUCTURES

9.3.1 Introduction

For obtaining reliable seismic response behaviour the principles concerning choice of form, materials, and failure mode control discussed in Section 5.3 apply to steel structures, while further factors specific to steel are discussed below.

Designing for failure mode control requires consideration of the structural forms used, with all the forms discussed in Section 5.4 being appropriate for steel, i.e.:

(1) Moment-resisting frames;
(2) Framed tube structures;
(3) Structural walls;
(4) Concentrically braced frames;
(5) Eccentrically braced frames;
(6) Hybrid (or composite) structures.

In designing these structural forms with failure mode control in mind, in addition to the discussion in Section 5.3.8 it should be noted that the essential objectives are

(a) Beams should fail before columns (unless extra column strength is provided);
(b) Premature instability failure modes should be suppressed;
(c) An appropriate degree of ductility (toughness) should be provided (see also as discussed in Section 7.2.3.2).

If a structure is designed to have 'limited ductility', i.e. of $\mu = 2$ or less, and hence designed to resist higher code loadings as discussed in Section 6.6.7.3(i), the element design criteria of most non-seismic steel codes are likely to be adequate. For structural elements of fully ductile structures, i.e. where $2 \leqslant \mu \leqslant 6$, the design details must ensure that full plastic deformations can be obtained. The rotation of plastic hinges under strong seismic shaking is considerably in excess of those envisaged by non-seismic steel codes, and more stringent stability requirements are needed to maintain section capacity.

While hot rolled steel sections are generally preferable to *cold formed sections* because the latter have limited ductility, cold formed sections may be used in earthquake resistant structures provided that the appropriate measures are taken. In the same way that timber (which is brittle) may be protected by ductile connections so as to form ductile structures (Chapter 10), so may relatively brittle cold formed steel sections be joined by ductile steel plates at appropriate locations[10] to ensure that ductile failure modes occur.

The collapse mechanisms of the so-called *plastic design method* may not always be consistent with the objectives of seismic failure mode control, but it may be used as long as the objectives of the latter are met and excessive lateral sway is avoided. A New Zealand study group on steel structures has also recommended[6] that the plastic design method should not be used for structures having more than four mass levels because of the unreliability of failure mode control.

9.3.2 Material quality of structural steel

For the construction of reliable earthquake-resistant ductile steel frames the basic steel material must, of course, be of good quality. While steels suitable for seismic resistance are found amongst those produced for general structural purposes, not all normal structural grades are sufficiently ductile. The main properties required are as follows;

Property	Requirements
(1) Adequate ductility;	Aseismic requirements
(2) Consistency of mechanical properties;	
(3) Adequate *notch* ductility;	General requirements
(4) Freedom from laminations;	
(5) Resistance to lamellar tearing;	
(6) Good weldability.	

9.3.2.1 Ductility

Ductility may be described generally as the post-elastic behaviour of a material (Section 6.6.2). For steel it may be expressed simply from the results of elongation tests on small samples, or more significantly in terms of moment-curvature of hysteresis relationships, as discussed later in this chapter.

Steels manufactured in various countries may have sufficient ductility, and earthquake codes of practice often recommend suitable steels. In California, for example, the UBC[11] requires that steels conform to the latest edition of the following ASTM Specification;

A36 (Structural carbon steel);
A441 High-strength, manganese–vanadium);
A500 (Grades B, C), (Cold formed carbon steel);
A572 (Grades 42, 45), (High strength, columbium or vanadium).
A588 (High strength multiple alloy).

Cold formed steel sections have only limited ductility and hence should only be used as noted in Section 9.3.1.

9.3.2.2 Consistency of mechanical properties

In economically designing so that beams fail before columns it is desirable that the maximum and minimum strengths of members are as nearly equal in magnitude as possible. This means that the standard deviation of strengths should be as small as possible. While it is satisfactory for non-seismic design, it is unfortunate for earthquake-resistant design that steel manufacturers have been more concerned with simply achieving their minimum guaranteed yield strengths, than in producing consistent ultimate strengths.

9.3.2.3 Notch ductility

Notch ductility is a measure of the resistance of a steel to brittle fracture and is a separate property from that of general ductility discussed in Section 9.3.2.1. Adequate notch ductility is required in *all* structural steelwork, not only in seismic areas of the world. It is generally expressed as the energy required to fracture a test piece of particular geometry. Three widely used tests are the Charpy V-notch test, the Izod test, and Charpy keyhole test. The results, although quantitative, are generally empirical and are not comparable between tests and between materials. While many national steel standards specify levels of resistance to brittle fracture, American standards do not.

9.3.2.4 Laminations

Laminations are large areas of unbonded steel found in the body of a steel plate or section. This implies a layering of the steel with little structural connection between the layers. The laminated areas originate in the casting and cropping procedures for the steel ingots, and may be as much as several square metres in extent. Steel may be screened ultrasonically for lamination before fabrication, and some guidance may be found on this procedure in a British Standards Institution publication.[12]

9.3.2.5 Lamellar tearing

It should first be pointed out that lamellar tearing should not be confused with laminations, the two being different phenomena.

Lamellar tearing is a tear or stepped crack which occurs under a weld where sufficiently large shrinkage stresses have been imposed in the through thickness direction of susceptible material. It commonly occurs in **T**-butt welds and in corner welds and is caused by inclusions which act as 'perforations' in the steel. Lamellar tearing has been discussed in some detail by Farrar and Dolby[13] and Jubb.[14] Unfortunately no non-destructive method of screening for susceptibility to lamellar tearing is as yet available. The usual method of checking is by measuring the ductility in a through plate tensile test.

Electric arc steelmaking incorporating vacuum degassing can produce steels with reduced (but not eliminated) susceptibility to lamellar tearing—although at some extra cost.[14] The risk of lamellar tearing can also be reduced by the use of suitable welding techniques and details.[3,13,14]

9.3.2.6 Weldability

Weldability may be considered simply as the capacity of the parent metal to be joined by sound welds. The weld metal should be able to closely match the properties of the parent plates, and few material defects should arise. To some extent weldability will be assured by the use of steels produced according to major national standards, such as those referred to in Section 9.3.2.1, or British Standard 4360, 'Weldable structural steels'. The weldability of a steel is often assessed by means of a formula based on the chemical analysis of the steel. Such methods determine the preheating temperature necessary to avoid hydrogen cracking.[15] In general the higher the tensile strength of the steel, the lower is its weldability.

9.4 STEEL BEAMS

In this section the behaviour and design of beams acting primarily in bending will be considered. In most beams axial forces are small enough to be neglected, but where large axial forces may occur column design procedures should be employed.

9.4.1 Moment–curvature relationships for steel beams under monotonic loading

For the adequate aseismic design of the steel beams, and the associated connections and columns, the moment–curvature or moment–rotation relationship should be known. A long stable plastic plateau is required which is not terminated too abruptly by lateral or local buckling effects, such as indicated by terminating at points *A*, *B*, and *C* in Figure 9.2. The curves terminating at *D* and *E* are typical of the desired behaviour achieved by well designed beams under moment gradient and uniform moment respectively. The moments in Figure 9.2 have been normalized in terms of the plastic moment capacity:

$$M_p = Sf_y \tag{9.2}$$

where S = the plastic modulus of the section; and
f_y = the characteristic yield stress of the steel.

Moment gradient is the usual loading condition to be considered with plastic hinges forming at the ends of beams in laterally loaded frames. The localization of high stresses produced by the moment gradient causes strain-hardening to occur during plastic rotation, resulting in an increase in moment capacity above

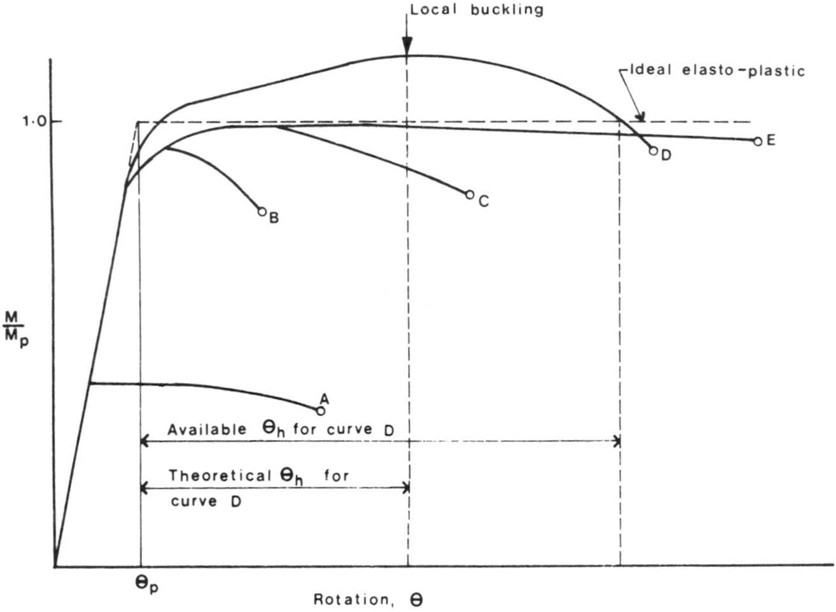

Figure 9.2 Behaviour of steel beams in bending

the ideal plastic moment *M* (curve *D* in Figure 9.2). Strain-hardening may increase the plastic moment by as much as 40 percent.[16] Local buckling and lateral buckling arising from plastic deformation of the compression flanges generally produces a reduction of moment capacity in the later stages of rotation,[17] as illustrated in curve *D* of Figure 9.2.

In order to predict the rotation capacity of a plastic hinge the following expressions presented by Lay and Galambos[18] for the monotonic inelastic hinge rotation θ_h (Figure 9.3) of a beam under *moment gradient* may be used:

$$\theta_h = 2.84\epsilon_y(\beta - 1)\frac{b}{d}\frac{t_f}{t_w}\left(\frac{A_w}{A_f}\right)^{1/4}\left(1 + \frac{V_1}{V_2}\right) \tag{9.4}$$

where b = flange width,
d = overall depth of section,
t_f = flange thickness,
t_w = web thickness,
A_f = flange area,
A_w = web area,
V_1 *and* V_2 = absolute values of shears acting either side of the hinge; arranged so that $V_1 \leqslant V_2$,

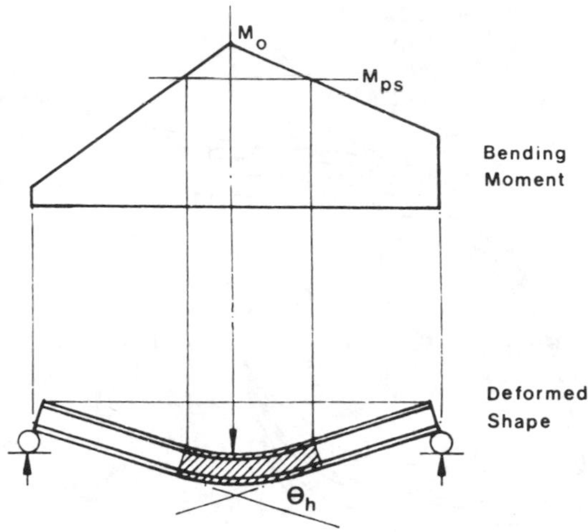

Figure 9.3 Beam under moment gradient with plastic hinge deformations and the hinge rotation θ_h of equation (9.4) as defined by Lay and Galambos[18]

β = ratio of strain at onset of strain-hardening to strain at first yield,
ϵ_y = strain at first yield.

θ_h represents a substantial proportion of the total rotation capacity of the beam (Figure 9.2). For the American section 10WF25 (A36 steel), equation (9.4) predicts that $\theta_h = 0.07$ radians.[18] It should be noted that equation (9.4) incorporates simplifications which lead to underestimations of θ_h of 20 percent or more.

The degree to which the plasticity of a section is utilised in rotation may be expressed by the rotation capacity R, which is a ratio of the plastic hinge rotation to the rotation at or near first yield. Under monotonic loading the rotation capacity is a function only of the beam section properties and its lateral supports, and decreases as some inverse function of the slenderness ratio l/r_y. Using the definition

$$R_1 = \theta_h/\theta_p - 1 \tag{9.5}$$

where $\theta_p = M_p l/EI$, Takanashi *et al.*[19] have shown that for typical Japanese beam section that R_1 exceeds 10 for l/r_y less than about 40, and $R_1 = 2$ for l/r_y of about 100.

A similar alternative to equation (9.5), often used in the literature and referred to later here, calculates the rotation capacity from

$$R = \frac{\theta_h}{\theta_y} \tag{9.6}$$

where θ_y is the elastic rotation between the far ends of the beam segment up to the formation of the hinge.

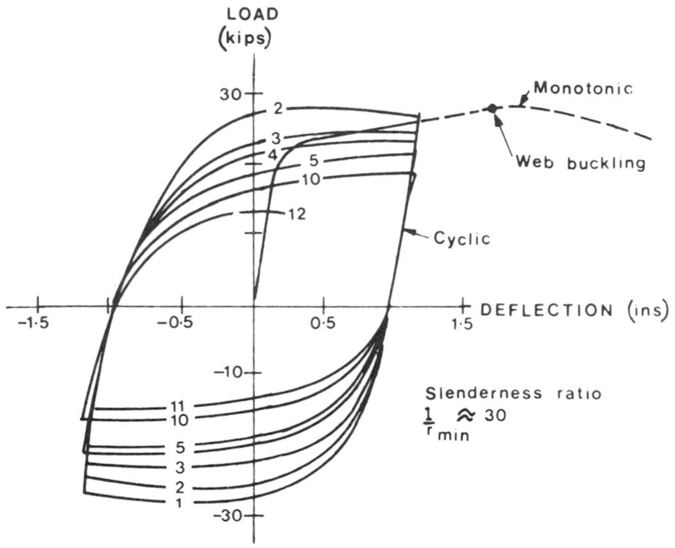

Figure 9.4 Hysteresis loops for a steel beam tested under moment gradient (after Vann *et al.*[20])

The rotation capacity of plastic hinges may be subject to reduction under cyclic loading, as discussed below.

9.4.2 Behaviour of steel beams under cyclic loading

In steel frames designed to make good use of inelastic resistance in earthquakes several reversals of strain of 1.5 percent or more may have to be withstood. As discussed in Section 9.2, stable repetition of the monotonic ductile capacity of beams, as measured by θ_h or R in equations (9.4) and (9.6), may not be possible under cyclic loading to higher strains. The hysteretic degradation of strength observed by Vann *et al.*[20] was mainly due to web buckling, but flange buckling and lateral torsional buckling, plus low cyclic fatigue can also have similar effects.

The rate at which strength degradation occurs is, of course, significant to design. It is of interest that Vann *et al.*[20] found that the strength of an American W8 × 13 I-section (Figure 9.4) had degraded to 72 percent of its plastic moment after 11 load cycles. This behaviour would be acceptable for full ductile design in New Zealand where 20 percent degradation after four load cycles is permitted (Section 6.6.2).

9.4.3 Design of steel beams

As may be concluded from the above discussion, the key factor in maintaining beam strength under seismic loading is the provision of stiffness or restraints

Table 9.2. Maximum width to thickness ratios for steel members (from Walpole and Butcher[21])

		Category 1 Parts of members requiring full ductility	Category 2 Parts of members requiring limited ductility	Category 3 Parts of members requiring elastic behaviour
Flanges and plates in compression with one unstiffened edge (e.g. I or [flanges)	$\frac{b_1\sqrt{(F_Y)}}{t_f}$	120	136	256
Flanges of welded box sections in compression	$\frac{b_2\sqrt{(F_Y)}}{t_f}$	500	512	560
Flanges of rectangular hollow sections	$\frac{b_2\sqrt{(F_Y)}}{t_f}$	350	420	635
Webs under flexural compression	$\frac{b_1\sqrt{(F_Y)}}{t_w}$	1000	1120	1340
Webs under uniform compression	$\frac{b_1\sqrt{(F_Y)}}{t_w}$	500	512	560

Table 9.3. Spacing of lateral restraints for steelwork (adapted from Walpole and Butcher[21])

	Parts of members requiring full ductility $R=24$, $\alpha=0.75$		Parts of members requiring limited ductility $R=10$, $\alpha=1.0$	
Flange length L_y where the compression flange is fully yielded	$L_y \geqslant 480a$	$L_y < 480a$	$L_y \geqslant 640a$	$L_y < 640a$
Spacing of braces within length L_y	$\leqslant 480a$	one brace required	$\leqslant 640a$	one brace required
Spacing to brace adjacent to length L_y	$\leqslant 720a$	$\leqslant 720a$	$\leqslant 960a$	$\leqslant 960a$

Notes: (1) Parts of members responding elastically should be braced according to allowable stress rules.
(2) Symbols are defined in Section 9.5.2.3.

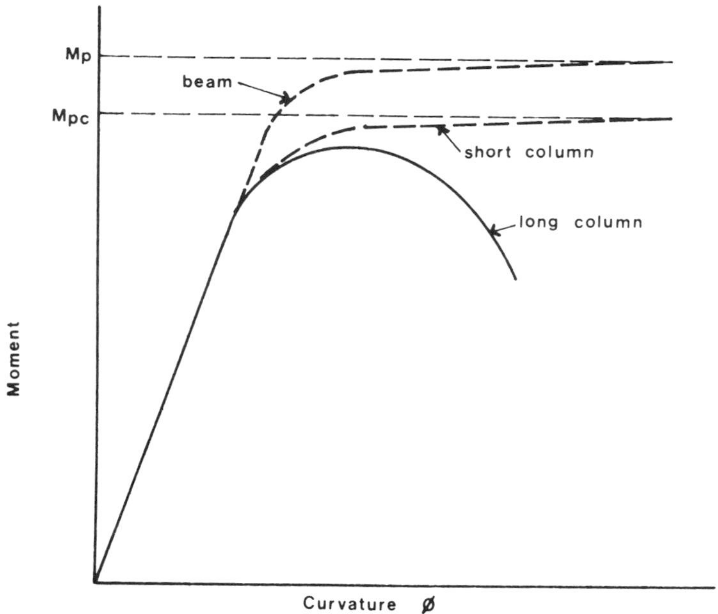

Figure 9.5 Typical moment–curvature relationships for short and long columns compared with pure beam behaviour

to control local and lateral buckling. New Zealand proposals[21] set out in Tables 9.2 and 9.3 show the increasing stability needs with increasing ductility demands.

For calculating the local buckling stiffness requirements, in Table 9.2 the symbols are defined as follows:

b_1 = the outstand of a flange or stiffener beyond the web;
b_2 = the width of a flange between two adjacent lines of connections to other parts providing support;
d_1 = the web distance clear between flanges, or between the tension flange and a horizontal stiffener where provided;
t_f = the mean thickness of the flange;
t_w = the thickness of the web.

The background to the required spacing of lateral restraints (given in Table 9.3) is as described for columns in Section 9.5.2.3.

9.5 STEEL COLUMNS

9.5.1 Monotonic and hysteretic behaviour of steel columns

Columns are often required to resist appreciable bending moments as well as axial forces. The moment-curvature relationships for the so-called 'beam–column'

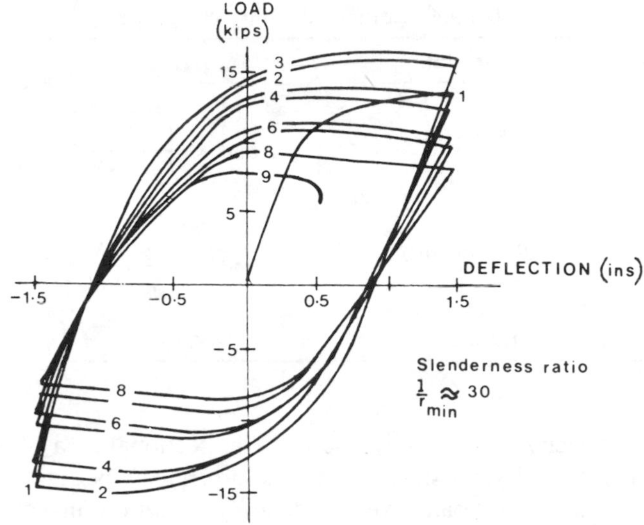

Figure 9.6 Hysteresis loops for a steel member under cyclic bending and with a constant axial force of $N = 0.3\ N_y$ (after Van *et al.*[20])

are similar to those for beams under uniform moment, except that the capacity is reduced below the beam plastic moment M_p by the presence of axial load, as shown in Figure 9.5 and Section 9.5.2.5.

As indicated in Figure 9.5, the full plastic moment M_{pc} of columns may not be developed because of local buckling or lateral torsional buckling as for beams.

Although columns should generally be protected against inelastic cyclic deformations by prior hinging of the beams, some column hysteretic behaviour is likely in strong earthquakes in most structures, and even with beam hinging mechanisms column hinges (or pins) are required at the lowest point in columns, as shown in Figure 5.5(b).

The behaviour of steel columns under cyclic bending is similar to that of beams without axial load, except that the axial force added to the bending moment concentrates the yielding in the regions of larger compressive stress. This leads to a more rapid decay of load capacity owing to more extensive buckling, as may apparently be inferred by comparing Figure 9.6 with Figure 9.4. Second-order bending ($P \times \Delta$ effect) may also be important in the inelastic range.

Design recommendations which allow for the effects of axial load and restraint are discussed below.

9.5.2 Design of steel columns

9.5.2.1 Column ductility

The degree of ductility which a column should be designed to supply is a function of the level of axial compression N expressed as a fraction of its yield

Table 9.4. Column ductility categories

Column state	N/N_y
Fully ductile	<0.5 and $<\dfrac{1+\beta-\lambda}{1+\beta+\lambda}$
Limited ductility	<0.7 and $<\dfrac{1+\beta-\lambda}{1+\beta+\lambda}$
Elastic	<1.0

compression capacity $N_y = A_s f_y$, where A_s is the sectional area of the member and f_y is the specified yield stress. The recommended[22] permissible range of N/N_y for three different design levels of ductility are set out in Table 9.4, from which it can be seen that fully ductile column hinging (for, say, $\mu = 4$ to 6, see Section 9.3.1) is appropriate only at relatively low values of N/N_y. In Table 9.4,

β = the ratio of end moments (of the same sense) with the numerically larger moment in the denominator (i.e. β ranges from $+1$ for double curvature, through 0 for one end pinned, to -1 for single curvature);

$$\lambda = \frac{l}{\pi r}\sqrt{\left(\frac{f_y}{E}\right)} \qquad (9.6)$$

where r = the radius of gyration about the same axis as the applied moment; and
l = the actual length of the column.

9.5.2.2 *Effective lengths of steel columns*

In earthquake resistant design special considerations regarding effective length of columns arise through the effects of inelasticity and drift limitations. The limits on drift imposed for some types of structure by earthquake codes impose a lateral stiffness constant for protecting non-structural components, which also assists in the control of structural stability. This effect has been acknowledged in California[23] and New Zealand,[22] with proposals that the effective length factor may be taken as $K = 10$ in the plane of seismic bending forces in frames designed to their drift limit (Kl = the effective length).

Alternatively, if the effective lengths of columns are to be calculated it should be noted that effective length factors for elastic and inelastic columns will be the same only when the column acts as an independent member. In other cases, i.e. when an inelastic column is part of a continuous frame, its effective length should be calculated appropriately,[22] suitable simplified methods having been described by Yura,[24] Le Messurier,[25] and Wood.[26]

9.5.2.3 Lateral buckling of steel columns

Columns designed to respond elastically in earthquakes may be designed to the normal non-seismic rules for lateral restraint against buckling, but extra precautions are required for the development of limited or full ductility as set out in Table 9.3.

As shown in the table, appropriate levels of rotation capacity R, as defined by equation (9.6), are $R=10$ for structures of limited ductility ($R=10$ is normally used for non-seismic plastic design), and $R=24$ for fully ductile structures. The spacings in Table 9.3 are based upon the length $640\alpha a$, where

$$\alpha = \frac{1.5}{\sqrt{(1+R/8)}} \tag{9.8}$$

$$a = r_y / \sqrt{(f_y)} \tag{9.9}$$

and L_y is the length of column (or beam) over which the compression flange is fully yielded, which may be taken[22] as occurring where

$$M > 0.85\ M_{pc} \qquad \text{if } N/N_y < 0.15 \tag{9.10}$$

$$\text{or } M < 0.75\ M_{pc} \qquad \text{if } N/N_y \geqslant 0.15 \tag{9.11}$$

When using Table 9.3 for beams, M_p should, of course, be used instead of M_{pc} in equation (9.9).

9.5.2.4 Local buckling of steel columns

The section geometry limits for controlling local buckling of steel columns are the same as for beams as given in Table 9.2, the symbols being defined in Section 9.4.3.

9.5.2.5 Forces in struts

The maximum compressive load capacity of struts not subjected to bending may be taken as

$$N_{ac} = \frac{A_s F_{ac}}{0.6} \tag{9.12}$$

where F_{ac} is the maximum compressive stress as a function of the slenderness ratio, calculated on a permissible stress basis, and A_s is the sectional area of the member.

9.5.2.6 Combined axial load and moment

At a support, the capacity of I-section members may be found from the interaction formulae given in the literature.[22,27,28] At the time of writing such formulae were undergoing review,[22,27] those given below being from Butterworth and Spring.[22]

(1) *Bending about the strong axis*

$$\text{For } N/N_y < 0.15, \qquad M = M_p \tag{9.13}$$

$$\text{For } N/N_y \geqslant 0.15, \qquad \frac{N}{N_y} + 0.85\frac{M}{M_p} \leqslant 1.0 \tag{9.14}$$

(2) *Bending about the weak axis*

$$\text{For } N/N_y < 0.4, \qquad M = M_p \tag{9.15}$$

$$\text{For } N/N_y \geqslant 0.4, \qquad \left(\frac{N}{N_y}\right)^2 + 0.85\frac{M}{M_p} \leqslant 1.0 \tag{9.16}$$

Bending about both axes

$$\text{For } N/N_y < 0.15, \qquad \frac{M_x}{M_{px}} + \frac{M_y}{M_{py}} \leqslant 1.0 \tag{9.17}$$

For $N/N \geqslant 0.15$,

$$\frac{N}{N_y} + 0.85\left[\frac{M_x}{M_{px}} + \frac{M_y}{M_{py}}\right] \leqslant 1.0 \tag{9.18}$$

Away from supports other section capacity rules for combining axial load and moment should be used, such as given in references 22, 27, and 28.

9.5.2.7 Shear in columns

In the unusual circumstances where the shear stress is high in a column, i.e. $V/V_p = 2/3$ or more, the interaction formula for moment, axial load and shear given by Neal[29] is appropriate:

$$\frac{M}{M_p} + \left(\frac{N}{N_y}\right)^2 + \frac{(V/V_p)^4}{[1-(N/N_y)^2]} \leqslant 1.0 \tag{9.19}$$

where V_p is as defined in equation (9.21).

9.6 STEEL FRAMES WITH DIAGONAL BRACES

Diagonally braced frames are now commonly discussed under two classifications, depending on whether the braces create perfect triangulation or not, namely:

(1) Concentrically braced frames (CBFs); and
(2) Eccentrically braced frames (EBFs).

Reliable ductility under cyclic loading is much more readily obtained from EBFs than CBFs, as will be evident from the following discussion.

9.6.1 Concentrically braced steel frames

9.6.1.1 Introduction

The general characteristics of CBFs have been described in Section 5.4.4, to which some further points are noted here. As illustrated in Figure 9.7, the braces

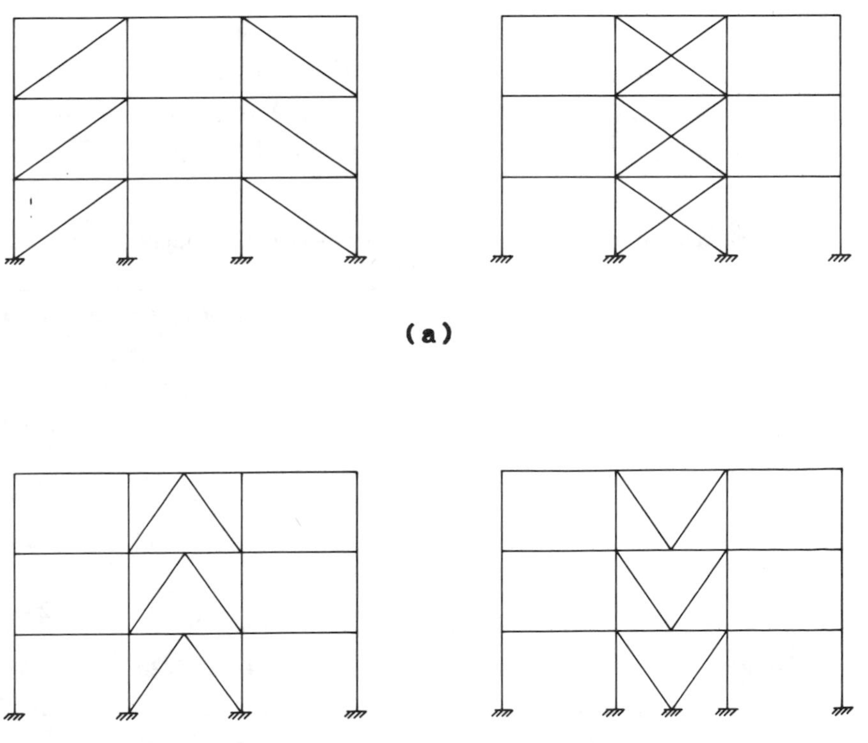

Figure 9.7 Typical concentrically braced frames with recommended symmetry of bracing for (a) Z- and X-braces and (b) V- or K-braces

may be arranged in either of two ways; (1) the braces may lie along thc lines between beam–column joints (Z- or X-braces), or (2) pairs of braces may meet at a point along a beam (V-braces, also called K-braces). In case (1) the diagonals of Z-bracing should be arranged in opposing pairs (as in Figure 9.7), as recommended in California[30] and New Zealand,[31] rather than all sloping in the same direction, so as to avoid the larger residual sway deflections that occur in asymmetrically braced frames. V-bracing is likely to be inferior to X-bracing, because the resistance of V-bracing to storey shears will be governed by the compression resistance of the braces.

The effects of instability in reducing the strength of braces under cyclic loading include those described for columns (Section 9.5.1), but are more serious because of additional effects of tensile yielding and because braced are generally more slender than columns. Various workers[32–34] have described the rapid loss in strength, stiffness, and energy absorption capacity that occurs under inelastic cyclic loading of V- and X-braced frames.

Further insight into the behaviour of CBFs may be obtained from studies of offshore platforms, such as by Marshall *et al.*[35]

9.6.1.2 Design of concentrically braced frames

For CBFs having negligible bending strength, values of the deflection ductility factor μ that are considered[31] to be reliably achievable are given in Table 9.5.

Table 9.5. Design values of μ for CBFs having negligible bending strength (after Walpole[31])

		Bracing slenderness $\frac{Kl}{r}\sqrt{\left(\frac{f_y}{250}\right)}$		
	No. of	<40	41–80	81–135
Bracing type	storeys	μ	μ	μ
Pairs of braces between beam–column joints	1	5.0	3.9	3.0
	2	4.5	3.5	2.4
	3	4.0	3.0	1.8
V-bracing	1	3.7	2.4	1.5
	2	3.0	1.8	1.2
	3	2.7	1.5	1.0

The value of μ reduces as the bracing slenderness increases, and μ also reduces with numbers of storeys (i.e. mass levels) such that not more than three mass levels were recommended for this type of construction. The appropriate earthquake loading may be obtained from μ as described in Section 6.6.7.3(i).

CBFs having appropriate moment resisting action will have better seismic reliability for the same number of mass levels than the structures described in Table 9.5, and more than three mass levels would be appropriate, depending on the moment resistance available. Insufficient research data exist to formulate reliable design rules for such structuring.

If the bracing is made from threaded rods it is recommended[31] that the structure be treated as elastically responding, i.e. it should be designed using $\mu = 1$.

Concentrically braced frames are well suited to protection from earthquakes using base isolation techniques (e.g. Figure 5.11) partly because they are relatively low-period structures (Section 5.5.2). Also, a suitable location for energy-dissipating devices is in the diagonals either as discussed in Section 5.5.7.2 or by simply deliberately bending the diagonals so that they must form moment hinges. The use of such devices would greatly increase the number of mass levels which could be built with acceptable reliability compared with the restrictions noted above.

9.6.2 Eccentrically braced steel frames

9.6.2.1 Introduction

The general characteristics of EBFs have been described in Section 5.4.5, to which some further points are noted here. As illustrated in Figure 9.8, EBFs are formed by deliberately creating eccentricities, e, with Z- and K-braces (V-braces) such that moments and shears exist in the short length of the beams known by terms such as the *link beam* or *shear link*. The ductility of this link beam may be utilized to obtain reliable seismic ductile response of the frame as a whole.

Referring to Figure 9.8, the following comparative merits are noted:

Figure 9.8(a) is well suited to short column spacings;
Figure 9.8(b) is the best for column safety, as the link (with its heavy welds) is located away from the columns;
Figure 9.8(c) minimizes the rotation angle θ for a given lateral displacement.

As noted earlier, one of the advantages of braced frames is the reduction in lateral drift compared with moment-resisting frames (MRF). This is illustrated by the graph of lateral stiffness (Figure 9.9) for a rectangular frame plotted as a function of the bracing eccentricity e, which varies from zero (a CBF) to unity (a MRF).

Considerable research has been done in California[36–39] on the response of I-section shear links and eccentric Z-bracing subject to cyclic loading, and the design recommendations given below are based mainly on this work.

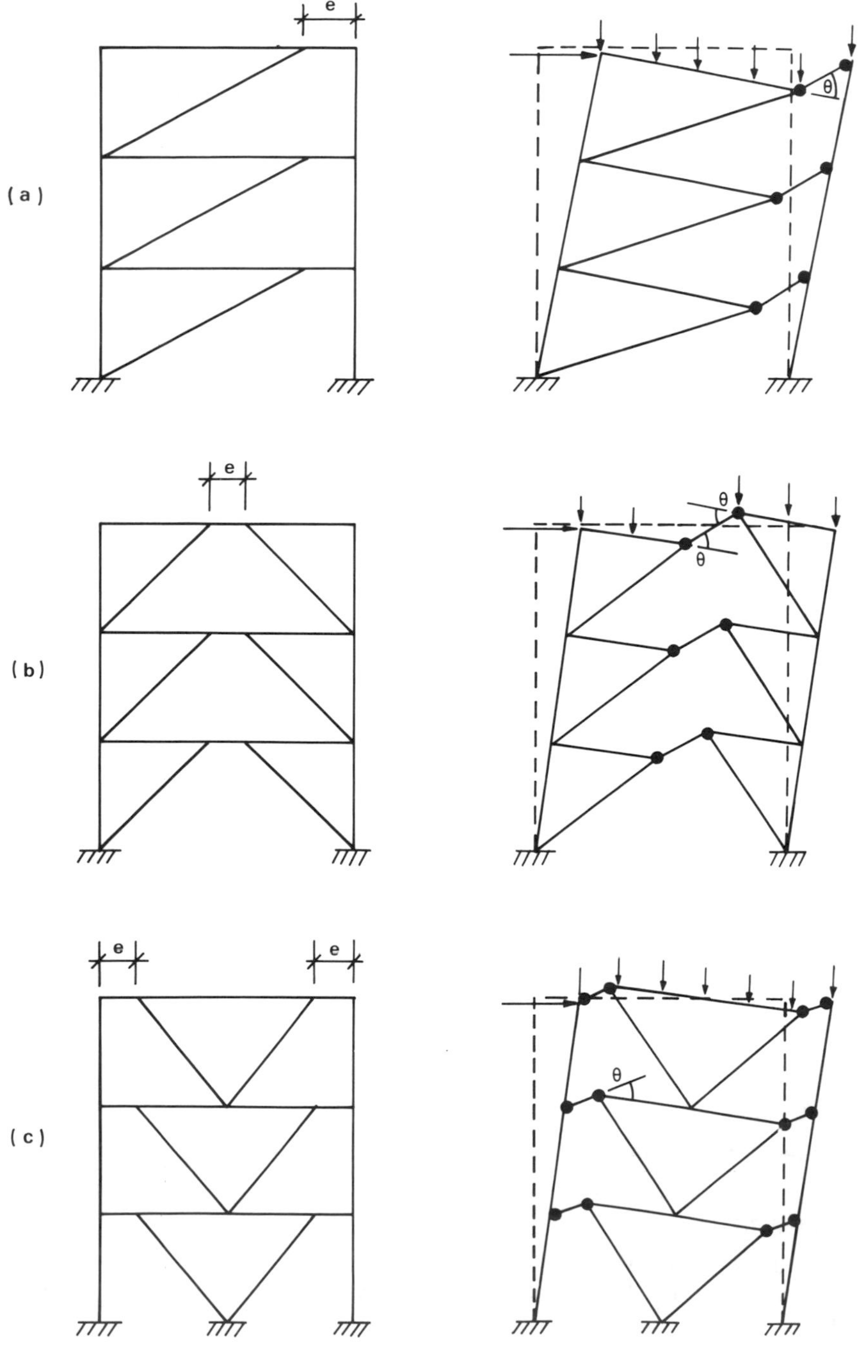

Figure 9.8 The three preferred eccentrically braced frames showing kinematics of deformation

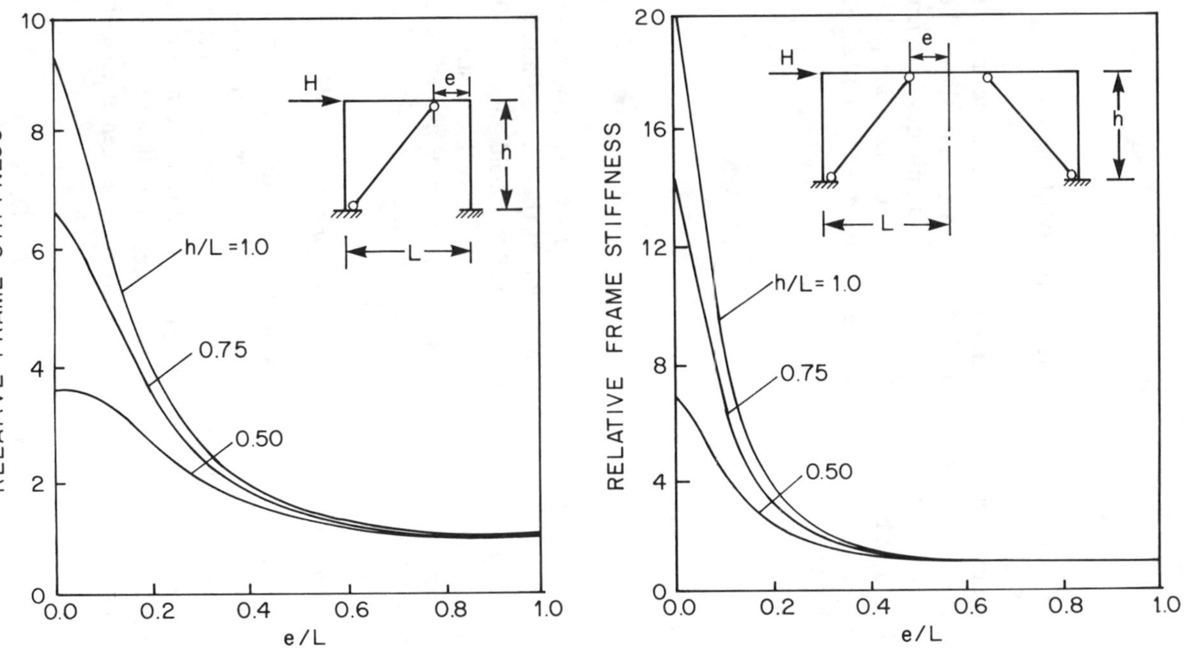

Figure 9.9 Stiffness of braced frames varying with eccentricity (after Hjelmstad and Popov[39])

9.6.2.2 *Design of ductile link beams in EBFs*

Because of the magnitude of shear stresses in the link beam, Sidwell[40] notes that the shear V and axial force N should comply with von Mises criterion that

$$\left[\frac{N}{N_y}\right]^2 + \left[\frac{V}{V_p}\right]^2 \leqslant 1.0 \tag{9.20}$$

where $N_y = A_s f_y$ is the axial load yield capacity of the beam,
and $V_p = 0.55\ d t_w f_y$ (9.21)

Because of the high ductility performance of link beams in inelastic shear it appears advantageous to create short links which yield in shear (shear links), rather than longer ones where moment yielding dominates (moment links).[38] A shear link is created by ensuring that the web of the link beam yields in shear, by taking the length of the link beam, e, not to exceed the maximum length b^* of a shear hinge,

$$b^* = 2M_p^*/V_p^* \tag{9.22}$$

where $M^* = f_y(\mathrm{d} - t_f)(b - t_w)t_f$ (9.23)

$$V_p^* = \frac{f_y}{\sqrt{3}}(d - t_f)t_w \tag{9.24}$$

A link longer than b^* forms a *moment link.*

Because there is a limit to the rotation capacity of plastic hinges (e.g. see equation (9.4)) it has been recommended[40] that the rotation angle θ between the link and the remainder of the beam (Figure 9.8) should not exceed 0.06 radian for link beams of clear length e up to $1.6M_p/V_p$. Also web stiffeners should be provided along the link beam at spacings not exceeding $24t_w$. For $1.6M_p/V_p < e \leqslant 2.4M_p/V_p$, θ should not exceed 0.03 radian.

9.6.2.3 *Design of other components of EBFs*

For failure mode control it is necessary to ensure that the remainder of the frame is sufficiently stronger than the link beams. This requires consideration of the potential overstrength of the link beam based on an upper limit on its yield strength rather than the characteristic strength, and the cracking strength of the floor at the link zone should be minimized and allowed for in the frame design. Procedures for designing these members have been given elsewhere.[40,41]

9.7 STEEL CONNECTIONS

9.7.1 Introduction

Connections as well as members should be designed to conform to the failure mode controls for the structure concerned. Thus, unless a connection is required to yield prior to the adjacent members as part of an energy-absorbing scheme, as is sometimes done with holding-down bolts, it is usual to design each connection to carry greater loads than the members entering it. In addition, the panel zones of beam–column joints should have stiffness appropriate to the assumptions made in the analysis of frame response.

Connections should also be designed to make fabrication and erection of the framework as simple and quick as possible. They should not be too sensitive to factory or field tolerances, and should minimize the use of highly skilled crafts. Connections should also permit adequate inspections to be made at the time of construction as proper quality control of fabrication processes, particularly welding, is, of course, essential. Important aspects of workmanship are discussed elsewhere.[3]

Butt welding, fillet welding, bolting, and rivetting may be employed for aseismic connections, either individually or in combination. As fully bolted or rivetted connections tend to be very large and expensive, fully welded connections or a combination of welding and bolting are most frequently used. Bolts have the advantage of contributing more damping to frames than welding (Section 9.2).

9.7.2 Behaviour of steel connections under cyclic loading

Compared with beam and column elements, relatively few cyclic load tests have been carried out on steel connections, and conclusive design criteria are not yet available for seismic conditions. Popov and Pinkney[42] tested five types of joint, two involving minor-axis bending of the column and three involving major-axis bending of the column. The latter three joints (Figure 9.10) were of the following types;

(1) A butt-welded joint;
(2) A fillet-welded joint using flange plates;
(3) A joint using high-strength bolts and flange plates.

In the tests it was found that the butt-welded joints were superior to the other two types in terms of total energy absorption. In the bolted joints the hysteresis loops were reduced in area considerably by slippage, although the use of smaller than normal oversize holes reduced this effect. All the joints sustained loads in excess of their design limit values until the onset of cracking.

In tests on connections using fully welded and flange welded-web bolted joints by Popov and Stephen,[43] very large increases in bending strength (up to 69 percent) due to strain-hardening were observed.

The comparative cyclic load behaviour of bolted connections in the snug tightened and fully tightened conditions has not been well established. In tests by Popov and Pinkney,[42] although the degree of tightness is not clear, some pinching of the hysteresis loops indicates the effect of slip on the faying surfaces. The extent of pinching was reduced for holes drilled only 0.4 mm oversize instead of the 1.6 mm oversize, which is standard in the USA.

292

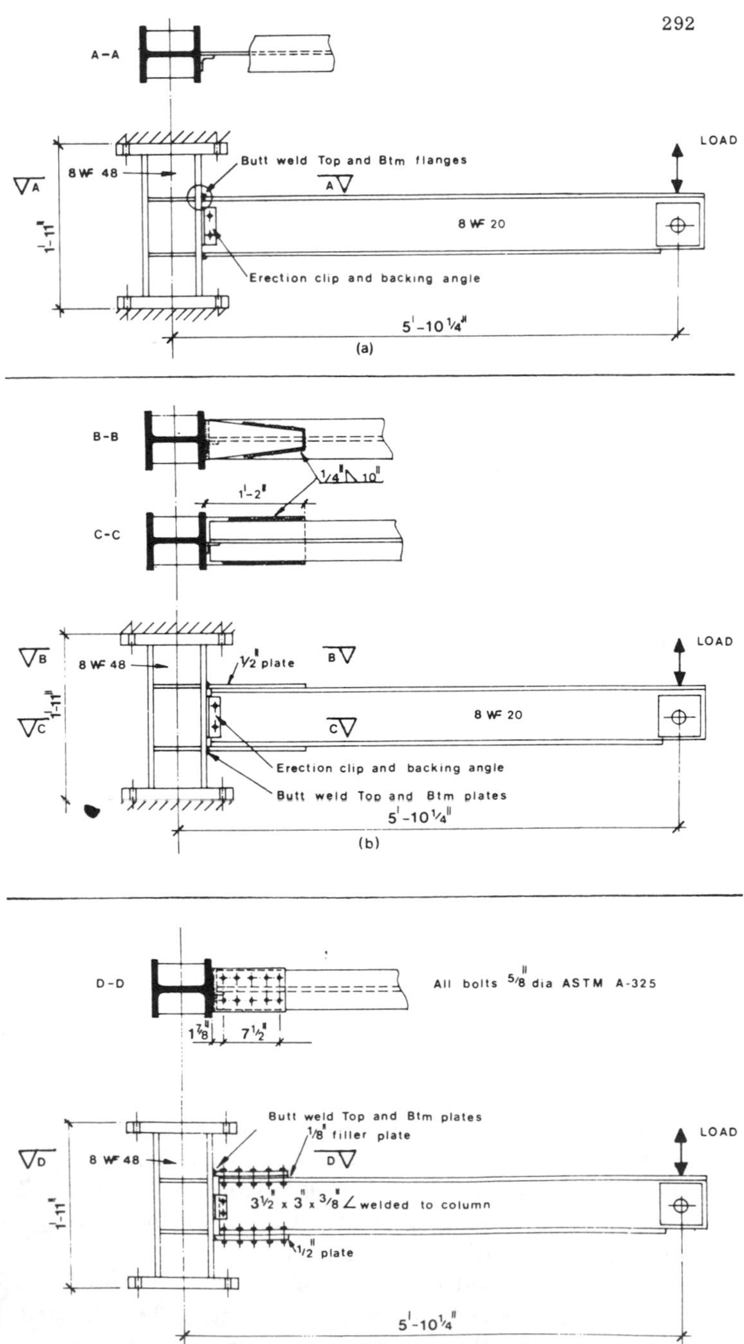

Figure 9.10 Beam–column connections with major axis column bending tested by Popov and Pinkney.[42] (a) Butt-welded beam–column joint; (b) fillet welded beam–column joint; (c) bolted beam–column joint

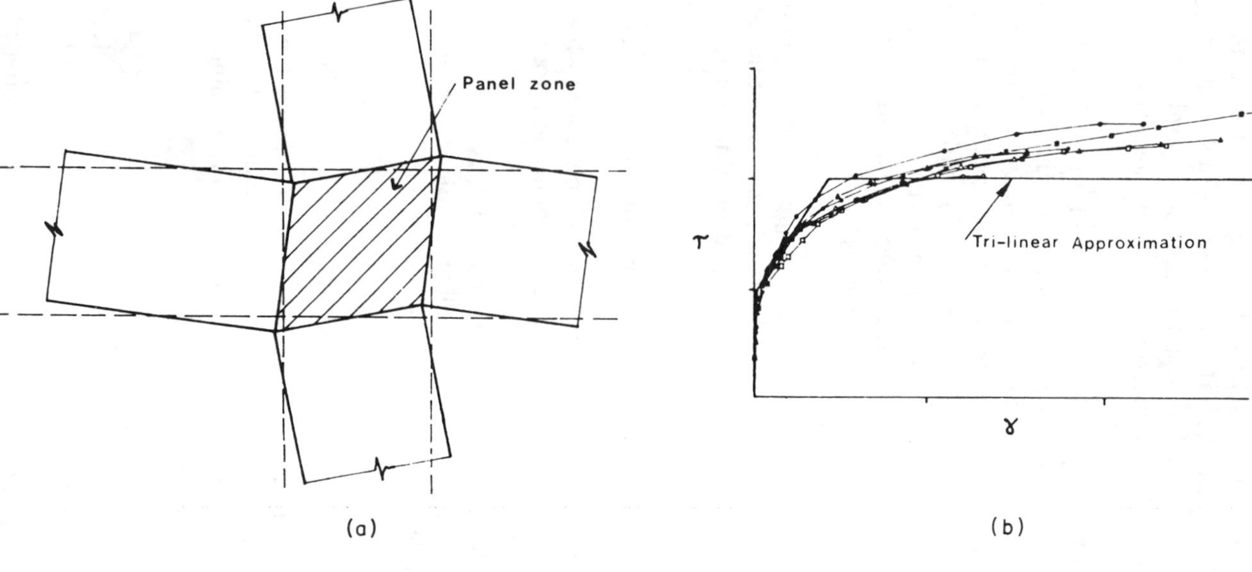

Figure 9.11 Idealized shear deformation of beam–column panel zones

9.7.3 Deformation behaviour of steel panel zones

The panel zone of a connection between two members is the intersection zone common to the two members. This zone is assumed to deform in shear as indicated in Figure 9.11(a). Kata and Nakao[44] have suggested a tri-linear relationship between the shear stress and shear strain as a good approximation to the results of their monotonic tests on Japanese **H**-Section connections (Figure 9.11b).

Although little testing has been done on the deformation characteristics of panel zones, specially under cyclic loading, it has been demonstrated[45] that the deformation of beam–column connections may contribute up to about one third of the interstorey deflection in multi-storey buildings, and of this deformation about half may arise from the shear deformation of the panel zone itself.[46]

The large influence of panel zone behaviour on overall frame strength and stiffness has also been indicated by Kato and Nakao.[44,47] If bilinear hinges were assumed in the panel zones, the ultimate shear resistance of the frame was formed only in the frame members.

However, in fully ductile frames it appears to be economic to have some yielding in the panel zones ($\mu \approx 3$) as well as in the beam plastic hinges, as has been proposed in the draft revisions to the Californian earthquake code.

9.7.4 Design of steel connections for seismic loading

The previous sections give the background to the following introduction to the exacting task of providing well-detailed connections for aseismic steelwork. In addition, it is noted that all the components of connections should be arranged to give a smooth stress flow between members, so that stress raising notches and sharp re-entrant angles should be avoided. Fuller recommendations on connection design have been given elsewhere.[31,48,49]

9.7.4.1 Design forces for connections

As discussed earlier, connections are usually designed to be stronger than the adjacent members, the strength of which should be based on some probability that the actual strength will exceed the guaranteed minimum strength. Typical increases are indicated below:

Guaranteed minimum f_y	*Average f_y*
250 MPa	$1.15 f_y$ min.
350 MPa	$1.10 f_y$ min.

Allowance should also be made for the increase in strength beyond yield point due to strain hardening. Combining both those effects, it is recommended[48] that the design forces for connections in *fully ductile* structures should be derived using $1.5 f_y$, and in structure of *limited ductility* $1.35 f_y$ should be used. Such design forces need not exceed those applicable if the structure were designed to be elastically responding.

There is also concern[48] that connections may be subjected to larger forces than given by the analysis due to unpredictable movements of the structure, and it was therefore recommended that connections be able to withstand the following minimum forces:

50 percent of the member strength in compression or tension ($0.5A_s f_y$). This requirement is severe and need only be applied when the design axial forces are significant;
30 percent of the member strength in flexure ($0.3Zf_y$);
10 percent of the member strength in shear ($0.15A_v f_y$).

9.7.4.2 *Welding*

Full penetration *butt welds* are the best means of load transfer, while partial penetration butt welds should not be used in areas of stress reversal. *Fillet welds* are also acceptable for load transfer provided that a variety of design controls are practised. For example, the throat thickness should not be less than half the thickness of the plate being welded. Nicholas[48] and McKay[3] describe aseismic design rules for welds which should be used in conjunction with normal practice for welding.

9.7.4.3 *Bolting*

The design and performance of bolted connections is affected by the following factors:

(1) The size of the hole and method of protection, i.e. the hole should be snug on the bolt and/or the bearing stress should not exceed f_y under ultimate load conditions.
(2) The conditions of the faying (mating) surfaces affect the frictional load transfer, and the effect of different paint systems should be considered.
(3) The threaded portion of the bolt may lie in the shear zone. On smaller contracts it may be impractical to avoid this.
(4) The bolt-tightening procedure affects the design method because fully tightened bolts are more reliable under seismic loading than snug tightened bolts.

More detail on bolting is given by Nicholas[48] and Walpole.[49]

9.7.4.4 *Cleats*

Cleats should be treated as sub-members in their own right and should be designed for effects such as eccentricity of the applied forces, buckling, bearing, punching, and splitting.

9.7.4.5 Beam–column joints

Beam–column joints are obviously one of the most common types of connection in steelwork, and the principles given above apply to them. Allowance needs to be made for reduction in section due to bolt holes, and for the stiffness that may be required for local stress effects in webs and flanges both within the panel zone and adjacent to it. Within the panel zone (Figure 9.11) the shear strength may be found using the von Mises criterion for yielding in the form given in equation (9.20), where N_y is the compressive yield capacity of the column, and

$$V = \frac{M_{b1} + M_{b2}}{d_b} - V_{c1} + V_{c2} \tag{9.25}$$

where d_b is the depth of the beam.

It will often be necessary to use doubler plates to increase the web area to comply with equation (9.20). Alternatively, diagonal stiffeners may be used (Figure 9.11), in conjunction with horizontal stiffeners, to reduce the shear force V acting on the web by

$$V = f_y A_{st} \cos \beta \tag{9.26}$$

where A_{st} and f_y are the area and specified yield stress of the diagonal stiffener, respectively, and β is the angle of the stiffener to the horizontal.

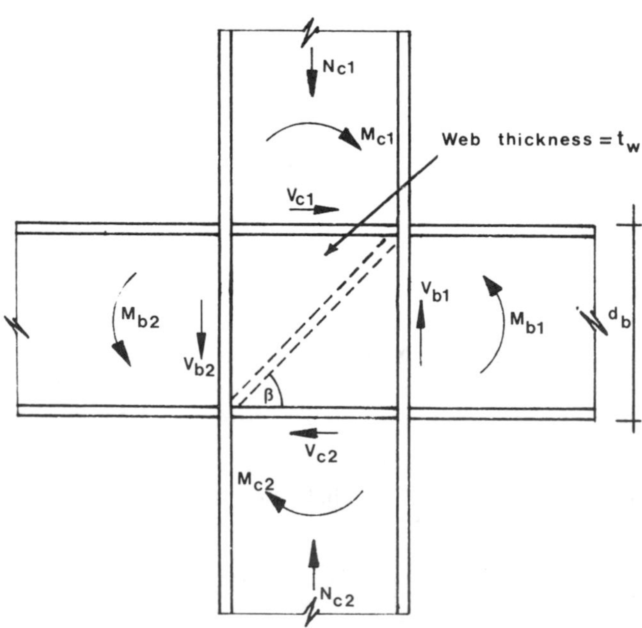

Figure 9.12 Forces acting on a typical panel zone

The aseismic design of beam–column joints has been discussed in more detail by Walpole.[49]

9.7.4.6 Connections in diagonally braced frames

In addition to the above principles, special considerations relating to failure mode control arise in the design of connections in diagonally braced frames, as discussed elsewhere.[31,40,41]

9.8 COMPOSITE CONSTRUCTION

In many steel structures, particularly multi-storey buildings, the steel acts compositely with concrete which is used for floors or fire protection of columns. Obviously the concrete may add strength and stiffness to the steel frame, but for failure mode control it may also imply additional forces to be dealt with in any given member, increasing the overstrength demands from adjacent members.

The principles outlined in the previous sections of this chapter apply to composite construction, but special aseismic requirements exist, as discussed by Clifton.[50]

9.9 STEEL STRUCTURAL WALLS

Recent studies[51,52] suggest that structural walls sheathed with steel plate may be an economical and effective means of resisting earthquakes in both new and strengthened structures. Webb stiffeners to control plate buckling appear not to be necessary.[51]

REFERENCES

1. Seismic design of Steel Structures Study Group, Sections A-K, *Bull. NZ Nat. Soc. for Earthq. Eng.*, **18**, No. 4, 323–405 (1985).
2. Bertero, V. V., and Popov, E. P., 'Effect of large alternating strains on steel beams', *J. Structural Division, ASCE*, **91**, No. ST1, 1–12 (1985).
3. McKay, G. R., 'Materials and workmanship', Section K of Ref. 1, 400–5.
4. Kaneta, K., Kohzu, I., and Nishizawa, H., 'Cumulative damage of welded beam-to-column connections in steel structures subjected to destructive earthquakes', *Proc. 8th World Conf. on Earthq. Eng., San Francisco*, **VI**, 185–92 (1984).
5. Krawinkler, H., 'Damage assessment in steel structures subjected to severe earthquakes', *Proc. 8th World Conf. on Earthq. Eng., San Francisco*, **IV**, 887–94 (1984).
6. Patton, R. N., 'Analysis and design methods', Section B of Reference 1, 329–36.
7. Symonds, P. S., 'Viscoplastic behaviour in response of structures to dynamic loading', in *Behaviour of Materials under Dynamic Loading* (Ed. N. J. Huffington), ASME, New York (1965).

8. Udagawa, K., Takanashi, K., and Kato, B., 'Effects of displacement rates on the behaviour of steel beams and composite beams', *Proc. 8th World Conf. on Earthq. Eng., San Francisco*, **VI**, 177–84 (1984).
9. Wakabayashi, M., Nakamura, T., Iwai, S., and Hayashi, Y., 'Effects of strain rate on the behaviour of structural members subjected to earthquake force', *Proc. 8th World Conf. on Earthq. Eng., San Francisco*, **IV**, 491–8 (1984).
10. Clifton, G. C., 'Cold formed sections', Section J of Reference 1, 397–9.
11. International Conference of Building Officials, *Uniform Building Code*, ICBO, Whittier, California (1985), Chapter 27.
12. British Standards Institution, 'Methods for ultrasonic testing and specifying quality of ferritic steel plate', BS 5996: 1980.
13. Farrar, J. C. M., and Dolby, R. E., *Lamellar Tearing in Welded Steel Fabrication*, The Welding Institute, Cambridge (1972).
14. Jubb, J. E. M., 'Lamellar tearing', Welding Research Council, USA, *Bulletin 168* (1968).
15. Coe, F. R., *Welding Steels without Hydrogen Cracking*, The Welding Institute, Cambridge (1973).
16. Erasmus, L., 'The mechanical properties of structural steel sections and the relevance of these properties to the capacity design of structures', *Trans. Institution of Prof. Engineers, New Zealand*, **11**, 3/CE, 105–11 (1984).
17. Welding Research Council and the American Society of Civil Engineers, 'Plastic design in steel—a guide and commentary', *ASCE Manual No. 41* (2nd end), ASCE, New York (1971).
18. Lay, M. G.,, and Galambos, T. V., 'Inelastic beams under moment gradient', *J. Structural Division, ASCE*, **93**, No. ST1, 389–99 (1967).
19. Takanashi, K., Udagawa, K., and Tanaka, H., 'Failure of steel beams due to lateral buckling under repeated loads', *Symposium on Resistance and Ultimate Deformability of Structures Acted on by Well Defined Loads*, IABSE, Lisbon, Preliminary Report, 163–9 (1973).
20. Vann, W. P., Thompson, L. E., Walley, L. E., and Ozier, L. D., 'Cyclic Behaviour of rolled steel members', *Proc. 5th World Conf. on Earthq. Eng., Rome*, **1**, 1187–93 (1973).
21. Walpole, W. R., and Butcher, G. W., 'Beam design', Section C of Reference 1, 337–43.
22. Butterworth, J. W., and Spring, K. C. F., 'Column design', Section D of Reference 1, 344–50.
23. Structural Engineers Association of California, *Recommended lateral force requirements and commentary: Draft Revision* (1984).
24. Yura, J. A., 'The effective length of columns in unbraced frames', *ASIC Eng J.*, **8**, No. 2, 37–42 (1971).
25. Le Messurier, W. J., 'A practical method of second order analysis', *AISC Eng. J.*, **14**, No. 2, 49–67 (1977).
26. Wood, R. H., 'Effective lengths of columns in multi-storey buildings', (in three parts), *The Structural Engineer*, **52**, Nos 7–9 (1974).
27. Chen, W. F., and Lui, E. M., 'Stability design criteria for steel members and frames in the United States', *J. Construct. Steel Research*, **5**, 31–74 (1985).
28. Chen, W. F., and Atsuta, T., *Theory of Beam-columns* (2 vols), McGraw-Hill, New York (1977).
29. Neal, B. G., 'The effect of shear and normal forces on the fully plastic moment of an I beam', *J. Mech. Eng. Sci.*, **3**, (1961).
30. Nordensen, G. J. P., 'Notes on the seismic design of steel concentrically braced frames', *Proc. 8th World Conf. on Earthq. Eng., San Francisco*, **V**, 395–402 (1984).
31. Walpole, W. R., 'Concentrically braced frames', Section E of Reference 1, 351–5.

32. Wakabayashi, M., Matsui, C., Minanui, K., and Mitani, I., 'Inelastic behaviour of steel frames subjected to constant vertical and alternating horizontal loads', *Proc. 5th World Conf. on Earthq. Eng., Rome*, **1**, 1194–7 (1973).
33. Popov, E. P., and Black, G. R., 'Steel struts under severe cyclic loading', *Proc. ASCE*, **107**, No. ST9, 1857–81 (1981).
34. Asteneh-Asl, A., Goel, S. C., and Hanson, R. D., 'Cyclic behaviour of double angle bracing members with bolted connections', *Proc. 8th World Conf. on Earthq. Eng., San Francisco*, **VI**, 249–56 (1984).
35. Marshall, P. W., Gates, W. E., and Anagnostopoulos, S., 'Inelastic dynamic analysis of tubular offshore structures', *Proc. Offshore Technology Conference, Houston, Texas, Paper OTC 2908*, 235–46 (1977).
36. Roeder, C. W., and Popov, E. P., 'Cyclic shear yielding wide-flange beams', *J. Structural Division, ASCE*, **104**, No. EM4, 763–80 (1978).
37. Roeder, C. W., and Popov, E. P., 'Eccentrically braced steel frames for earthquakes', *J. Structural Division, ASCE*, **104**, No. ST3, 391–412 (1978).
38. Kasai, K., and Popov, E. P., 'On seismic design of eccentrically braced steel frames', *Proc. 8th World Conf. on Earthq. Eng., San Francisco*, **V**, 387–94 (1984).
39. Hjelmstad, K. D., and Popov, E. P., 'Seismic behaviour of active beam links in eccentrically braced frames', *Report No. UBC/EERC-83/15*, Earthquake Engineering Research Center, University of California, Berkeley (1983).
40. Sidwell, G. K., 'Eccentrically braced frames', Section F of Reference 1, 355–9.
41. Structural Engineers Association of California, *Draft provisions for eccentrically braced frames* (1985).
42. Popov, E. P., and Pinkney, R. B., 'Cyclic yield reversal in steel building connections', *J. Structural Division, ASCE*, **95**, No. ST3, 327–53 (1969).
43. Popov, E. P., and Stephen, R. M., 'Cyclic loading of full size steel connections', American Iron and Steel Institute, *Steel Research for Construction Bull.*, No. 21 (1972).
44. Kato, B., and Nakao, M., 'The influence of the elastic plastic deformation of beam-to-column connections on the stiffness, ductility and strength of open frames', *Proc. 5th World Conf. on Earthq. Eng., Rome*, **1**, 825–8 (1973).
45. Teal, E. J., 'Structural steel seismic frames—drift ductility requirements', *Proc. 37th Annual Convention, Structural Engineers Association of California* (1968).
46. Surtees, J. O., and Mann, A. P., 'End plate connections in plastically designed structures', *Conf. on Joints in Structures*, Institution of Structural Engineers and the University of Sheffield (1970).
47. Kato, B., 'A design criteria of beam-to-column joint panels', *Bull. NZ Nat. Soc. for Earthq. Eng.*, **7**, No. 1, 14–26 (1974).
48. Nicholas, C. J. A., 'Connection design', Section G of Reference 1, 360–8.
49. Walpole, W. R., 'Beam-column joints', Section H of Reference 1, 369–80.
50. Clifton, G. C., 'Composite design', Section I of Reference 1, 381–96.
51. Canon, T. J., and Dean, R. G., 'Structural steel plated shear walls', *Proc. 8th World Conf. on Earthq. Eng., San Francisco*, **V**, 1237–44 (1984).
52. Kulak, G. L., 'Behaviour of unstiffened steel plate shear walls', *Proc. Pacific Structural Steel Conference, Auckland*, **1**, 43–57 (1986).

Chapter 10

Timber Structures

10.1 INTRODUCTION

Timber structures have a well-deserved reputation for high resistance to earthquakes. This is due to a number of factors, particularly the high strength-to-weight ratio of timber and also its enhanced strength under short-term loading and the ductility of its steel fastenings such as nails and bolts. However, despite these qualities timber has traditionally been used mainly in domestic construction, and has been the least used of the main structural materials for engineered structures even in those earthquake areas where timber is a plentiful resource. This apparent reluctance to use timber for engineered structures has probably been due mainly to the difficulty of making structural connections and to reservations about fire resistance. However, a change in attitudes was marked by the success of the 1984 Pacific Timber Engineering Conference held in Auckland, New Zealand, and by the surge of research and development in aseismic design of timber which started in New Zealand in the late 1970s.

In earthquakes[1–4] the main causes of inadequate performance of timber construction have been as follows:

(1) Large response on soft ground;
(2) Lack of integrity of substructures;
(3) Asymmetry of the structural form;
(4) Insufficient strength of chimneys;
(5) Inadequate structural connections;
(6) Use of heavy roofs without appropriate strength of supporting frame;
(7) Deterioration of timber strength through decay or pest attack;
(8) Inadequate resistance to post-earthquake fires.

Within certain limitations, means are available for dealing with all these aspects of earthquake resistance of timber construction, as discussed below.

10.2 SEISMIC RESPONSE OF TIMBER STRUCTURES

It has been variously reported[4,5] that timber buildings suffer more earthquake damage when sited on soft ground than when on hard ground. The reasons for this occurrence are uncertain. For instance, the possible role of resonance is obscure. As most traditional one- and two-storey timber buildings have been domestic in nature with fundamental periods of vibration in the range 0.1 to 0.6 s, resonance with the ground would seem likely on very thin rather than thick layers of soft ground. After heavy shaking a timber building will loosen at the joints and its natural periods are likely to lengthen, but the manner of its vibration is uncertain and it may not have well-defined modes in which resonance can occur.

It is possible that timber houses on soft ground are weakened by seasonal ground movements, making them more vulnerable to earthquakes. Also the differential earthquake ground movements in softer soils are larger than in firm ones and this is likely to affect timber buildings with light foundations more than stiffer forms of construction. If timber buildings are to be built on soft ground in an earthquake area, extra measures should be taken to ensure structural integrity, particularly at foundation level.

The response of timber structures to earthquakes depends on the combined response properties of the components, i.e. the timber and the connections. As shown by the typical monotonic stress–strain curves in Figure 10.1, timber is ductile in compression and brittle in tension, so that members failing in axial tension or in bending of the parent timber do so in a brittle manner. Column behaviour will be ductile or brittle, depending on the ratio of compression to

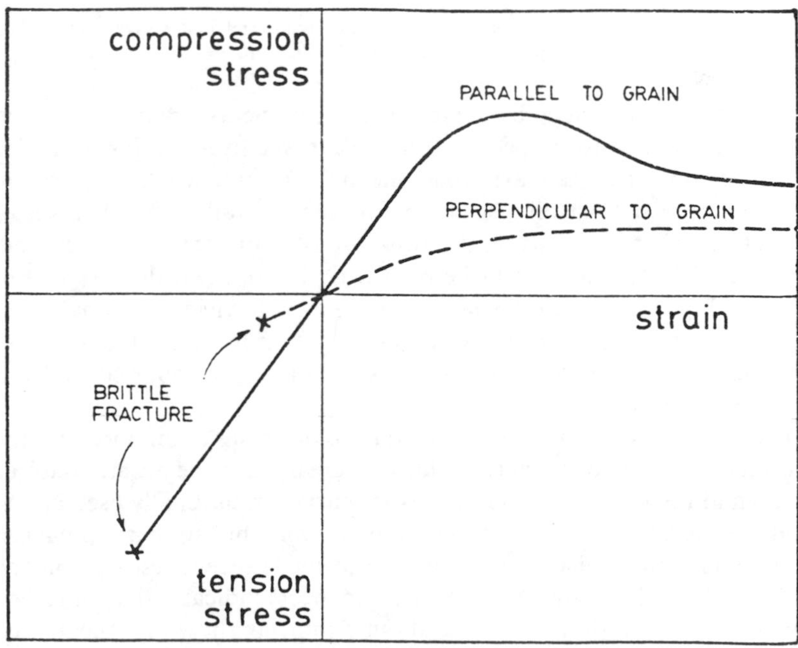

Figure 10.1 Stress–strain relationships for timber (after Buchanan[32])

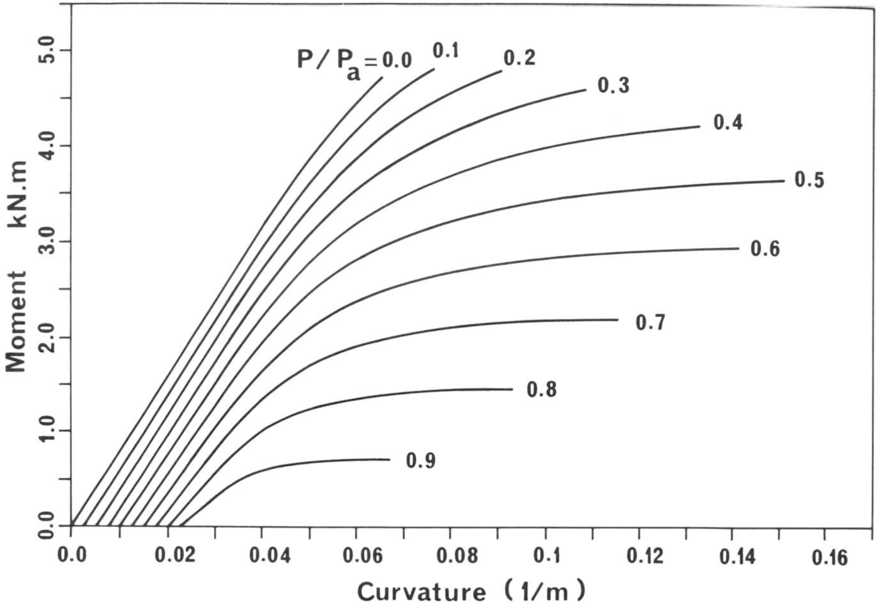

Figure 10.2 Typical moment–curvature relationships for commercial quality sawn timber (Canadian spruce-pine-fir) (after Buchanan[6])

bending stress, as indicated by the monotonic moment-curvature relationships found by Buchanan[6] for a species group of average strength (Figure 10.2). For strong green timber the zero axial load line in Figure 10.2 would be more curved, as more compression yielding occurs before tension failure finally takes place.

The effect of connections on the response of timber structures depends on the nature of the fastening and whether the timber or the fastening governs the behaviour. For example, steel nails may be detailed to yield so that nailed timber structures can have ductile response, as shown by the hysteresis loops for particle board sheathed walls subjected to slow cyclic loading as reported by Thurston and Flack[7] (Figure 10.3).

It is well established that the *duration of loading* has an important effect on the ultimate strength of timber, the latter increasing as the duration decreases. As shown in Figure 10.4 by the 'Madison' curve (traditionally used in US and Canadian codes), it was originally thought that this strength enhancement continued into the dynamic load range, but more recent research on shorter durations of loading (down to 1/20 s) by Spencer[8] indicates that enhancement ceases for load durations less than about 5 min, as shown by the Johns and Madsen[9] curve in Figure 10.4. This would imply that slow cyclic load tests reflect the actual dynamic response behaviour correctly, rather than under-estimating it, as suggested by the Madison curve. However, the strength of timber

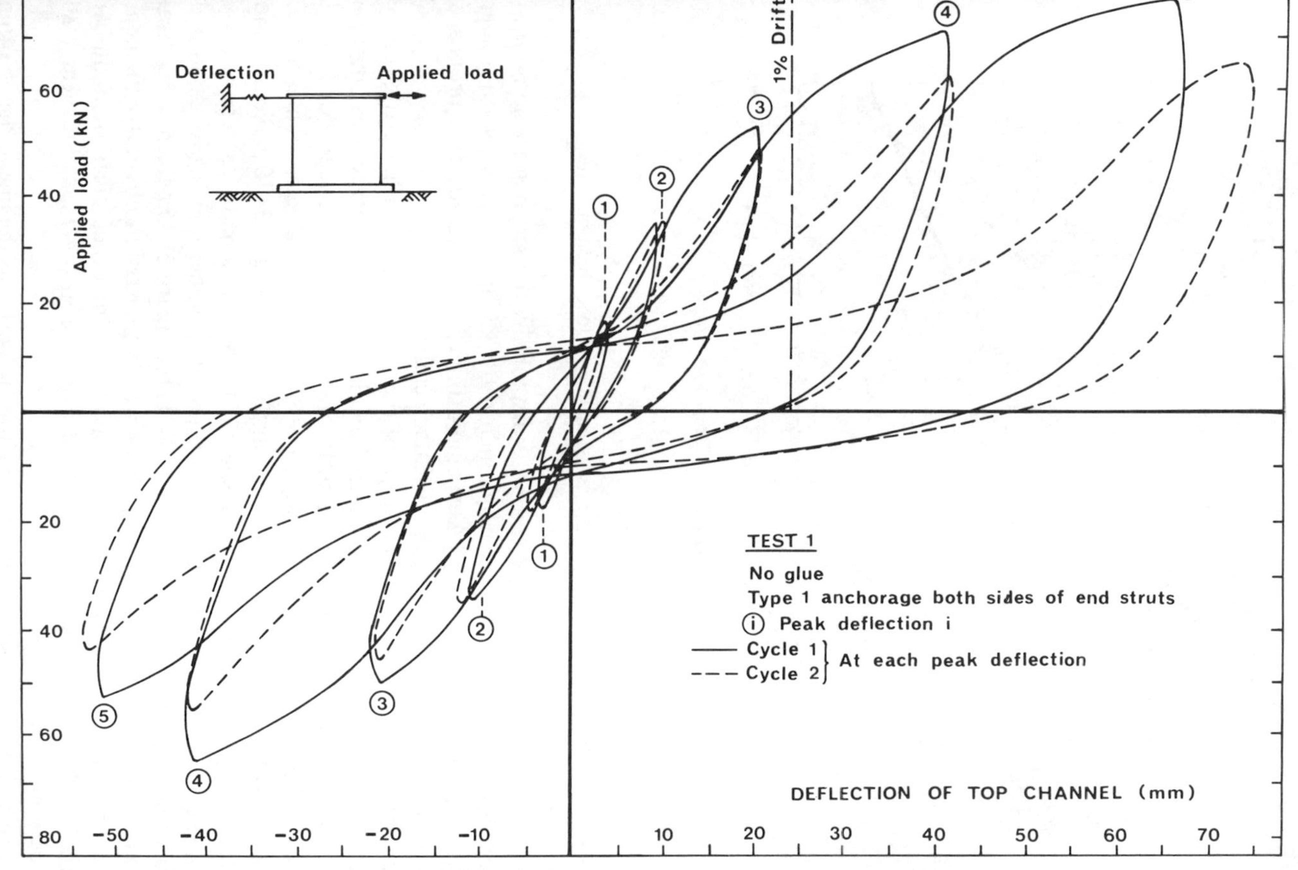

Figure 10.3 Hysteretic behaviour in a cyclic load test on a 2.4 m square particle board sheathed wall (after Thurston and Flack[7])

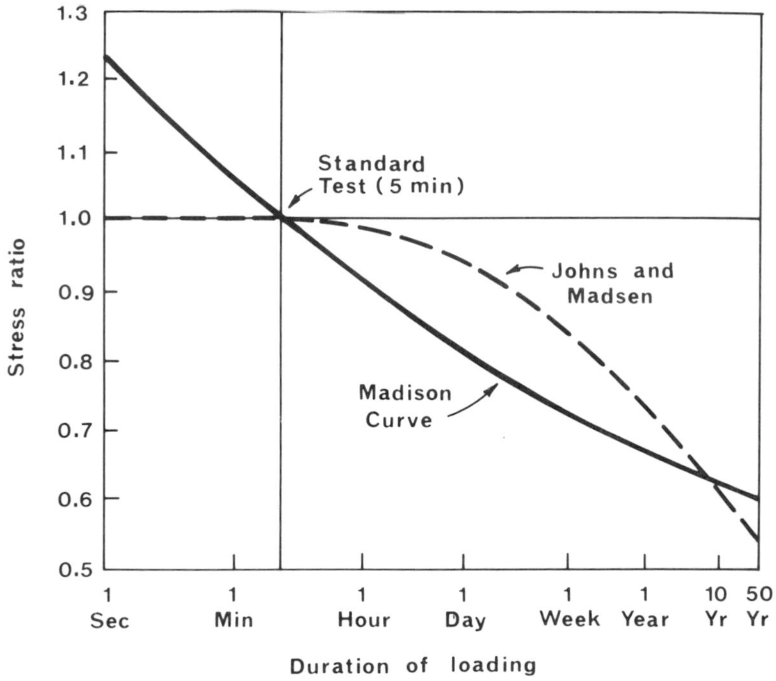

Figure 10.4 Relative strength of timber as a function of load duration (after Buchanan[32])

under earthquake loads is, of course, substantially higher than for long-term dead loads, and an enhancement factor of 1.7 times the strength under permanent load as used in the Australian timber code is also being proposed for a revision of the New Zealand timber code.[10]

Hysteretic behaviour with fat loops of the type shown in Figure 10.3 imply that the equivalent viscous *damping* available in such structures may be considerable. For example, from tests on sheathed diaphragms Medearis[11] found the equivalent viscous damping to be 8–10 percent regardless of amplitude. This figure bears comparison with the range of damping of 3–10 percent found for timber houses with plywood sheathed walls in Japan, as reported by Sugiyama.[12] The latter also found that traditional timber housing in Japan had must higher damping at 7–25 percent. Thus the figure of 15 percent of critical damping in Table 6.17 for timber shear wall construction obviously includes additional damping which is often available from other parts of a timber building, but which was apparently not available in the Japanese shear wall houses noted above.

The hysteretic behaviour shown in Figure 10.3 exhibits the phenomenon often referred to as *pinching* or *pinched loops*, where there is reverse curvature on rising and falling arms of the loop, and sometimes the loop is thinner in the middle than near its ends. Pinching is typical of hysteresis of many forms of timber construction where nail yield occurs, and also occurs in bolted construction, as reviewed by Dowrick.[13]

10.3 RELIABLE SEISMIC BEHAVIOUR OF TIMBER STRUCTURES

For obtaining reliable seismic response behaviour the principles concerning choice of form, materials, and failure mode control discussed in Section 5.3 apply to timber structures. Regarding the form (configuration) of timber structures, the remarks made in Sections A.3.2, A.4.2, and A.4.8 should also be noted.

Designing for failure mode control requires consideration of the structural form used, and of the forms discussed in Section 5.4 the main options for earthquake load resistance in timber construction are:

(1) Sheathed walls (shear walls);
(2) Moment-resisting frames;
(3) Concentrically braced frames;
(4) Hybrid moment-resisting/braced frames (e.g. see Figure 10.11).

The failure mode of each of these forms can be made ductile, generally by using the ductility of the steel connections of holding-down bolts (Figure 10.3). As this requires that the timber is designed to be stronger than the connections, the lateral load design may be governed by earthquakes even in cases when the wind base shear exceeds the earthquake base shear (as is often the case for timber structures).

Further discussion of failure mode control is given in the following sections on different member types.

10.4 FOUNDATIONS OF TIMBER STRUCTURES

The principles for design of timber structures are, of course, the same as those for other materials, as discussed in Section 6.4, but some details specific to timber construction need to be observed. For example, for the design of foundations of low-rise timber buildings the guidance given in Sections A.3.4 and A.4.6 may be appropriate.

The holding down of timber structures to concrete foundations has been prone to problems which occur wherever disparate materials are connected, and holding-down fastenings of the type shown in Figure 10.10 should be designed for strength and ductility to ensure that the desired ductile failure mode occurs.

The type of foundation provided by pole construction (Figure 10.5) overcomes some of the weaknesses of orthodox substructures to timber building, as the poles themselves provide vertical continuity and the pole frameworks develop the necessary resistance to horizontal forces.[14–16] For some species of timber,

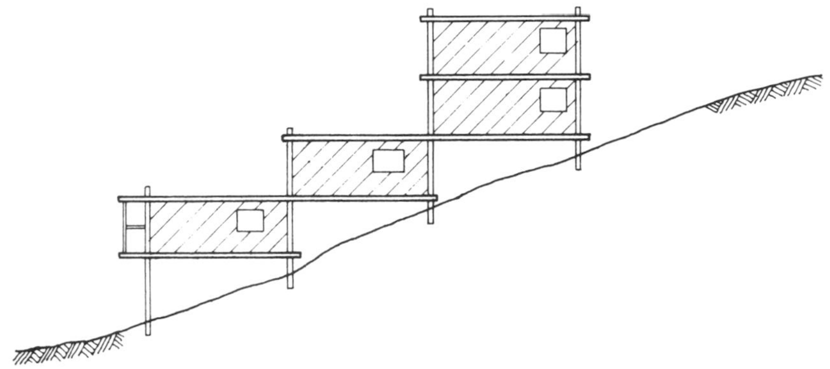

Figure 10.5 Pole frame apartments as built at Lugunda Beach, California[17]

preservative treatment such as tantilizing will, of course, be essential for durability below ground.

10.5 TIMBER-SHEATHED WALLS (SHEAR WALLS)

Most timber buildings derive their strength and stiffness from shear panels or diaphragms which may constitute walls, floors, ceilings, or roof slopes. Individual shear elements are built up from planks, plywood, metal, plaster, or other sheeting which is fixed to the basic timber framework by nails, screws, or glue. The effectiveness of different types of wall or diaphragm for resisting in-plane shears depends on:

(1) Its overall size and shape;
(2) The size, shape and position of any apertures;
(3) The nature of the timber framework;
(4) The nature and disposition of the diagonal or sheeting members;
(5) The connections between elements (3) and (4).

Regarding walls in particular, a useful study of some of the above factors was carried out by the US Forest Products Laboratory in 1946, the results of which are shown in Figure 10.6. The superiority of plywood for the panelling compared with diagonal or strip boarding is obvious in Figure 10.6. More recent tests[7] show that modern composite timber panels (particle board) are similar to plywood panels in this respect. While Figure 10.6 shows that glued construction has higher strength and rigidity than nailed construction, much more recent cyclic load tests in New Zealand[17] did not confirm that an increase in stiffness occurs, but, more important, they showed this form of construction to be very brittle at failure load.

The design data for shear walls given in the 1985 Uniform Building Code[18] needs reviewing in the light of recent studies. Extensive research in New Zealand into plywood sheathed walls[19,20] using cyclic and dynamic testing and non-linear dynamic analyses has been providing a basis for improved design procedures[19–21] which ensure ductile behaviour. This is done by providing extra

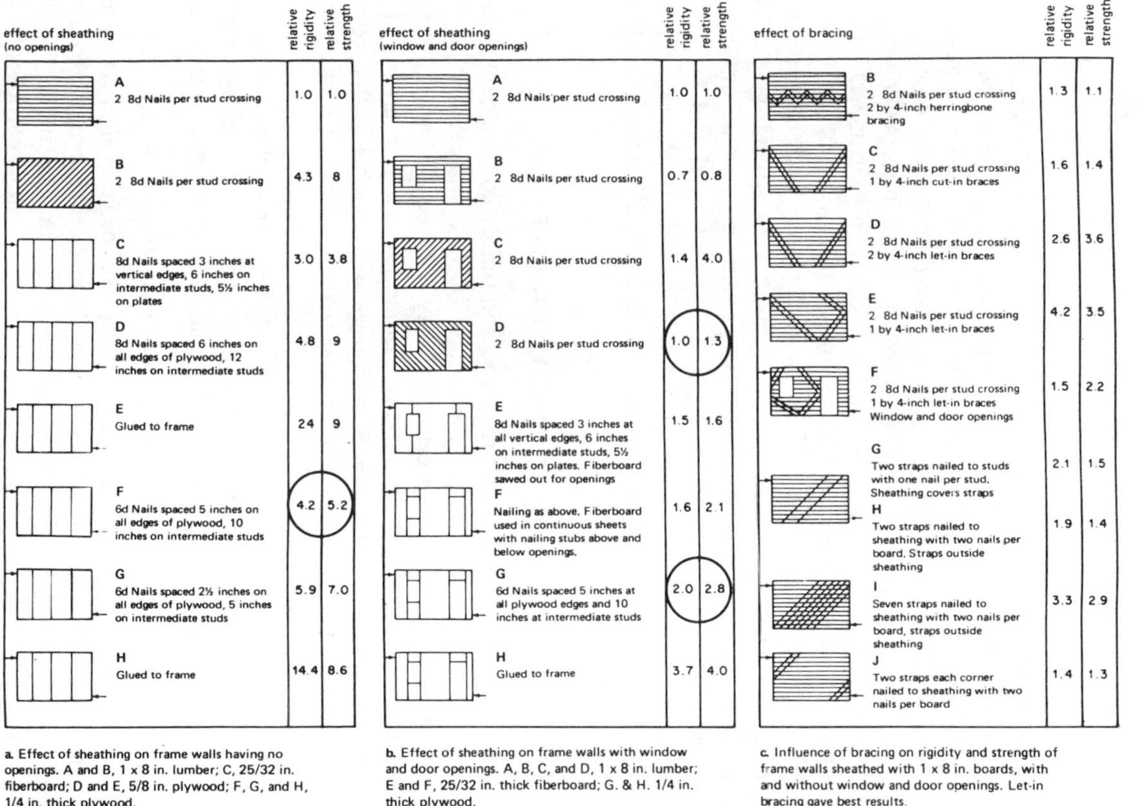

Figure 10.6 Tests on timber-framed walls with various forms of sheeting and fixing carried out by the US Forest Products Laboratory

strength in the framing chords and holding-down fixings so that yielding occurs in the sheathing nails only.

Further, it is noted that openings in sheathed walls require special attention, especially in highly stressed walls,[21] and a simplified analysis method has been proposed by Dean *et al.*[22]

From Figure 10.6 it is also clear that diagonals are much more effective when continuous between opposite framing members of a panel, rather than when broken by apertures. In domestic buildings it is, of course, common for only one or two diagonals to be used within any individual wall unit, and such diagonals should clearly be inclined between 30 degrees and 60 degrees to the horizontal for greatest effectiveness. In timber which is likely to split, nail holes near the ends should be predrilled slightly smaller than the nail diameter. In framing up shear panels, care should be taken that the perimeter members, and diagonals if used, are made from sound timbers. The framing members for door and window apertures should similarly be of good-quality timber.

External timber framed walls are often clad with plaster, and the earthquake performance of such walls has been greatly improved using expanded metal lath. In Japan and California expanded metal is now commonly used in conjunction with cement, lime, and sand plaster (mixes of between 1:1:3 and 1:1:4$\frac{1}{2}$ are found to be durable), the lime reducing the brittleness of the plaster. The application of two or three coats of plaster, giving a total thickness of about 20 mm, is normal practice.

10.6 TIMBER HORIZONTAL DIAPHRAGMS

Horizontal diaphragms such as floors or roofs in timber construction require more design consideration than corresponding concrete or steel diaphragms. This occurs largely because of the greater flexibility of timber diaphragms, which may render invalid the usual simplifying assumption made in analysis that horizontal diaphragms are rigid, and which may lead to troubles with excessive deflections, as shown by Figure 10.7. Thus, stiffness rather than strength may be the controlling design criterion.

There are two alternative (and contrasting) design approaches that may be adopted[23] for diaphragms, both of which relate to the high degree of non-linearity which readily occurs in diaphragms under design earthquake loading.

Design approach No. 1—Suppression of non-linearity

The distribution of loading through the diaphragm to the vertical structure and the control of deflections can be more easily and more reliably predicted if the non-linearity in the nail deformations is suppressed sufficiently so that the diaphragm may be considered effectively linear elastic for design purposes. In the majority of buildings this will be the preferable and economical design procedure.

Design approach No. 2—Recognition of inelasticity

In this approach the inelastic (non-linear) behaviour of the diaphragm at design loads is allowed for explicitly in calculating the loads, load distribution, and

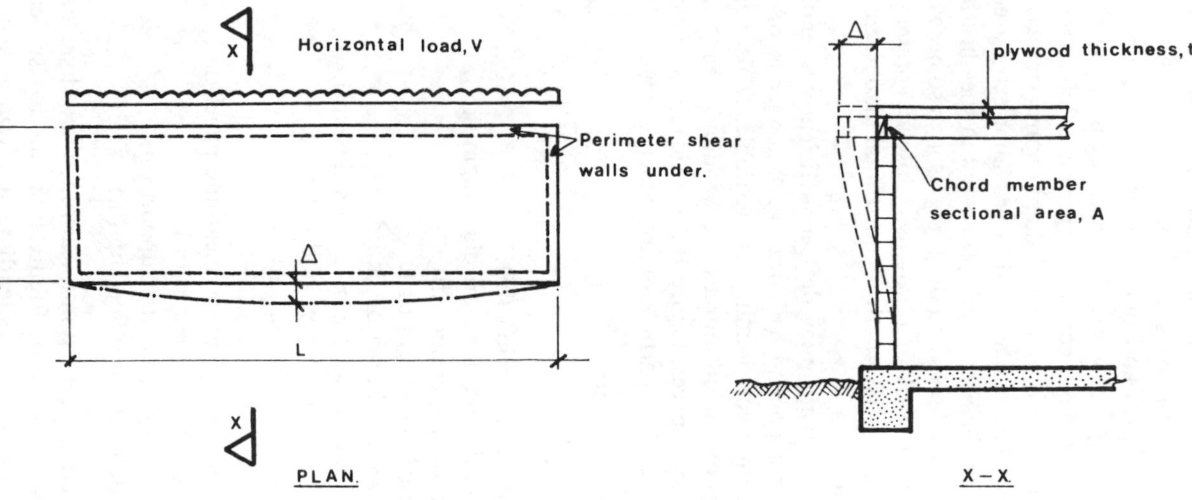

Figure 10.7 A typical horizontal timber diaphragm showing the effect on supporting walls of deflections under horizontal loading

deflections. The energy absorption and ductility of the diaphragm implied by the inelasticity may be utilized to reduce the loadings generated within the diaphragm (e.g. by the use of reduced code structural factors permitted by higher ductility). This results in a reduction in the estimated forced transferred from the diaphragm to the supporting structure.

For the strength design of horizontal diaphragms, the forces are usually found using the girder analogy,[23,24] where the sheathing is the 'web' of the girder and top plate of perimeter wall or a continuous perimeter joist is the flange. However, because of the great depth of typical diaphragms, strength is not generally problematical. Brittle failure is avoided by providing extra strength in the chords (flanges), so they yielding (and failure) occurs in the sheathing nails only.

Excessive deflection of plywood diaphragms is to some extent controlled by limiting the aspect ratio of diaphragms. Closely representing international practice, New Zealand[10] the length of horizontal diaphragms may not exceed 4.5 times the width.

As horizontal diaphragms may deflect sufficiently to endanger supporting or attached wall components (Figure 10.7), a means of calculating their deflections under in-plane loading is desirable. This deflection involves contributions from four main sources, i.e. bending, shear, nail slip, and splice slip (in flange or chord splices, Figure 10.9), which for a single-span diaphragm sheathed with plywood or particle board panels is given[23,24] by

$$\Delta = \frac{5WL^3}{192EAB^2} + \frac{WL}{8GBT} + \frac{Le_n}{4}\left(\frac{1}{h} + \frac{1}{b}\right) + \frac{\Sigma\Delta_c x_s}{2B} \tag{10.1}$$

where Δ = horizontal deflection (m),
W = total horizontal load on diaphragm (kN),
L = length of diaphragm (m),
B = width of diaphragm (m),
A = cross-sectional area of chord (m^2),
E = modulus of elasticity of chords (kN/m^2),
G = shear modulus of plywood (kN/m^2),
t = thickness of plywood (m),
e_n = nail deformation (m),
h = length of an individual sheathing panel (m),
b = width of sheating panel (m),
x_s = distance from splice to support (m),
Δ_c = individual flange splice slip (m).

The nail deformation e_n, is a non-linear function of load level, nail type, and sheathing thickness as shown by Figure 10.8, and is also dependent on other factors such as the nature of the sheathing and framing and the surface of the nail.[26] Thus it is clearly desirable to obtain e_n from the results of tests matching the components to be used. For example, in New Zealand for local plywood and medium density particle board sheathing, the recommended nail slip at design load level is 0.9 mm, regardless of sheathing thickness.

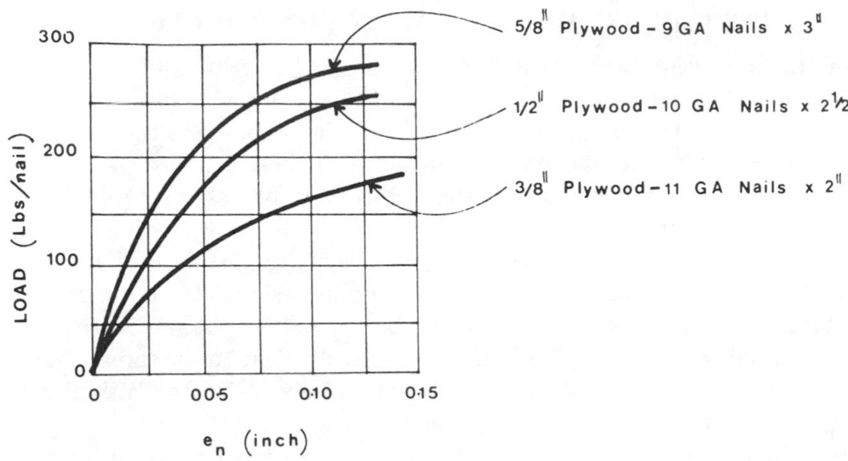

Figure 10.8 Nail deformation in plywood diaphragms framed on to Douglas fir as derived in the USA[25]

Finally it is noted that design problems arise from large openings in diaphragms, as discussed elsewhere.[22–24]

10.7 TIMBER MOMENT-RESISTING FRAMES AND BRACED FRAMES

The most common form of *moment-resisting frame* in timber is the portal frame. Because of their high flexibility, and hence long period of vibration, in many cases the design of timber portals may be governed by wind loads and by deflections rather than by seismic strength. Multi-storey moment-resisting frames in timber are unlikely to reach more than about two-storeys in height, without excessive beam and column sizes or without adding other means of providing horizontal stiffness, because of the high inter-storey drifts that would occur[27] and the difficulties of making such structures cost-competitive.[28]

As reviewed elsewhere by the author,[13] ductility is obtained in moment-resisting frames by yielding in the connections, which may be of three types:

(1) Steel side plates, with yielding in the nails;
(2) Steel side plates, necked so that yield occurs in the plate;
(3) Plywood side plates, with yielding in the nails.

Braced frames in timber construction may be created by traditional timber braces or by steel (e.g. Figure 10.11). It appears that double bracing is better than single bracing, because of progressive deflection in one direction only in the latter asymmetrical case, as shown by the lopsided hysteretic behaviour found by Sakamoto *et al.*[29]

10.8 CONNECTIONS IN TIMBER CONSTRUCTION

Connections between timber members may be formed in the timber itself, or may involve glue, nails, screws, bolts, metal straps, metal plates, or toothed metal connectors. Under earthquake loading, joints formed in the timber are inferior to most other forms of joint. In light timber construction such as smaller dwellings, the use of metal nail plates (Hurricane braces) or toothed steel connectors is now widespread.

The nailed joints, the nail load, size, and spacing require careful attention. A nail driven parallel to the timber grain should be designed for not more than two thirds of the lateral load which would be allowed for the same size of nail driven normal to the grain. Nails driven parallel to the grain should not be expected to resist withdrawal forces. Edge or end distance of nails should not be less than half the required nail penetration.

The effectiveness of nails in enhancing seismic response behaviour improves with increasing thickness of (1) side plates in moment-resisting connections and (2) sheathing of walls.[20] More research, however, is required before the effect of thickness can be fully described in simple design rules.

In diaphragms, perimeter framing may need jointing capable of carrying the longitudinal forces arising from wind or seismic loading. A simple method of connection is shown in Figure 10.9.

Connections between shear walls and foundation or between successive storeys of shear walls must be capable of transmitting the horizontal shear forces and the overturning moments applied to them. Details which are considered good practice in California for these connections are illustrated in Figure 10.10, but obviously both details on the right-hand side of the figure are capable of resisting only small overturning tensions.

For some comments on the connections of timber roof diaphragms to walls of other materials the reader should refer to Sections 10.6 and A.3.5.

Pole frame buildings are usually jointed using bolts, steel straps and clouts (Figure 10.11) as described in detail elsewhere.[15,16] An effective means of obtaining resistance to lateral shear forces is to create moment-resisting triangles at the knees of portals (Figure 10.11) using steel rods as the diagonal member.

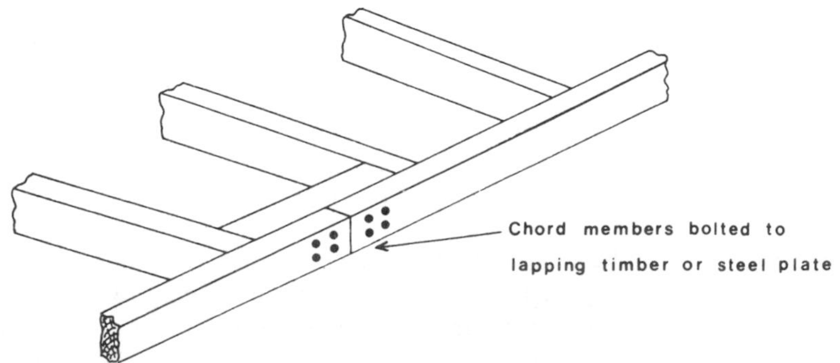

Figure 10.9 Method of jointing chord members of timber diaphragms

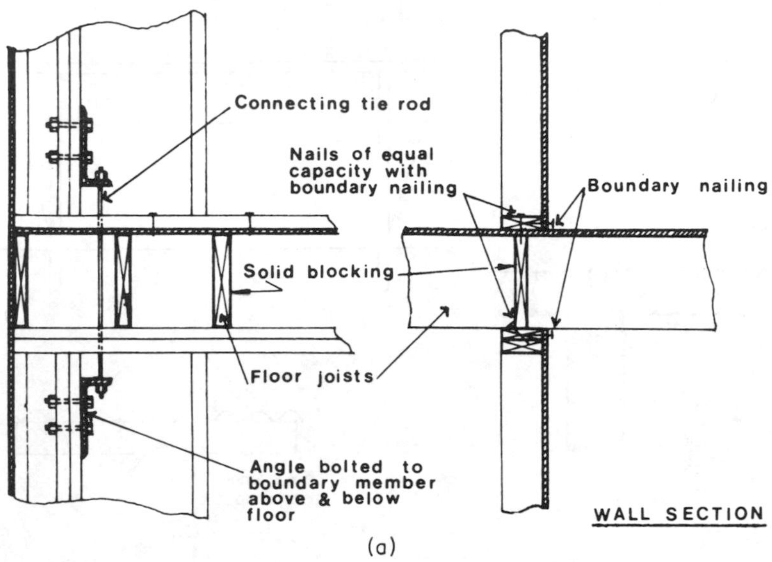

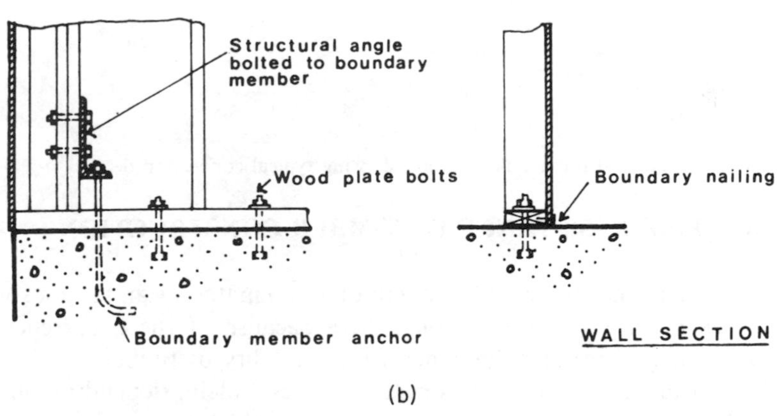

Figure 10.10 Connection details for plywood shear walls.[30] (a) Interstorey connections in timber buildings; (b) connection of timber members to concrete foundations

As noted above, the reliable performance of timber structures in earthquakes depends heavily on appropriate detailing of connections. As this subject is currently developing fast, the reader should refer to recent publications on the particular types of connection being designed, some of which are referred to throughout this chapter.

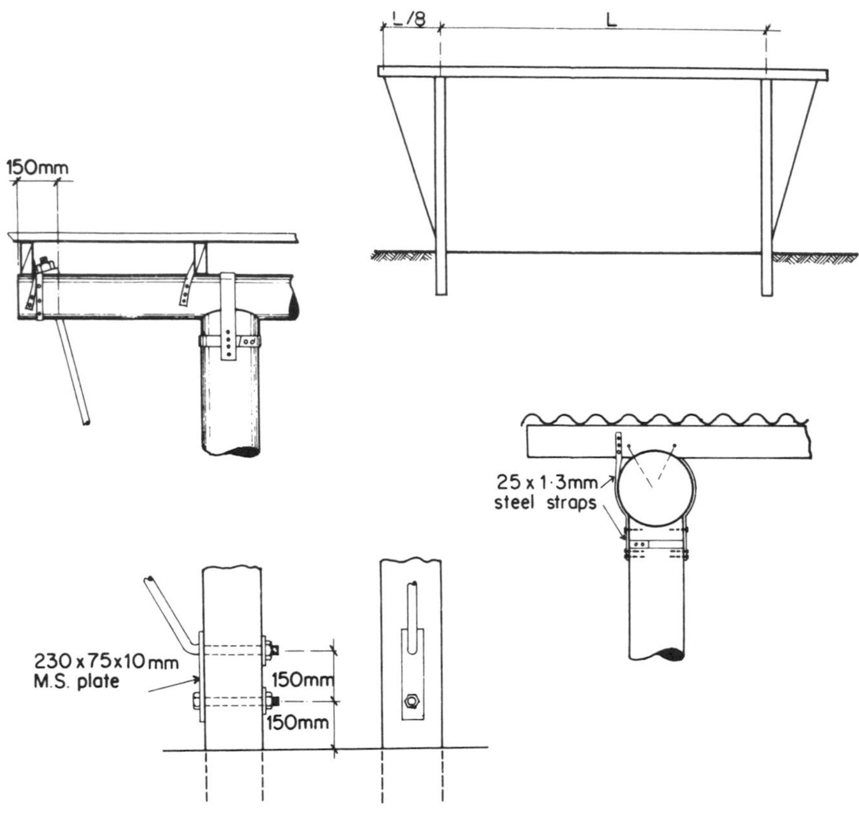

Figure 10.11 Tied rafter pole building showing typical connection details[14]

10.9 FIRE RESISTANCE OF TIMBER CONSTRUCTION

Although fire resistance is a problem common to all materials and is not solely related to seismic areas it is mentioned here because of the occurrence of earthquake-induced fires and the general flammability of timber.

The fire resistance of timber construction varies widely, depending on the thickness of the timbers used. Pole frames and other thick timbers such as used in moment-resisting frames have relatively low fire risk because of their large volume-to-surface area ratio. Surface charring is relatively shallow and protects the interior of the member from the flame. Apparently badly burned structures of heavy timber have been found to be strong enough to continue in service. Such behaviour in fire is likely to be superior to equivalent unprotected steelwork, and its fire risk merits are becoming recognized for insurance purposes.

While the above ability of thick timbers to withstand fire is well understood, less is known about the effect of connections on fire resistance. A few fire tests have been carried out on certain types of connection, as reviewed by Smith,[31] who also discusses the means of fire protection of steel plate connections.

Because of its obvious flammability, light timber construction should only be used where low fire rating is acceptable. Various chemical fire retardants have been marketed in some parts of the world in an attempt to improve the resistance of flammable construction, but their value for timber construction seems limited, as they appear to increase the time to ignition by only a few minutes.

REFERENCES

1. Falconer, B. H., 'Preliminary comments on damage to buildings in the Managua earthquake', *Bull. NZ Soc. for Earthq. Eng.*, **1**, No. 2, 61–71 (1968).
2. 'The San Fernando California earthquake of February 9, 1971', *NBS Report 10556*, US Dept of Commerce, National Bureau of Standards, March (1971).
3. Earthquake Engineering Research Laboratory, California Institute of Technology, 'Engineering features of the San Fernando earthquake', *Bull. NZ Soc. for Earthq. Eng.*, **6**, No. 1 22–45 (1973).
4. Soil Research Team, Earthquake Research Institute, 'Earthquake damage and subsoil conditions as observed in certain districts in Japan', *Proc. 2nd World Conf. on Earthq. Eng., Japan*, **1**, 311–25 (1960).
5. Tsai, Zuei-Ho, 'Earthquake and architecture in Japan', *Proc. 3rd World Conf. on Earthq. Eng., New Zealand*, **III**, V–114 (1965).
6. Buchanan, A. H., 'Strength model and design methods for bending and axial load interaction in timber members', *PhD thesis*, Dept of Civil Engineering, University of British Columbia (1984).
7. Thurston, S. J., and Flack, P. F., 'Cyclic load performance of timber sheathed bracing walls', New Zealand Ministry of Works and Development, Central Laboratories, *Report No. 5–80/10* (1980).
8. Spencer, R. A., 'Rate of loading effects on bending for Douglas-Fir lumber', *First International Conf. on Wood Fracture*, Banff, Canada (1978).
9. Johns, K. C., and Madsen, B., 'Duration of load effects in lumber, Parts I,II,III', *Canadian J. Civil Eng.*, **9**, No. 3, 502–36 (1982).
10. Standards Association of New Zealand, 'Code of practice for timber design', *NZS 3603–1981*.
11. Medearis, K., 'Static and dynamic properties of shear structures', *Proc. International Symp. Effects of Repeated Loading on Materials and Structures, RILEM—Inst. Ing.*, Mexico, **VI** (1966).
12. Sugiyama, H., 'Japanese experience and research on timber buildings in earthquakes', *Proc. Pacific Timber Engineering Conf., Auckland*, **III**, 431–8 (1984).
13. Dowrick, D. J., 'Hysteresis loops for timber structures', *Bull. NZ Nat. Soc. for Earthq. Eng.*, **19**, No. 2, 143–52 (1986).
14. New Zealand Timber Research and Development Association, 'Pole frame buildings', TRADA, *Timber and Wood Products Manual*, Section 16–1 (1972).
15. New Zealand Timber Research and Development Association, 'Engineering design data for softwood poles', TRADA, *Timber and Wood Products Manual*, Section 26–2 (1986).
16. New Zealand Timber Research and Development Association, 'Pole type construction', TRADA *Timber and Woods Products Manual*, Section 2f-1 (1973).
17. Yap, K. K., 'Slow cyclic testing of shear panels bonded with adhesives', NZ Ministry of Works and Development, *Central Laboratories Reports No. 5–85/14* (1985).
18. International Conference of Building Officials, *Uniform Building Code*, ICBO, Pasedena, California (1982).

19. Stewart, W. G., *PhD Thesis*, Civil Engineering Dept, University of Canterbury, New Zealand (in progress 1986).
20. Dean, J. A., Stewart, W. G., and Carr, A. J., 'The seismic behaviour of plywood sheathed shearwalls', *Bull. NZ Nat. Soc. for Earthq. Eng.*, **19**, No. 1, 48–63 (1986).
21. Dowrick, D. J., and Smith, P. C., 'Timber sheathed walls for wind and earthquake resistance', *Bull. NZ Nat. Soc. for Earthq. Eng.*, **19**, No. 2, 123–34 (1986).
22. Dean, J. A., Moss, P. J., and Stewart W., 'A design procedure for rectangular openings in shear walls and diaphragms', *Proc. Pacific Timber Engineering Conf., Auckland*, **II**, 513–18 (1984).
23. Smith, P. C., Dowrick, D. J., and Dean, J. A., 'Horizontal timber diaphragms for wind and earthquake resistance', *Bull. NZ Nat. Soc. for Earthq. Eng.*, **19**, No. 2, 135–42 (1986).
24. Applied Technology Council, California, *Guidelines for the design of horizontal wood diaphragms*, ATC-7 (1981).
25. Timber Engineering Company, *Timber Design and Construction Handbook*, F. W. Dodge Corporation, New York (1956).
26. Collins, M. J., 'Design data for nailed joints in shear', *Bull. NZ Nat. Soc. for Earthq. Eng.*, **19**, 1986 (in press).
27. Moss, P. W., 'Seismic performance of a multistoreyed timber frame having moment-resisting nailed joints', *Proc. Pacific Timber Engineering Conf., Auckland*, **II**, 559–68 (1984).
28. Thurston, S. J., and Garrett, I. J., 'Design of a ductile three storey aseismic laminated timber frame building', *Proc. 3rd South Pacific Conf. on Earthq. Eng., Wellington*, **2**, 269–82 (1983).
29. Sakamoto, I., Ohashi, Y., and Shibata, M., 'Some problems and considerations on aseismic design of wooden dwelling houses in Japan', *Proc. 8th World Conf. on Earthq. Eng., San Francisco*, **V**, 669–76 (1984).
30. New Zealand Timber Research and Development Association, 'Plywood design for seismic areas', TRADA, *Timber and Wood Products Manual*, Section 5f-1 (1973).
31. Smith, P. C., 'Timber joints in fire', *The New Zealand J. of Timber Construction*, 22–6 (May 1985).
32. Buchanan, A. H., 'Wood properties and seismic design of timber structures', *Proc. Pacific Timber Engineering Conf., Auckland*, **II**, 462–9 (1984).

Chapter 11

Earthquake resistance of services and equipment

11.1 SEISMIC RESPONSE AND DESIGN CRITERIA

11.1.1 Introduction

This chapter sets out to advise engineers on the earthquake-resistant design of services components and installations. Much of the background information on the earthquake problem is contained in other chapters of this book or in the literature of structural engineering and seismology. Up till about 1970 only a comparatively small effort had been made in this field by services engineers on their own account. Since the 1971 San Fernando California earthquake,[1,2] in which about 10 percent of the total cost of damage was attributed to damage to mechanical and electrical equipment, there has been a growing awareness that equipment needs its own specialist aseismic design and detailing. The following points are worthy of attention:

(1) Aseismic design of equipment is a problem of dynamics, which cannot be treated adequately with equivalent static methods alone.
(2) Earthquake accelerations applied in the design of equipment generally should be much larger than the corresponding values used in the design of the buildings housing the equipment.
(3) The response spectrum method (See Section 11.1.4) provides ready-worked solutions of the equations of motion, and is a powerful aid to understanding the true dynamic nature of the earthquake problem.
(4) In many cases a high level of earthquake resistance can be provided at relatively small extra cost.[3]

11.1.2 Earthquake motion—accelerograms

Strong-motion earthquakes are most commonly recorded by accelerographs, which produce accelerograms which are a plot of the variation of acceleration in a given direction with time. A widely used accelerogram is that obtained at El Centro, California, during the Imperial Valley earthquake of 18 May 1940. Figure 1.4 shows the north-south accelerogram of this earthquake, with a peak acceleration of $0.33g$. By integration of the acceleration record, the ground velocity was deduced, showing a maximum value of 34 cm/s. Similarly, by integrating the velocity, the displacement of the ground was inferred, showing a maximum of 21 cm. The record of acceleration in the east-west direction was similar, with a maximum value of $0.22g$. The vertical component showed considerably more rapid variations, reaching a maximum of $0.2g$.

More detailed discussions of earthquake motion are given earlier in this book, especially in Chapter 4.

11.1.3 Design norms—design earthquakes

It is usual for a design earthquake to be adopted in a given region for certain types of construction. The design earthquake is defined as the worst earthquake officially likely to occur in that region with a given average return period (say 100 years), and may be specified in terms of peak ground accelerations, a response spectrum, or an earthquake (Richter) magnitude. Although there is a risk of a worse real earthquake occurring, the standard for officially acceptable minimum risk is set by the design earthquake. Individual structures or equipment items may be designed to some authorized or discretionary fraction (greater or less than unity) of the design earthquake, depending on the design levels of risk and ductility.

Design earthquakes are discussed more fully in Chapter 4.

11.1.4 The response spectrum design method

A direct analytical approach to the problem of earthquake strength is to make a mathematical model of the structure and subject it to accelerations as recorded in actual earthquakes. Many structures approximate to the single degree-of-freedom model shown in Figure 11.1, where a mass is supported by a spring and is connected to a damping device. If linear material behaviour is assumed, the ratio of spring stiffness to horizontal shear is constant. For mathematical convenience the damping force is usually taken as proportional to velocity, which is generally a satisfactory approximation.

Figure 6.29 is a typical 'response spectrum' diagram. It shows the maximum acceleration response to a given earthquake motion of a linear single degree-of-freedom structure, with any fundamental period in the range $T=0$ to $T=4.0$ s. Note the considerable reduction in response resulting from an increase in damping.

As individual earthquakes give different irregular reponses dependent on local ground conditions, a design criterion is sought by averaging the response curves for a number of earthquakes (Figure 11.2).

The curves in Figure 11.2 clearly show that for earthquakes recorded on firm ground:

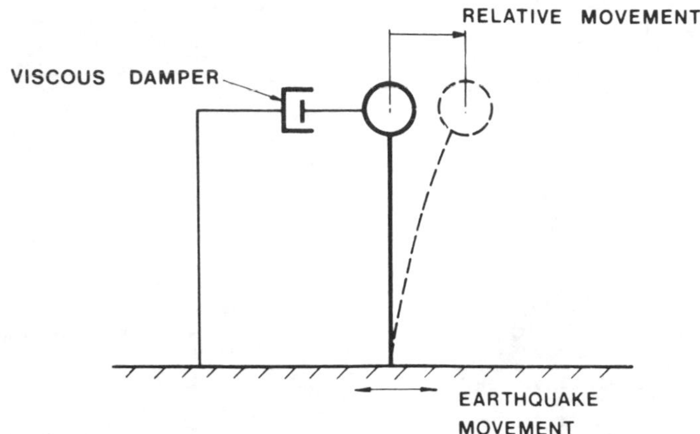

Figure 11.1 Single-degree-of-freedom model for studying earthquake response of equipment

(1) Extremely rigid structures (i.e. period less than 0.05 s) experience peak accelerations not much larger than the maximum applied acceleration;
(2) Structures with small flexibility (periods of 0.2–0.6 s) act as mechanical amplifiers and experience accelerations up to four or five times the peak applied (ground) acceleration;
(3) Structures with large flexibilities experience accelerations less than the peak applied acceleration;
(4) Structures with strong damping, whatever their natural period, experience greatly reduced response to ground motion.

11.1.5 Comparison of design requirements for buildings and equipment

The principles underlying the choice of earthquake design loads and level of material response are, of course, the same for buildings and equipment, depending on the consequences of any given response and their acceptability (Section 1.3). Thus where equipment and buildings have the same risk characteristics, they should be designed as follows:

(1) They should be designed to the same design earthquake, except as modified by the building (Section 11.1.6); and
(2) They should be designed to the same response (stress) levels, so that they will generally be damaged to a comparable degree in a given event.[4]

However, in some cases it is acceptable for a building to be deformed into the post-elastic range, using its ductility, while some of its contents may have to respond elastically to remain operational or safe. Thus vital services may be designed to more stringent earthquake criteria than the buildings housing

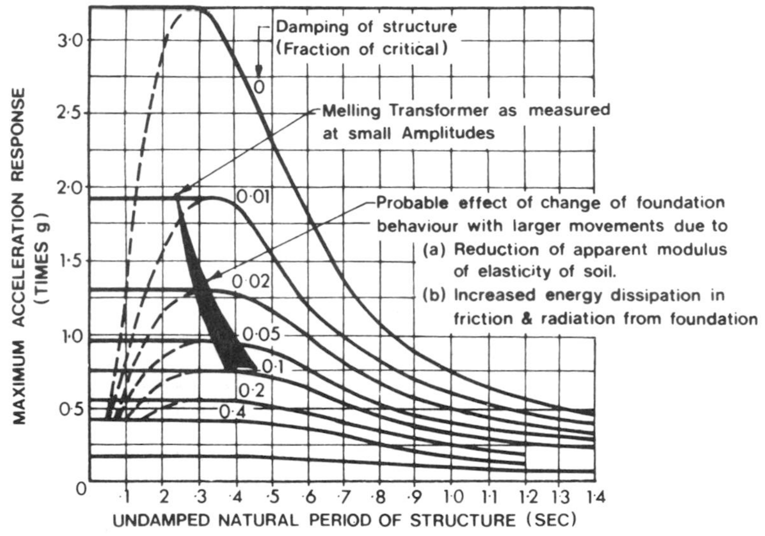

Figure 11.2 Response of 30-tonne transformer on concrete pad on soft subsoil to design earthquake (After Hitchcock,[3] reproduced from *New Zealand Engineering* by permission of the New Zealand Institution of Engineers)

them, e.g. safety equipment, such as used for firefighting or emergency ventilation. Also dangerous substances should not be released, such as gas, steam, or toxic chemicals.

The typical difference between traditional design loads for structures designed to be fully ductile ($\mu = 4$ to 6) and those designed to remain elastic ($\mu = 1$) may be seen by comparing the bottom curve in Figure 11.2 with the upper family of curves. Obviously there will be a large difference in design loads for equipment required to remain elastic and the building housing the equipment, if that building is designed to respond in a fully ductile manner. The question of the effect of ductility μ on the design loads is discussed in Section 6.6.7.3(i).

The continuation of electricity supply is a major factor in the success of emergency plans after earthquakes. A notable failure of electricity supply equipment occurred in the 1971 San Fernando, California, earthquake, where there was \$30 million damage to the newly built \$110 million Pacific Intertie Electric Converter Station.[1]

11.1.6 Equipment mounted in buildings

Equipment mounted in a building should be designed to withstand the earthquake motions to which it will be subjected by virtue of its dynamic relationship to the building. The design of the equipment and its mountings should take into account the dynamic characteristics of the building, both as a whole and in part.

A building tends to act as a vibration filter and transmits to the upper floors mainly those frequencies close to its own natural frequencies. Thus on the upper floors there will be a reduction in width of the frequency band of the vibrations affecting the equipment. As a rough guide, the fundamental period of a flexible building may be taken as $0.1\,N$s, where N is the number of storeys, but individual parts of the structure such as floors (on which the equipment is mounted) may have lower fundamental periods. Also the magnitude of the horizontal accelerations will generally increase with height up the building; and hence amplification of the ground accelerations usually takes place.

The accurate prediction of the vibrational forces occurring in equipment mounted in a building is a complex dynamical problem which at present is only attempted on major installations such as nuclear power plants. In ordinary construction a simpler approach has to suffice, such as the response spectrum technique. With such methods, however, it is difficult to make realistic allowance for the filtering and amplification characteristics of the building.

(See also Section 11.2.2.2.)

11.1.7 Material behaviour

11.1.7.1 Failure modes

As noted in Section 11.1.5, whether a material behaves elastically or inelastically has a great influence on the seismic response (or loading). It is thus of great importance to understand whether a material is brittle or what degree of ductility may be obtained, and for maximum reliability the failure modes need to be controlled, as discussed for general structures in Section 5.3.8. Some further points relating to equipment are noted below.

(i) Brittle materials

Whereas structural engineers generally try to avoid the use of brittle materials, electrical engineers have no choice but to use one of the most brittle of all materials, namely porcelain, in most of their structures. Because there is no ductility, any failure of such a material is total. Therefore seismic design accelerations for these structures should be ten to twenty times those used for ordinary buildings. Diagonally braced structures carrying heavy loads, even though made of steel, may also have to be designed for large accelerations to avoid sudden failure by buckling of struts.

(ii) Ductility

For massive rigid bodies such as transformers all the energy imparted by an earthquake has to be absorbed in holding-down bolts or clamps, which are very small compared with the mass of the whole structure. Thus if reliance is to be placed on the ductility of these fastenings to justify using reduced seismic accelerations for elastic design or for protecting the transformer, considerable knowledge of the post-yield behaviour and energy absorbing capacity of the fastenings is essential.

Reference to the behaviour of steel (Chapter 9), particularly to steel connections in Section 9.7, is relevant to the design of holding-down fastenings.

11.1.7.2 Damping

In the design of buildings, as only a fairly small amount of damping is readily available, survival in a large earthquake depends largely on post-yield energy dissipation and ductility. On the other hand, with much electrical equipment the provision of high damping becomes a practical possibility because of the smaller masses involved. Such damping may be in the form of rubber pads, stacks of Belleville washers, or true viscous damping units.

The beneficial effect of damping on seismic response is apparent in Figures 11.2 and 11.5.

11.1.8 Cost of providing earthquake resistance for equipment

Many smaller items of electrical equipment can withstand horizontal accelerations of $1.0g$ as currently designed and installed. Even a 10 tonne transformer with the height of its centre of gravity equal to the width of its base could be secured against a horizontal acceleration of $1.0g$ with four 20 mm diameter holding-down bolts without exceeding the yield stress—an inexpensive protection for such a valuable piece of equipment.

Provided the nature of earthquake loading is understood and taken into account from the beginning of a design, earthquake resistance can often be obtained at little cost.[3] The introduction of additional earthquake strength into an existing design is bound to be more expensive.

11.2 SEISMIC ANALYSIS AND DESIGN PROCEDURES FOR EQUIPMENT

Seismic analysis techniques and design procedures are the same in principle for equipment as those described for general structures elsewhere in this book, particularly Sections 1.4 and 6.6. The following discussion, which considers the main points specific to equipment, separately describes procedures using dynamic and equivalent-static analyses.

In either case the basic considerations are, of course, the same. For example, vertical seismic accelerations of a similar size to the horizontal accelerations occur, and may exceed gravity in a severe earthquake (Section 2.4.2.1). Thus as the motions reverse in direction, equipment may need to survive nett vertical accelerations ranging from about zero to $2.0g$, prior to considering the dynamic response of the equipment itself. Such accelerations would obviously greatly affect the stability of equipment. For example, friction between the base and the floor could not be relied on to locate the equipment horizontally.

Further background on the aseismic design of equipment may be found in a paper by Schiff regarding electric power substations[5] and in the New Zealand code for equipment of various types in buildings.[4]

11.2.1 Design procedures using dynamic analysis

11.2.1.1 Single degree-of-freedom structures

For structures which can be thought of as having effectively only one mode of vibration in a given direction, the following simple *response spectrum* design procedure will usually prove to be both easy to carry out and seismically realistic:

(1) Ascertain the natural period of vibration in the direction being studied (by calculation or by measurement of similar structures). This should be done for its condition as installed, including the effects of supports and foundations.
(2) Determine an appropriate value of equivalent viscous damping by measurement of similar structures or by inference from experience, or by calculation if special dampers are provided.
(3) Read the acceleration response to the standard earthquake from the appropriate spectrum, e.g. Figure 11.2. For natural periods less than 0.3 s, the maximum value for the damping concerned should generally be used, unless there is convincing proof that the structure concerned will remain very rigid throughout strong shaking, i.e. that it will always retain a natural period less than 0.1 s. This latter condition is very difficult to prove, and should not normally be used.
(4) Combine the stresses from this earthquake loading with other stresses such as those from dead loads, and working pressures, including short-circuit loads, but not with stresses due to wind loads.
(5) For ductile structures, design to meet total loadings with stresses not exceeding normal working stresses, or such higher stresses as may be shown to meet the specification for survival in a major earthquake.
(6) For brittle structures, design to meet total loading with stresses that allow a factor of safety of at least 2.0 on the guaranteed breaking load of brittle components, or at least 2.5 if the breaking load is not based on statistically adequate information.

11.2.1.2 Multi-degree-of-freedom structures

For structures that have more than one mode of vibration two main methods of dynamic analysis exist as described below:

(1) A *response spectrum* technique similar to that described in the preceding section can be used, but it is more complex in that the responses due to a number of modes must be combined (Section 1.4).

(2) A more powerful dynamic analysis involves the application of *time-dependent forcing functions* directly to structures, rather than using response spectra. The equations of motion for the structure are solved using *full modal analysis* or *direct integration*, as described in Sections 1.4 and 6.6.7 of this book. This approach is being increasingly used in the dynamic analysis of a wide range of engineering structures, and it has been used by some manufacturers of high-voltage circuit breakers.

11.2.2 Design procedures using equivalent-static analysis

Where dynamic analysis is not used, it is desirable to establish suitable equivalent-static forces expressed as coefficients of gravity, C_p, such that the horizontal base shear, F, acting on the equipment is

$$F = C_p W_p \tag{11.1}$$

where W_p is the weight of the item of equipment concerned (Section 6.6.7.1).

Such coefficients should preferably be determined only for structures that fall into well-defined groups within which dynamic characteristics do not vary greatly. This has been done for a wide range of types of equipment in some codes as discussed later in Section 11.2.2.5. Preferably each equipment group should have its coefficients derived from fundamental principles in such a way as to cover reasonable variations from the chosen dynamic characteristics. Hitchcock[3] suggested three such groups: (1) basemounted free-standing equipment, (2) equipment mounted on suspended floors, and (3) equipment that would fail in a brittle manner. The following discussion of these groups is largely based on Hitchcock's paper.

11.2.2.1 Base-mounted free-standing equipment

Transformers are the chief members of this group of equipment. Figure 11.3 shows the forces acting on such equipment; the graph shows how the calculated holding-down force, expressed as a fraction of the weight, varies with the maximum acceleration experienced and with the ratio of height of centre of gravity to effective width of base.

This graph shows the weakness of the once widely used 0.25g earthquake specification, under which any equipment with a height of centre of gravity less than twice the effective base width would have a calculated holding-down requirement of zero. When such equipment is subjected to strong earthquake accelerations, which generally will be larger than 0.25g, the required holding-down forces will be much greater than zero.

As an example of equipment in this group, consider the power transformer mounted on a concrete pad shown in Figure 11.4. Some field measurements with small amplitude vibrations in the transverse direction gave the damping as 0.9 percent of critical, and the fundamental period as 0.24 s. Plotting these

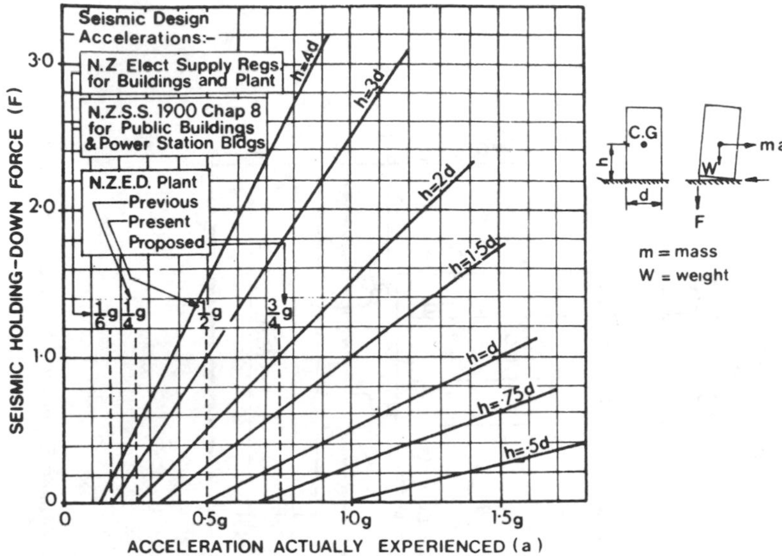

Figure 11.3 Relationship between seismic holding-down bolt loads, applied accelerations, and geometry of base-mounted free-standing equipment (after Hitchcock,[3] reproduced from *New Zealand Engineering* by permission of the New Zealand Institution of Engineers)

values of damping and period on Figure 11.2 shows that the acceleration response of this transformer to the proposed New Zealand standard earthquake would be nearly 2.0g if the period and the damping remain unchanged. It is known, however, that when foundations rock in this manner, the subsoil properties may be modified; its modulus of elasticity (and hence natural frequency) decreases while the energy dissipated per cycle (and hence equivalent damping) increases.

It has been estimated (Table 6.4) that the equivalent damping of rocking foundations can reach about 10 percent of critical as compared with about 20 percent for foundations moving vertically without rocking. As the overall equivalent damping factor for this example will probably lie in the range 2–10 percent, it can be seen from Figure 11.2 that this particular transformer would be subjected to a peak acceleration of 1.3g to 0.75g in an earthquake corresponding to this response spectrum.

Therefore the equivalent-static design method for basemounted freestanding equipment (including transformers and fastenings) should be as follows:

(1) If the natural period of vibration of the equipment as finally installed on its foundations is not known or is known to be larger than 0.1 s, then a design acceleration of 0.7g should be used, in conjunction with normal working stresses and with properly designed ductile material behaviour in the weakest part of the fixings.

(2) If the natural period of vibration of the equipment as finally installed on its foundations can be shown to be less than 0.1 s (and to remain so for

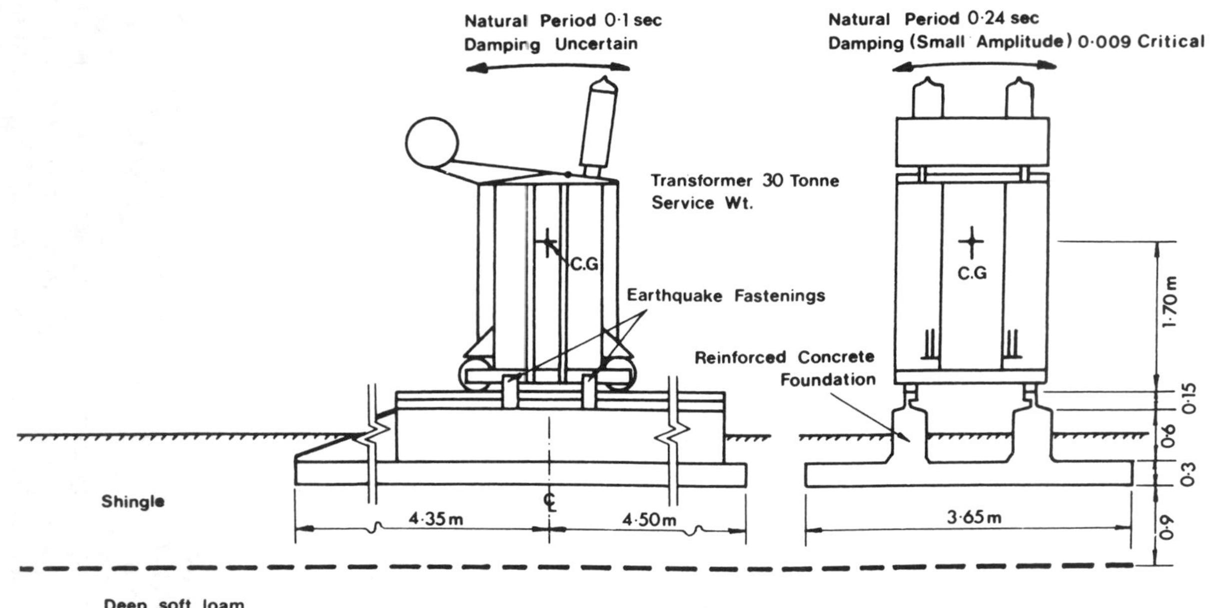

Figure 11.4 110/33 kV transformer on concrete pad (after Hitchcock,[3] reproduced from *New Zealand Engineering* by permission of the New Zealand Institution of Engineers)

accelerations up to 0.4g) then a design acceleration of 0.4g should be used, in conjunction with normal working stresses and properly designed ductile material behaviour. As mentioned previously, it is very difficult to be sure that the lower portion of the response spectrum for very small values of the period (T) can be safely used, and this provision should seldom be applied in practice.

In a study[6] of the dynamic response of transformers it was found that the equivalent-static force representative of the dynamic loads was best made by a single force applied at the top, this force varying depending on the type of mountings and whether shear or moment was being considered.

11.2.2.2 Equipment mounted on suspended floors of buildings or other structures

Equipment mounted in buildings is generally subjected to modified earthquake effects; this can mean amplification especially in the upper floors of buildings. This amplification of the ground and building motions is worse when the building and the equipment resonate, i.e. when they have equal periods of vibration. Fortunately damping between the building and the equipment can be used to drastically reduce amplification, as it is not always possible to avoid the resonance effect. Table 11.1 illustrates the effect of resonance and damping as obtained in a simple analysis by Shibata *et al.*[7]

Any rule of thumb for seismic design should require all equipment items in a building above ground floor to be designed and fastened for higher excitations than on the ground, as reflected in the difference in the provisions for single-storey and multi-storey buildings in the New Zealand code.[4,8]

When closer design modelling is required, such as when resonance may occur, the excitation at any given floor level may be found from the dynamic characteristics of the building. This is conveniently done in the form of floor response spectra.[9,10]

11.2.2.3 Equipment that would fail in a brittle manner in earthquakes

Hitchcock[3] reports on some tests on the dynamic characteristics of porcelain-supported equipment, some ground-mounted, some supported on concrete posts.

Table 11.1. Response of plant mounted in buildings. Plant resonant with building with fundamental periods in range 0.2–0.4 s

Fraction of critical damping for building	0.07	0.07	0.07	0.07
Friction of critical damping for plant item	0.007	0.02	0.1	0.2
Peak response of plant item to 1940 El Centro earthquake	8–10g	5–7g	3g	2g

The natural periods of vibration were found to be the range 0.2–0.4 s, corresponding to the peak of the response spectra in Figure 11.2. As the damping ranged from 0.018–0.006 of critical, the expected response varied from about 1.5g greater than 2.0g. Most of the items of equipment involved had strengths appreciably less than those required to withstand such accelerations. In order to deal with this situation Hitchcock[3] suggested three alternative procedures as follows:

(1) Provide the required strength with factors of safety of the order of 2–3 to cover uncertainties in the assessment of the strength of brittle materials.
(2) Provide ductile components that yield early enough to prevent the brittle components reaching breaking load.[11]
(3) Provide additional damping. This solution is quite practicable when dealing with small masses of electrical equipment.[12]

Equipment used in an electrical installation must in general be suitably rigid to avoid variations in clearance between live parts, and to limit the amount of flexibility to be provided in electrical connections. In fact any acceptable structure of equipment is unlikely to have a period longer than about 0.4 s. From Figure 11.2 it can be seen that for periods less than about 0.4 s acceleration is taken as constant for any given damping. Hence the relationship between acceleration response and damping can be plotted as in Figure 11.5.

Assuming that the New Zealand design earthquake is to be used, the following design rules for this type of brittle equipment may be adopted.

(1) If the amount of damping in the equipment is not accurately known the equivalent-static acceleration for equipment that fails in brittle components under horizontal loading should be 1.5g. This should be used as a factor of safety of 2.0 on the guaranteed breaking strength of the brittle portions, and with ordinary working stresses in the ductile parts of the structure.
(2) Alternatively, if satisfactory evidence is available of the amount of damping inherent in the equipment, the seismic coefficient may be that read from Figure 11.5 for that amount of damping.

These rules would be suitable for the design of standard items of equipment installed in any part of a seismic country, because any type of foundation, from extremely rigid to highly flexible, could be used without invalidating the underlying assumptions.

11.2.2.4 Design examples

For examples of design calculations using the preceding equivalent-static design coefficients on a variety of electrical equipment, readers are referred to the appendices of Hitchcock's paper[3] and reference 4.

11.2.2.5 Code seismic coefficients for equipment

The coefficient C_p for finding the horizontal base shear defined by equation (11.1) has not been widely codified. Examples for a variety of equipment are given in Table 11.2, as specified in New Zealand,[4,8] which is an area of moderate seismic risk.

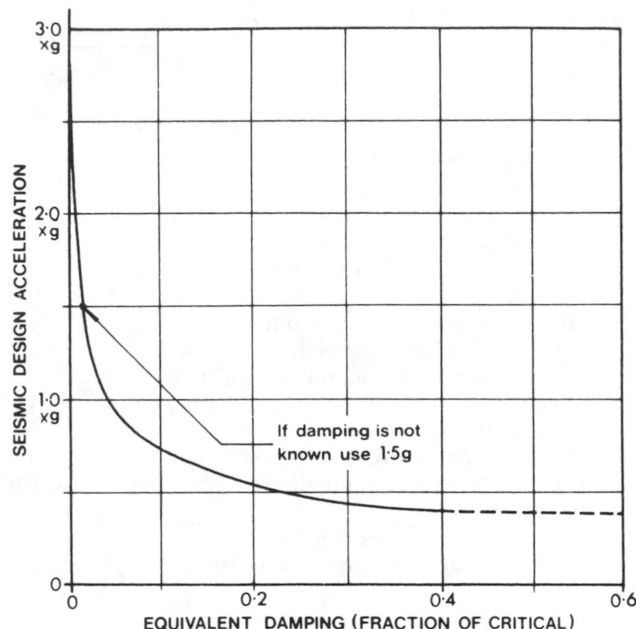

Figure 11.5 Seismic design accelerations for brittle structures with fundamental period less than 0.4 s (after Hitchcock,[3] reproduced from *New Zealand Engineering* by permission of the New Zealand Institution of Engineers)

Table 11.2. Seismic coefficients for equipment in buildings, as required in New Zealand[4,8]

Item	Part of portion	$C_{p\ max}$	$C_{p\ max}$
7	Towers not exceeding 10% of the mass of the building. Tanks and full contents, not included in item 8 or item 9; chimneys and smoke stacks and penthouses connected to or part of the building except where acting as vertical cantilevers:		
	(a) Single-storey buildings where the height to depth ratio of the horizontal force resisting system is:		
	(i) Less than or equal to 3	0.2	
	(ii) Greater than 3	0.3	
	(b) Multi-storey buildings where the height to depth ratio of the horizontal force resisting system is:		
	(i) Less than or equal to 3	0.3	
	(ii) Greater than 3	0.5	

(continued)

Table 11.2. *(continued)*

Item	Part of portion	$C_{p\ max}$	$C_{p\ min}$
8	Containers and full contents and their supporting structures; pipelines, and valves:		
	(a) For toxic liquids and gases, spirits, acids, alkalis, molten metal, or poisonous substances, liquid and gaseous fuels including containers for materials that could form dangerous gases if released:		
	(i) Single-storey buildings	0.6	0.5
	(ii) Multi-storey buildings	1.3	0.9
	(b) Fixed firefighting equipment including fire sprinklers, wet and dry riser installations, and hose reels:		
	(i) Single-storey buildings	0.5	0.3
	(ii) Multi-storey buildings	1.0	0.6
	(c) Other:		
	(i) Single-storey buildings	0.3	0.2
	(ii) Multi-storey buildings	0.7	0.4
9	Furnaces, steam boilers, and other combustion devices, steam or other pressure vessels, hot liquid containers; transformers and switchgear; shelving for batteries and dangerous goods:		
	(i) Single-storey buildings	0.6	0.5
	(ii) Multi-storey buildings	1.3	0.9
10	Machinery; shelving not included in item 9; trestling, bins, hoppers, electrical equipment not specifically included in other items 8,9 or 11; other fixtures:		
	(i) Single-storey buildings	0.3	0.2
	(ii) Multi-storey buildings	0.7	0.3
11	Lift machinery, guides, etc., emergency standby equipment	0.6	
12	Connections for items 8 to 11 inclusive shall be designed for the specified forces provided that the gravity effects of dead and live loads shall not be taken to reduce these forces		
13	Suspended ceilings including attached equipment, lighting and attached partitions, see clause 3.6.5	0.6	
14	Communications, detection or alarm equipment for use in fire or other emergency:		
	(i) Single-storey buildings	0.5	0.3
	(ii) Multi-storey buildings	1.0	0.6

The severity of the requirements compared with those for buildings is clear; for example, pipework for sprinkler systems in normal-use multi-storey buildings in Zone A of New Zealand should be designed for horizontal accelerations of up to 1.0*g*.

Such equivalent-static force values should be adequate unless equipment with low damping has a natural period of vibration close to one of the important periods of the building (see also Section 11.2.2.2).

11.3 ASEISMIC PROTECTION OF EQUIPMENT

11.3.1 Introduction

As discussed by Blackwell,[13] there are two main problems affecting the protection of many types of equipment. The first concerns movements, and the second energy absorption. Both problems are worsened if resonance or quasi-resonance exists. Movements can be dealt with in either of the following ways:

(1) By preventing serious relative displacement during an earthquake by anchoring the components of the installation to the building structure;
(2) By accommodating the relative movements of components without fracture of pipelines, ducts, cables and other connections. (These relative movements may result from movements of either the building fabric or the mechanical services components themselves.)

The energy absorption problem means dealing with the seismic stresses occurring in the equipment, its mountings, and its fastenings to the structure. The equipment may have to be strengthened, and damping devices may have to be fitted. Mountings should not be made too strong because, apart from the expense, this may cause the equipment to fail somewhere else. Also the resulting lower period of vibration sometimes leads to higher stresses; for example, it can be seen from Figure 11.2 that if equipment with a natural period of 1.4 s and 5 percent damping is stiffened so that its natural period decreases to 0.5 s, the inertial force will have increased three times.

It would be impracticable to make connections at positions of maximum sway on equipment whose natural period is above 1.0 s. Fitting limit stops might overcome this problem but such stops would have to be designed to limit shock loading (Section 11.3.3). Energy is absorbed by the deformation of fasteners, springs, and rubber mountings, but, if the materials have little natural damping, the deformation remains within the limits of elasticity, dissipation of energy may be insufficient for earthquake protection. In this respect springs and even rubber mountings may prove unsatisfactory. Hydraulic or friction dampers could be added to increase the energy absorption but this would be expensive and require detailed design.

Simpler methods of absorbing energy are usually possible, including the plastic deformation of supports and holding-down bolts, and the frictional work done when units slide about on the floor. Once plastic deformation has taken place, bolts will be slack on the return movement and this is when floor friction is useful. Floor friction is free from back-lash and shock effects apart from the deceleration at the end of the slide, and is generally free of cost; the unit of course must be designed not to tip over.

When fastenings are designed for plastic deformation, they should be proportioned and sized so that the stresses are evenly distributed throughout the whole volume of the material, because the amount of energy dissipated is directly proportional to the stress developed and to the volume of material developing stress. Fastenings should be free of weak links or stress concentrations which would result in early fracture of the fastening without much dissipation of energy.

11.3.2 Rigidly mounted equipment

11.3.2.1 Boilers, calorifiers, control panels, batteries, air-conditioners, kitchen equipment, and hospital equipment

(1) The first requirement is to prevent the equipment sliding across the floor. The coefficient of friction rarely exceeds 0.3, and the effectiveness of friction can be greatly reduced by the upwards component of the earthquake, so friction alone is unlikely to be sufficient. Mounting on bituminous-felt or lead would increase the friction and may be sufficient for less important equipment. The use of a suitable glue with neoprene pads would also give increased security against sliding.

(2) The next requirement is to ensure stability against overturning. In the first instance a simple geometric calculation will show whether the equipment is inherently stable or not. This is a function of the base width and the height of centre of mass of the equipment (Section 11.2.2.1 and Figure 11.3). If the horizontal acceleration is $0.6g$ (the maximum required for boilers in single-storey buildings in New Zealand, see Table 11.2), the holding-down force would be zero if the centre of gravity of the equipment is not higher then 0.84 times the width of the base.

(3) Where overturning stability cannot be obtained from geometric considerations, the equipment will have to be fastened to the building structure in some way. If this is done by holding-down bolts fixed into the floor, the bolts should be the weakest part of the system so that they protect the equipment by yielding first. This is particularly desirable when the equipment itself is not very strong. Fine-thread bolts with a length not less than ten times the diameter should be used, and the thread should be designed such that the ultimate strength of the bolt based on the thread root area exceeds the yield strength of the gross bolt area (see also Section 11.3.1). Restraint against overturning can in some cases be obtained by fastening the top of the equipment to walls or columns; but the walls in particular must be seen to be strong enough for this purpose.

(4) Pipework and electrical wiring connections are vulnerable and therefore must be strong. It would also be wise to allow some flexibility in the pipes and wires away from the equipment in case of relative seismic movement between the items on either side of the connections.
(5) Doors to control-panels should be hinged to prevent them being dislodged in earthquakes; loose covers can fall against live contacts, shorting out the equipment.
(6) Mercury switches should be avoided, as should essential instruments that have heavy movable components likely to break away from their supports.
(7) Boilers with extensive brickwork are undesirable as it is very difficult to reinforce the fire brick.

11.3.2.2 Chimneys

Chimneys should be subjected to a thorough seismic structural design (Appendix A.2). Lightweight double wall sheet-metal flues should be used where possible and prefabricated stacks should be avoided or used with great care.

11.3.2.3 Tanks

As well as considerations of sliding and overturning as discussed above, the following points are peculiar to tanks:

(1) Corrugations of copper tanks are liable to collapse with subsequent failure of the bottom joint. This can be remedied by making a stronger joint and possibly reducing the number of corrugations. Alternatively, welded stainless steel tanks can be used to increase the tank strength while retaining corrosion resistance.
(2) Where there will be a possibility of a tank sliding, severance of the connections can be avoided by flexibility in the pipes (ten diameters on each side of a bend should be adequate) and by provision of strong connections between the pipes and the tank. The bottom connection can be strengthened by passing it right through the tank and welding it at each end. The top connection could be similarly treated unless a large arm ball valve were required, when extra strengthening at the connection would be satisfactory.
(3) Suspended tanks should be strapped to their larger systems, and provided with lateral bracing.
(4) Because of the build-up of surface waves in liquid during earthquakes, some protection against liquid spillage may be desirable. This may be either in the form of a lid, or a spill tray with a drain under the tank. The effects of pressures on the tank due to the liquid oscillation may have to be taken into account in the design of larger tanks.

For further discussion of tanks, see Section A.2.6.1.

11.3.3 Equipment mounted on isolating or energy-absorbing devices

11.3.3.1 Introduction

In 1970 Blackwell[13] described flexible mountings as falling into groups relating to the predominant motions of short-period earthquakes, as follows:
Group 1. Mountings with a natural period less than the predominant earthquake period (i.e. below about 0.07 s)—felt, cork, and most rubber mountings would usually come into this category. Provided the mountings will not permit sliding to occur (e.g. by gluing) and the connections to the equipment are reasonably flexible, no further mounting precautions should be necessary.
Group 2. Mountings with a natural period corresponding to the predominant earthquake periods (i.e. above about 0.07 s). Spring mountings fall into this category, such as those used for low-speed fans, engines, compressors, and possibly electric motors. As resonance is likely, some method must be provided which limits the movement and transfers the forces directly to the floor instead of through the mounts. Steel rods or angles would be suitable and should be designed to yield at the design acceleration. An example of such a fastener is shown in Figure 11.6. Note the covering round the rod to reduce shock loading. Alternatively the rods could be replaced by multiple-strand steel wire. The flexibility normally provided in pipe, duct, and electric wire connections would be adequate for an earthquake.

11.3.3.2 Developments in isolation and energy-absorption methods

The successful development of seismic isolation and energy absorption devices for major structures, as described in Section 5.5, has been paralleled by similar applications for equipment. For example, as recommended about 1970 by Hitchcock and others[3,11,12] brittle equipment can be protected by energy-absorbing supports (Section 11.2.2.3).

Subsequent developments of the principles of Section 5.5 include the use on large industrial boilers supported from the top on laterally flexible hangers,[14,15] together with the use of energy absorbers.[14] This permits the use in seismic regions of equipment that is not in itself designed for earthquakes. The same approach has also been used for seismic protection of nuclear installations such as reactor vessels[16] and much other equipment[16–20] such as piping or computers.

11.3.4 Light fittings

Pendant fittings can have a wide range of natural frequencies. Wire-supported fittings may not fail but could swing and smash if brittle. Heavy fittings and brittle materials for supports should be avoided, as should any combination of low damping and low fundamental periods (in the range 0.2–1.0 s). In many cases the aseismic design of lighting will be related to that of suspended ceilings when the light fittings are attached to them (see also Table 11.2 and Section 12.4.2).

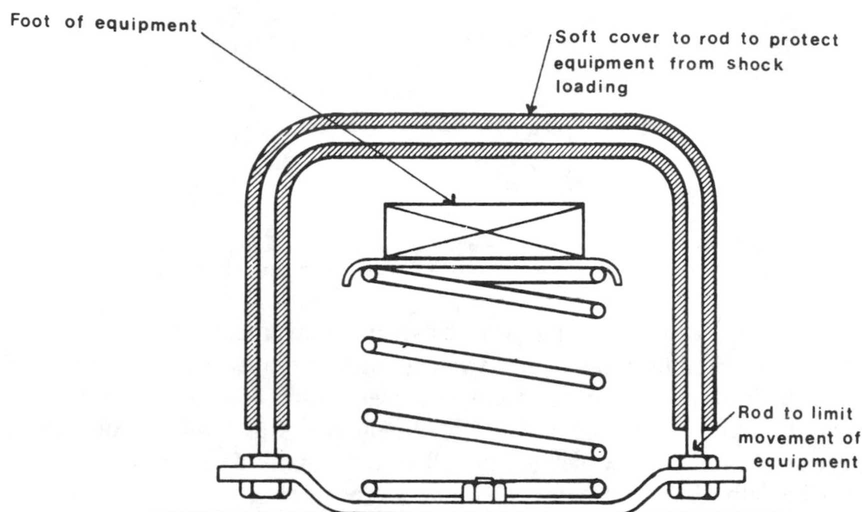

Figure 11.6 Detail of a flexible mounting with a resilient restraint against excessive movements (after Blackwell,[13] reproduced from *New Zealand Engineering* by permission of the New Zealand Institution of Engineers)

11.3.5 Ductwork

Ductwork is usually quite strong in itself, and despite relatively flexible hangers, it is usually susceptible to earthquake damage only where it crosses seismic movement gaps in buildings. At these points flexible joints should be provided which are long enough to take up the seismic movements. Canvas joints may be suitable, asbestos if there is a fire risk, or lead-impregnated plastic if noise is a problem. Wherever possible, seismic movement gaps in buildings should not be crossed. It may be possible to locate fire walls at seismic movement gaps and to design pipe and duct systems to be separate on each side of the gap, thus avoiding crossing the gap as well as keeping the number of systems down to a minimum.

The other most vulnerable position in ductwork is at its connection to machines (e.g. fans). At these positions flexible duct connections should be installed in a semifolded condition with enough material to allow for the expected differential deflection between the machines and the ductwork.

Duct openings and pipe sleeves through walls or floors should be large enough to allow for the anticipated movement of the pipes and ducts.

11.3.6 Pipework

11.3.6.1 Flexibility requirements

Flexibility is required in pipework to allow for building and equipment movement. Seismic flexibility requirements are different from those for

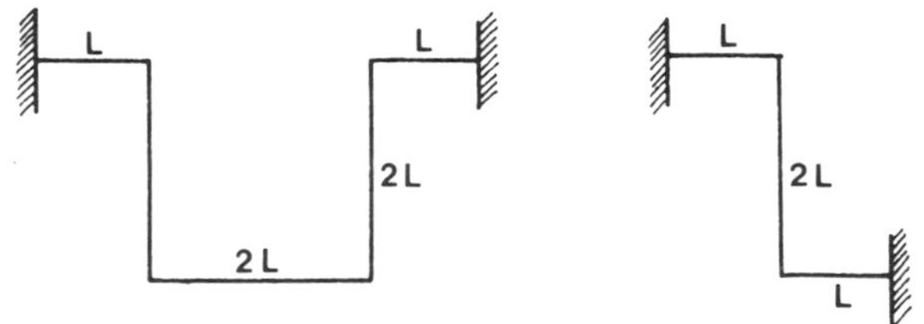

Figure 11.7 Suggested pipe arrangements for crossing movement gaps (reproduced from *New Zealand Engineering* by permission of the New Zealand Institution of Engineers)

accommodating thermal expansion, as seismic movements take place in three dimensions. Sliding joints or bellows cannot be used as they do not have the required flexibility and introduce a weakness which could cause early failure without making use of the ductility of the pipework as a whole. Accordingly, those expansion joints which are installed to accommodate thermal expansion must be fully protected from earthquake movements.

The movements should be taken up by bends, off-sets or loops which have no local stress concentrations and which are so arranged that if yielding occurs there will not be any local failure. Note that short-radius bends can cause stress concentrations. Anchors adjacent to loops must also be strong, and connections to equipment must be able to resist the pipe forces caused by earthquake movements. Connections using screwed nipples and some types of compression fittings should be avoided, unless they can be arranged so as to be unaffected by seismic movements.

U-bends and Z-bends as shown in Figure 11.7 can be used to obtain flexibility. The dimensions L should be determined by calculation, so as to give safe stresses in the pipes and at the supports for the applied seismic movements.

Where the laying of pipes across seismic movement gaps in buildings cannot be avoided, details as shown in Figures 11.7, 11.8, or 11.9 can be used. Such

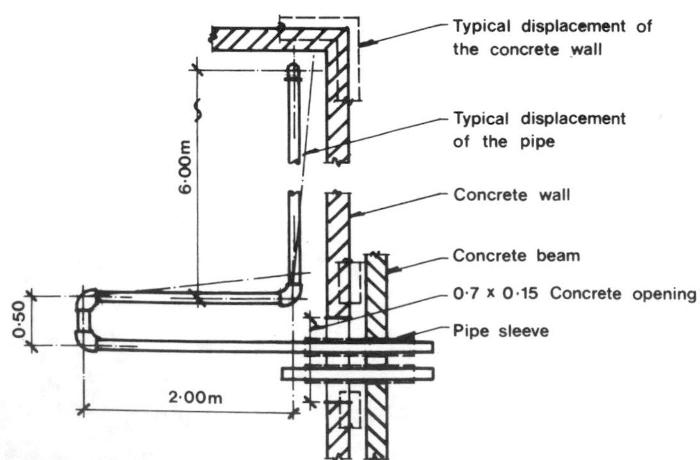

Figure 11.8 Plan view of pipework crossing a seismic movement gap (after Berry[21])

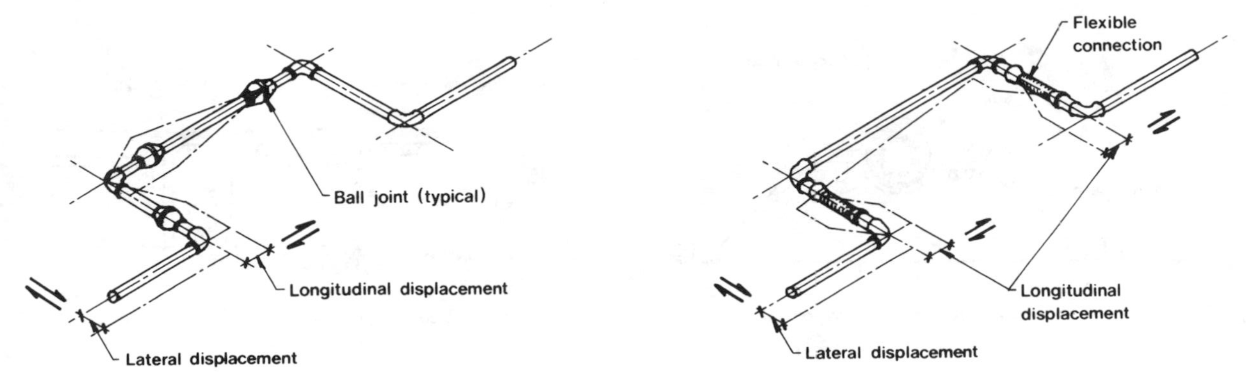

Figure 11.9 Pipework details for crossing seismic movement gaps where space limitations prevent the use of pipe loop shown in Figure 11.8 (after Berry[21])

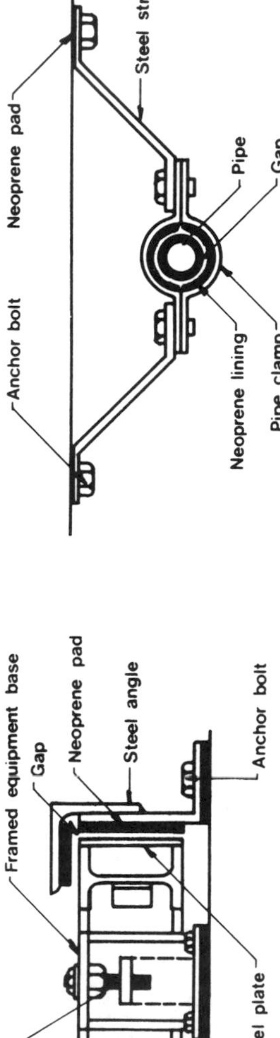

Figure 11.10 Combined earthquake mountings and vibration isolation for machine bases and pipework (after Berry[21])

crossings should be made at the lowest floor possible, in order to minimize the amount of movement which has to be accommodated.

Pipework should be tied to only one structural system. Where structural systems change, the relative deflections are anticipated, flexible joints should be provided in the pipework to allow for the same amount of movement. Suspended pipework systems should have consistent degrees of freedom throughout. For example, branch lines should not be anchored to structural elements if the main line is allowed to sway. If pipework is allowed to sway, flexible joints should be installed at equipment connections.

For further information on installation details applicable to any piping systems reference may be made to the Standards of the US National Board of Fire Underwriters for earthquake protection to fire sprinkler systems.

11.3.6.2 Methods of supporting pipework

Simple hangars will allow the pipe to swing like a pendulum. With usual support spacings, pipes will have a fundamental period of about 0.1 s if sideways movement is prevented at every support, and the period will increase to 0.2 s with twice this spacing, and to about 1.0 s with three times the spacing. The latter two periods are very close to common building periods and the resulting resonance would cause large movements, considerable noise and possible failure. This can be avoided by the provision of horizontal restraints or by the use of two hangars in a V-formation.

As noted in Section 11.3.3, specially designed energy absorbing supports for pipework are an aseismic design alternative that may have design advantages in some circumstances, especially in critical facilities.

REFERENCES

1. Housner, G. W., and Jennings, P. C., 'The San Fernando earthquake', *Earthquake Engineering and Structural Dynamics*, **1**, No. 1, 5–32 (1972).
2. US Department of the Interior and US Department of Commerce, 'The San Fernando California earthquake of February 9th, 1971', (A preliminary report), Geological Survey Professional Paper 733, US Govt, Washington (1971).
3. Hitchcock, H. C., 'Electrical equipment and earthquakes', *New Zealand Engineering*, **24**, No. 1 (1969).
4. Standards Association of New Zealand. 'Specification for seismic resistance of engineering systems in buildings', NZS 4219: 1983.
5. Schiff, A. J., 'Seismic design practice for electrical power substations', *Proc. 8th World Conf. on Earthq. Eng., San Francisco*, **VIII**, 181–8 (1984).
6. Chandrasekeran, A. R., and Singhal, N. C., 'Behaviour of a 220 kV Transformer under similated earthquake conditions', *Proc. 8th World Conf. on Earthq. Eng., San Francisco*, **VIII**, 149–56 (1984).
7. Shibata, H., Sata, H., and Shigeta, T., 'Aseismic design of machine structures', *Proc 3rd World Conf. on Earthq. Eng., New Zealand*, **2**, II–552 (1964).
8. Standards Association of New Zealand, 'Code of Practice for general structural design and design loadings for buildings', NZS 4203: 1984.

9. Asfura, A., and Der Kiureghian, A., 'A new floor response spectrum method for seismic analysis of multiply supported secondary systems', *Report No. UCB/EERC-84/04*, Earthquake Engineering Research Center, University of California, Berkeley (1984).
10. Igusa, T., and Der Kiureghian, A., 'Generation of floor response spectra including oscillator–structure interaction', *Earthquake Engineering and Structural Dynamics*, **13**, No. 5, 661–76 (1985).
11. Gilmour, R. M., and Hitchcock, H. C., 'Use of yield ratio response spectra to design yielding members for improving earthquake resistance of brittle structures', *Bull. NZ Soc. for Earthq. Eng.*, **4**, No. 2, 285–93 (1971).
12. Winthrop, D. A., and Hitchcock, H. C., 'Earthquake design of structures with brittle members and artificial damping by methods of direct integration', *Bull. NZ Soc. for Earthq. Eng.*, **4**, No. 2, 294–300 (1971).
13. Blackwell, F. N., 'Earthquake protection for mechanical services', *New Zealand Engineering*, **25**, No. 10, 271–5 (1970).
14. Hollings, J. P., and P., Sharpe R. D., and Jury, R. D., 'Earthquake performance of a large boiler', *Proc. 8th European Symposium on Earthquake Engineering, Portugal* (1986).
15. Lopez, G., Makimoto, Y., Mii, T., Mitsuhashi, K., and Pankow, B., 'Analytical–experimental dynamic analysis of the El Centro power plant Unit No. 4 steam generator. . .', *Proc. 8th World Conf. on Earthq. Eng., San Francisco*, **VII**, 189–96 (1984).
16. Spencer, P., 'The design of steel energy-absorbing restrainers and their incorporation into nuclear power plants for enhanced safety (Vol. 5): Summary Report', *Report No. UCB/EERC-80/34*, Earthquake Engineering Research Center University of California, Berkeley (1980).
17. Schneider, S., Lee, H. M., and Godden, W. G., 'Behaviour of a piping system under excitation: (etc.)', *Report No. UCB/EERC-80/03*, Earthquake Engineering Research Center, University of California, Berkeley (1982).
18. Mainante, J. A., *Seismic Mountings for Vibration Isolation*, Wiley-Interscience, New York (1984).
19. Chang, I. K., 'Dynamic testing and analysis of improved computer/clean room raised floor system', *Proc. 8th World Conf. on Earthq. Eng., San Francisco*, **V**, 1175–8 (1984).
20. Fujita, T., 'Earthquake isolation technology for industrial facilities—Research, development and applications in Japan', *Bull. NZ Nat. Soc. for Earthq. Eng.*, **18**, No. 3, 224–49 (1985).
21. Berry, O. R., 'Architectural seismic detailing', *State of the Art Report* No. 3. Technical Committee No. 12, Architectural–Structural Interaction IABSE–ASCE, International Conference on Planning and Design of Tall Buildings, Lehigh University (1972), (Conference Preprints, Reports Vol. 1a–12).

Chapter 12

Architectural detailing for earthquake resistance

12.1 INTRODUCTION

A large part of the damage done to buildings by earthquakes is non-structural. For instance, in the San Fernando, California, earthquake of February 1971, a total of $500 million worth of damage was done to the built environment of which over half was non-structural. The importance of sound anti-seismic detailing in earthquake areas should need no further emphasizing. The choice of a suitable structural form is crucial (Section 5.3.7).

Buildings in their entirety should be tailored to ride safely through an earthquake and the appropriate relationship between structure and non-structure must be logically sought. For the effect of non-structure on the overall dynamic behaviour of a building see Section 5.3.8.2, where the question of full separation or integration of infill panels into the structure is discussed.

Architectural items such as partitions, doors, windows, cladding and finishes need proper seismic detailing; many non-seismic construction techniques do not survive strong earthquake motion as they do not provide for the right kinds or size of movements. Detailing for earthquake movements should, however, be considered in conjunction with details for the usual movements due to live loads, creep, shrinkage, and temperature effects. As with so many other problems, it is worth saying that good planning can provide the right framework for practical aseismic details.[1]

An ironic example of the inadequacy of a non-structural item comes from the San Fernando earthquake; a modern firestation withstood the earthquake satisfactorily with regard to its structure, but the main doors were so badly jammed that all the fire engines were trapped inside.

Unfortunately there is little literature available giving specific guidance on aseismic architectural detailing. Indeed few countries seem to have Codes of Practice on this subject, though there are helpful clauses in the City of Los Angeles building bylaws. Little basic research had been done in this area and it appears that architects in earthquake areas to date have largely relied on details considered to be 'good practice', without discussing their experience.[2] We are forced to start almost from square one, observe what goes wrong with architectural details in earthquakes, and try to prevent repetitions.[3] The

San Fernando and more recent carthquakes caused numerous failures, many on photographic record,[4] from which we can learn.

There are some signs of awakening interest, at least in California and New Zealand; for example, Arnold and Reitherman's work on architectural configuration,[5] Durkin's studies of causes of deaths and injuries in earthquakes such as in the March 1985 San Antonio, Chile, earthquake,[6] and the US–New Zealand review of design and construction practice for architectural elements in earthquakes.[7,8] In the latter the need for architectural aseismic detailing to receive its due attention has been highlighted by Hopkins *et al.*,[8] a need which has been responded to by an experimental research programme[9,10] on the dynamic response behaviour of non-structural partitions and suspended ceilings started in California in the 1980s.

12.2 NON-STRUCTURAL INFILL PANELS AND PARTITIONS

12.2.1 Introduction

The recommendations of this section should be applied in conjunction with normal design considerations regarding creep, shrinkage, and temperature effects which overlap, but are generally less exacting than the seismic design requirements for infill panels.

In earthquakes all buildings sway horizontally producing differential movements of each floor relative to its neighbours. This is termed storey drift (Figure 12.1), and is accompanied by vertical deformations which involve changes in the clear height h between floors and beams.

Any infill panel should be designed to deal with both these movements. This can be done by either (1) integrating the infill with the structure or (2) separating

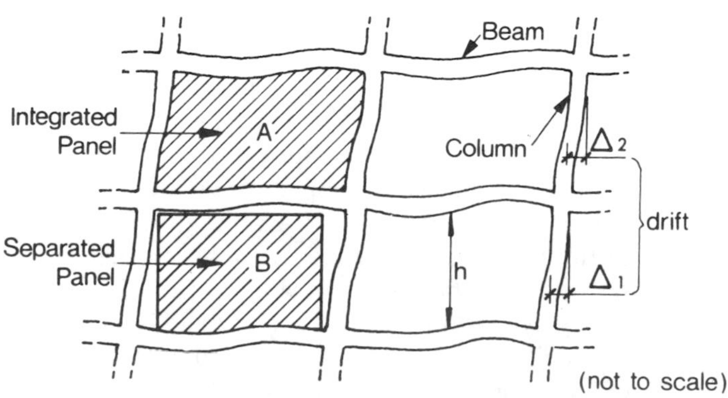

Figure 12.1 Diagrammatic elevation of structural frame and non-structural infill panels

the infill from the structure. A discussion of both systems of constructing infill panels follows, while further guidance on the aseismic effectiveness of some types of partitions may be found in Reference 10.

12.2.2 Integrating infill panels with the structure

In this case the panels will be in effective structural contact with the frame such that the frame and panels will have equal drift deformations (Panel A in Figure 12.1). Such panels must be strong enough (or flexible enough) to absorb this deformation, and the forces and deformations should be computed properly. Where appreciably rigid materials are used the panels should be considered as *structural* elements in their own right as discussed in Sections 6.6.6 and 8.5. Reinforcement of integrated rigid walls is usually necessary if seismic deformations are to be satisfactorily withstood.

Integration of infill and structure is most likely to be successful when very flexible partitions are combined with a very stiff structure (with many shear walls). Attention is drawn to the fact that partitions not located in the plane of a shear wall may be subjected to deformations substantially different from those of the shear wall. This is particularly true of upper-storey partitions.

Light partitions may be dealt with by detailing them to fail in controlled local areas thus minimizing earthquake repairs to replaceable strips (Figure 12.2).

Finding suitably flexible construction for integral infill may not be easy, especially in beam and column frames of normal flexibility. These may experience a storey drift of as much as 1/100 of the storey height in an earthquake.

12.2.3 Separating infill panels from the structure

(See Figure 12.1, Panel B.) For important structural reasons this method of dealing with non-structural infill is likely to be preferable to integral construction when using flexible frames in strong earthquake regions. The size of the gap

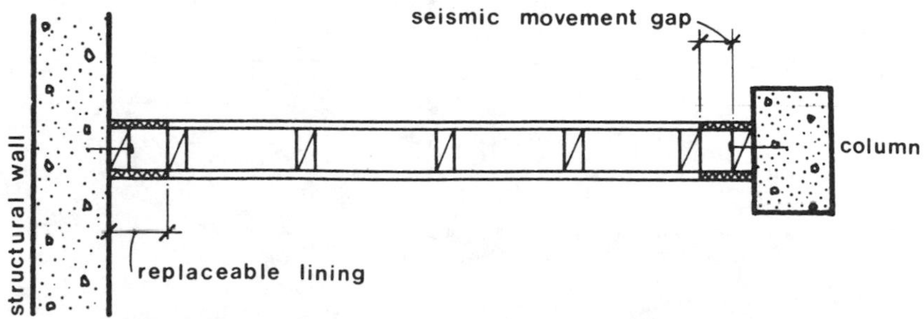

Figure 12.2 Lightweight partition detailed so that earthquake hammering by the structure will damage limited end strips only

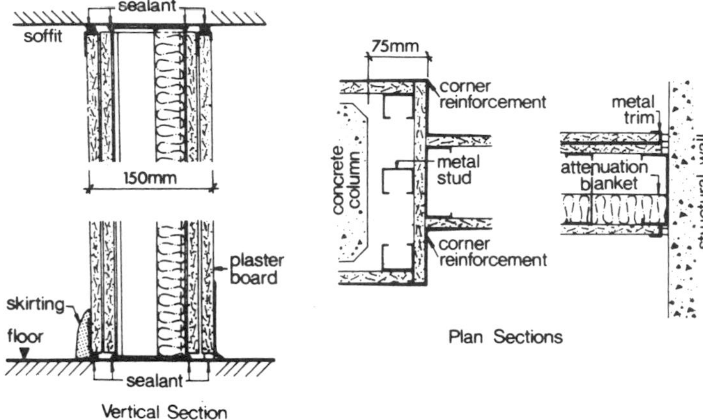

Figure 12.3 Light partition details for small seismic movements (i.e. suitable for stiff-framed buildings or small earthquakes)

between the infill panels and the structure is considerably greater than that required in non-seismic construction. In the absence of reliable computed structural movement, it is recommended that horizontal and vertical movements of between 20 mm and 40 mm should be allowed for. The appropriate amount will depend on the stiffness of the structure, and the structural engineer's advice should be taken on this.

This type of construction has two inherent detailing problems which are not experienced to the same extent in non-seismic areas. First, awkward details may be required to ensure lateral stability of the elements against out-of-plane forces. Second, soundproofing and fireproofing of the separation gap is difficult. Moderate soundproofing of the movement gap can be achieved with cover plates or flexible sealants, but where stringent fireproofing and soundproofing requirements exist, the separation of infill panels from the structure is inappropriate. Designers should be careful in the choice of so-called 'flexible'

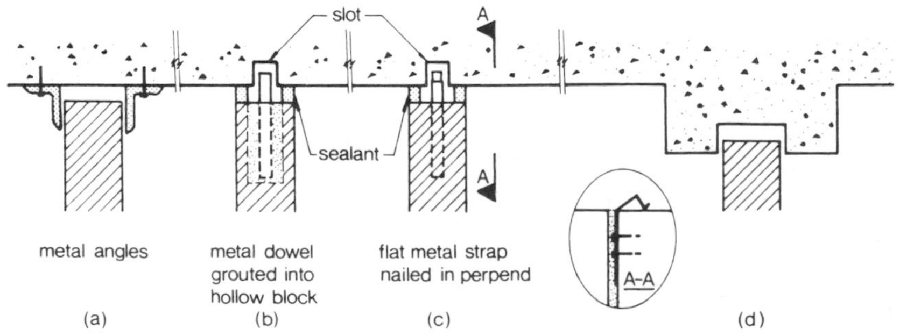

Figure 12.4 Separated stiff partitions: top details for lateral stability of brick or block walls (see Section 12.2.2)

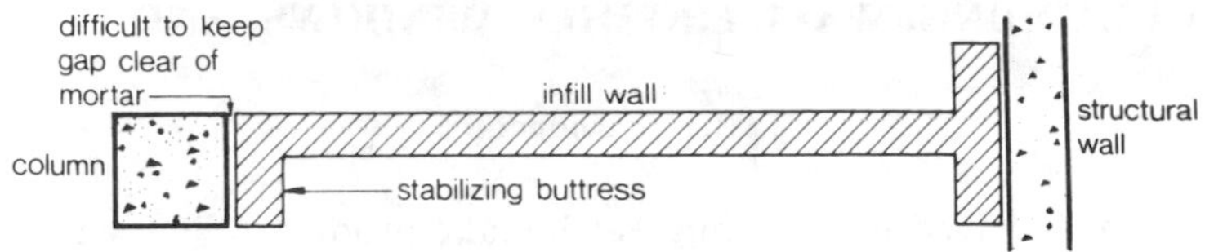

Figure 12.5 Separated stiff partition: plan view of stabilizing buttress systems

materials in movement gaps; the material must be not only sufficiently soft but also permanently soft. Both polysulphide and foamed polyethylene are *not* flexible enough (or weak enough) in this situation.

It is in fact difficult to find a suitable material; Mono-Lasto-Meric is both permanently and sufficiently soft, but is not suitable for gap widths exceeding 20 mm. Foamed polyurethane is probably the best material from a flexibility point of view and will provide modest sound-insulation, but may have little fire resistance. A fire-resistant possibility is Declon 156, a polyester/polyurethane foam which intumesces in fire conditions.

Figures 12.3 to 12.6 show some details used for separated infill panels. Note that great care has to be taken during both detailing and building to prevent the gaps being accidentally filled with mortar or plaster. Figure 12.6 shows a detail which helps prevent plaster bridging the gap. Further details suitable for small seismic movements may be found elsewhere.[11]

12.2.4 Separating infill panels from intersecting services

Where ducts of any type penetrate a full-height partition, the ducts should not be tied to the partition for support. Support should occur on either side of the partition from the building structure above. If the opening is required to be sealed because of fire resistance or acoustics, the sealant should be of a resilient non-combustible type to permit motion of the duct without affecting the partition of duct. It is important for both seismic and acoustic considerations that the duct be independently supported by hangers and horizontal restraints from the building structure.

Further discussion of ducts is to be found in Section 11.3.6, and for some remarks on the required properties of gap sealants around ducts, see discussion on infill panels in Section 12.2.3.

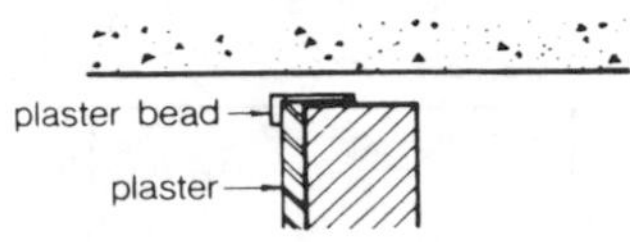

Figure 12.6 Plastering detail to ensure preservation of gap between partition and structure

12.3 CLADDING, WALL FINISHES, WINDOWS, AND DOORS

12.3.1 Introduction

The problems involved in providing earthquake-proof details for these items are the same in principle as those for partitions as discussed in the preceding section. Their in-plane stiffness renders them liable to damage during the horizontal drift of the building, and the techniques of integral or separated construction must again be logically applied.

12.3.2 Cladding and curtain walls

Precast concrete cladding is discussed in Section 7.4. Suffice it here to point out that in flexible buildings, non-structural precast concrete cladding should be mounted on specially designed fixings which ensure that it is fully separated from horizontal drift movements of the structure. Brick or other rigid cladding should be either fully integral and treated like infill walls (Sections 6.6.6, and 8.5), or should be properly separated with details similar to those for rigid partitions (Figure 12.4, 12.5) or for parapets such as shown on Figure 12.7.

External curtain walling may well be best dealt with as fully-framed prefabricated storey-height units mounted on specially-designed fixings capable of dealing with seismic movements in a similar way to precast concrete cladding, as mentioned above.

12.3.3 Weather seals

Weather seals that may be damaged in severe earthquakes should be accessible and suitable for replacement.

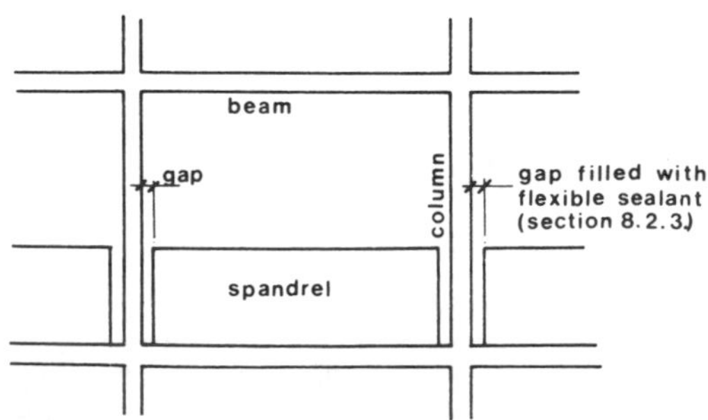

Figure 12.7 Detail of external frame showing separation of spandrel or parapet from columns to avoid unwanted interaction

12.3.4 Wall finishes

Brittle or rigid finishes should be avoided or specially detailed on any walls subjected to shear deformations, i.e. drift as applied to panel A, Figure 12.1. This applies to materials such as stone facings or most plasters. In Japan it is recommended that stone facings should not be used on walls where the storey drift is likely to be more than 1/300.

Brittle veneers such as tiles, glass, or stone should not be applied directly to the inside of stairwells, escalators, or open wells. If they must be used, they should be mounted on separate stud walls or furrings. Preferably the stairwells should be free of material which may spall or fall off and thus clog the exit way or cause injury to persons using the area.

Heavy ornamentation such as marble veneers should be avoided in exit lobbies. If a veneer of this type must be used, it should be securely fastened to structural elements using appropriate structural fastenings to prevent the veneers from spalling off in the event of seismic disturbance.

Plaster on separated infill panels must be carefully detailed to prevent its bridging the gap between panel and structure (Figure 12.6) as this may cancel the purpose of the gap, resulting in damage to the plaster, the infill panel and the structure.

12.3.5 Windows

It is worth observing that in the 1971 San Fernando, California, earthquake, which caused $500 million worth of damage, glass breakage cost more than any other other single item.

Window sashes should be separated from frame action except where it can be shown that no glass breakage will result. If the drift is small, sufficient protection of the glass may be achieved by windows glazed in soft putty (Figure 12.8) where the minimum clearance c all round between glass and sash is such that

$$c > \frac{\Delta_w}{2[1+(h/b)]} \tag{12.1}$$

The failure mode of hard putty glazed windows tends to be of the explosive buckling type and should be used only where sashes are fully separated from the structure, for example when glass is in a panel or frame which is mounted on rockers or rollers as described in Section 7.4. Further discussion of window behaviour in earthquakes may be found elsewhere.[12,13]

12.3.6 Doors

Doors which are vital means of egress, particularly main doors of highly populated and emergency service buildings, should be specially designed to

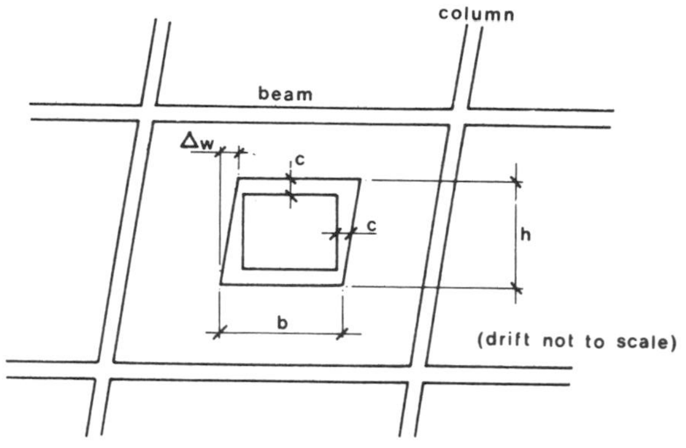

Figure 12.8 Detail of external frame with window glazing set in soft putty

remain functional after a strong earthquake. For doors on rollers, the problem may not be simply a geometric one dealing with the frame drift Δ, but may also involve the dynamic behaviour of the door itself.

12.4 MISCELLANEOUS ARCHITECTURAL DETAILS

12.4.1 Exit requirements

Every consideration should be given to keeping the exit ways clear of obstructions or debris in the event of an earthquake. As well as the requirements for wall finishes and doors outlined in Sections 12.3.4 and 12.3.6, the following points should be considered.

Floor covers for seismic joints in corridors should be designed to take three-dimensional movements, i.e. lateral, vertical, and longitudinal. Special attention should be given to the lateral movement of the joints.

Free-standing showcases or glass lay-in shelves should not be placed in public areas, especially near exit doors. Displays in wall-mounted or recessed showcases should be tied down so that they cannot come loose and break the glass front during an earthquake. Where this is impracticable tempered or laminated safety glass should be used for greater strength.

Pendant-mounted light fixtures should not be used in exit ways. Recessed or surface-mounted independently supported lights are preferred.

12.4.2 Suspended ceilings

In seismic conditions ceilings become potentially lethal. Individual tiles or plaster may jar loose from the supports and fall. Ceiling-supported light fixtures may loosen and drop out, endangering persons below. Thus alternatives to the

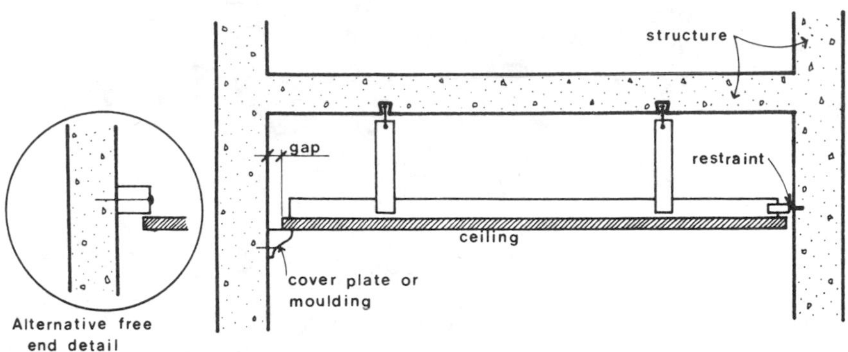

Figure 12.9 Details at periphery of suspended ceilings to prevent hammering and excessive movement

standard ceiling construction procedures should be considered. A thorough review of the seismic hazard from suspended ceilings and detailing recommendations has been given by Clarke and Glogau,[14] while studies on dynamic response behaviour have been made by Rihal and Granneman.[10]

The horizontal components of seismic forces to which a ceiling may be subjected can be allowed for in several ways. A dimensional allowance should be made at the ceiling perimeter for this motion so as to minimize damage to the ceiling where it abuts the walls: one way of doing this is to provide a gap and a sliding cover (Figure 12.9). Some ceiling suspension systems need additional horizontal restraints at columns and other structural elements, such as diagonal braces to the floor above, in order to minimize ceiling motion in relation to the structural frame.[14] This will reduce hammering damage to the ceiling and tiles will be less likely to fall out. The suspension system for the ceiling should also minimize vertical motion in relation to the structure.

Lighting fixtures which are dependent upon the ceiling system for support should be securely tied to the ceiling grid members. If such support is likely to be inadequate in earthquakes, the light fixtures should be supported independently from the building structure above. Diffuser grilles, if required for the air supply system, should also be hung independently.

In seismic areas, a lay-in T-bar system for ceiling construction should be avoided if at all possible, as its tiles and lighting fixtures drop out in earthquakes. In both the 1964 Alaska and the 1971 San Fernando earthquakes, the economical (and therefore popular) exposed tee grid suspended ceilings suffered the greatest damage. Evidently the differential movement between the partitions and the suspended ceilings damaged the suspension systems, and as the earthquake progressed the ceilings started to sway and were battered against the surrounding walls. This damage was aggravated when the ceilings supported lighting fixtures, and in many instances the suspension systems were so badly damaged that the lighting fixtures fell.

The need for independent support and lateral bracing of lighting fixtures mounted in suspended ceilings requires further study. The City of Los Angeles

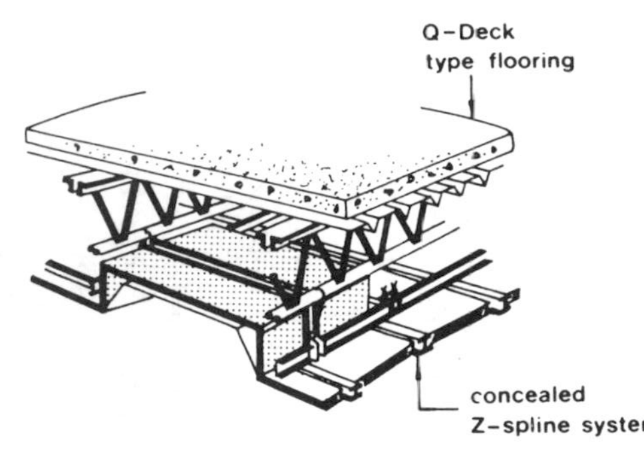

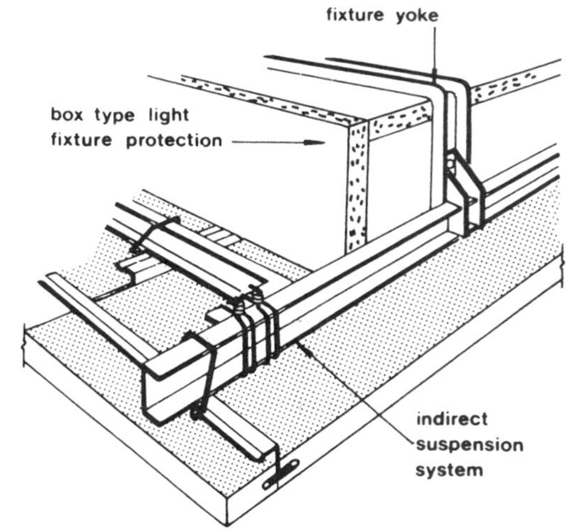

Figure 12.10 Two details of suspended ceiling construction providing movement restraint and secure tile fixing (after Berry[3])

has a regulation which stipulates minimum requirements for ceiling suspension systems supporting acoustic tile ceilings and lighting fixtures. It requires that ceiling suspension systems be designed to support a minimum load of 2.5 pounds per square foot of ceiling area, except that if the suspension system also supports lighting fixtures, this requirement is increased to 4 pounds per square foot. It also stipulates that the lighting fixtures shall not exceed 50 percent of the ceiling area and that they be fastened to the web of the load-carrying member. It does not, however, require independent support of the lighting fixtures or any lateral bracing.

Damage to ceilings can also occur where sprinkler heads project below the ceiling tiles. One way of minimizing this problem is to mount the heads with a swivel joint connection so that the pipe may move with the ceiling. Figures 12.9 and 12.10 give suggestions for seismic detailing of suspended ceilings.

12.4.3 Landscape elements

An interesting feature of the 1971 San Fernando earthquake was that many of the free-standing items of landscape furniture were found upside down after the earthquake. In order for this to have occurred, the items of furniture must have been subjected to horizontal forces equal to their own weights, throwing them about dangerously. This suggests that heavy items of movable landscape furniture should be secured to the ground in strong motion areas in order to prevent personal injury.

12.4.4 Window-washing rigs

Window-washing rigs should be restrained close to the building against earthquake forces as well as wind. Curtain wall mullions may have to accommodate this additional load, or instead of structural mullion guides attached to the building structure, a spring-loaded roller arrangement may be included on the window-washing rig provided the building structure has projecting fins between which the window-washing rig can ride. The roof-mounted carriage of the window-washing rig should also be secured to the building structure against seismic forces.

REFERENCES

1. Brandenburger, J., 'Internal details that permit movement', *Symposium on Design for Movement in Buildings*, The Concrete Society, London, 14 October 1969.
2. Housner, G. W., and Jennings, P. C., 'The San Fernando California earthquake', *Earthquake Engineering and Structural Dynamics*, **1**, No. 1, 5–32 (1972).
3. Berry, O. R., 'Architectural seismic detailing', *State of the Art,* Report No. 3. Technical Committee No. 12, Architectural-Structural Interaction. IABSE–

ASCE International Conference on Planning and Design of Tall Buildings, Lehigh University (1972) (Conference Preprints, Reports Vol. 1a–12).

4. US Department of Commerce, National Bureau of Standards, 'The San Fernando, California, earthquake of February 9, 1971', *NBS Report* 10556, March (1971).
5. Arnold, C., and Reitherman, R., *Building Configuration and Seismic Design*, John Wiley and Sons, New York (1982).
6. Durkin, M. E., and Aroni, S., 'The cause of casualties and the response of the care system in the March 3rd 1985 Chile earthquake', Presented at Fourth Chilean Conf. on Seismology and Earthq. Eng., Viña del Mar, Chile (1986).
7. Arnold, C., Hopkins, D. C., and Elsessor, E., *Architectural detailing for seismic damage control*, Report to the National Science Foundation, USA, 1986.
8. Hopkins, D. C., Massey, W. E., and Pollard, J. L., 'Architectural elements in earthquakes, a review of design and construction practice in New Zealand', *Bull. NZ Nat. Soc. for Earthq. Eng.*, **18**, No. 1, 21–40 (1985).
9. Rihal, S. S., and Granneman, G., 'Experimental investigation of dynamic behaviour of building partitions and suspended ceilings during earthquakes', *Proc. 8th World Conf. on Earthq. Eng., San Francisco*, **V**, 1135–40 (1984).
10. Rihal, S. S., and Granneman, G., *Dynamic behaviour of non-structural partitions and ceiling systems during earthquakes*, Research report submitted to National Science Foundation, USA, September (1983).
11. Toomath, S. W., 'Architectural details for earthquake movement', *Bull. NZ Soc. for Earthq. Eng.*, **1**, No. 1, (1968).
12. Bouwkamp, J. G., 'Behaviour of window panels under in-plane forces', *Bull. Seism. Soc. Amer.*, **51**, No. 1, 85–103 (Jan. 1961).
13. Osawa, Y., Morishita, T., and Murakami, M., 'On the damage to window glass in reinforced concrete buildings during the earthquake of April 20, 1965', *Bull. Earthquake Research Institute, University of Tokyo*, **43**, 819–27 (Dec. 1965).
14. Clarke, W. D., and Glogau, O. A., 'Suspended ceilings: the seismic hazard and damage problem and some practical solutions', *Bull. NZ Nat. Soc. for Earthq. Eng.*, **12**, No. 4, 292–304 (1979).

Appendix A

Earthquake resistance of specific structures

The previous twelve chapters describe the different phases of the design process illustrated in the figure in the introduction to this book and provide the design basis for a wide variety of structures and elements thereof. However, some types of structure have problems of earthquake resistance peculiar to themselves, and are conveniently dealt with in chapters specific to the type of structure concerned. Hence the appendix provides design guidance for several types of structure, and should be used in conjunction with the previous more general chapters of this document.

A.1 EARTHQUAKE RESISTANCE OF BRIDGES

A.1.1 Introduction

Until the late 1960s reports of serious earthquake damage to bridges were relatively few compared to those of other structures. Since then earthquakes in California, Papua–New Guinea and New Zealand have added considerably to our knowledge of the seismic response of bridges.[1–5] These earthquakes have demonstrated that bridges are vulnerable to differential longitudinal, lateral and vertical movements at piers and abutments. Dealing with these relative movements gives major design problems at junctions between horizontal and vertical members and within the supports themselves; the deck members are generally only modestly affected by earthquake stresses.

Perhaps the other main lesson from earthquake damage has been realizing the importance of exercising control over the failure modes, i.e. ensuring that energy dissipation occurs in desired locations.

Ongoing research has been contributing to great advances in the aseismic design of bridges, and international pooling of experience and identification of problems has assisted this process.[6]

A.1.2 Reliable seismic behaviour of bridges

For obtaining reliable seismic response behaviour of bridges we should apply the principles concerning choice of form, materials, and failure modes (discussed below) together with the more general factors discussed in Section 5.3 wherever they are appropriate to bridges.

Table A.1. Advantages and disadvantages of various configurations of bridge structure (after Chapman[7])

	Advantages	Disadvantages
1. *Multiple simply supported spans* (a) *All spans separate but restrained by:* *Longitudinally:*		
(i) Fixing to piers via shear keys and/or linkage bolts at each end of each span.	Good integrity in earthquake. Tolerant to differential settlement. Longitudinal forces can be shared between piers	Provisions required for allowing superstructure to shorten
(ii) Fixing to piers via shear keys only at one end of each span with freedom to slide at the other.	Superstructure shortening effects create no problems	Precautions necessary to prevent free span end from leaving pier—hydraulic shock absorbers may serve this purpose.
(iii) Linkage bolts to adjacent spans—all spans interconnected and sliding over intermediate piers. Restraint at one abutment.	Superstructure shortening effects create no problems	Unpredictably large horizontal inertia forces at abutment, plus small vertical reaction, likely to require heavy anchorage system
Transversely: (i) Fixing to piers and abutments via concrete or steel shearkeys		
(b) In situ *concrete deck cast continuously on top of simple spans for full bridge length. Horizontal restraint by:* *Longitudinal:*		
(i) Fixing to some or all piers via shear keys; sliding over remainder.	All advantages of 1a(i) above plus: Deck joints eliminated. Unequal transverse seismic response of piers can be redistributed	As 1a(i) above plus: Detail required to avoid slab damage being caused by differential vertical movement of beam ends

(ii)	Restraint at one abutment—sliding over intermediate piers	As 1a(iii) above	As 1a(iii) above
	Transversely:		
(i)	Fixing to some or all piers via concrete or steel shear keys	Superstructure may be free at abutment	
(c)	In situ *concrete deck, beam ends and diaphragms cast to create live load moment continuity between spans. Horizontal restraint by:*		
	Longitudinally:		
(i)	Fixing to some or all piers via shear keys (non-moment connection); sliding over remaining piers	As 1b(i) above except for differential settlement	As 1a(i) above plus: Lacks much tolerance to differential settlement
(ii)	Fixing to some or all piers via monolithic (moment resisting) connection; sliding over remaining piers	As 1c(i) above plus: added redundancy should give more security against collapse	As 1c(i) above plus: Increased pier stiffness leads to increased effects of shortening
(iii)	Restraint at one abutment—sliding over intermediate piers	As 1a(iii) above	As 1a(iii) above
	Transversely:		
(i)	Fixing to some or all piers via shear keys—two or more bearings; or—single bearing	Use of alternative depends on stabilizing from other piers and torsional strength of superstructure	
(ii)	Fixing to some or all piers via monolithic connection		Transverse deflection of pier leads to rotation of top of single stem pier—induces torsion in s/structure and uplift at ends of curved bridge
2	*Continuous spans of various types (slabs, box girders, etc.)*	These, for seismic purposes, are similar to 1(c) above	

A.1.2.1 Choice of structural form for bridges

As with buildings, the choice of structural form can have considerable bearing on seismic performance and costs. Unfortunately important non-seismic factors may conflict with purely seismic considerations when selecting the form of the superstructure, and compromises must be made.[7] For example, it is desirable for earthquake resistance purposes to make the superstructure as continuous and redundant as possible, but deck-shortening effects can cause larger pier moments in monolithic construction. On the other hand, if the deck is arranged to slide over the piers, difficulties arise in providing a satisfactory anchorage structure, as the horizontal inertia forces are then concentrated at fewer supports. The optimum solution for a given structure will depend on achieving a balance between pier heights, span lengths, and foundation problems.

The advantages and disadvantages of various solutions for average road bridge and elevated motorway construction are summarized in Table A.1, and have been further discussed by Chapman.[7]

A.1.2.2 Failure mode control for bridges

Ductility at any location will improve the behaviour near failure, but the most cost-effective degree of ductility supplied will not necessarily be high (Section 7.2.3.2) and will in any case vary, depending on factors such as the overall geometry of the bridge. The number (if any) and location of energy-dissipation points will need to be decided, the energy being dissipated by either plastic hinge zones or specific energy-dissipating devices. The repairability of bridges is an important aspect of aseismic design, and it is good practice to structure the bridge so that potential damage zones such as plastic hinges are reasonably accessible, e.g. not deeply buried below ground or water.

The above factors have been used in New Zealand[8] to determine the maximum degree of ductility, in terms of the deflection ductility factor μ (Section 6.6.7.3i), that should be assigned in various circumstances, as set out in Table A.2 and illustrated in Figure A.1. In Table A.2 a *ductile structure* is one that behaves effectively in an ideal elastoplastic manner, while a *partially ductile structure* behaves such that after yield there is a significant upward slope in the force-displacement curve up to the design displacement. Short bridges (up to 80 m long) with relative displacement provisions at the abutments, where the inertia forces are taken directly into the ground at the abutments, are considered 'locked in' to the ground motions so that their effective period of vibration is $T=0$.

In the footnote to Table A.2, 'Type L1 limited ductility' refers to structures where the ductility *demand* is limited (i.e. $1 \leqslant \mu \leqslant 6$) because its proportions are such that its yield strength exceeds the minimum specified load. 'Type L2' structures simply have limited ductility simply because of their detailing.

Table A.2. Design deflection ductility factor, μ maximum allowable values for bridges in New Zealand[8]

Energy-dissipation category:	μ
Ductile (type D), or partially ductile structure (type P1), in which plastic hinges form at design load intensity, above ground or water level	6
Ductile (type D), or partially ductile structure (type P1), in which plastic hinges form in reasonably accessible positions, e.g. less than 2 m below ground, but not below water level	4
Ductile (type D), or partially ductile structure (type P1), in which plastic hinges are inaccessible, forming more than 2 m below ground or below water level, or at a level not precisely predictable	3
Partially ductile structure (type P2)	3
Spread footings designed to rock	3
Hinging in raked piles in which earthquake load induces large axial forces	2
Structure 'locked in' to the ground ($T=0$)	1
Structure of limited ductility (type L3, elastic)	1

Note: The ductility factor for structures of limited ductility (types L1 and L2) is to be determined from actual member properties

A.1.3 Seismic analysis of bridges

For bridges both dynamic and equivalent-static analyses are used, the respective advantages and limitations of which are discussed in Chapter 6. For important bridges, dynamic analysis may be desirable, and where foundations are constructed in, or driven through, softer soils a dynamic response analysis of the site may be considered essential.

In conjunction with the above criteria a sensible condition requiring dynamic analysis is for physical situations where it is not appropriate to represent the structure as a single degree of freedom oscillator,[8] such that equivalent-static load analysis will be unreliable. Such situations include:

(1) Bridges where the mass of the piers (including added masses) is significant compared to the mass of the deck.
(2) For transverse analysis, where the stiffness varies significantly between adjacent piers of abutments (including the effects of foundations and bearings).
(3) Bridges where other conventional piers and abutments comprise the seismic load resisting system, e.g. suspension, cable-stayed, and arch bridges, or when energy-dissipating devices or uplifting piers are used.

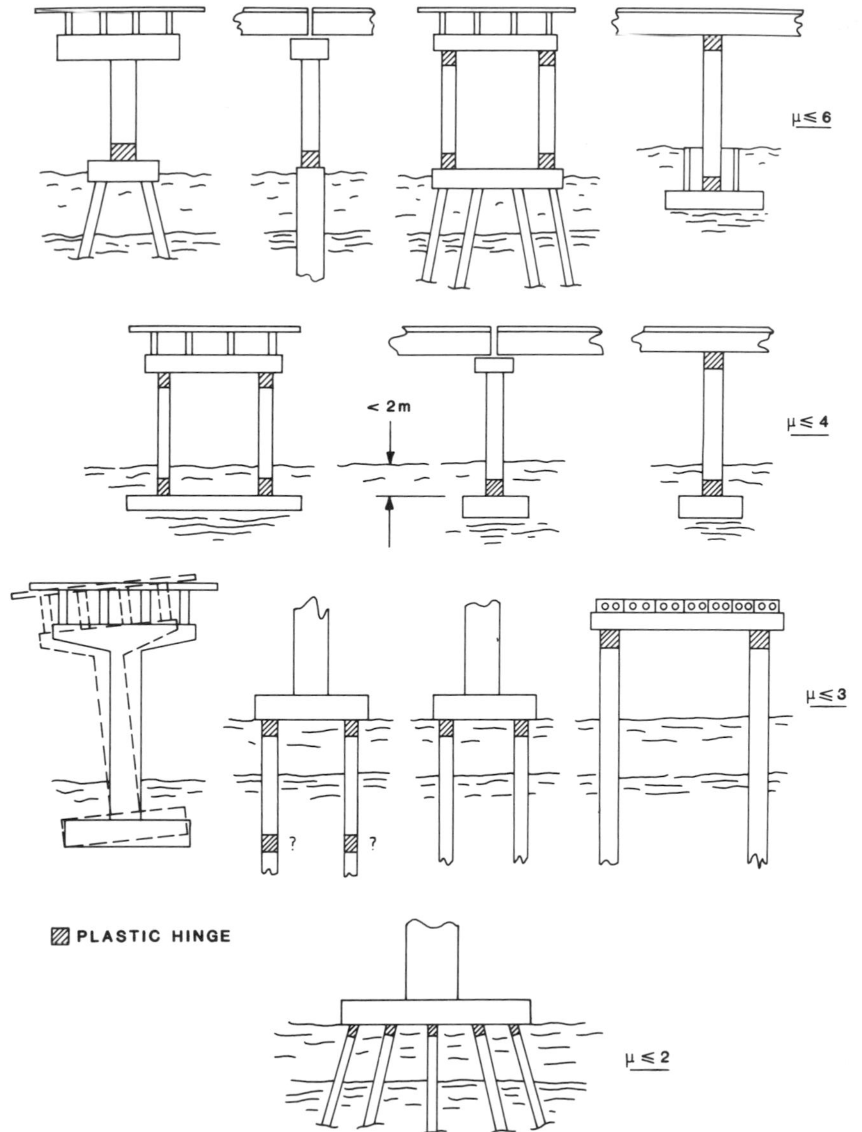

Figure A.1 Examples of maximum values of lateral deflection ductility factor μ allowed for bridges by Table A.2 (after reference 8)

Although carried out more as a research project than as a design exercise, a study reported by Penzien[9] illustrates the type of information which may be derived from a response analysis of soft soils. A similar but simpler study carried out for the design of a bridge at Tamaki, Auckland, has been described by Parton *et al.*[8] The piles for this bridge were assumed to deform as much as the soil, as predicted by the site response analysis, and were reinforced to resist corresponding curvatures. This involved the possibility of inelastic behaviour near the top of the pile. Two opposing views on building foundations in similar ground conditions (soft clay) near San Francisco have been described elsewhere.[11,12]

For the dynamic analysis of special bridges such as long multiple-span bridges, suspension and tied cantilever bridges, and complex elevated roadways, special computer programs are required.[13–15]

In any bridge analysis a realistic degree of damping must be allowed for. The overall damping of most ordinary bridges in earthquakes may be taken as about 5 percent of critical.[16,17] Allowance for foundation damping in softer soils is problematical, and needs careful consideration (Sections 6.2 and 6.3). In the design of long-span steel bridges the damping may be nearer 2 percent of critical, depending on the influence of concrete road surface elements.

Finally, it is noted that the design earthquake loadings should take account of the deflection ductility factor μ, determined as discussed in Section A.1.2.2. In elastic analyses this may involve reducing the elastic ($\mu = 1$) response spectrum loadings by the reduction factor R described in Section 6.6.7.3(i).

A.1.4 Strength and ductility of bridges

The main overall design criterion is to prevent partial or total collapse in strong earthquakes; bridges on roads which are strategic for relief and/or economic reasons should remain open at least to light traffic at all times. Various categories of safety desirability can readily be worked out in a given area, and some local authorities stipulate them.

Local regulations governing earthquake design of bridges exist in various placed, examples of which are those for Japan,[18,19] New Zealand[8,20] or those of the California Department of Transportation.

Because of the desirability of locating plastic hinges in accessible locations (Section A.1.2.2) buried members such as piles should commonly be designed to remain elastic, in which case they will be designed for strength rather than ductility. In some cases this will not prove feasible or economical, and piles then should be designed as ductile in a similar manner to columns. For example, the New Zealand concrete code[21] gives recommendations for the ductile design of concrete piles.

Appropriate strength and ductility may be provided by using the design methods discussed for concrete, steel or timber structures as appropriate (Chapters 7,9, and 10) while recommendations specific to substructures (piers, abutments, and foundations) may be found in Chapter 6 and in the specialist literature, notably reference 22.

A.1.5 Superstructure forces on abutments

The stiff non-ductile nature of abutments compared with the piers means that they would generally carry most of the lateral seismic loads if permitted to do so.

Even allowing for the higher damping of the abutments this may not be feasible, and the actual response of the abutments is in any case difficult to predict because of the difficulty of assessing how much soil acts with the abutment.[7] It is therefore good practice, on structures with two or more piers and with continuous deck diaphragm action, to separate superstructure from abutments transversely and to carry all transverse loads on the piers.

For shorter bridges there has been growing interest[23] in the performance of bridges with integral deck and abutments. The design criteria for this form of construction were still being studied in the mid-1980s, but a limit of 80 m in the length of such 'locked in' bridges for temperature stress reasons has been recommended as noted in Section A.1.2.2.

Care should be exercised in minimizing horizontal rotations about a vertical axis, and such torque should be resisted by bending rather than torsion in the piers. Where the pier and deck geometry is controlled by other considerations and torsion is high, the abutments may have to be used to resist the lateral forces.

A.1.6 Movement joints and horizontal linkage systems

Large amounts of seismic movement may have to be allowed for at movement joints in bridges as discussed in Section A.1.1 above; in strong ground motion adjacent structural components may undergo large inelastic oscillations out of phase. The minimum clearance provided at the time of construction between major structural elements should be[8]

$$C = \frac{T}{3} - S + \sqrt{(E_1^2 + E_2^2)} \tag{A.1}$$

where T is the temperature-induced movement from the median temperature position, S is the long-term shortening, and E_1 and E_2 are the out-of-phase seismic movements of the adjacent members separated by the movement gap under consideration.

Although it is usually straightforward to provide bearings even for this order of movement, it is expensive to provide the corresponding movement joints in the deck surface. It may be reasonable on economy grounds to provide for a small fraction of this total movement in a deck joint, and allow for the remaining movement by accepting a concentrated zone of secondary damage. This may be in the form of a sacrificial 'knock up' wedge of concrete.

In order to ensure structural integrity during earthquakes, i.e. that decks will not separate from their vertical support, horizontal linkage systems should be provided, such as shear keys, linkage bolts, or hinged deck slabs. Elastomeric bearings which have shear dowels top and bottom are not a substitute for linkage.

A.1.7 Holding-down devices

It is recommended[8] that holding-down devices should be provided when the nett vertical reaction under design earthquake conditions is less than 20 percent of the dead load reaction, and they should have a strength equal to at least 10 percent of the dead load reaction which would exist if the span were simply supported.

Problems arise in allowing for seismic movements, and at the same time avoiding damage to holding-down bolts. Some protection may be provided by using long bolts which will yield in flexure rather than shear, and rubber packing is sometimes provided under washers. Access to nuts and bolts for repair purposes may also be desirable.

A.1.8 Energy-isolating and dissipating devices for bridges

The beneficial application of these devices to reducing the design forces and increasing the reliability of bridges in earthquakes, has been discussed in Section 5.5, notably the use of lead–rubber bearings (Section 5.5.3) and isolation using uplifting (stepping) piers (Section 5.5.5).

A.2 CHIMNEYS AND TOWERS

A.2.1 Introduction

Towers and industrial chimneys pose a series of specialist design and construction problems generally related to their height and slenderness. They are vulnerable to earthquakes because they usually have only one line of defence, the failure of any one part of the structure resulting in spectacular failure. The earthquake resistance of these structures is discussed briefly in this section.

A.2.2 Reliable seismic behaviour of chimneys and towers

For obtaining reliable seismic response behaviour of chimneys and towers we should apply the principles of form, materials, and failure modes discussed in Section 5.3, noting that simple cantilever structures obviously will be much simpler to consider, but more vulnerable, than the more redundant structural forms discussed in Section 5.4.

A variety of forms of chimney and tower construction has been used in earthquake areas, including simple cantilevers, guyed structures, chimneys with supporting towers, and structurally combined multiple chimneys. The adoption of the latter two forms has the advantage of increasing the redundancy of the structures and hence decreasing their seismic vulnerability. The structures

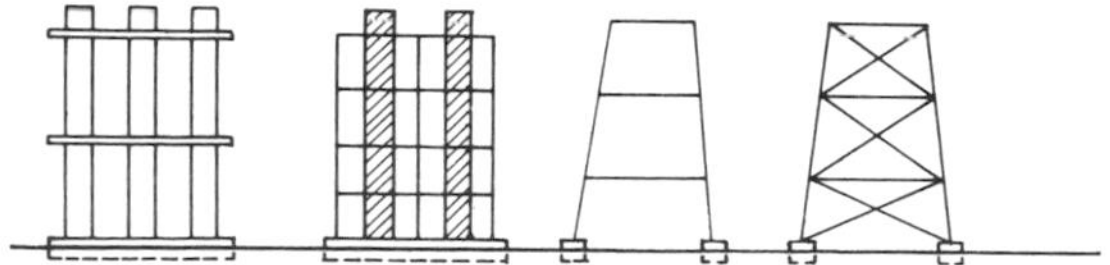

Figure A.2 Typical examples of chimneys and towers utilizing structural frame action (not to scale)

may be prismatic, or taper or step with height, while the inverted pendulum form is implicit for elevated water tanks (Figures A.2, A.3). Whereas stack-like structures are basically shells or tubular members, towers or supporting structures may also involve braced and unbraced frameworks.

Heights of well over 100 m (300 ft) are common for chimneys and towers in seismic areas, and steel, reinforced concrete or prestressed concrete are used in their construction.

A.2.3 Seismic analysis of chimneys and towers

For chimneys and towers of moderate size a dynamic earthquake analysis is highly desirable. Equivalent static loadings of codes of practice are not well suited to modelling higher mode effects which can be significant in slender structures. The controlling design criterion may be deformation rather than stress in the case of chimney linings, and wind loading may govern the design in shear or moment or both in some structures.

Unlike in building structures, it is seldom feasible to use the concept of ductility to make chimneys and towers more economical, as one plastic hinge will usually be sufficient to induce partial or total collapse. The need for elastic rather than inelastic behaviour is reflected in the relatively high loadings required by most codes for this type of structure. Only in multi-redundant supporting frames or chimney groups is ductility likely to be safely usable.

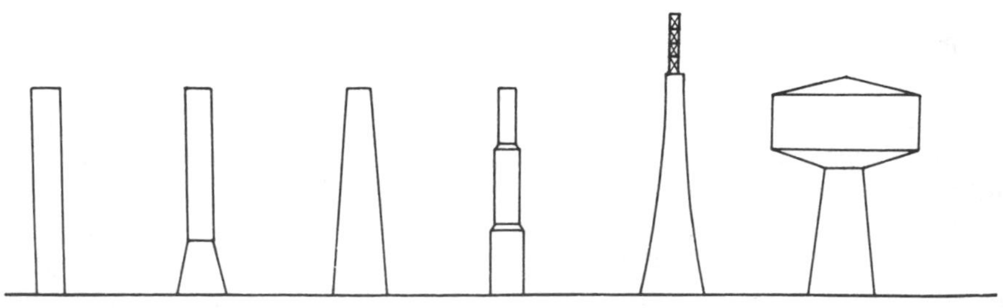

Figure A.3 Some typical free-standing chimneys and stack-like towers (not to scale)

A.2.4 Framed chimney and tower structures

A few examples of the many types of framed chimney and tower structures are shown in Figure A.2. Seismic analysis may be carried out as for normal building frames, and plane frame or space frame dynamic analysis may be appropriate.

Diagonally braced towers usually have slender members which are assumed to carry zero load in compression. The seismic response of lightly braced 10-storey single bay frames has been studied by Goel and Hanson,[24] who found that elastic analysis with or without viscous damping did not represent the dynamic behaviour of their frames when considerable yielding occurred in most of the members. They reported maximum ductility ratios of about seven in the bracing and five in the columns. The inference may be drawn that where only modest inelasticity is permitted in braced towers during the design earthquake, elastic analysis may provide sufficiently reliable design criteria (cf. Section 6.6.7.3(i)). In any case a conservative design is warranted because of the vulnerability of these structures to accidental torsions arising from asymmetrical yielding of the diagonal bracing. Diagonally braced structures are further discussed in Section 9.6.

Inverted pendulum structures are discussed specifically in Section A.2.6.

A.2.5 Free-standing chimneys and stack-like towers

A variety of chimney and tower structures fall into this category, as illustrated in Figure A.3. They range from simple prismatic cantilevers to tapered inverted pendulums. In this section free-standing chimneys and towers with relatively uniform distributions of mass with height will be considered, inverted pendulums being discussed in Section A.2.6. For the type of structure under consideration, Newmark and Rosenblueth[25] considered it important to take into account the very high harmonics, particularly when the fundamental period is so long that the design acceleration spectrum is hyperbolic over several periods of vibration. In the design of a 200 m high chimney for the Hsieh Ho power station in Taiwan in which the author was involved (unpublished report, 1978) the fourth mode gave the largest contribution to the bending moment near the top of the windshield, while the fifth mode was the most important in the liner at the same location. The fundamental period of the chimney was $T = 2.1$ s. This is slightly contrary to the suggestions by others[26–8] that no significant errors arise through considering only the first three or four modes of vibration of chimneys.[26,27] Response spectrum analyses, in which the total response is taken as the square root of the sum of the squared modal responses, appear suitable for chimneys.[25,28] In any case, the desirability of dynamic analysis for chimneys is not disputed. Any such analysis should incorporate the effects of bending and shear deformations, soil–structure interaction, rotational inertia, and gravitational loading including the P-delta effect, as all of these effects may be important.

Since inelastic behaviour is undesirable, these structures should be designed to remain largely elastic in strong earthquakes, and hence elastic analysis will be appropriate. As with any sizeable structure without any redundancy, it is important to ensure reliability by designing the structure to remain elastic during the normal design earthquake and to check survivability under a rarer event (Section 5.2) as the latter condition in some cases may control the design.

A.2.5.1 Reinforced concrete chimneys

A description of the response spectrum analysis of eight reinforced concrete chimneys up to 250 m in height has been given by Rumman,[28] who used seven earthquake inputs and a structural damping of 5 percent of critical. Rumman found that three or four modes should be taken in order to achieve satisfactory accuracy with the response spectrum technique, but more would be safer, see Section A.2.5.

Reporting in a more general paper on the design of reinforced concrete chimneys, Maugh and Rumman[26] pointed out that as the seismic moments and shears are inversely proportional to the damping, particular care should be exercised in choosing the value of this parameter. Unfortunately they gave no specific guidance on this point. A damping value of 5 percent of critical seems commonly taken for reinforced concrete chimneys, and this value is reasonably appropriate for their behaviour in the elastic range (Section 6.6.4). However, the effect of the lining on the overall damping of the chimney should be considered.

The ACI[29] recommends that the design of chimneys be based on a modal response analysis, and that they should withstand at least the equivalent static forces described below. In order to make some allowance for the effect of higher modes, the ACI suggests that 15 percent of the total horizontal shear be applied at the top of the chimney. The remainder of the shear is distributed vertically so that the force F_x at any height h_x above the base is given by

$$F_x = \frac{(V - V_H)\upsilon_x h_x}{\Sigma \upsilon_x h_x} \tag{A.2}$$

where V is the total seismic base shear, V_H is that fraction of V applied at the top, and υ_x is the weight of the segment of the chimney at height h_x.

For deriving the base shear V,

$$V = ZUCW \tag{A.3}$$

where Z corresponds to the American seismic zones;[30]

$$C = 0.1/T^{\frac{1}{3}}; \tag{A.4}$$

U = risk-related use factor varying from 1.3 to 2.0;
W = total weight of lining (excluding lining if not supported by the shell).

Hence the base shear V is a function of the fundamental period T of the chimney. For chimneys on a rigid base, the ACI[29] gives a dimensionally inconsistent empirical formula for T equivalent to

$$T = \frac{0.49\,H^2}{(3D_b - D_t)\sqrt{(E)}} \sqrt{\left(\frac{m_1}{m}\right)} \tag{A.5}$$

where H = height of chimney (m),
D_b = external diameter at the top (m),
D_t = external diameter at the top (m),
E = modulus of elasticity (N/mm^2),
m_1 = total mass of the chimney including linings etc.,
m = total mass of the chimney structure only.

Other formulae for calculating the fundamental periods of chimneys have been proposed, such as that by Housner and Keighley[31] for tapered cantilevers. This work also presents formulae for the second and third mode periods. In an unpublished work[32] Mitchell developed a method suitable for computing the fundamental periods of cylindrical, tapered, and step-tapered chimney structures.

A useful comparative analysis of the above three methods of computing T has been made by Rinne.[33]

The design moment at any level of the chimney may be found from

$$M_x = J_x[0.15V(H - h_x) + \Sigma F_h(h - h_x)] \tag{A.6}$$

where $$J_x = J + (1 - J)(h_x/H)^3 \tag{A.7}$$

$$J = 0.6/T^{\frac{1}{3}}, \qquad (0.45 \leqslant J \leqslant 1) \tag{A.8}$$

and F_h is the lateral force applied at level h.

It should be pointed out that in the above discussion no account has been taken of the effect of subsoil flexibility or gravity effects. The significance of soil–structure interaction is discussed in Section 6.3, where an example of a stack-like tower was cited for which the fundamental period was $T = 1.2$ s for a rigid base and $T = 3.0$ s for a soft soil base. In slender chimneys gravity effects during seismic deformation may also be significant.

A.2.5.2 Steel chimneys

The analytical considerations for steel chimneys follow lines similar to those for reinforced concrete discussed above. Blume[27] provides an interesting discussion of the dynamic analysis of a number of steel chimney and tower structures which were damaged in the Chilean earthquakes of May 1960. He came to a similar conclusion to that regarding concrete chimneys quoted above (Section A.2.5.1) that the first three modes were significant. This does not conform to the contention of Newmark and Rosenblueth[25] as discussed in Section A.2.5.

According to two reports[27,34] the damping of steel chimneys is about 1 to 2 percent of critical, including the effect of the lining. In some instances it may be feasible to reduce the seismic (and wind) response which would otherwise occur, by introduing special structural damping devices such as studied by Johns *et al.*[35] An increase in damping from 2 to 4 percent, for example, would be of considerable benefit.

Where a fully computerized dynamic analysis is not envisaged, the fundamental period of a cantilever chimney of uniform section (or any similar structure) may be derived from

$$T = 1.79 H^2 \sqrt{\left(\frac{v_g}{EI_g}\right)} \tag{A.9}$$

where v_g = weight per unit height (kN/m),
g = acceleration of gravity (m/s^2),
H = height (m),
E = modulus of elasticity (kN/m^2),
I = moment of inertia of cross section (m^4).

For chimneys with a flared base, the fundamental period may be found from

$$T = 2\pi \sqrt{\left(\frac{0.08\Delta}{g}\right)} \tag{A.10}$$

where Δ = the calculated deflection in metres at the top of the chimney due to 100 percent of its weight applied as a lateral load.

The structural contribution of the lining to the stiffness of the steel shell should be considered; the effect of gunited linings, for example, on the period of vibration may be considerable.[27,33] Lining should not be considered effective in buckling or yield resistance unless specially designed for composite action with the shell. A chimney shell with a critical buckling stress below the yield point is undesirable as there is little energy absorption capacity after buckling.

As mentioned above, little reliance should be placed on ductile behaviour in chimneys and towers. This is particularly true for steel chimneys where yield can rapidly develop into a secondary failure. Blume,[27] however, made the single concession that holding-down bolts and connections should have yield capacity above the concrete foundation surface, as this affords protection to the chimney itself through the energy absorbed in bolt stretching and controlled rocking. The principle of designing holding-down bolts to yield prior to the remainder of the structure is now a widely used seismic protection procedure for many different types of structure.

Finally the remarks in Section A.2.5.1 regarding soil–structure interaction should be noted.

A.2.6 Inverted pendulum structures

Inverted pendulums consist of tower or column structures with a large concentrated mass at the top, and occur commonly in forms such as canopies, observation platforms, elevated restaurants and water towers. They may have one or more vertical supports which in some cases form frameworks (Figures A.3, A.4). The large mass at the top makes such structures especially vulnerable to earthquakes because of the accompanying horizontal inertia forces and the so-called $P \times \Delta$ effect (Section 6.6.7.3(ii)). For this reason most codes of practice are appropriately even more conservative for inverted pendulums than for other chimneys and towers.

The seismic bending moments at the tops of columns may govern the design of the columns and of parts of the structure above.[36] Asymmetry of live load and unintended asymmetry of structural mass distribution may induce significant moments about horizontal and vertical axes.

In these structures the previously mentioned unsuitability of inelastic behaviour is emphasized by the large mass at the top. Considering a simple inverted pendulum Newmark and Rosenblueth[25] found that the column design moment including gravity effects was almost double the moment determined without taking gravity into account.

A.2.6.1 Elevated liquid containers

Because of hydrodynamic effects it is convenient to consider elevated water tanks and other liquid containers as a special case of the inverted pendulum. These structures may be either supported on a single vertical member or a framework (Figure A.4). In either case the conclusion of both Blume *et al.*[37] and Boyce[38] may be applied, namely that elevated tanks should be modelled as two-degree-of-freedom structures. Boyce demonstrated this with observations of a real water tower, and also showed that large errors are involved in using a single-degree-of-freedom model (Figure A.5).

If the water is completely contained to prevent vertical motion of the water surface (sloshing), the water tower may be treated as a normal inverted pendulum (Section A.2.6). Sloshing will usually act as damping, and may result in a useful reduction in seismic response of the structure compared with the contained liquid case. However, sloshing may damage the roof of the tank or cause spillage of toxic or other liquids (Section 11.3.2.3).

The hydrodynamics of sloshing is mathematically complex,[39] but a simplified dynamic analysis has been suggested by Housner[40] as a result of a study of the great damage to elevated water tanks which occurred in the Chilean earthquakes of May 1960. However, for design office purposes the most convenient dynamic analysis for elevated tanks which has so far been developed is that of Blume *et al.*[37] This work presents graphs enabling the rapid determination of the complex constants used in the hydrodynamic equations.

It should also be pointed out that some computer programs have also been written for the dynamic analysis of elevated water tanks, such as that by

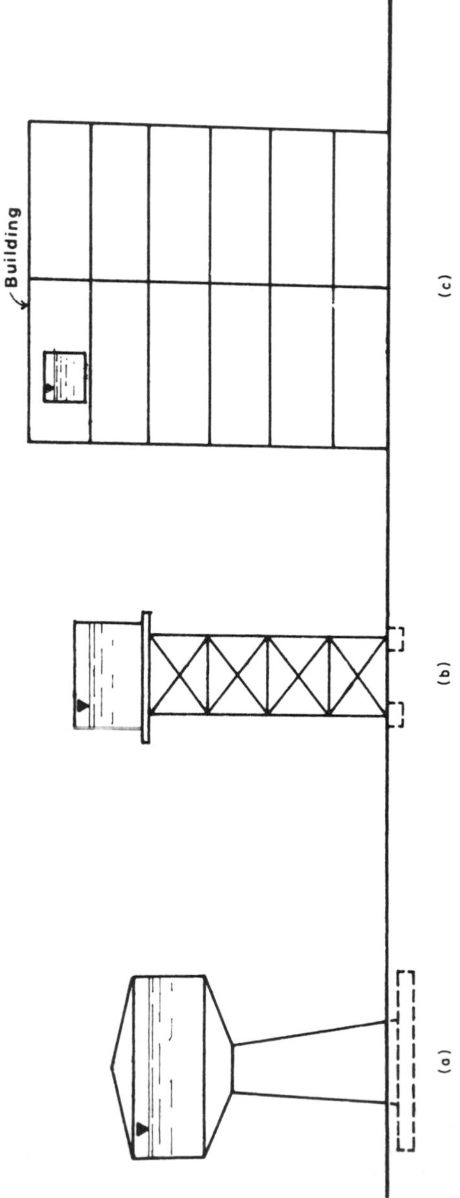

Figure A.4 Typical elevated water tank structures

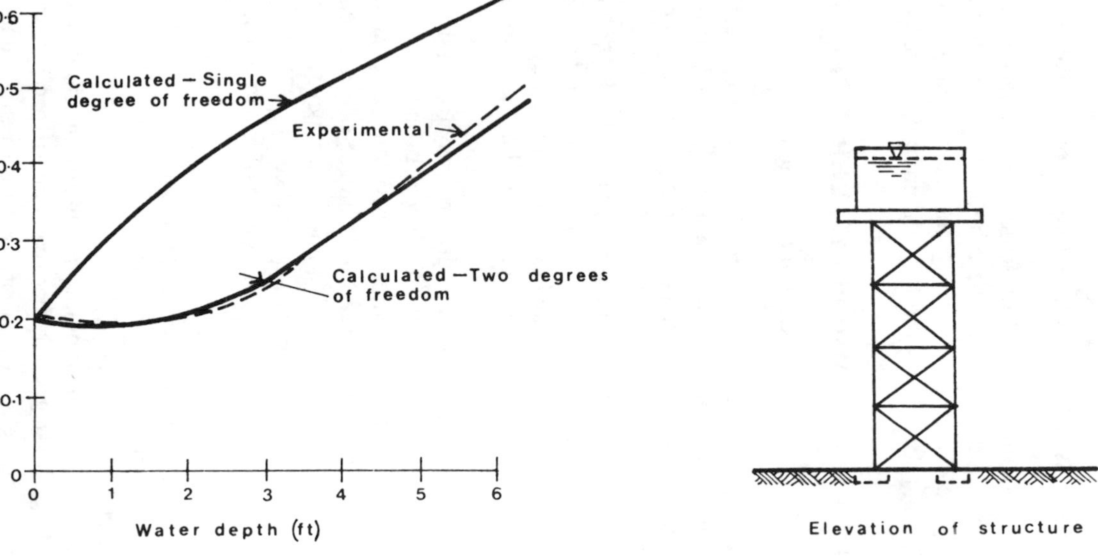

Figure A.5 Comparison of measured and calculated periods of vibration for an elevated water tank (after Boyce[38])

Shepherd[41] for a three-storeyed cross-braced supporting tower. Computer programs written for offshore oil platforms could also be used.

Further information on the design of liquid containers has been conveniently compiled by a New Zealand study group on the seismic design of storage tanks.[42]

A.2.7 Energy-isolating and dissipating devices for chimneys and towers

In Section 5.5 the application of these devices has been discussed for reducing the design forces and increasing the reliability of a wide range of structures. In particular, the reader is referred to the discussion of the use of uplift and energy dissipators for a free-standing chimney discussed in Section 5.5.5.

A.3 LOW-RISE COMMERCIAL–INDUSTRIAL BUILDINGS

A.3.1 Introduction

This large rather ill-defined class of buildings has several earthquake resistance problems peculiar to itself which are not readily treated in the more general earlier chapters of this document. The buildings referred to here are generally of one- or two-storeys and are used for a wide variety of purposes such as warehouses, light manufacturing, shops, supermarkets, and entertainments. They represent a considerable proportion of the annual investment in new buildings in many countries; in the USA for example, about half the total value of new building construction in recent years has been spent on commercial–industrial buildings.

However, because these buildings are not easy to classify as structural types or to analyse reliably, and because they are only low-rise, they are often inadequately dealt with by earthquake resistance regulations and by design and construction practice. Hence they tend to suffer in earthquakes disproportionately to their monetary value and occupancy, as occurred for example in the San Fernando, California, earthquake of 1971.[43–45]

In addition to the above reasons for failure of low-rise buildings, the following have been noted elsewhere:[46]

(1) Inability of the structure to act as a unit;
(2) Inadequate bracing;
(3) Eccentric stiffness elements;
(4) Mixed material construction or non-uniform distribution of ductile elements, both leading to far greater than calculated forces on certain elements;
(5) Inadequate connection of superstructure to the foundations.

A.3.2 Reliable seismic behaviour of low-rise commercial–industrial buildings

For obtaining reliable seismic response behaviour we should apply the principles of form, materials, and failure modes discussed in Section 5.3, some points specific to low-rise commercial–industrial buildings being noted below.

The control of failure modes is frequently made particularly difficult because of use of widely different materials for different components such as walls, floors, or roofs, and the consequent difficulties of designing effective connections. Also non-structural elements (partitions) seem frequently to be disproportionately stiff and destroy the designed structural resistance pattern. Spencer and Tong[47] have proposed a procedure to ensure that inelastic behaviour occurs in the desired locations for a one-storey precast concrete building.

A.3.3 Seismic analysis

The unsatisfactory earthquake performance of commercial–industrial buildings may often be largely attributed to unsatisfactory seismic analysis. This may result in serious underestimations of the strength required in connections or the significance of asymmetries. To overcome this more use should be made of dynamic analysis. In buildings with stiff walls and diaphragm roof construction (Section 10.6), three-dimensional finite-element dynamic analysis will readily demonstrate the vulnerable features. Even coarse meshed elastic analysis will help provide the ground rules for this type of aseismic construction.

In a study of the non-linear dynamic response of single storey tilt-up wall construction Adham *et al.*[48] found lumped mass parameter models to be satisfactory, and also found that panel–diaphragm interaction effects result in design moments for the panels and forces in the panel-roof connections being higher than those specified by normal US design procedures.

A.3.4 Foundations for low-rise commercial–industrial buildings

The provisions of suitable foundations for earthquake resistance of low-rise construction *on soft ground* is a basic engineering problem for commercial–industrial buildings. Because the foundation requirements for gravity and wind loading are minimal in such buildings, the extra cost for providing protection at source against differential ground movements is large compared with that for taller structures.

It should be recognized that considerable amounts of differential horizontal and vertical movement may be imposed on a low-rise structure provided with economical foundations, even when following the recommendations for shallow foundations suggested in Section 6.4.2. Such movements should be allowed for in the superstructure by providing suitable continuity of articulation especially at roof level (Section A.3.5).

A.3.5 Connections in low-rise commercial–industrial buildings

Commercial–industrial buildings structured entirely in one material, or in compatible materials such as masonry and *in situ* concrete, are generally more effective in earthquakes than buildings comprised of a mixture of materials, and should be dealt with as described in the appropriate sections of Chapters 7 to 10.

Many buildings of the type under consideration are built with different members in different materials. For example, the walls may be of precast concrete, reinforced bricks, or concrete blocks, while roofs or floors may consist of steel trusses, laminated timber beams, and plywood elements. Such heterogeneous construction has proved to be particularly vulnerable at the connections between dissimilar materials.[43]

In the 1971 San Fernando earthquake the commonest failures according to Bockemohle[43] were as follows;

(1) Separation of plywood roof diaphragms from the supporting timber ledgers on the walls;
(2) Separation of roof girders from the tops of walls, columns, or pilasters;
(3) Inadequate continuity of roof chords at dowels and laps;
(4) Inadequate shear transmission through roof diaphragms into shear walls in buildings of irregular plan form.

A connection detail between timber roofs and concrete or masonry walls which performed well in the San Fernando earthquake[43] and in subsequent tests[49] is indicated in Figure A.6(a). An alternative detail using a steel angle ledger is shown in Figure A.6(b).

The Los Angeles City Building Department requires joist anchors that are spaced at not more than 1.2 m, and that transmit a minimum horizontal force of 4.4 kN/m.

A number of details for connection of masonry walls to different roof and floor constructions, some of which are appropriate for earthquake resistance, have been given by the ATC[50] and Amrhein.[51]

As mentioned above, care is also necessary to ensure the integrity of connections between girders and wall columns or pilasters. Holding-down bolts must be adequately embedded, and the shear resistance of the column section should be provided by adequate transverse reinforcing links particularly in the region immediately below the girder support.

Finally, the need for failure mode control (Section A.3.2) should be emphasised when considering the design of connections, the following factors needing attention:

(1) Use of yielding connections may be made to protect structural elements;
(2) Yielding elements should be distributed throughout the structure so as to produce the desired failure mode;
(3) Connections that are expected to undergo inelastic deformations must have adequate ductility.[47]

A.4 LOW-RISE HOUSING

A.4.1 Introduction

It is difficult to say exactly what low-rise housing comprises, because there is not precise definition of the term 'low-rise'. However, for the purposes of this book it may be said to include one- and two-storey houses generally, and sometimes may also refer to three- or four-storey construction. We are

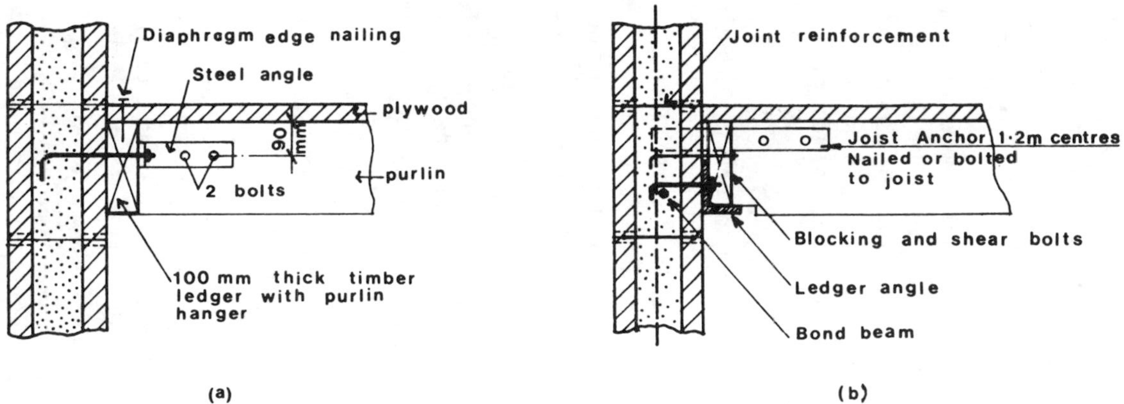

Figure A.6 Joist anchor connections between roof and wall. (Part (b) reproduced by permission of the Masonry Institute of America)

considering housing for the design of which a structural engineer's direct involvement is usually marginal or non-existent.

The provision of adequate earthquake-resistant housing poses a considerable world problem. Every year sees much damage, homelessness, and loss of life due to the effect of earthquakes on housing, particularly in developing countries. Low-rise buildings are especially vulnerable because of the consequent lack of engineering design and use of lower grades of construction technology. These drawbacks are worsened in developing countries, which also suffer from having to use less suitable materials in the masonry range, and there is a desperate need to provide even the most nominal level of earthquake resistance to millions of such homes.

Because of the lack of engineering design the responsibility lies on the architect, and the governmental building supervisor (if they are involved) and on the builders themselves, to produce earthquake-resistant construction. The problem is to use the available materials to the best advantage by choosing a sound structural form and by using building details which provide maximum structural continuity.

In seismic regions with reasonably advanced technologies, architects and builders are usually assisted by having to comply with building regulations which have an earthquake engineering basis. For example, in New Zealand there are codes of practice for timber and masonry buildings *not* requiring specific engineering design.[52,53] Such regulations are naturally written for standard house forms, and buildings on soft or sloping ground or those using a mixture of building materials are likely to warrant specific engineering design consideration.

In less advanced countries, little engineering-based guidance may be available for the builder of low-rise dwellings. The problem of selecting construction standards appropriate to the local technology is difficult.[54,55] More should be done to discourage dangerous practices and to foster the use of seismically successful vernacular construction details. For example, in Latin America a form of construction called *quincha*, consisting of a timber and cane lattice plastered with mud, has proved remarkably effective in earthquakes.[55]

From the foregoing brief discussion it will be evident that there are great difficulties, including economic ones, in the way of achieving satisfactory performance of houses in earthquakes. This situation is emphasised by the reviews of vulnerability of houses in the USA[44,46] and New Zealand,[56,58] despite aseismic building standards being high on a world scale in both these countries. The reasons for failure of houses are often the same as those for low-rise commercial–industrial buildings listed in Section A.3.1. Ideally, in order to improve the situation the principles of reliable seismic behaviour given in Section 5.3, together with the utilization of the strength and ductility of the materials used (Chapter 7 to 10), should be incorporated into house-building practice. For more immediate practical assistance the essential features of aseismic design of some types of low-rise housing are discussed below, while advice regarding strengthening of existing houses has been given by Cooney.[57]

A.4.2 Symmetry in plan of low-rise housing

The building's resistance against horizontal forces should be derived from walls providing reasonably symmetrical resistance in two orthogonal directions in plan (Figure A.7(a)). If one facade only consists mainly of window and door apertures, horizontal diaphragm action at eaves level should be capable of transferring the resulting earthquake torque to the end walls at right-angles to that facade (Figure A.7(b)). It should be noted that because of the inherent high torsional flexibility of buildings with essentially only three resisting walls, this type of construction is forbidden by the Uniform Building Code[30] when using masonry walls and timber roof diaphragms; some short elements resisting horizontal shear must be introduced into the window facade. The resulting reduced torsions will nevertheless need to be distributed through a horizontal diaphragm. Damage arising from excessive asymmetry occurred in the San Fernando earthquake, as shown in Figures 27 and 28 of the paper by the California Institute of Technology.[45]

A.4.3 Apertures in walls

Apertures for doors and windows require care in positioning and detailing. In masonry the positioning of apertures is particularly important as discussed in Section 8.3.2. Lintels in heavy materials need careful detailing against falling during earthquakes. If it is structurally necessary for a wall to act as a whole, the effect of apertures on the integrity of the wall should be considered.

A.4.4 The strength and stiffness of walls

The strength and stiffness of *timber walls* required to act in shear in their own plane are greatly enhanced by use of panelled linings of timber such as plywood or particle board (Section 10.5) or metal cladding.[59] Alternatively, diagonal bracing, preferably symmetrical, is also effective.

Masonry walls and masonry veneers are often unreinforced in traditional low-rise housing, and suffer great damage in earthquakes accordingly.[44,57] The very

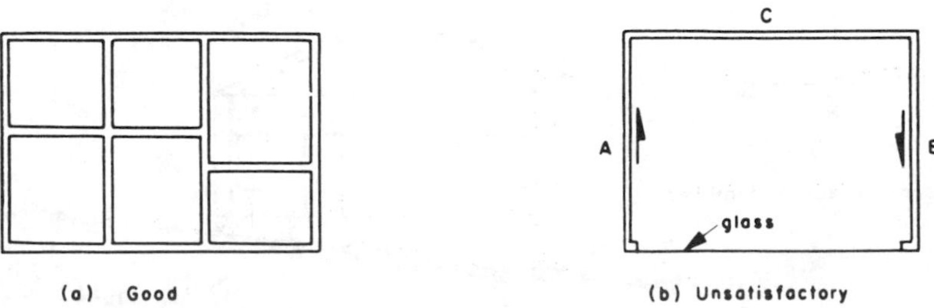

Figure A.7 Schematic plans showing layout of shear walls in low-rise housing

beneficial effect of even minimal reinforcing has been demonstrated, as discussed for concrete masonry in Section 8.2, and for clay bricks or adobe construction, as shown in Latin American Research.[60] A minimum recommendation is therefore to reinforce the perimeter of major wall elements with a vertical steel bar (10 mm diameter) at each end, and horizontal reinforcing bands at top and bottom, as described in the next section.

A.4.5 Horizontal continuity

Horizontal continuity at floor and roof levels should be provided by special connections or lapping reinforcement, and such continuity should go around facade corners.

A.4.6 Foundations for low-rise housing

Foundation problems in low-rise housing are similar to those expressed in Section A.3.4 for commercial–industrial buildings. Also, in timber housing the substructure between the footings and the first occupied floor tends to have inadequate horizontal shear resistance, and sidesway damage (Figure A.8) occurs in earthquakes.[44,45,58] Pole frame construction as illustrated in Figure 10.5 readily overcomes this problem.

Another common failing has been that the timber structure is inadequately connected to the concrete foundation blocks or strips. The detail shown in Figure 10.10(b) for example, should be provided with adequate bolts.

A.4.7 Roofs of heavy construction

Roofs of heavy construction are a great menace, causing large loss of life in earthquakes. In some unerdeveloped areas massive earth and masonry roofs are the norm, but less heavy tiles can also be dangerous. Where this type of construction is unavoidable, appropriate measures should be taken to ensure the integrity of the roof during earthquakes. Apart from proper vertical support,

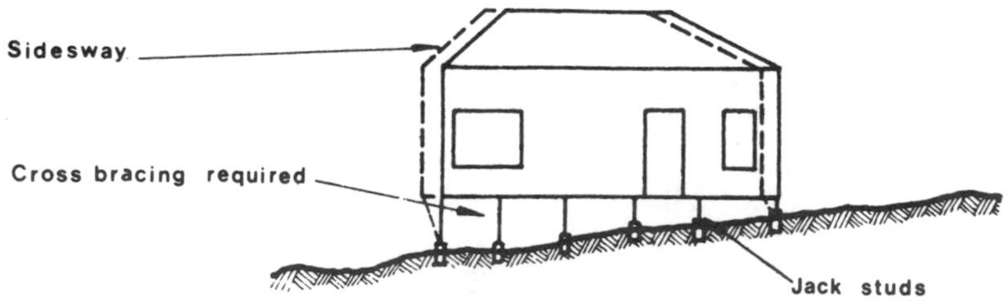

Figure A.8 Substructure in timber stud construction requiring extra horizontal shear strength

horizontal diaphragm action at eaves level to prevent spreading and collapse of the roof is particularly valuable.

A.4.8 Chimneys and decorative masonry panels

Elements which are stiffer and heavier than the rest of the building cause a great deal of damage in earthquakes. Concrete and masonry chimneys in basically timber houses are particularly vulnerable.[44,58] In many cases the ideal solution would be to make the stiff elements structurally independent of the rest of the building, but difficulties arise in detailing the movement gaps. Otherwise the stiff and the flexible elements should be much more strongly tied together than has been common practice in the past. When a timber structure is tied to a stiff element, the latter becomes a major horizontal shear resisting element for the whole building, and the building should be designed accordingly.

A.5 IMPROVING THE EARTHQUAKE RESISTANCE OF EXISTING STRUCTURES

In an acknowledgement of the enormous worldwide seismic risk, in particular to human life and historical monuments, represented by existing older structures which have far less earthquake resistance than modern standards require, philosophies and techniques for strengthening (retrofitting) such structures are being developed in many countries. Because of the vast variety of existing structures, the development of general rules of real use is difficult and to a large extent each structure must be approached as a strengthening problem on its own merits. Some of the factors which need consideration are as follows:

(1) The form of the structure and non-structure, and the need for change, e.g. to create symmetry;
(2) The materials used in the existing construction;
(3) The permissible visual and functional effect of the strengthening;
(4) The desired further design life;
(5) The desired seismic resistance;
(6) The acceptable damage to the existing fabric in the design event;
(7) The parts requiring strengthening and the problems of access thereto, e.g. piles;
(8) The degree to which ductile failure modes are required. (Significant ductility is not reliably achievable at reasonable cost in many older constructions, particularly of masonry, or may imply heavy damage to the existing fabric.)
(9) The extent to which other components are to be upgraded as well as the strength, e.g. architectural features and building services;
(10) Continuance of normal function during the strengthening works;
(11) Costs.

Depending on the above factors, significant seismic resistance can be obtained for most structures for only a small fraction (5–30 percent) of their replacement cost, while the long-term upgrading of historical buildings or monuments may exceed the cost of their replacement (where that is meaningful).

Because the means of strengthening are so diverse, ranging from simply installing a few steel ties to complete base-isolation, no attempt to outline them will be made here, but the principles of earthquake resistance discussed in this book apply. The reader is referred to the literature, such as the papers from the 8th World Conference on Earthquake Engineering[61] or from the New Zealand National Society for Earthquake Engineering.[62] The latter organization has also produced recommendations for the strengthening of buildings,[63] while Cooney[57] has described simple means for strengthening houses.

REFERENCES

1. Wood, J. H., and Jennings, P. C., 'Damage to freeway structures in the San Fernando earthquake', *Bull. NZ Soc. for Earthq. Eng.*, **4**, No. 3, 347–76 (Dec. 1971).
2. Tseng, W. S., and Penzien, J., 'Seismic response of highway overcrossings', *Proc. 5th World Conf. on Earthq. Eng., Rome*, **1**, 942–51 (1973).
3. Ellison, B. K., 'Earthquake damage to roads and bridges, Madang, R. P. N. G. Nov. 1970', *Bull. NZ Soc. for Earthq. Eng.*, **4**, No. 2, 243–57 (April 1971).
4. Hollings, J. P., and Fraser, I. A. N., 'Earthquake damage to three railway bridges 1968 Inangahua earthquake', *Bull. NZ Soc. for Earthq. Eng.*, **1**, No. 2, 22–48 (Dec. 1968).
5. Wilson, J. B., 'Notes on bridges and earthquakes', *Bull. NZ Soc. for Earthq. Eng.*, **1**, No. 2, 92–7 (Dec. 1968).
6. Applied Technology Council, California, 'Comparisons of United States and New Zealand seismic design practices for highway bridges', ATC-12 (1982).
7. Chapman, H. E., 'Earthquake resistant design of bridges and the New Zealand Ministry of Works bridge design manual', *Proc. 5th World Conf. on Earthq. Eng., Rome*, **2**, 2242–51 (1973).
8. New Zealand Ministry of Works and Development, 'Highway bridge design brief', Amendment of Section 2.4 Earthquake resistant design, Draft Copy (January 1985).
9. Penzien, J., 'Soil–pile foundation interaction', in *Earthquake Engineering* (Ed. R. L. Wiegel), Prentice-Hall, Englewood Cliffs, NJ (1970), Chap. 14, pp. 349–81.
10. Parton, I. M., and Melville Smith, R. W., 'Effect of soil properties on earthquake response', *Bull. NZ Soc. for Earthq. Eng.*, **4**, No. 1, 73–93 (March 1971).
11. ASCE 'Building foundation for soft clay, earthquake area—Mat foundation', *Civil Engineering, ASCE*, **44**, No. 2, 56–7 (Feb. 1974).
12. ASCE 'Building foundation for soft clay, earthquake area—Pile foundation', *Civil Engineering, ASCE*, **44**, No. 2, 58–9 (Feb. 1974).
13. Tezcan, S. S., and Cherry, S., 'Earthquake analysis of suspension bridges', *Proc. 4th World Conf. on Earthq. Eng., Chile*, **II**, A3, 125–40 (1969).
14. Arya, A. S., and Thakkar, S. K., 'Earthquake response of a tied cantilever bridge', in 'Earthquake engineering', *Proc. 3rd European Symposium on Earthquake Engineering, Sofia*, 343–53 (1970).

15. Tseng, W. S., and Penzien, J., 'Linear and non-linear seismic analysis computer programs for long multiple highway bridges', *Report No. EERC 73-20*, Earthquake Engineering Research Center, University of California, Berkeley (1973).
16. Katayama, 'Dynamic characteristics of bridge structures', *Lecture notes*, presented at the International Institute of Seismology and Earthquake Engineering, Tokyo (1972).
17. Charleson, 'The dynamic behaviour of bridge substructures', *Report*, to Road Research Unit of New Zealand National Roads Board (1970).
18. Standards of Aseismic Civil Engineering Constructions in Japan published in *Earthquake regulations—a world list, 1973* International Association for Earthquake Engineering, Tokyo (1973).
19. JSCE, *Earthquake resistant design for civil engineering structures, earth structures and foundations in Japan*, compiled by the Japan Society of Civil Engineers, 1977.
20. Papers resulting from deliberations of the Society's Discussion Group on the Seismic Design of Bridges, *Bull. NZ Nat. Soc. for Earthq. Eng.*, **13**, No. 3, 226–309 (1980).
21. Standards Association of New Zealand, 'Code of practice for the design of concrete structures', NZS 3101, Parts 1 and 2: 1982.
22. Priestley, M. J. N., and Park, R., 'Strength and ductility of bridge substructures', *Road Research Unit Bulletin 71*, National Roads Board, New Zealand (1984).
23. Wolde-Tinsae, A. M., Greimann, L. F., and Johnson, B. V., 'Performance of integral bridge abutments', *IABSE Proceedings* P-58/83, 17–29 (1983).
24. Goel, S. C., and Hanson, R. D., 'Seismic behaviour of multistorey braced steel frames', *Proc. 5th World Conf. on Earthq. Eng., Rome*, **2**, 2934–43 (1973).
25. Newmark, N. M., and Rosenblueth, E., *Fundamentals of Earthquake Engineering*, Prentice-Hall, Englewood Cliffs, NJ, (1971).
26. Maugh, L. C., and Rumman, W. S., 'Dynamic design of reinforced concrete chimneys', *J. ACI*, **64**, No. 9, 558–67 (Sept. 1967).
27. Blume, J. A., 'A structural-dynamic analysis of steel plant structures subjected to the May 1960 Chilean earthquakes', *Bull. Seism. Soc. Amer.*, **53**, No. 2, 439–80 (Feb. 1963).
28. Rumman, W. S., 'Earthquake forces in reinforced concrete chimneys', *J. Structural Division, ASCE*, **93**, No. ST6, 55–70 (Dec. 1967).
29. American Concrete Institute, *Specification for the design and construction of reinforced concrete chimneys*, (ACI307-79), American Concrete Institute (1979).
30. International Conference of Building Officials, *Uniform Building Code*, ICBO, Pasadena, California (1985).
31. Housner, G. W., and Keightley, W. O., 'Vibrations of linearly tapered cantilever beams', *Trans. ASCE*, **128**, Part 1, 1020–48 (1963).
32. Mitchell, W. W., 'Determination of the period of vibration of multi-diameter columns by the method based on Rayleigh's principle', *Unpublished report*, for the Engineering Department of the Standard Oil Company of California, San Francisco, (1962).
33. Rinne, J. E., 'Design of earthquake resistant structures: towers and chimneys', in *Earthquake Engineering* (Ed. R. L. Wiegel), Prentice-Hall, Englewood Cliffs, NJ, (1970) pp. 495–505.
34. Kircher, *et al.*, 'Seismic analysis of oil refinery structures', *Proc. 2nd US Nat. Conf. on Earthq. Eng.*, Stanford, California, 127–36 (1979).
35. Johns, D. J., Britton, J., and Stoppard, G., 'On increasing the structural damping of a steel chimney', *Earthquake Engineering and Structural Dynamics*, **1**, No. 1, 93–100 (July-Sept. 1972).
36. Rascón, O. A., 'Effectos sismicos en estructuras en forma de péndulo invertido', *Ref. Soc. Mex. Ing. Sísm.*, **3**, No. 1, 8–16 (1965).

37. Blume, J. A., and Associates, *Earthquake Engineering for Nuclear Reactor Facilities*, J. A. Blume and Associates, San Francisco (1971), (particularly pp. 111–23).
38. Boyce, W. H., 'Vibration tests on a simple water tower', *Proc. 5th World Conf. on Earthq. Eng., Rome*, **1**, 220–5 (1973).
39. Housner, G. W., 'Dynamic analysis of fluids in containers subjected to accelerations', Appendix F in *Nuclear Reactors and Earthquakes*, US Atomic Energy Commission, TID—7024 (1963).
40. Housner, G. W., 'The dynamic behaviour of water tanks', *Bull. Seism. Soc. Amer.*, **53**, No. 2, 381–7 (Feb. 1963).
41. Shepherd, R., 'The seismic response of elevated water tanks supported on cross braced towers', *Proc. 5th World Conf. on Earthq. Eng., Rome*, **1**, 640–49 (1973).
42. Priestley, N. J. M. (Ed.), 'Seismic design of storage tanks', *New Zealand Nat. Soc. for Earthq. Eng.*, Wellington (1986).
43. Bockemohle, L. W., 'Earthquake behaviour of commercial-industrial buildings in the San Fernando valley', *Proc. 5th World Conf. on Earthq. Eng., Rome*, **1**, 76–81 (1973).
44. National Bureau of Standards, 'The San Fernando California earthquake of February 9, 1971', *NBS Report 10556*, US Department of Commerce (March 1971).
45. Earthquake Engineering Research Laboratory, California Institute of Technology, 'Engineering features of the San Fernando earthquake', *Bull. NZ Soc. for Earthq. Eng.*, **6**, No. 1, 22–45 (March 1973).
46. Gupta, A. K. (Ed.), 'Seismic performance of low rise buildings', Proc. Workshop at Chicago, 1980, American Society of Civil Engineers, New York (1981).
47. Spencer, R. A., and Tong, W. K. T., 'Design of a one story precast concrete building for earthquake loading', *Proc. 8th World Conf. on Earthq. Eng., San Francisco*, 653–60 (1984).
48. Adham, S. A., Ewing, R. D., and Agbabian, M. S., 'Mitigation of seismic hazards in tilt-up wall buildings', *Proc. 8th World Conf. on Earthq. Eng., San Francisco*, 637–44 (1984).
49. Briasco, E. *Joist Anchors vs. Wood Ledgers*, Los Angeles Department of Building and Safety (1971).
50. Applied Technology Council, 'Guidelines for the design of horizontal wood diaphragms', ATC-7, California (1981).
51. Amrhein, J. E., *Reinforced Masonry Engineering Handbook*, Masonry Institute of America, Los Angeles (3rd edn) (1978).
52. Standards Association of New Zealand, 'Code of practice for light timber buildings not requiring specific design', NZS 3604: 1981.
53. Standards Association of New Zealand, 'Masonry buildings not requiring specific design (means of compliance)', NZS 4229: 1986.
54. Flores, R., 'An outline of earthquake protection criteria for a developing country', *Proc. 4th World Conf. on Earthq. Eng., Chile*, **III**, J4, 1–14 (1969).
55. Evans, F. W., 'Earthquake engineering for the smaller dwelling', *Proc. 5th World Conf. on Earthq. Eng., Rome*, **2**, 3010–13 (1973).
56. Cooney, R. C., and Fowkes, A. H. R., 'Houses in New Zealand—What will happen?', in *Large Earthquakes in New Zealand*, The Royal Society of New Zealand, Miscellaneous Series No. 5, 101–9 (1981).
57. Cooney, R. C., 'Strengthening houses against earthquakes', Building Research Association of New Zealand, *Technical Paper P37* (1982).
58. Cooney, R. C., 'The structural performance of houses in earthquakes', *Bull. NZ Nat. Soc. for Earthq. Eng.*, **12**, No. 3, 223–37 (1979).
59. Tracey, W. J., 'A simulated earthquake test of a timber house', *Bull. NZ Soc. for Earthq. Eng.*, **2**, No. 3, 289–94 (Sept. 1969).
60. de Beeck, M. S., and San Bartolome, A., 'Relevant masonry projects carried out in the structures laboratory at the Catholic University of Peru', *Proc. 8th World Conf. on Earthq. Eng., San Francisco*, **VI**, 823–30 (1984).

61. Various authors, Section 2, pp. 441–708, *Proceedings 8th World Conf. on Earthq. Eng., San Francisco*, (1984).
62. Various authors, 'Case studies: Earthquake risk buildings', *Bull. NZ Nat. Soc. for Earthq. Eng.*, **16**, No. 1, 60–79 (1983); 16, No. 2, 162–78 (1983); and 17, No. 1, 57–68 (1984).
63. New Zealand National Society for Earthquake Engineering, 'Recommendations and guidelines for classifying, interim securing and strengthening earthquake risk buildings', NZNSEE, Wellington (1985).

Appendix B

Miscellaneous Information

B.1 MODIFIED MERCALLI INTENSITY SCALE

I. Not felt except by a very few under exceptionally favourable circumstances.

II. Felt by persons at rest, on upper floors, or favourably placed.

III. Felt indoors; hanging objects swing; vibration similar to passing of light trucks; duration may be estimated; may not be recognized as an earthquake.

IV. Hanging objects swing; vibration similar to passing of heavy trucks, or sensation of a jolt similar to a heavy ball striking the walls; standing motor cars rock; windows, dishes, and doors rattle; glasses clink and crockery clashes; in the upper range of IV wooden walls and frames creak.

V. Felt outdoors; direction may be estimated; sleepers wakened, liquids disturbed, some spilled; small unstable objects displaced or upset; doors swing, close, or open; shutters and pictures move; pendulum clocks stop, start, or change rate.

VI. Felt by all; many frightened and run outdoors; walking unsteady; windows, dishes and glassware broken; knick-knacks, books, etc., fall from shelves and pictures from walls; Furniture moved or overturned; weak plaster and masonry D* cracked; small bells ring (church or school); trees and bushes shaken (visibly, or heard to rustle).

VII. Difficult to stand; noticed by drivers of motor cars; hanging objects quiver; furniture broken; damage to masonry D, including cracks; weak chimneys broken at roof line; fall of plaster, loose bricks, stones, tiles, cornices (also unbraced parapets and architectural ornaments); some cracks in masonry C*; waves on ponds; water turbid with mud; small slides and caving in along sand or gravel banks; large bells ring; concrete irrigation ditches damaged.

VIII. Steering of motor cars affected; damage to masonry C or partial collapse; some damage to masonry B*; none to masonry A*; fall of stucco and some masonry walls; twisting and fall of chimneys, factory stacks, monuments, towers and elevated tanks; frame houses moved on foundations if not bolted down; loose panel walls thrown out; decayed piling broken off; branches broken from trees; changes in flow or temperature of springs and wells; cracks in wet ground and on steep slopes.

IX. General panic; masonry D destroyed; masonry C heavily damaged, sometimes with complete collapse; masonry B seriously damaged; general damage to foundations; frame structures if not bolted shifted off foundations; frames racked; serious damage to reservoirs; underground pipes broken; conspicuous cracks in ground; in alluviated areas sand and mud ejected, earthquake fountains and sand craters appear.

X. Most masonry and frame structures destroyed with their foundations; some well-built wooden structures and bridges destroyed; serious damage to dams, dikes and embankments; large landslides; water thrown on banks of canals, rivers, lakes, etc.; sand and mud shifted horizontally on beaches and flat land; rails bend slightly

XI. Rails bent greatly; underground pipelines completely out of service.

XII. Damage nearly total; large rock masses displaced; lines of sight and level distorted; objects thrown into the air.

* Masonry A, B, C, and D as used in MM scale above.

Masonry A: Good workmanship, mortar, and design; reinforced, especially laterally, and bound together by using steel, concrete, etc., designed to resist lateral forces.

Masonry B: Good workmanship and mortar; reinforced, but not designed in detail to resist lateral forces.

Masonry C: Ordinary workmanship and mortar; no extreme weaknesses like failing to tie in at corners, but neither reinforced nor designed against horizontal forces.

Masonry D: Weak materials, such as adobe; poor mortar; low standards of workmanship; weak horizontally.

B.2 QUALITY OF REINFORCEMENT FOR CONCRETE

The following notes provide some amplification of the points on reinforcement made in Section 7.2.5.

For adequate earthquake resistance, suitable quality of reinforcement must be ensured by both specification and testing. As the properties of reinforcement vary greatly between countries and manufacturers, much depends on knowing the source of the bars, and on applying the appropriate tests. Particularly in developing countries the role of the resident engineer is crucial. Even in California there has been concern amongst designers[1] at ‘the lack of quality control provided by the present ASTM Standards and the lack of uniformity in the reinforcement presently available’.

In order to obtain satisfactory ductility and control of plastic hinge mechanisms the following points require consideration.

(*a*) *Minimum yield stress*

An adequate minimum yield stress (or 0.2 percent proof stress) may be ensured by specifying steel to an appropriate standard, such as BS 4449,[2] or BS 4461,[3] ASTM A615,[4] or ASTM A706.[5]

(*b*) *Variability of yield stress*

For economical design control of structural collapse mechanisms the variability in the yield stress should be smaller than is generally obtainable in practice. For example, the ACI[6] tries to exert some control on the variability of steel to ASTM A615 by requiring that the actual yield stress should not exceed the minimum specified yield stress (characteristic strength) by more than 124 N/mm^2 (18 000 psi). This nominal control in the scatter of yield values is essentially a compromise with manufacturing economy, the design preference being for much less variability, say about half the above value.

(*c*) *Higher strength steels*

Grades of steel with characteristic strength in excess of 415 N/mm^2 (60 000 psi) are not recommended in some earthquake areas, e.g. California and New Zealand. This is because higher strengths generally imply decreased ductility (a shorter yield plateau), but where adequate ductility is proven by tests, somewhat higher strengths may be used where regulations permit. For example, steels to BS 4461[3] with characteristic strengths of 460 N/mm^2 appear satisfactory (Item (d) below). Hot rolled bars of similar strength are also available but problems have been encountered with their ductility.

(*d*) *Cold worked steels*

Cold worked steels are effectively excluded from use in a number of earthquake countries. For example, only steels to ASTM A706 and ASTM A615 (hot rolled) are recommended.[6] In California it is also recommended that the ultimate tensile stress should not be less than 1.33 times the actual yield stress of the bar. This requirement is ostensibly to ensure adequate post-elastic energy absorption capacity in the relatively brittle American steels, but this capacity is as well provided by various other steels with better elongation characteristics such as the British steels,[2,3] but which would have less difference between the ultimate and yield points. In fact for analytical purposes elastoplastic behaviour is more convenient. Hence the 1.33 ratio criterion given above should not be applied to adequately ductile steels such as the British ones referred to above.

Park[7] has stated the usual reasons for avoiding cold worked steels:

(1) It is commonly held that cold worked bars are too brittle for seismic loading conditions. This is not true for all cold worked bars; for example, steel to BS 4461[3] is at least as ductile as most hot rolled steels. The tests

Table B.1

Bar size			Elongation on $5.65\sqrt{(S_0)}$ gauge length (%)					
US		British	Mild Steel		High yield steel			
Bar no. ($\frac{1}{8}$ in)	Dia. (mm)	Dia. (mm)	ASTM A615	BS 4449	ASTM A706	BS A615	BS 4449	4461
3	9.5		20		25	16		
4	12.7		19		22	14		
5	15.9		18		20	13		
		16						
6	19.1		15		19	12		
7	22.2				15	10		
8	25.4			22	14	10	14	12
9	28.7				13	8		
10	32.3				13	8		
11	35.8				12	7		
		40						
14	43.0				10	7		
		50						
18	57.3				9	6		

mentioned below in items (f), (g), (i), and (j) would have to be passed to ensure adequate ductility.

(2) The lack of a yield plateau in cold worked steel is considered a disadvantage. This is certainly analytically inconvenient in that it adds further complications to the determination of plastic hinge positions and to post-elastic behaviour generally. This objection to cold worked steel is obviously rather idealistic considering the many other simplifications in seismic design, and for members nominally without plastic hinges in the seismic collapse condition (such as columns, or floor slabs) it is invalid.

(e) *Substitution of higher grades of steel*

The contractor should not be permitted to use other than the specified grade of steel in the members of moment resisting frames, as this is likely to cause dangerous changes to the collapse mechanism of the structure. Substitution of higher strength steel in beams is particularly undesirable.

(f) *Elongation tests—general ductility*

The most basic measure of ductility of reinforcement is its elongation at failure. Steels with good elongation behaviour are more likely to perform well in other tests of ductility discussed below than steel with low elongation. Reinforcements complying with BS 4449, BS 4461, ASTM A706, and ASTM A615 have moderate to good elongation values, as compared in Table B.1. The American results have been converted to a gauge length of $5.65\sqrt{(S_0)}$ using the Oliver formula[8] in order to conform to modern international practice; this ensures geometric similarity and allows direct comparison of elongations for different specimen diameters.

It can be seen from Table B.1 that the ASTM elongation requirements are less stringent than the British ones for mild steel and for high yield bars of larger diameter.

(g) Bend tests

Bend tests are most important for ensuring sufficient ductility in reinforcement in the bent condition. It is important to use test conditions appropriate to the minimum diameters permitted on site. Because some bars to ASTM A706 and ASTM A615 require large minimum bends in construction (Table B.3), the mandrel diameters used for bend tests on American steels are greater than for British steels in some cases (Table B.2).

Table B.2. Mandrel diameters for bend tests (around 180 degrees unless otherwise stated)

Bar Number ($\frac{1}{8}$ inch)	Diam. (mm)	Mild Steel ASTM A615	Mild Steel BS 4449	High Yield Steel ASTM A615	High Yield Steel ASTM A706	High Yield Steel BS 4449	High Yield Steel BS 4461
3, 4, 5	10–16	4ϕ	2ϕ	4ϕ	3ϕ	3ϕ	3ϕ
6, 7, 8	19–25	5ϕ	2ϕ	5ϕ	3ϕ	3ϕ	3ϕ
9, 10, 11	29–36		2ϕ	8ϕ	6ϕ	3ϕ	3ϕ
14	43			10ϕ*	8ϕ	3ϕ	3ϕ

Notes: ϕ is the bar diameter; * around 90 degree bend only.

(h) Minimum bend radius

The minimum bend radius should be chosen to suit the basic ductility of the steel. As discussed in (f) above, some bars to ASTM A706 and ASTM A615 are not required to be as ductile as British steels; this is taken into account in that larger minimum bend diameters are required in American practice, as shown in Table B.3.

Table B.3

Bar number ($\frac{1}{8}$ in)	Diam. (mm)	Minimum mandrel diameter for bends USA	Minimum mandrel diameter for bends Britain
3–8	10–25	6ϕ	6ϕ
9, 10, 11	29–36	8ϕ	6ϕ
14, 18	43, 57	10ϕ	6ϕ

(i) Resistance to brittle fracture

The brittle fracture problem arises because all carbon steels (and most other types) undergo a transition from ductile to brittle behaviour as the temperature is reduced. This property is strain rate sensitive. At slow rates of loading a steel can behave in a ductile manner, while at the same temperature but a higher loading rate it could fail with nil deformation. Stress concentrations also increase the risk of brittle fracture.

Although reinforcing steels are not normally assessed for resistance to brittle fracture, this is likely to be important when service conditions include shock loading (earthquakes) and low temperatures (say, below 3–5°C). The desirability of testing will be a matter of judgement, depending on the type of structure, the climate, and the seismic risk. For example, important structures such as North Sea oil platforms should always be checked for brittle fracture despite the low seismic risks involved. Heated buildings in Iran, where the winters may be cold, may not merit such tests, unless important structural elements are exposed to the weather.

The brittle fracture characteristics of reinforcing bars may be assessed from their ductile/brittle transition curves, as obtained from a series of impact tests (such as the Charpy test) carried out on standard notched specimens at various temperatures. Although this simple test provides only a fairly crude check on brittle fracture, a reasonable minimum requirement would be that the steel should have a notch toughness of 27 Joules (20 ft lb) measured in the Charpy test at the minimum service temperature.[9]

(*j*) *Strain–age embrittlement*

This form of embrittlement results from cold working steel bars and subsequent ageing which tends to raise the transition temperature from ductile to brittle behaviour. The cold working may either be part of the manufacturing process; for example, steel to BS 4461,[3] or will occur in subsequent bending, strain–age embrittlement may be minimized by suitable controls in the steel-making processes, particularly of the free nitrogen content of the steel.

Strain–age embrittlement shows itself as otherwise unexpected fractures which generally occur following impact load at the bends. Its existence if detected by the *rebend test*, a generally suitable procedure for which may be found in BS 4449[2] or BS 4461. This should be followed if bending practice on site is in accordance with UK practice. If the bending practice of other countries is to be followed, the mandrel size for the rebend test should be the same as that for the appropriate bend test. Whereas this test is often optional in non-seismic areas, it should be applied to all batches of steel for earthquake-resistant structures.

(*k*) *Weldability*

Few high-tensile reinforcing bars are readily weldable. This is particularly true of hot rolled high-tensile bars which generally get their high-tensile strength from increased carbon content. This renders the heat affected zone of the weld liable to serious embrittlement and cracking. General information on this subject is given clsewhere.[10]

Some reinforcement suppliers are capable of given detailed technical advice on the welding of their product, but in general a competent welding engineer or an organization such as the Welding Institute should be consulted. Testing will probably be necessary.

(*l*) *Galvanizing*

A reinforcing bar can be embrittled by other processes. For example, galvanizing of pre-bent reinforcement has been shown to embrittle it by accelerated strain–ageing. Hence coatings such as galvanizing should not be applied to reinforcing bars without giving special consideration to possible embrittling effects.

B.3 STATISTICAL METHODS FOR PROBABILITY STUDIES

B.3.1 Introduction

Probability studies by definition involve statistics, and a brief outline of some elementary statistical operations used in earthquake probability problems is given below. Further explanation of these and related matters may be found in standard references on statistical methods.[11,12] For further reading on probability and statistics as applied to earthquake problems, reference may be made to Benjamin and Cornell,[13] Newmark and Rosenblueth[14] and Lomnitz.[15]

B.3.2 Definitions of some statistical terms

The following terms may be usefully defined here before their use in the subsequent text.

Variance σ^2, is the mean square deviation from the mean. If a set of values (a finite population) consists of n observations x_i, whose mean is μ, the *deviation* of each observation is $x_i - \mu$, and the variance is written

$$\sigma^2 = \frac{\sum_{i=1}^{n} (x_i - \mu)^2}{n} \tag{B.1}$$

If we are considering only a limited number (a sample) out of a population, then the variance is defined as A^2 such that

$$S^2 = \frac{\sum_{i=1}^{n} (x_i - \bar{x})^2}{n-1} \tag{B.2}$$

where $\bar{x}$ is the mean of the sample. It is important to choose the correct variance, either equation (B.1) or (B.2), except, of course, where n is large and the difference between the equations becomes negligible.

Standard deviation σ is the square root of the variance and is therefore written

$$\sigma = \sqrt{\left[\frac{\sum_{i=1}^{n} (x_i - \mu)^2}{n}\right]} \tag{B.3}$$

Distributions. A set of observations may be arranged in various frequency distribution forms. Earthquakes are generally considered to be randomly occurring phenomena and hence have what is commonly termed a *normal distribution* described by the so-called Gauss function

$$p = Ce^{-h^2X^2} \tag{B.4}$$

where X = deviation from the mean;

p = probability of occurrence of this deviation;

$C = \dfrac{1}{\sigma\sqrt{(2\pi)}}$, a constant equal to the maximum height of the curve (Figure B.1)

$h = \dfrac{1}{\sigma\sqrt{(2)}}$, the precision constant determining the spread of the curve.

Hence equation (B.4) can be written

$$p = \frac{e^{-X^2/2\sigma^2}}{\sigma\sqrt{(2\pi)}} \tag{B.5}$$

The standard deviation σ determines the horizontal spread of the distribution curve, and for many purposes it is convenient to use σ as a unit of deviation from the mean, i.e. let

$$Z = \frac{1}{\sigma\sqrt{(2\pi)}}\, e^{-Z^2/2} \tag{B.6}$$

Call $F(Z)$ the area under the curve between the mean and $Z = Z$. The areas under the normal distribution curve $F(Z)$ corresponding to deviations in steps of one standard deviation are written on Figure B.1.

Significance level represents the probability of drawing an erroneous conclusion. When observations are normally distributed, $\{1\text{-}2F(Z)\}$ represents the probability of a value falling outside the range. For example, when

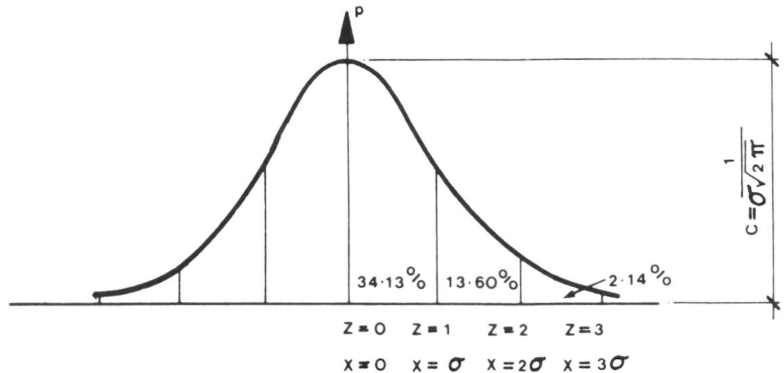

Figure B.1 Area under the normal probability curve

$Z = 1.96$, $\{1\text{-}2F(Z)\} = 0.05$, and it is said that the level of significance is 5 percent.

Confidence level represents the probability of drawing a correct conclusion. It is described by $2F(Z)$ expressed as a percentage. For example, when $Z = 1.96$, $2F(Z) = 0.95$, i.e. there is a 95 percent probability of a value falling within the range $\mu \pm 1.96\sigma$, and the confidence level is 95 percent.

Confidence limits are defined as $(\mu - Z\sigma)$ and $(\mu + Z\sigma)$, the interval between them being the *confidence interval*.

Degrees of freedom v may be defined as the number of independent observations that can be hypothesized. For example, in determining a least squares line, two dependent observations exist, namely the slope and the intercept. Hence if n observations (of M, say) exist, then there are $v = n - 2$ degrees of freedom.

B.3.3 Establishing relationships from data of seismicity observations

From a set of earthquake data for a given site a typical relationship that we may wish to obtain is the magnitude-frequency relationship as discussed in Section 2.3.3. If the magnitude M is plotted against $\log N$, where N is the number of earthquakes of magnitude M or greater per year, a scatter of points is obtained (Figure B.2).

First let us make the usual assumption of the linear relationship discussed in Section 2.3.3, i.e.:

$$\log N = A - bM \tag{2.11}$$

As well as obtaining the 'best' values of A and b, it is desirable to be able to:

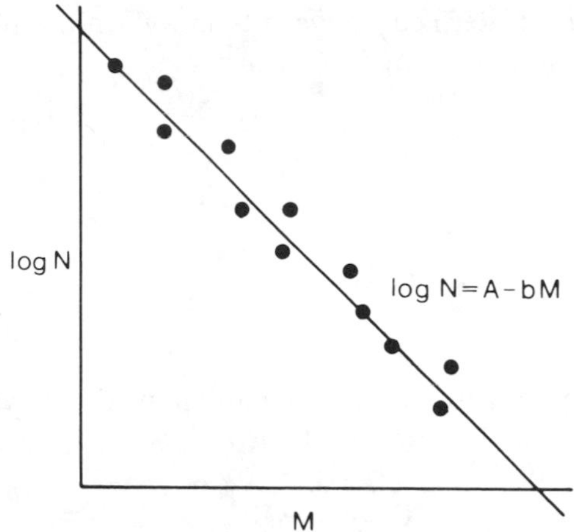

Figure B.2 Typical plot of magnitude against frequency of occurrence

(1) Evaluate how well the line fits the data; and also estimate
(2) What confidence we can have in the 'best' fitting line.

In order to do both of these things, the statistical operations discussed in Sections B.3.4 and B.3.5 should be carried out.

However, first we must fit the 'best' straight line to the data by the method of least squares. This line is called the *linear regression line*, the principle of which is that the most probable position of the line is such that the sum of the squares of deviations of all points from the line is a minimum.

Considering the line shown in Figure B.3,

$$y = a + bx \tag{B.7}$$

the deviations ϵ_i are measured in the direction of the y-axis. The underlying assumption is that x has either negligible or zero error (being assigned), while y is the observed or measured quantity. The observed y is thus a random value from a population of values of y corresponding to a given x.

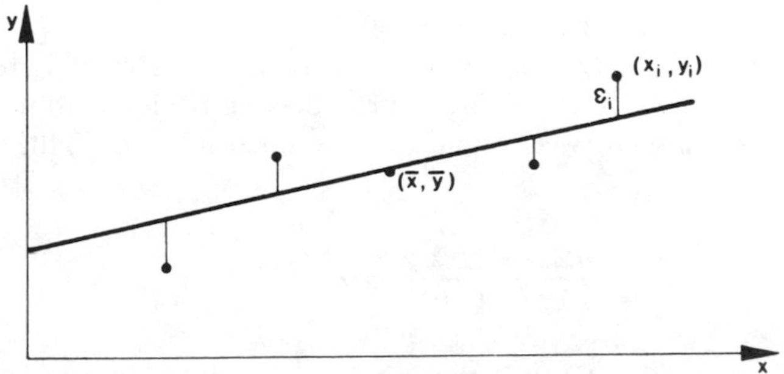

Figure B.3 Fitting the 'best' straight line through data points

The regression line as defined above is found when the sum of the squares of the deviations is a minimum, i.e. when

$$P = \Sigma\epsilon_i^2$$

is a minimum, i.e.

$$P = \Sigma\{y_i - (a + bx_i)\}^2$$

is minimized.

By differentiating, P is found to be a minimum for values of a and b such that the equation of the line of best fit is

$$y = \frac{\Sigma x^2\Sigma y - \Sigma x\Sigma xy}{n\Sigma x^2 - (\Sigma x)^2} + \frac{n\Sigma xy - \Sigma x\Sigma y}{n\Sigma x^2 - (\Sigma x)^2}\, x \qquad \text{(B.8)}$$

Equation (B.8) is the regression line of y upon x.

If the properties of the variables are reversed such that y is assigned and x is observed then the regression line of x upon y is found to be described by equation (B.9):

$$x = \frac{\Sigma x\Sigma y^2 - \Sigma y\Sigma xy}{n\Sigma y^2 - (\Sigma y)^2} + \frac{n\Sigma xy - \Sigma x\Sigma y}{n\Sigma y^2 - (\Sigma x)^2}\, y \qquad \text{(B.9)}$$

The two equations (B.8) and (B.9) coincide at the centroidal point $(\bar{x},\bar{y})$ of the data, but generally will have differing slopes (Figure B.4).

B.3.4 Estimating goodness of fit—correlation

Having fitted the 'best' straight line to the data by the regression analysis outlined above, the 'goodness to fit' of the line to the data should be ascertained. Although the regression line has the most appropriate slope and passes through the centroid of the data, it may still be an inappropriate description of the data, such as in the case illustrated in Figure B.5.

A convenient test for 'goodness to fit' is to measure what is termed the *correlation* between the two variables. It can be seen from equation (B.8) that if there is no correlation between y and x (if y is independent of x) the coefficient of x is zero, i.e.:

$$\frac{n\Sigma xy - \Sigma x\Sigma y}{n\Sigma x^2 - (\Sigma y)^2} = 0 \qquad \text{(B.10)}$$

Similarly, from equation (B.9) if there is no correlation between x and y the coefficient of y is zero, i.e.:

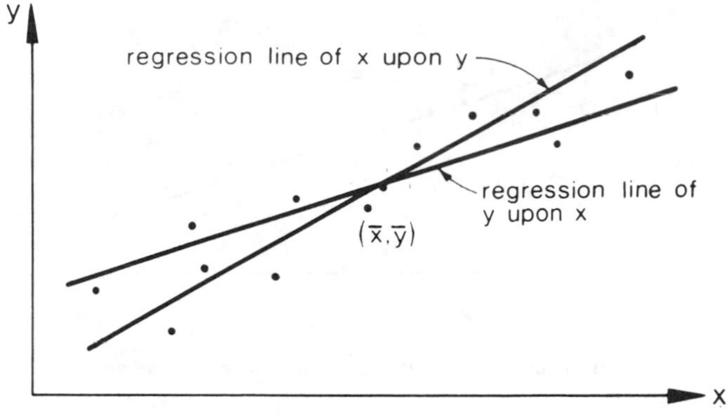

Figure B.4 Regression lines based on the two alternative variables x and y

$$\frac{n\Sigma xy-\Sigma x\Sigma y}{n\Sigma y^2-(\Sigma y)^2}=0 \tag{B.11}$$

If there is no correlation between the two variables being studied, the product of the slopes given by equations (B.10) and (B.11) is zero, i.e.:

$$r^2=\frac{n\Sigma xy-\Sigma x\Sigma y}{n\Sigma x^2-(\Sigma x)^2}\cdot\frac{n\Sigma xy-\Sigma x\Sigma y}{n\Sigma y^2-(\Sigma y)^2}=0 \tag{B.12}$$

Conversely, when there is perfect correlation, i.e. all the points lie exactly on each of the two regression lines, the lines coincide; their slopes are the same and hence from equation (B.12):

$$r^2=1 \tag{B.13}$$

A measure of the correlation of two variables is thus given by r, the *correlation coefficient* and

$$r=\frac{n\Sigma xy-\Sigma x\Sigma y}{\sqrt{[\{n\Sigma x^2-(\Sigma x)^2\}\{n\Sigma y^2-(\Sigma y)^2\}]}} \tag{B.14}$$

or simply, writing $X=x-\bar{x}$, and $Y=y-\bar{y}$,

$$r=\frac{\Sigma XY}{\sqrt{[\Sigma X^2\Sigma Y^2]}} \tag{B.15}$$

The correlation coefficient lies in the range $0\leqslant r\leqslant 1$. If r is positive, y increases with increasing x, and if r is negative, y decreases with increasing x. If $r=\pm 1$ perfect correlation exists, i.e. all the data points lie exactly on one straight line.

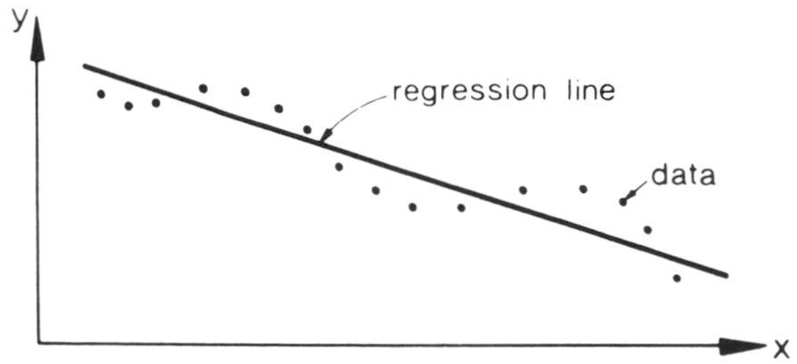

Figure B.5 Illustration of difference between a 'best' straight line and 'goodness to fit'

In order to assess the meaning of a calculated value of r, Table B.4 should be consulted. If the calculated (absolute) value of r is equal to or greater than the appropriate value of r in Table B.4, we conclude that correlation exists with a level of significance (Section B.3) equal to or better than that implied by the table. The level of significance represents the probability of our having drawn a wrong conclusion, i.e. it represents the probability that the relationship of the points to a straight line arose by chance and not because there was a real linear relationship. In most seismicity studies we would choose a 5 percent level of significance. If the calculated value of r is less than the appropriate value of r in Table B.4, the correlation between the data and the best-fitting line is worse than that desired, i.e. there is smaller probability that we have found a valid straight line.

Table B.4. Values of Correlation Coefficient r. (From Crow, Davis and Maxfield, *Statistical Manual*, Dover, 1960)

$P(|r|)$

0 $|r|$ 1

	5 percent level of significance				1 Percent level of significance				
	Total number of variables				Total number of variables				
ν	2	3	4	5	2	3	4	5	ν
1	0.997	0.999	0.999	0.999	1.000	1.000	1.000	1.000	1
2	0.950	0.975	0.983	0.987	0.990	0.995	0.997	0.998	2
3	0.878	0.930	0.950	0.961	0.959	0.976	0.983	0.987	3
4	0.881	0.881	0.912	0.930	0.917	0.949	0.962	0.970	4
5	0.754	0.836	0.874	0.898	0.874	0.917	0.937	0.949	5
6	0.707	0.795	0.839	0.867	0.834	0.886	0.911	0.927	6
7	0.666	0.758	0.807	0.838	0.798	0.855	0.885	0.904	7
8	0.632	0.726	0.777	0.811	0.765	0.827	0.860	0.882	8
9	0.602	0.697	0.750	0.786	0.735	0.800	0.836	0.861	9
10	0.576	0.671	0.726	0.763	0.708	0.776	0.814	0.840	10

continued

Table B.4. *continued*

	5 percent level of significance				1 Percent level of significance				
	Total number of variables				Total number of variables				
v	2	3	4	5	2	3	4	5	v
11	0.553	0.648	0.703	0.741	0.684	0.753	0.793	0.821	11
12	0.532	0.627	0.683	0.722	0.661	0.732	0.773	0.802	12
13	0.514	0.608	0.664	0.703	0.641	0.712	0.755	0.785	13
14	0.497	0.590	0.646	0.686	0.623	0.694	0.737	0.768	14
15	0.482	0.574	0.630	0.670	0.606	0.677	0.721	0.752	15
16	0.468	0.559	0.615	0.655	0.590	0.662	0.706	0.738	16
17	0.456	0.545	0.601	0.641	0.575	0.647	0.691	0.724	17
18	0.444	0.532	0.587	0.628	0.561	0.633	0.678	0.710	18
19	0.433	0.520	0.575	0.615	0.549	0.620	0.665	0.698	19
20	0.423	0.509	0.563	0.604	0.537	0.608	0.652	0.685	20
21	0.413	0.498	0.552	0.592	0.526	0.596	0.641	0.674	21
22	0.404	0.488	0.542	0.582	0.515	0.585	0.630	0.663	22
23	0.396	0.479	0.532	0.572	0.505	0.574	0.619	0.652	23
24	0.388	0.470	0.523	0.562	0.496	0.565	0.609	0.642	24
25	0.381	0.462	0.514	0.553	0.487	0.555	0.600	0.633	25
26	0.374	0.454	0.506	0.545	0.478	0.546	0.590	0.624	26
27	0.367	0.446	0.498	0.536	0.470	0.538	0.582	0.615	27
28	0.361	0.439	0.490	0.529	0.463	0.530	0.573	0.606	28
29	0.355	0.432	0.482	0.521	0.456	0.522	0.565	0.598	29
30	0.349	0.426	0.476	0.514	0.449	0.514	0.558	0.591	30
35	0.325	0.397	0.445	0.482	0.418	0.481	0.523	0.556	35
40	0.304	0.373	0.419	0.455	0.393	0.454	0.494	0.526	40
45	0.288	0.353	0.397	0.432	0.372	0.430	0.470	0.501	45
50	0.273	0.336	0.379	0.412	0.354	0.410	0.449	0.479	50
60	0.250	0.308	0.348	0.380	0.325	0.377	0.414	0.442	60
70	0.232	0.286	0.324	0.354	0.302	0.351	0.386	0.413	70
80	0.217	0.269	0.304	0.332	0.283	0.330	0.362	0.389	80
90	0.205	0.254	0.288	0.315	0.267	0.312	0.343	0.368	90
100	0.195	0.241	0.274	0.300	0.254	0.297	0.327	0.351	100
125	0.174	0.216	0.246	0.269	0.228	0.266	0.294	0.316	125
150	0.159	0.198	0.225	0.247	0.208	0.244	0.270	0.290	150
200	0.138	0.172	0.196	0.215	0.181	0.212	0.234	0.253	200
300	0.113	0.141	0.160	0.176	0.148	0.174	0.192	0.208	300
400	0.098	0.122	0.139	0.153	0.128	0.151	0.167	0.180	400
500	0.098	0.109	0.139	0.153	0.128	0.151	0.167	0.180	400
1,000	0.062	0.077	0.088	0.097	0.081	0.096	0.106	0.116	1,000

The critical value of r at a given level of significance, total number of variables, and degrees of freedom, v, is read from the table. If the computed $|r|$ exceeds the critical value, then the null hypothesis that there is no association between the variables is rejected at the given level. The test is an equal-tails test, since we are usually interested in either positive or negative correlation. The shaded portion of the figure is the stipulated probability as a level of significance.

Table B.5. Distribution of t (Table B.5 is taken from Table III of Fisher and Yates: *Statistical Tables for Biological, Agricultural and Medical Research*, published by Longman Group Ltd, London (previously published by Oliver and Boyd, Edinburgh), and by permission of the authors and publishers)

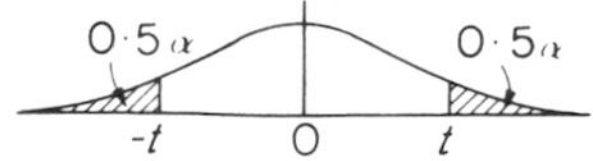

Degrees of freedom ν	Probability α			
	0.10	0.05	0.01	0.001
1	6.314	12.706	63.657	636.619
2	2.920	4.303	9.925	31.598
3	2.353	3.182	5.841	12.941
4	2.132	2.776	4.604	8.610
5	2.015	2.571	4.032	6.859
6	1.943	2.447	3.707	5.959
7	1.895	2.365	3.499	5.405
8	1.860	2.306	3.355	5.041
9	1.833	2.262	3.250	4.781
10	1.812	2.228	3.169	4.587
11	1.796	2.201	3.106	4.437
12	1.782	2.179	3.055	4.318
13	1.771	2.160	3.012	4.221
14	1.761	2.145	2.977	4.410
15	1.753	2.131	2.947	4.073
16	1.746	2.120	2.921	4.015
17	1.740	2.110	2.898	3.965
18	1.734	2.101	2.878	3.922
19	1.729	2.093	2.861	3.883
20	1.725	2.086	2.864	3.850
21	1.721	2.080	2.831	3.819
22	1.717	2.074	2.819	3.792
23	1.714	2.069	2.807	3.767
24	1.711	2.064	2.797	3.745
25	1.708	2.060	2.787	3.725
26	1.706	2.056	2.779	3.707
27	1.703	2.052	2.771	3.690
28	1.701	2.048	2.763	3.674
29	1.699	2.045	2.756	3.659
30	1.697	2.042	2.750	3.646
40	1.684	2.021	2.704	3.551
60	1.671	2.000	2.660	3.460
120	1.658	1.980	2.617	3.373
∞	1.645	1.960	2.576	3.291

This table gives the value of t corresponding to various values of the probability α (level of significance) of a random variable falling inside the shaded areas in the figure, for a given number of degrees of freedom ν available for the estimation of error. For a one-sided test the confidence limits are obtained for $\alpha/2$.

First, calculate the correlation coefficient using equation (B.15):

$$r = \frac{\Sigma XY}{\sqrt{[\Sigma X^2 \Sigma Y^2]}}$$

$$= \frac{2(9+25+63+120+99)}{2(2^2+3^2+5^2+7^2+9^2+10^2+11^2)\times 2(2^2+3^2+5^2+7^2+9^2+12^2)}$$

$$= \frac{379}{389}$$

i.e. $r = 0.974$

To enter Table B.4, the number of degrees of freedom equals the number of observations minus 2, that is:

$$\begin{aligned} v &= n-2 \\ &= 14-2=12 \end{aligned}$$

For a 1 percent level of significance and $v = 12$, from Table B.4, $r = 0.661$. As the calculated value of r exceeds this value, then the line fits the data with a level of significance better than 1 percent, i.e. the data fits the regression line very well.

The data are linearly distributed but with quite large scatter from the line. The meaning of this scatter is studied in the following section.

B.3.5 Confidence limits of regression estimates

Having found the regression line through the observed data and having determined the 'goodness of fit' by the correlation factor, we must still discover what confidence we can have in this estimate of the equation to the line. A commonly used confidence test is the so-called *t-test*. For the desired level of significance and the appropriate number of degrees of freedom we find t such that the true value of some parameter y lies in the range

$$y \pm tS_y$$

The probability of being wrong is equal to the level of significance of the value of t. S_y is the standard deviation of the observed values of y. Values of t are tabulated in Table B.5. The confidence limits are thus defined as $(y + tS_y)$ and $(y - tS_y)$. For a regression line we can determine the confidence limits for such

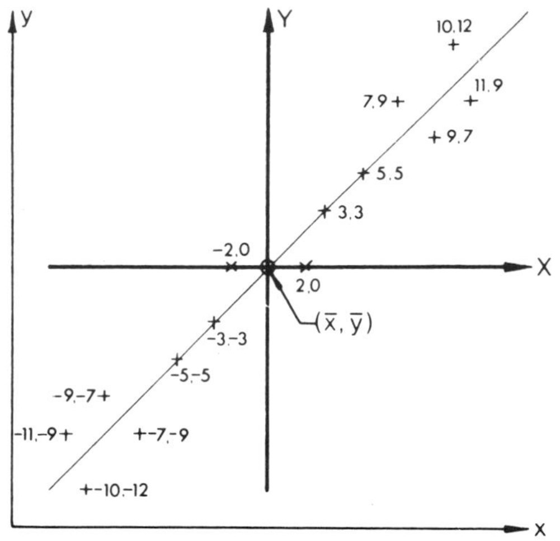

Figure B.6 Data points and regression line for Example B.1

parameters as the mean $\bar{y}$, the slope b, or a future individual value of y_i related to a specific value of x. These confidence limits are illustrated in the following example.

Example B.2—*Confidence limits*

We will refer to the data of the previous example as plotted in Figure B.6.

(*a*) *Confidence limits for the mean, $\bar{y}$.* The variance of y:

$$S_y^2 = \frac{\epsilon_i^2}{\nu} = \frac{10 \times 2^2}{12} = \frac{10}{3}$$

The variance of $\bar{y}$:

$$S_{\bar{y}}^2 = \frac{S_y^2}{n} = \frac{10}{3 \times 14} = \frac{5}{21}$$

From Table B.5 taking $\nu = 12$ and a 5 percent level of significance

$$t = 2.179$$

Therefore the confidence limits for $\bar{y}$ are

$$\begin{aligned} &\bar{y} \pm tS_{\bar{y}} \\ = \; &\bar{y} \pm 2.179\sqrt{(\tfrac{5}{21})} \\ = \; &\bar{y} \pm 1.063 \end{aligned}$$

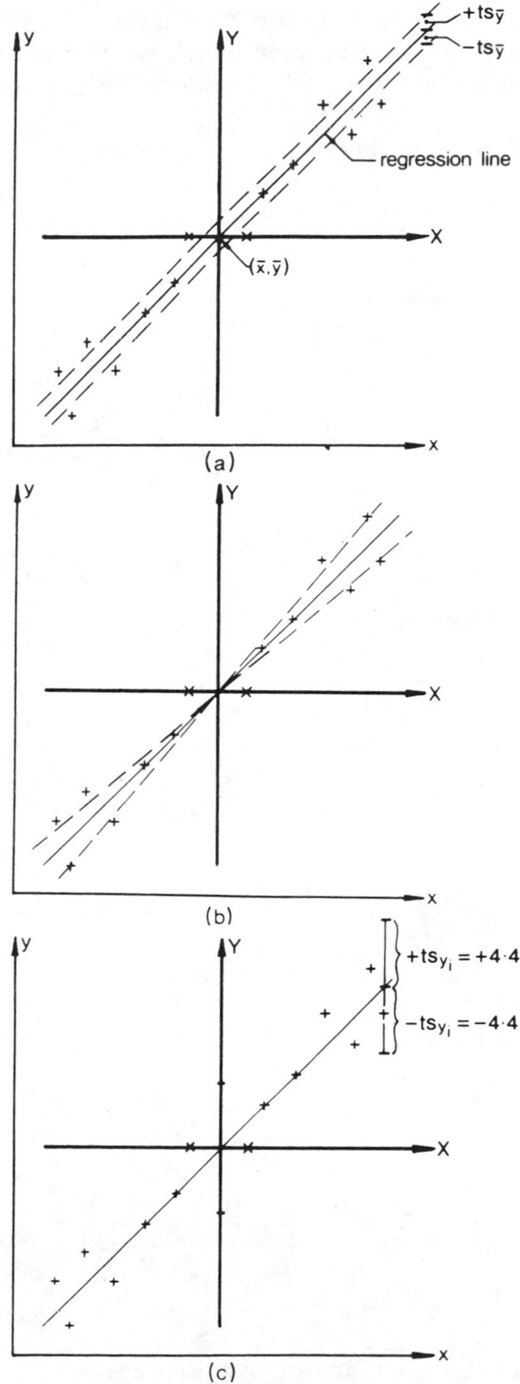

Figure B.7 Confidence limits for regression estimates. (a) Confidence limits for $\bar{y}$; (b) confidence limits of slope; (c) confidence limits for y

There is a 95 percent probability that $\bar{y}$ lies in this range. Since the regression line must pass through the centroidal point, an error in $\bar{y}$ leads to a constant error in y for all points on the line as shown in Figure B.7(a).

(b) Confidence limits for the slope, b. The variance of the slope b:

$$S_b^2 = \frac{S_y^2}{\Sigma(x-\bar{x})^2} \tag{B.16}$$

$$= \frac{S_y^2}{\Sigma X^2}$$

$$= \frac{10}{3} \times \frac{1}{2(2^2+3^2+5^2+7^2+9^2+10^2+11^2)}$$

$$S_b^2 = \frac{5}{3\times 389}$$

As before, for a 5 percent level of significance

$$t = 2.179$$

Therefore the confidence limits for b are

$$b \pm tS_b$$

$$= 1.00 \pm 2.179\sqrt{\left(\frac{5}{3\times 389}\right)}$$

$$= 1.00 \pm 0.143$$

Again there is a 95 percent probability that b lies in this range (Figure B.7):

$$S_{y_i}^2 = S_y^2\left(1+\frac{1}{n}+\frac{X^2}{\Sigma X^2}\right) \tag{B.17}$$

$$= \frac{10}{3}\left(1+\frac{1}{14}+\frac{11^2}{2\times 389}\right)$$

$$= 4.09$$

As before, for a 5 percent level of significance

$$t = 2.179$$

Therefore the corresponding confidence limits for y_i(or Y_i) at $X=11$ are

$$Y_i \pm tS_{y_i} = Y_i \pm 2.179\sqrt{(4.09)}$$

$$= Y_i \pm 4.407$$

Note that the confidence limits for Y_i at $X=-11$ equal those at $X=+11$. Similarly, the confidence limits for y_i at $X=0$ can be found by substituting $X=0$ in equation (B.17). This gives the confidence limits for y_i (or Y_i) at $X=0$ as

$$Y_i \pm 4.118$$

Again there is a 95 percent probability that $Y_{(X=0)}$ lies in this range. These confidence limits are illustrated in Figure B.7(c).

The confidence limits found in Example B.2 show there is appreciable scatter from the regression line, despite the fact that the correlation coefficient $r=0.974$ indicates that the data fit the straight line very well. Part (c) of Example B.2 shows how we can estimate a probable maximum value of magnitude M, or ground acceleration a, for a given return period from seismicity data such as shown on Figure B.2.

For examples of the use of statistics on earthquake acceleration data, see Figure 2.23 and its source reference.

B.4 DETERMINING PROBABILITIES OF GROUND-MOTION CRITERIA

B.4.1 The basic method

The application of probability theory to the analysis of seismic risk and hazard is well established, notable contributions having been made by Cornell,[16] Cornell and Vanmarcke,[17] Esteva and Villaverde,[18] and McGuire.[19]

The estimation of the probability of occurrence of an earthquake effect, i, at a given site requires the consideration of a number of factors, i.e. uncertainty in attenuation, randomness of occurrence, size, and location of the events. First, let us consider the uncertainties in attenuation. As discussed in Section 2.4.2.1, the attenuation of peak ground accelerations, velocities, and displacements may be expressed in the form of equation (2.16). This form may also be used for spectral amplitudes, so that the equation for all these ground-motions criteria may be written either as

$$i = b_1' \; 10^{b_2 M} \; (r+b_4)^{-b_3} \; \epsilon' \tag{B.18}$$

or

$$\log i = b_1 + b_2 M - b_3 \log (r+b_4) + \epsilon \tag{B.19}$$

The values of the empirically derived constants b_1 to b_4 in the above equations found in the literature vary considerably. For example, Cornell[16] used early forms of the attenuation equation for a, v, and d in which $b_4=0$, while Esteva and Villaverde[18] used different values of b_4 for each parameter a, v, S_a, and S_v, and McGuire[19] used $b_4=25$ for all site effects. Non-zero values of b_4 are intended to prevent unbounded effects at sites close to the source. The terms ϵ' and ϵ are residual terms accounting for derivations of individual observations from the predicted mean value.

The attenuation relationship is found by performing a regression analysis on the data, usually using the logarithmic form of the relationship, (i.e. equation (B.19)). The parameters b_2 and b_3 thus evaluated are used directly in equation (B.18) to predict i, whereas b_1' should be obtained from

$$b_1' = 10^{b + \frac{1}{2} s_e} \tag{B.20}$$

where s_e is the standard error of the estimate of the predictions of log i. Equation (B.20) is obtained[19] from the relationship between the mean and variance of a normally distributed variable and the mean of the corresponding log-normally distributed variable.

Next we will discuss the uncertainty due to randomness in occurrence, size, and location of earthquakes. As discussed in Section 2.3.3, the seismicity of an area is generally expressed in the form

$$\log N = A - bm \tag{B.21}$$

or

$$N = 10^{A - bm} \tag{B.22}$$

where N is the number of earthquakes of magnitude m or greater per year per unit source area (or length) and A and b are constants for the locality concerned.

Any given site may be affected by a number of source regions having their own particular A,b values. From equation (B.22) the number of earthquakes of interest can be found as those lying in the magnitude range m_0 to m_1, where m_0 means the minimum magnitude of engineering and $m_1 = M_{max}$ the maximum magnitude event (Section 2.3.3). It then follows that the cumulative distribution of magnitude of earthquakes is given by

$$F_M(m) = k[1 - 10^{-b(m - m_0)}] \qquad m_0 \leqslant m \leqslant m_1 \tag{B.23}$$

where

$$k = 1/[1 - 10^{-b(m_1 - m_0)}] \tag{B.24}$$

It follows from equation (B.20) that the probability density function for magnitude is

$$f_M(m) = bk\ 10^{-b(m - m_0)} \qquad m_0 \leqslant m \leqslant m_1 \tag{B.25}$$

The cumulative distribution of maximum effect i at a given site due to an earthquake of random size and location may be expressed by

$$F_I(i) = \int_{m_0}^{m_I} \int_r P[\mathrm{I} \leqslant i|m,r]\, f_{\mathrm{m}}(m) f_{\mathrm{R}}(r)\; dm\; dr \tag{B.26}$$

where $P[I \leqslant i|m,r]$ is the probability that the maximum effect I is less than i given m and r and $f_{\mathrm{M}}(m)$ is the probability density function for magnitude given by equation (B.25). The probability distribution function for distance $F_{\mathrm{R}}(r)$ is dependent on the geometric nature of the source, i.e. on the spatial distribution of possible earthquake foci. The most common source geometries considered are point sources, line sources (repeating faults), and horizontal planes or area sources in which there is an equal probability of events occurring at any point.

For a line source, consider a site situated symmetrically in relation to a fault deemed to be a line source at depth H below the horizontal plane EFS as shown in Figure B.8. Assuming that events are equally likely to occur at any point on the line, the location variable X is uniformly distributed on the interval $(0,l/2)$. The cumulative probability distribution $F_{\mathrm{R}}(r)$ of R is thus

$$F_{\mathrm{R}}(r) = P[R \leqslant r] = P[R^2 \leqslant r^2]$$

$$= P[X^2 + d^2 \leqslant r^2] = P[|X| \leqslant \sqrt{(r^2 - d^2)}]$$

$$F_{\mathrm{R}}(r) = 2\sqrt{(r^2 - d^2)}/l \qquad d \leqslant r \leqslant r_0 \tag{B.27}$$

Therefore the probability density function of R is

$$f_{\mathrm{R}}(r) = \frac{dF_{\mathrm{R}}(r)}{dr}$$

$$= \frac{2r}{l\sqrt{(r^2 - d^2)}} \qquad d \leqslant r \leqslant r_o \tag{B.28}$$

Substituting the density functions for M and R, i.e. equations (B.25) and (B.28), in equation (B.26), and using the lognormal distribution of equation (B.19) for i, equation (B.26) becomes

$$F_I(i) = \int_{m_0}^{m_I} \int_{d}^{r_0} \phi \left\{ \frac{\log v - \mu}{\sigma} \right\} bk\; 10^{-b(m - m_0)}\; \frac{2r}{l\sqrt{(r^2 - d^2)}}\; dm\; dr \tag{B.29}$$

where

$$\phi \frac{1}{\sqrt{(2\pi)}} \int_{-\infty}^{x} \mathrm{e}^{-u^2/2}\, du$$

is the cumulative distribution function for the standard normal distribution, and μ and σ are the mean and standard deviation of log i determined by the regression for an earthquake of magnitude m and r.

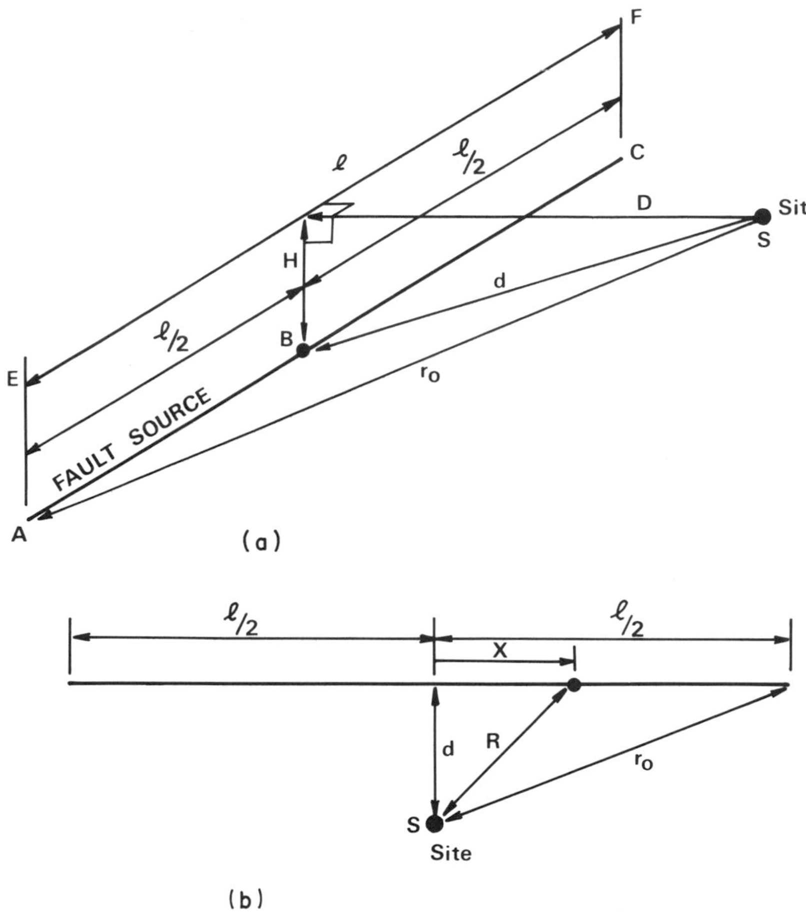

Figure B.8 Model of linear earthquake source. (a) Perspective; (b) plane ACS

It is noted that the probability that effect i will be exceeded is obtained from equation (B.29) using

$$p_i = P[I > i] = 1 - F_I(i) \tag{B.30}$$

As the probability p_i applies to any earthquake occurring within the source region, if on average N earthquakes per year occur in the source region, then the average annual probability of i being exceeded at the site is

$$p_D = p_i N \tag{B.31}$$

and the average return period of i being exceeded is

$$T_R = 1/p_D = 1/p_i N \tag{B.32}$$

B.4.2 Probabilistic enhancement from attenuation variability

The attenuation of seismic effects with distance, as required for the above calculations, may be taken in the form of either of a deterministic function or as a probabilistic statement. Much larger effects (e.g. spectral accelerations) are often calculated for a given return period when the probabilistic form of the attenuation relationship is used than when a deterministic approach is taken using the same mean relationship with no variability. This increase arises because, in the probabilistic model, high levels of shaking are associated not only with the mean levels from rare earthquakes but also with the upper tail motions of much more common smaller events. The increase in the calculated effect for a given return period, or the corresponding increase in mean occurrence rate for a given level of effect, is sometimes referred to a 'probabilistic enhancement'. Because of the large scatter in attenuation data, this enhancement is sometimes considerable. For example, the effect seems to be worse for New Zealand than for USA data.[20,21]

In the case of an unbounded magnitude model, Peek[21] shows the derivation of the enhancement factors, A_z and B_z. Let

$$p_i{}^* = A_z p_i \tag{B.33}$$

where $p_i{}^*$ is the corrected probability of occurrence;
p_i is the probability of occurrence based on the geometric mean of the attenuation data (no scatter in the model);

$$A_z = \exp\left[\tfrac{1}{2}\left\{\frac{\sigma b}{b_2}\right\}^2\right] \tag{B.34}$$

where σ is the standard deviation of $ln\ i$.

and letting $$i^* = B_z i \tag{B.35}$$

where i^* is the corrected earthquake effect;
i is the earthquake effect for a given return period based on the geometric mean of the attenuation data (no scatter in the model); and

$$B_z = \exp\left[\frac{\sigma^2 b}{2b_2}\right] \tag{B.36}$$

For typical values of b, b_2 and σ, Figure B.9 shows that the magnitude of A_z and B_z may easily exceed 2.0. Figure B.9 is based on the assumption that the magnitude range is unbounded. This approximation does not always predict higher accelerations for a given return period than 'exact' estimates based on direct use of probabilistic theory incorporating variability and using a finite maximum magnitude.[22]

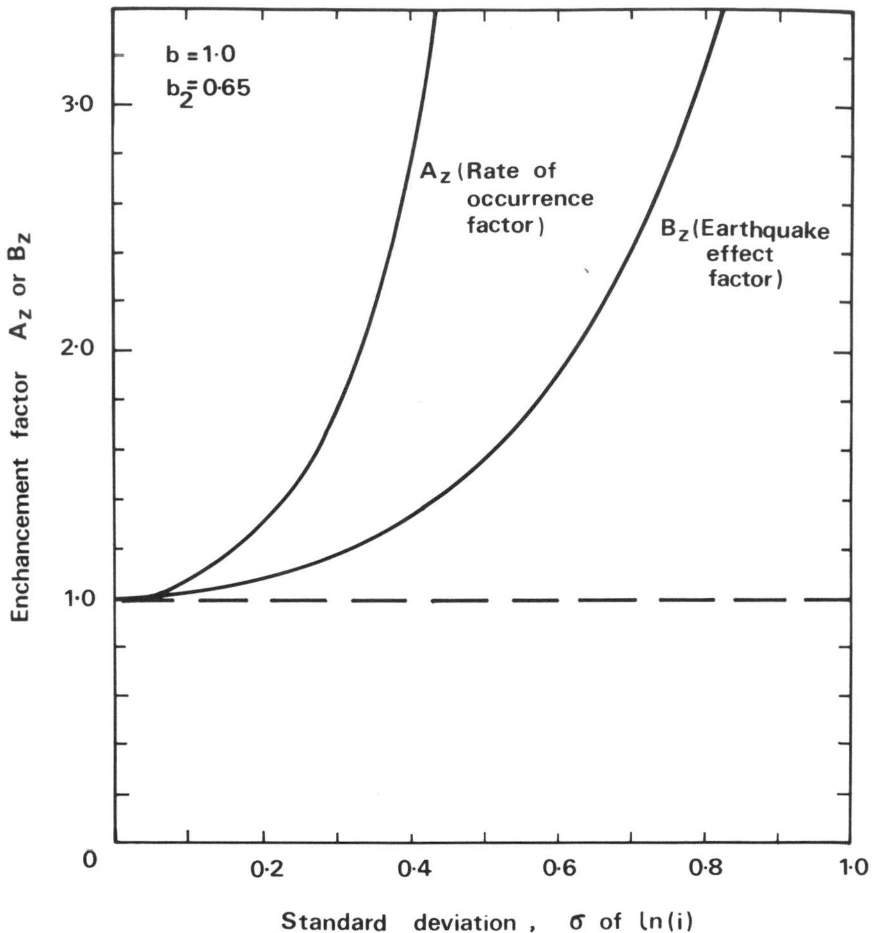

Figure B.9 Typical enhancement factors, A_z and B_z, to allow for variability in attenuation data (adapted from Peek[21])

Further discussion of the effects of variability of attenuation data have been given elsewhere.[19–20]

REFERENCES

1. Seismology Committee, SEAOC, *Recommended lateral force requirements and commentary*, Structural Engineers Association of California (1973).
2. British Standards Institution, 'Hot rolled steel bars for the reinforcement of concrete', BS 4449 (1978).
3. British Standards Institution, 'Cold worked steel bars for the reinforcement of concrete', BS 4461 (1978).

4. American Society for Testing Materials, 'Deformed and plain billet-steel bars for concrete reinforcement', ASTM A615 (1984).
5. American Society for Testing and Materials, 'Low-alloy deformed bars for concrete reinforcement', ASTM A706 (1984).
6. American Concrete Institute, 'Building code requirements for reinforced concrete', ACI 318–83.
7. Park, R., 'Steel in earthquake resistant structures', *Metallurgy in Australasia, Proc. 27th Annual Conference of the Australian Institute of Metals*, 136–42 (1974).
8. British Standards Institution, 'Methods for converting elongation measurements for steel: carbon and low alloy steels', BS 3894, Part 1 (1965).
9. Boyd, G. M., *Brittle Fracture in Steel Structures*, Butterworths, London (1970).
10. American Welding Society, 'Standard welding code—Reinforcing steel', AWS D1.4.
11. Neville, A. M., and Kennedy, J. B., *Basic Statistical Methods for Engineers and Scientists*, International Textbook Co., Pennsylvania (1964).
12. Hald, A., *Statistical theory with engineering applications*, John Wiley and Sons, New York (1952).
13. Benjamin, J. R., and Cornell, C. A., *Probability, Statistics and Decision for Civil Engineers*, McGraw-Hill, New York (1970).
14. Newmark, N. M., and Rosenblueth, E., *Fundamentals of Earthquake Engineering*, Prentice-Hall, Englewood Cliffs NJ (1971).
15. Lomnitz, C., *Global Tectonics and Earthquake Risk*, Elsevier, Amsterdam (1974).
16. Cornell, C. A., 'Engineering seismic risk analysis', *Bull. Seism. Soc. Amer.*, **58**, No. 5, 1538–1606 (1968).
17. Cornell, C. A., and Vanmarcke, E. H., 'Analysis of uncertainty in ground motions and structural response due to earthquakes', *Research Report R69-24*, Dept of Civil Engineering, Massachusetts Institute of Technology (1969).
18. Esteva, L., and Villaverde, R., 'Seismic risk, design spectra and structural reliability', *Proc. 5th World Conf. on Earthq. Eng., Rome*, **2**, 2586–97 (1973).
19. McGuire, R. K., 'Seismic structural response risk analysis, incorporating peak response regressions on earthquake magnitude and distance', *Research Report R74-51*, Dept of Civil Engineering, Massachusetts Institute of Technology (1974).
20. Berrill, J. B., 'Seismic hazard analysis and design loads', *Bull. NZ Nat. Soc. for Earthq. Eng.*, **18**, No. 2, 139–50 (1985).
21. Peek, R., 'Estimation of seismic risk for New Zealand', *Research Report 80/21*, Dept of Civil Engineering, University of Canterbury, New Zealand (1980).
22. Bender, B., 'Incorporating acceleration variability into seismic hazard analysis', *Bull. Seism. Soc. Amer.*, **74**, No. 4, 1451–62 (1984).

Index

Accelerograms, 72
  as design earthquakes, 130
  of real earthquakes, 121, 130
  and response spectra, sources of, 120
  simulated, 121, 131, 134
Accelerograph, 72
Acceptable risk, 7, 113
Added mass of soil, 212, 222
Adhesion values for foundations, 234
Adobe, 361
Agadir 1960 earthquake, 131
Alaska 1964 earthquake, Great, 97
Allowable bearing pressure on soils, 232
Alluvium (*see* Soil)
Amplification (in soil) 190, 192, 198–9, 207, 208
Analysis (*see also* Seismic analysis)
  and material behaviour, method of, 275
  for structures, selection of method of, 274
Anatolian fault zone, 59
Architectural detailing, 443
Artificial accelerogram, 198 (*see also* Simulated earthquake)
Asperity model, 68
Attenuation, 67, 503
  model, 113, 114
  within soil, 198, 207, 208
Available ductility for r.c. members, 292
Averaged response spectra, 22, 119

Banco Central, Managua, 138
Bar bending, 315
Barrier model, 68
Base isolation, 152, 166 (*see also* Isolation)
  using flexible bearings, 168
Basins (geological), 38
Bauschinger effect, 259
Beam–column joints, 334, 398
Beam-hinging failure mechanisms, 158
Beams
  r.c., 330–5
  steel, 377
Bedrock
  depth to, 95, 102, 192
  effective (equivalent), 102, 192, 224
  motion, 131, 133, 197–8
Bend tests for reinforcement, 488
Benioff zone, 36
Bilinear (hysteresis model), 265–6
Body wave magnitude, 6
Bolting, 372, 397
Brittle fracture, resistance to, 488
Brittle materials, 276, 361, 423, 449
Braced frames, 413
Bridges, 126, 174, 176, 455
  strength and ductility of, 461
Bucharest 1977 earthquake, 56
Buckling of steel columns, 385
Buffer plates (tectonic), 37
Buildability, 147, 160

Caissons, 235
Cantilever walls, 303
Capacity design, 158, 160
Caracas 1967 earthquake, 95, 107, 152, 207
Cause of earthquakes, 4
Ceilings, suspended, 450
Characteristics of strong ground motion, 72
Chilean 1960 earthquake, 467, 469
Chimneys, 174, 435, 463–8, 479
  reinforced concrete, 466

Chimneys *(continued)*
steel, 467
Chinese earthquake catalogue, 117
Cladding, 350, 351, 448
Clay, sensitive (quick), 97, 231, 240
Cleats, 397
Clyde Dam, New Zealand, 139
Cold formed sections, 374
Cold worked steel, 486
Columns (r.c.), 327–9, 332
steel, 382
Compaction, degree of, 103
Composite construction, 399
Compressive underthrust faults, 42
Concentrically braced frames, 164, 387
Concrete quality, 316
Confidence limits, 492, 499
Configuration of construction, 148, 456
Confinement (*see also* Transverse steel)
effect on ductility, 297
reinforcing (columns), 327
Connections
in diagonally braced frames, 399
in industrial buildings, 473
in precast concrete, 340–9
in steelwork, 393, 396
in timber construction, 414
Consequences of earthquakes, 7
Construction
joints in structural walls, 309
materials, choice of, 156
Corner frequency f_o, 70
Correlation, 494
Cost
of construction, 11, 142, 172
of damage, 9
of earthquake resistance, 10, 424
Coupled walls, 164, 310, 311
Coupling beams, 311
Critical structures, 147
Cyclic loading behaviour (*see* Hysteresis)
Cyclic triaxial test, 108
Damageability, 263
Damping, 108, 186, 187, 218, 256, 372, 407, 424
ratios for structures, 266, 268
soil-structure system, 226
Dams, large, 139
Deflection
control, 265 (*see also* Drift *and* Deformation)
ductility factor, 170, 277, 292, 388, 458, 461
Deformation control, 154 (*see also* Deflection *and* Drift)
Density (*see* Mass densities)
Design
brief, 1
earthquakes, 113, 118
events, 87, 117
process, 1
Development (anchorage), 315
Diagonal braces, 387
Diaphragms, timber horizontal, 410
Directional effects, 77
Distant events, 87
Distribution in time of damaging earthquakes, 59
Djakarta, 55, 89
Doors, 449
Drift, 146 (*see also* Lateral drift)
limitations, 384 (*see also* Deflection control *and* Deformation control)
Duration
of loading, effect (on timber), 404
of strong motion, 82
Ductility, 262, 375, 423
adequate, 263, 375
bridges, 461
cantilever walls, 306
column, 383
demand, 292, 458
effect of axial load, 355
limited, 263
prestressed concrete, 353
ratio, section, 292
r.c. columns, 301

required, 291
Ductwork, 437
Duhamel integral, 19
Dynamic analysis, 273
of soil-structure systems, 209
Dynamic enhancement of yield stress, 373
Dynamic properties of soil, 182
Dynamics, theory of, 13

Earth pressures, seismic, 249
Earthquake
distribution in time and size, 59
occurrence processes, 64
response, 20 (*see also* Seismic response)
retaining structures, 248
source models, 67
Eccentrically braced frames, 164, 389
Economic consequences of earthquakes, 9
Effective length of steel columns, 384
El Centro 1940 earthquake, 14, 21, 258, 279
Elastic continuum analysis of piles, 243
Elastic response spectra, 122
Elastic seismic response of structures, 256
Elastoplastic hysteresis model, 265–6
Electrical equipment (*see* Equipment)
Embedment
effects in soil-structure systems, 230
of footings, 222
Energy
absorption, 262, 388, 436
dissipators, 165, 176, 177
isolating devices, 165
Epicentre, 73
Epicentral area (*see* Near field)
Epicentral distance, 72
Equations of motion, solution of, 19
Equipment, 419
aseismic protection of, 433
base-mounted free-standing, 426
brittle, 429
code seismic coefficients, 430–3
design requirement, 421
material behaviour in, 423
rigidly mounted, 434
on suspended floors, 429
Equivalent cantilever method (pile analysis), 244, 246
Equivalent radius (of footings), 187
Equivalent static force analysis, 272
Equivalent viscous damping, 196, 263
Exit requirements, 450

f_{max}, 70, 134
Failure mode control, 157, 160, 161, 165, 290, 363
for bridges, 458
Fault
activity, degree of, 43
displacements, probability of, 135
movements, designing for, 138
rupture length, area, displacement, 52–4
system, Nevis-Cardrona, 48
trace, 5
types, 42
Faults, 5
location of active, 40
risk and design considerations, 135
Field determination
of fundamental period of soil, 105
of shear wave velocity, 105
Field-tests for soil properties, 100, 101
Finite elements, 209, 222, 223
Fire resistance of timber construction, 416
Focal depth, effect of, 56, 195
Focal distance, 72
Focus, 5
Folds (geological), 38
Foundation
damping, 228
dashpot, 217
modelling, finite elements, 222
size effects, 127
spring stiffness, 211

Foundations, 231 (*see also* Substructure)
  aseismic design of, 231
  deep box, 234
  for industrial buildings, 473
  in liquefiable ground, 245
  for low-rise housing, 478
  piled, 235
  reinforced concrete, 317
  shallow, 233
  timber structures, 407
Fourier spectrum, 85, 135, 197
Framed tube structures, 163
Frequency
  content, 85
  domain, 197, 209, 223
Friction angles for foundations, 234
Fundamental period (*see also* Period of vibration)
  effective, 226
  for soil deposits, 192, 195
Function (of the project), 142

Gaussian white noise, 134
Gediz, Turkey 1970 earthquake, 207
Geology, 4, 31
  local, 94
Geophysical method, 105
Golden Gate Park, San Francisco 1957 earthquake, 130, 191, 198
Grabens, 38
Gravity retaining walls, 253
Greece–Turkey region, 9
Ground
  improvement techniques against liquefaction, 248
  motions, nature of, 67
Groundwater
  conditions, 101, 103
  discharge, 100

Half-space
  elastic, 209, 211, 215
  layered, 219–21
  theory, limitation of, 187
  viscoelastic, 209, 215
Hammering, 150, 445, 451 (*see also* Pounding)
Hazard
  function, 64
  refinement factor, 66, 138
Helena Montana 1935 earthquake, 130
Holding-down devices for bridges, 463
Horizontal linkage systems for bridges, 462
Housing, low-rise, 474
Hybrid
  half-space/finite element model, 223
  structural systems, 165
Hypocentre, 5
Hysteresis, 183, 186, 259–61, 263, 265, 276, 380, 383, 393, 404
Hysteretic behaviour, pinched loops, 393, 406

Imperial Valley 1940 earthquake (*see* El Centro)
Inelastic response spectra, 129
Infill panels
  effect on seismic response, 269
  non-structural, 160, 444–7
Infill walls, structural, 369
Inhomogeneous soil, 217, 218
Insurance, 12
Intensity, 5
Interaction of frames and infill panels, 269
Internal damping, 186, 187 (*see also* Material damping)
Interstorey drifts, 170
Intra-plate earthquakes, 37
Inverted pendulum structures, 469
Isolating devices, location of, 168
Isolation (*see also* Base isolation)
  equipment, 436
  from seismic motion, 166
  using flexible piles, 170
  using uplift, 173

Koyna Dam, India, 5

Laboratory tests for soils, 101, 107
Lamellar tearing, 376
Landslides (avalanches), 97
Lateral drift, 155 (*see also* Drift)
Lateral restraints, spacing of, 381
Lead–rubber bearings, 168
Light fittings, 436
Limit state design, 143
Limited ductility, 265, 291
Link beams in EBFs, 392
Liquefaction, 97, 101, 161, 203, 248
  of saturated cohesionless soils, 203
Local magnitude, 6
Longitudinal steel, 354
  in beams, 332
Low-rise construction, 156, 472–9
Lumped mass model
  for soil, 196
  for structures, 23

M_{max}, 52, 115 (*see also* Maximum magnitude)
Magnitude, 5
Magnitude–frequency relationship, 60
Magnitude versus fault
  displacement, 54
  rupture area, 53
  rupture length, 52
Managua 1972 earthquake, 135
Masonry, 13, 361
Mass densities, 189
Material damping, 186, 218 (*see also* Internal damping)
Maximum acceleration, 206
Maximum credible earthquake, 147
Maximum magnitude, 61, 504 (*see also* M_{max})
Mechanical equipment (*see* Equipment)
Mexico 1985 earthquake, 198
Mexico City, 95, 154, 195, 207
  Lake Zone of, 194
Michigan Basin, 39
Microtremor recording techniques, 107
Mid-plate earthquakes, 54
Modal responses combination, 27, 28
Mode
  shapes, 24
  superposition, 25
Models, mathematical, 265
Modified Mercalli intensity scale, 5, 484
Modulus of elasticity values for soils, 189
Moment magnitude, 7, 54, 75
Moment-resisting frames, 163, 413
Mononobe–Okabe equations, 250
Movement gaps, or joints, 150, 161, 437–41, 445, 448
Multi-degree-of-freedom systems, 22
Multiple events, 6
Muto slitted wall, 154, 164, 177, 271

Near-field ground motion, 74, 76
Near-field versus far-field effects, 129
Niigata Japan 1964 earthquake, 97, 203
Non-linear inelastic earthquake response, 28
Non-linear seismic response of structures, 258
Non-linear soil behaviour, 183, 218
Non-stationary random processes, 131
Non-structural infill panels, 444
Non-structure and failure mode control, 160
Normal faults, 42, 52, 54
Normal risk construction, 147
Normal mode analysis, 274
North Sea, 38, 97

Offshore structures (oil platforms), 224, 388

Pacoima Dam, 71, 85, 131
Panel zones, steel, 396
Parapets, 335
Particle
  on board, 405, 408
  size distribution, 107, 203
Partitions, 160, 444

P-delta effect, 154, 155, 279, 383, 469
Peak ground
  acceleration, 198
  acceleration versus Modified Mercalli intensity, 76
  displacement, 81
  motion, upper bound, 79
  motions, attenuation of, 72
  velocity, upper bound, 80
Peak horizontal acceleration, 79
Penetration resistance tests, 103
Period of vibration, 224, 256, 267
  of soil sites, 192, 195
Peruvian 1970 earthquake, 97, 160
Pile groups, 245
Piles, 235
  dynamic response, 239
  elastic continuum analysis, 243
  equivalent static lateral loads, 240
  lateral resistance, 241
Pipelines, 139
Pipework, 433, 437–41
Plastic design method, 375
Plastic hinges, 458
  mechanisms, 159
  rotation capacity, 379
Plasticity, 259
Platforms (geological), 38
Plumbing (*see* Equipment, Pipework *and* Services)
Plywood, 408
Poisson model, or process, 64, 67, 133, 137
Poisson's ratio for soils, 189
Porcelain, 276 (*see also* Brittle materials)
Post-yield behaviour, 145 (*see also* Non-linear *and* Hysteresis)
Pounding, 160, 369 (*see also* Hammering)
Precast concrete
  cladding, 350, 351
  structural, 339–49
Prestressed concrete, 352
Probability, 3, 4
  distribution, 491, 505
  distributions of ground motion criteria, 113, 503
  of occurrence of fault displacements, 135
  studies, 490

Quality of concrete, 316
Quality of reinforcement, 316, 485
Quality of structural steel, 375
Quincha, 476

Radiation damping, 186, 222, 226
Ramberg–Osgood model, 197, 265–6
Rate
  of loading, 372
  of strain energy release, 58
Reduction factor, for loading 277
Regression analysis, 73, 499, 504
Reinforced concrete design, 288
Reinforcement quality, 316
Relative density
  of soil, 202, 204
  test, 107
Reliability, 142, 143, 265
Reliable seismic behaviour
  bridges, 455
  chimneys and towers, 463
  concrete structures, 289
  industrial buildings, 472
  masonry, 363
  steel structures, 374
  timber structures, 407
Repairability, 159, 265
Repairable (structures), 146
Resonance, 256
Resonant column test, 110
Response spectra, 21
  averaged, 22, 119
  design earthquakes, 122
  elastic, 122
  inelastic, 129
  of real earthquakes, 121
  of simulated earthquakes, 121
  sources of, 120

special features of, 126
Response spectrum
analysis, 26
technique, 274
Response studies of regular soil sites, 198
Retaining walls, 254, 320
Retrofitting of existing structures, 479 (*see also* Upgrading)
Reverse faults, 42, 52, 54
Rise time, 70
Roofs of heavy construction, 478
Root-mean-square acceleration, 83
Rupture
surface, 67
velocity, 70

Safety criteria, 143
San Antonio Chile 1985 earthquake, 149, 150, 199
San Fernando California 1971 earthquake, 8, 150, 234, 418, 443
San Francisco 1957 earthquake, 130, 191, 198
San Francisco Bay area, 195
Seiches, 97
Seismic activity, 31
Seismic analysis
bridges, 459
chimneys and towers, 464
design procedures for equipment, 424
structures, methods of, 272
Seismic earth pressures, 249
Seismic event map, 56
Seismic gaps, 61, 63
Seismic hazard, 3, 31
Seismic moment, 6, 54
Seismic quiescence, 61
Seismic response, 13
masonry, 361
prestressed concrete, 352
reinforced concrete, 289
soils, 182
steel structures, 371
structures, elastic, 256
non-linear, 258
theory, 13
timber structures, 403
Seismic risk (definition), 3
Seismicity, 4, 31, 89
Seismicity model, 113, 114
Seismology, 5
Seismotectonics, global, regional, 31, 32, 37
Serviceability criteria, 143
Services, 419, 447 (*see also* Equipment)
Settlement of dry sands, 101, 202
Shear beam, 194, 215
Shear in columns (steel), 386
Shear modulus, 108, 183, 185
Shear strain, 185, 187
effect on damping and shear modulus, 183
Shear strength
of columns, 328
of structural walls, 308
Shear walls (*see under* Structural walls *and* Walls)
Shear wave velocity, 184, 194
Simulated earthquakes, 119, 121 (*see also* Artificial accelerogram)
Significant acceleration, 85
Single-degree-of freedom systems, 15
Site
characteristics, 94
investigations, 100
response to earthquakes, 190
soil conditions, 126
Skopje 1964 earthquake, 131
Slabs (r.c.), 335
Sloshing in water tanks, 469
Soft storey concept, 152
Soil
conditions, 94
layers, effect on bedrock excitation, 94–100, 190
models, 209
structure
interaction, 208, 224

Soil *(conditions)*
favourable, 230
systems, seismic response, of, 207
tests, 100
types, 189
Source parameters, 69, 125
Spatial distribution of earthquakes, 55
Spectra for different site conditions, 99
Splices (in reinforcement), 314
Springs and dashpots, 209
Sprinkler systems, 433
Standard deviation, 491
Standard penetration resistance, 103, 203
Stationary random processes, 133
Staircases, 335
Statistical methods, 490
Steel
beams
design of, 380
under cyclic loading, 380
under monotonic loading, 377
connections under cyclic loading, 393
cold worked, 486
higher strength, 486
structures, 371
Stiff structures versus flexible, 154
Stiffness
appropriate, 153
degradation, 259, 277
Strain–age embrittlement, 489
Strain-hardening, 259
Strain-release map, 57
Strain-softening, 259
Strength of an earthquake, 5
Strengthening of existing structures, 479 (*see also* Upgrading)
Stress drop (on faults), 68, 70
Strike-slip faults, 42, 52, 54
Structural form, 364
for bridges, 458
specific, 162
Structural walls, 163 (*see also* Walls)
flanged, 304
reinforced concrete, 302
steel, 399
Subduction zones, 32
Subgrade reaction method, 243
Substructure, 161 (*see also* Foundations)
Surface wave magnitudes, 6
Survivability, 146, 174
Symmetry, 148
of low-rise housing, 477

t-test (statistical), 499
Tanks, 137, 435
elevated, 46
underground, 234
Tectonics, 31
plates, 32
provinces, 38
Tilting (of strata), 40
Timber
sheathed walls, 408
structures, 402
Tokachi–Oki 1968 earthquake, 97
Topographical effects, 95, 199
Torsional (effects), 148–9, 477
Toughness, 160, 265
of concrete structures, 291
Towers, 463
Transverse steel, 334, 354 (*see also* Confinement)
Tsunami, 97

Uniform risk response spectra, 126
Upgrading, seismic, 11 (*see also* Strengthening of existing structures)
Upper bounds
to peak ground motion, 79
on peak ground velocity, 80

Variance, 490
Vertical shear beam model, 194
Vibration frequencies, 24
Viscoelastic finite elements, 223
Viscoelastic half space, 215

Viscous damper, 212
Void ratio, critical, 203
Volcanic activity, 4, 32
Von Mises criterion, 392
Vulnerability, 3, 10

Wall finishes, 449
Walls
  apertures in, 320, 322, 364, 477
  squat, 304
  strength and stiffness, 477
  structural r.c., 320–4
  timber sheathed, 408
Warping of strata, 40
Water content of soil, 97, 182, 254 (*see also* Groundwater *and* Liquefaction
  table, 103, 232
  tanks (*see* Tanks)
Weldability, 377, 489
Welding, 397
Western Montana, 39
White noise, 133–4
Windows, 449
Winkler spring method, 243
Workmanship, 147, 160

Yield stress, variability in steel, 486